中国农业科学院
兰州畜牧与兽药研究所科技论文集
（2012）

中国农业科学院兰州畜牧与兽药研究所　主编

中国农业科学技术出版社

图书在版编目（CIP）数据

中国农业科学院兰州畜牧与兽药研究所科技论文集（2012）/ 中国农业科学院兰州畜牧与兽药研究所主编．—北京：中国农业科学技术出版社，2015.3
ISBN 978－7－5116－1742－2

Ⅰ.①中… Ⅱ.①中… Ⅲ.①畜牧学－文集②兽医学－文集 Ⅳ.①S8－53

中国版本图书馆 CIP 数据核字（2015）第 004283 号

责任编辑 闫庆健 鲁卫泉
责任校对 贾晓红

出 版 者 中国农业科学技术出版社
北京市中关村南大街 12 号 邮编：100081
电　　话 (010)82109704(发行部) (010)82106632(编辑室)
(010)82109703(读者服务部)
传　　真 (010) 82106625
网　　址 http://www.castp.cn
经 销 者 各地新华书店
印 刷 者 北京昌联印刷有限公司
开　　本 880 mm×1 230 mm 1/16
印　　张 23.75 彩插 16 面
字　　数 599 千字
版　　次 2015 年 3 月第 1 版 2015 年 3 月第 1 次印刷
定　　价 80.00 元

《中国农业科学院兰州畜牧与兽药研究所科技论文集（2012）》

编　委　会

前　言

近年来，在中国农业科学院科技创新工程的引领下，中国农业科学院兰州畜牧与兽药研究所的科研水平和社会声誉获得快速提高，与尊重知识和尊重人才密不可分。实践证明，促进人才健康成长，要有良好创新环境的支持，更离不开科研人员和管理人员不断进行思考与探索，通过撰写论文进行思想和理论的升华。

目前，我所科研人员和管理人员不但有工作上的热情，更有对工作认识上的高度和对学科理解上的深度。他们在紧张繁忙的实践活动中，笔耕不辍，将自己的研究成果写成论文。这不单是科研人员和管理人员的工作总结、过程记录，更是他们智慧的结晶，最终成为研究所的一笔宝贵财富。

为了珍惜这笔财富，加强优秀论文的交流与传播，也为了记录他们的辛勤付出，营造更加浓厚的学术氛围，促进科研水平和管理水平的提升，切实推进研究所的科技创新，科技管理处等部门搜集了2012年研究所公开发表的论文172篇约70.7万字，并编印成《中国农业科学院兰州畜牧与兽药研究所科技论文集（2012）》。文集分为草业学科、畜牧学科、兽药学科、中兽医与临床兽医学科和其他等五部分，仅供参考和学习。由于时间仓促，可能还有论文未能收录，敬希鉴谅！

编　者

2014年12月

目　　录

畜牧学科

兽药学科

中兽医与临床兽医学科

草业学科

其　他

畜牧学科

A Comparative Study at two Different Altitudes with two Dietary Nutrition Levels on Rumen Fermentation and Energy Metabolism in Chinese Holstein Cows

Qiao G. H.[1], Shao T.[2], Yu C. Q.[3], Wang X. L.[1], Yang X.[1], Zhu X. Q.[1] and Lu Y.[1]

(1. Lanzhou Institute of Animal and Veterinarian Pharmaceutics Science, CAAS, Lanzhou 730050, China; 2. Institute of Ensilage and Processing of Grass, Nanjing Agricultural University, Nanjing, China; 3. Institute of Geographical Science and Resource, Chinese Academy of Sciences, Beijing 100081, China)

Summary: The object of this study was to investigate the effect of two altitudes (1 600 vs. 3 600m) with two nutritional levels [5.88MJ/kg dry matter (DM) vs. 7.56MJ/kg DM] on apparent total tract digestibility, rumen fermentation, energy metabolism, milk yield and milk composition in Chinese Holstein cows. Sixteen Chinese Holstein cows in their third lactation with close body weights, days in milk and milk yield were randomly divided into four groups, of which two were directly transferred from Lanzhou (altitude of 1 600m) to Lhasa (altitude of 3 600m). Four treatments (high plateau and high nutrition level, HA - HN; high plateau and low nutrition level, HA - LN; low plateau and high nutrition level, LA - HN; and low plateau and low nutrition level, LA - LN) were randomly arranged in a 2 × 2 factorial experimental design. Results indicated that the apparent total tract digestibility of a diet's DM, organic matter, crude protein, neutral detergent fibre and acid detergent fibre and DM intake were not affected by either altitude or nutrition level ($P > 0.05$). Milk protein percentage was higher for the diet with the high level of nutrition than for the diet with low nutrition level irrespective of altitude ($P < 0.05$). Percentages of milk fat and milk lactose were not affected by either altitude or nutrition level ($P > 0.05$). The metabolizable energy used for milk energy output was decreased by high altitude in comparison with that at low altitude ($P < 0.05$). No differences were observed in the live body weight or body condition score (BCS) of Chinese Holstein cows among all of the four treatments ($P > 0.05$).

Introduction

Chinese Holstein cows (♂ Holstein origin × ♀ Chinese yellow cattle) with an average 305 - day milk yield of 4 500kg are commonly used in the dairyproduction system in China (Qiao, 2011). With the rapid improvement in living conditions in Tibet, the dairy industry has undergone a very dramaticincrease. However, there is no good local breed of dairy cattle in Tibet. Therefore, Chinese Holstein cows were directly purchased from low – altitude areas in China and transferred to the Tibet plateau by some dairy farm owners. Particle semen from Chinese Holstein dairy bulls has also been used to obtain genetic improvements in Chinese yellow cattle. Currently, Chinese Holstein cows have become the predominant breed of dairy cattle in Tibet. From 1985 to 2010, the number of Chinese Holstein cows rapidly increased from 4. 09 million to 6. 95 million (Qiao, 2011).

Tibet is the sacred and pure land, and the altitude ranges from 2 700 to 4 500m above sea level. The climate in the area of Tibet is a typical plateau climate, where air pressure and oxygen concentrations are lower at sea level (approximately 60% as much as low plateau). Effects of high altitude on body metabolism in small to medium – sized mammals have been well documented in the past five decades. Some native species such as yaks (Han *et al.*, 2002; Dong *et al.*, 2004), and adapted species such as dogs (Forster *et al.*, 1981), sheep (Blaxter, 1978) and goats (Forster *et al.*, 1981) can adapt to a low – oxygen environment. The basal metabolism rate and nutrient digestibility of yaks remain unchanged at altitudes ranging from 2 270 to 4 500m, but the basal metabolism rate of yellow cattle was increased by increasing altitude (Han *et al.*, 2003). Responses of nutrient digestibility, energy metabolism, blood biochemical parameters and productive performance (milk yield and milk compositions) in dairy cows to an alpine altitude of 2 000m above sea level have been well documented recently (Bartl *et al.*, 2008; Bovolenta *et al.*, 2002a, b; Berry *et al.*, 2001a, b; Leiber *et al.*, 2004). Complex interactions of various environmental factors on the productive performances of dairy cattle were observed. Because these studies focused on the combination of environmental factors, the extent to which the diet type and environmental factors, particularly altitude, directly contributed to the effect of high altitude cannot be distinguished. Little is known regarding the effect of high altitude and diet type on the aforementioned aspects in Chinese Holstein cows.

Energy nutrition is the primary contributing factor that drives production in dairy cows. Current systems (NRC 2001, AFRC, INRA) for energy requirements of dairy cows were based on the data of *in vivo* studies that were carried out in lowland areas. However, these systems are probably invalid for use in high – altitude areas, such as the Tibet plateau (from 3 600 to 4 500m). According to a comparative investigation into dairy productive performance at two altitudes (Tibet, 3 600m vs. Gansu Province, 1 600m) with same chemical components in diet, the milk yield over 305days of cows was lower (< 3 000kg) in Tibet compared with that at lower – altitude area (Gansu Province, more than 4 600kg; Qiao, 2011). Therefore, this study was conducted to investigate the effects of high altitude and diet type on dry matter (DM) intake, the apparent total tract digestibility, rumen fermentation and the principle of energy partitioning in Chinese Holstein cows in Tibet plateau and to provide a theoretical basis and practical recommendations for dairy production.

Materials and methods

Experimental sites and cows

Sixteen Chinese Holstein cows in their third lactation with proximal milk yields ((17. 88 ±

5.12) kg), body condition scores (BCS) (2.57 ± 1.28), body weights ((525 ± 20) kg), DM intakes ((16.11 ± 1.01) kg) and days in milk ((76 ± 13) days) were strictly selected from the Yongdeng dairy farm in the Gansu Province. This dairy had a total of 946heads of Chinese Holstein lactating dairy cows, all of which were fed with total mixed rations in barns and were treated against parasites in the spring using Eprinomectin (Kanglong Co Ltd, Wuhan, China).

The sites selected for the experiment were the Dawashan dairy research farm of the Lanzhou Institute of Animal and Veterinarian Pharmaceutics Science in Lanzhou City (1 600m) and the dairy research farm of the College of Animal and Veterinary in Tibet (3 600m). Cows at two experimental sites were kept in barns to avoid radiation. Ambient temperatures of two experimental sites were close. The average daily ambient temperature at experimental sites during the experimental period ranged from 12.8 to 15.3℃ and 15.6 to 19.4℃ in Lhasa and Lanzhou respectively. Thus, this present experiment was designed to differentiate the effect of hypoxia on Tibet plateau from the other factors (radiation, ambient temperature) on cow's digestion, metabolism and productive performance. To minimize the risk of statistical type I errors, the cows were divided into four groups according to the principle that there is no significant difference in milk yield, body weight, BCS and DM intake between the four groups, as presented in Table 1. Each group contained four cows. Two groups were randomly selected and transferred to an experimental site in Tibet, and the other two were transferred to an experimental site in Lanzhou City.

Table 1 Body weight, milk yield, dry matter (DM) intake and days in milk in different groups*

Items	Group A	Group B	Group C	Group D	Mean	SEM
Body weight (kg)	523.25	524.75	527.00	527.50	525.63	1.64
BCS	2.54	2.61	2.58	2.53	2.57	0.04
DM intake (kg)	16.34	15.82	16.03	16.26	16.11	0.49
Milk yield (kg/day)	18.20	17.98	17.73	17.63	17.88	1.07
Days in milk	71.50	77.25	74.50	74.71	76.94	2.74

* There is no significant difference in body weight, body condition score (BCS), milk yield and days in milk between the four groups ($P > 0.05$).

Experimental design, management and diets

The diets selected, as presented in Table 2, were the classical diet types of the dairy production system in China, which are specific to small-scale dairy farms (5 ~ 20 heads, low nutrition) and large-scale dairy farms (over 30 heads, high nutrition). Alfalfa and Chinese wild rye in diets with high levels of nutrition and maize stover in diets with low levels of nutrition were firstly chopped to between 2.0 and 2.5cm and then rubbed into small pieces using an RC500machine (Keyang Co, Shandong, China). The diets were totally mixed by a mixing machine (EYH-1; Nanjing Equipment Co, Nanjing, China) and fed to the cows. Diets were formulated according to the National Research Council (NRC 2001) (Table 2). Feedstuffs used in each type of diet at each experimental site remained unchanged throughout the experiment.

The study was a 2 (altitude) × 2 (nutrition level) factorial experiment design. The two altitudes were 3 600m (high altitude, HA) and 1 600m (low altitude, LA). Each of the 16 cows was fed *ad libitum* in individual metabolism pens in barns at either high or low altitude with total mixed rations containing either a high nutrition level (HN) or a low nutrition level (LN). The four treatments were thus designated HA – HN, HA – LN, LA – HN and LA – LN for high – altitude and low – altitude offered diets containing high and low levels of nutrition respectively. Each treatment group contained four cows. Fresh water was available at any time throughout the day. Cows were fed once per day at 06:30. Daily feed refusals were discarded. Cows were milked twice daily at 06:00and 17:30, and milk yield was recorded. At each milking time, 300ml of milk was removed using a pipette. Morning milk samples and afternoon milk samples were pooled according to the cow and the ratio of morning milk yield to afternoon milk yield. Two drops of penicillin were added into each milk sample, and these were then stored in an ice box. The milk samples were divided into two parts for each cow, of which one was for the main composition analysis and the other was for energy content determination. Efforts were made to immediately return the milk for composition analysis to the laboratory.

Table 2 Experimental diets and chemical compositions (% dry matter)

Ingredients	High nutrition	Low nutrition
Alfalfa hay	5. 83	0
Chinese wild rye	11. 67	0
Maize stover	0	70
Maize silage	17. 5	0
Cracked maize	35	15. 5
Soybean meal	19. 1	8. 5
Wheat bran	9. 4	4. 5
Sodium chloride	0. 5	0. 5
Minerals premix*	0. 5	0. 5
Vitamin premix	0. 5	0. 5
Crude protein	19. 1	9. 7
Neutral detergent fibre	27. 2	59. 3
Acid detergent fibre	15. 5	36. 6
NFC	46. 1	23. 6
NE_L ‡ (MJ/kg)	7. 56	5. 88

NF_L ‡, Non – fibre carbohydrate.

* Per kilogram contains 270g Ca, 60g Mg, 40g P, 5g Zn, 4g Mn, 1. 5g Cu, 500mg I, 50mg Co, 15mg Se; and + Per kilogram contains 500IU vitamin E, 100 000IU vitamin D_3, 500 000IU vitamin A; ‡ Net energy for lactation, calculated according to NRC 2001. The other chemical components were the values determined in laboratory.

Experimental schedule and analysis

Two identical experimental protocols were followed at altitudes of 1 600 and 3 600m above sea level. The whole experiment was performed during a period of 2 months (60days, August 1to October 1). The average daily ambient temperature at experimental sites during the experimental period ranged from 12. 8 to 15. 3℃ and 15. 6 to 19. 4℃ in Lhasa and Lanzhou respectively. The first 53 days was the adaptive period, and the last 7 days was the experimental period. This schedule was chosen to give cows considerable time to achieve adaptation to high altitude, which was justified by scholar Leiber *et al.* (2004). During the last seven consecutive days, feed samples, feed refusals, faeces and urine were collected. Approximately 15% of the faecal samples, feed samples and feed refusals were analysed for DM, organic matter (OM), neutral detergent fibre (NDF) and acid detergent fibre (ADF) (AOAC, 1990). The analysis procedure for NDF strictly followed the method of Van Soest *et al.* (1991), without the use of sodium sulphite and with the inclusion of heat-stable α - amylase (Aldrich Sigma, A - 4551, Saint Louis, MO, USA). DM content was determined by putting the samples in an air - forced oven at 105℃ for 48 h. OM content was determined by putting samples in a muffle furnace at 550℃. The nitrogen (N) content of the samples was determined using Kjeldahl N ×6.25, and the composition of milk samples was analysed using a milk analyser (LactoStar, 4 000/SPI, Berlin, Germany). The energy contents of diets, refusals, faeces, urine and milk were determined using a calorimeter (Parr 6 300; Parr Instrument Co., Louisville, KY, America).

Diet energy metabolism

In brief, metabolizable energy (ME) intake (MJ/day) and milk energy output (MJ/day) were obtained from the cow production trail. Gross energy (GE) intakes were directly measured using a calorimeter (Parr 6 300; Parr Instrument Co. America). The ME intake was calculated from the DM intake (production trial) and the dietary ME concentration (energy metabolism study). The dietary ME concentration was estimated as the difference between the GE intake (digestibility trials), the energy outputs from faeces and urine (digestibility trials), and the methane energy output. Methane energy output was calculated according to the regression model of Holter and Young (1992). It is calculated as follows:

methane energy output (MJ/day) = [2.927 - 0.0405 × milk yield (kg/day) + 0.335 × milk fat (%) - 1.225 × milk protein (%) + 0.248 × dietary crude protein (CP) (% DM) - 0.448 × ADF (% DM) + 0.502 × forage ADF (% DM) + 0.0352 × digestibility of ADF (%)] × 4.2 + 420 × GE intake (MJ/day).

Sampling and analysis of ruminal fluid and blood

Ruminal fluid and blood samples were removed on the last three consecutive days of the experimental period. Ruminal fluid samples were removed via an oral tube at 4h after the morning feed and were then squeezed through four layers of cheesecloth. Ruminal fluid samples to be used for volatile fatty acid determination were acidified with 2ml of 25% metaphosphoric acid and centrifuged at 4℃, 3 000g for 15min. Supernatant was frozen at - 20℃ for subsequent analysis. Meanwhile, blood samples were removed from the jugular vein using blood sampling bottles coated with heparin on the inner wall. Volatile fatty acids (acetate, propionate and butyrate) in ruminal fluids were determined using the GC - 2010 (Japan, 30m × 0.32mm × 1.25mm capillary column, N_2 was the carrier gas). The ammonia nitrogen concentration of ruminal fluid was determined using the hypochlorite phenol method (Qiao *et al.*, 2010). The pH value of ruminal fluid was determined immediately using pH meter (Accumet pH meter; Fisher Scientific, Montreal, QC, Canada). Biochemical parameters of blood glucose in plasma were determined according to the instruction of commercial kits (YZB/0362 - 2006; Shanghai Rongsheng, Shanghai, China).

Live body weight was measured during the last 2 weeks. The mean difference in live body weight between the current week and the previous week was divided by 7. The value obtained was regarded as the daily live weight gain. BCS of the cattle were determined in the last week of experimental period. The BCS of each cow was determined using the method described by Mulvanny (1977), with five categories from 1 (very thin) to 5 (very fat).

Statistical analysis

All of the data obtained were subjected to a twoway (altitude vs. nutrition level) ANOVA procedure, using a 2 × 2factorial (altitudes vs. nutrition levels) experimental design (SAS 8.1). The data used were averaged for each individual cow. The statistical model used was $Y_{ij} = \mu + \alpha_i + \beta_i + \alpha_i \times \beta_i + c_k + e_{ijk}$, where μ is the overall mean, α_i is the altitude effect, β_i is the nutrition level effect, $\alpha_i \times \beta_i$ is the interaction of altitudes and nutrition levels, c_k is the cow effect and e is the random error. Significant differences between treatments were declared at $P < 0.05$ in this study using Duncan's multiple comparison.

Results

Effect of individual cow

Effect of individual cow did not reach significant level for all the parameters listed in the tables ($P > 0.05$).

Effects on DM intake and apparent total tract digestibility

The data on the effects of DM intake and apparent total tract digestibility in Chinese Holstein cows are shown in Table 3. The results indicated that DM intake and apparent total tract digestibility of the DM, OM, NDF, ADF and CP diets were not affected by either altitude or nutrition levels ($P > 0.05$, Table 3).

Table 3 Effect of altitudes and nutrition levels on dry matter (DM) intake and apparent total tract digestibility in Chinese Holstein cows

Items	HA		LA		SEM	Significance		
	HN	LN	HN	LN		Nutrition	Altitude	Interaction
DMI (kg/day)	16.2	16.1	16.5	14.7	0.31	NS[5]	NS	NS
Digestibility (%)								
DM	62.0	62.8	61.1	63.4	0.83	NS	NS	NS
OM	66.3	66.8	64.5	66.1	0.88	NS	NS	NS
CP	61.5	60.7	59.1	61.2	0.69	NS	NS	NS
NDF	54.3	55.9	53.8	55.6	1.29	NS	NS	NS
ADF	47.6	48.3	46.4	49.0	1.17	NS	NS	NS

* HA, high latitude; LA, low altitude; HN, high nutrition level; LN, low nutrition level; NS, not significant ($P > 0.05$); ADF, acid detergent fibre; CP, crude protein; NDF, neutral detergent fibre; OM, organic matter.

Effects on production performance

The data on the effects of altitudes and nutrition level on milk yield and milk compositions in Chinese Holstein cows are shown in Table 4. The results indicated that the milk yield was significantly decreased by either high altitude in comparison with low altitude ($P<0.05$) or diets with low nutrition levels ($P<0.05$). Percentages of milk fat and milk lactose were not affected by either nutrition levels or altitudes ($P>0.05$). Milk protein percentages were significantly increased by diets with high nutrition levels, irrespective of the altitude ($P<0.05$). Simultaneously, milk urea nitrogen levels were significantly increased by the diet with a high level of nutrition, irrespective of altitudes ($P<0.05$).

Table 4 Effect of altitudes and nutrition levels on milk yield and milk composition in Chinese Holstein cows

Items	HA		LA		SEM	Significance		
	HN	LN	HN	LN		Nutrition	Altitude	Interaction
Milk production, kg/day	17.0^{b}	12.9^{a}	22.6^{c}	14.4^{ad}	1.01	*	*	NS
Milk composition, %								
Fat	3.14	3.22	3.20	3.10	0.03	NS	NS	NS
Protein	3.05^{b}	2.93^{a}	3.10^{b}	2.91^{a}	0.02	*	NS	NS
Lactose	4.63	4.39	4.70	4.67	0.04	NS	NS	NS
Milk urea N, mg/dl	11.2^{a}	7.76^{b}	10.19^{a}	7.88^{b}	0.91	*	NS	NS

[*] HA, high latitude; LA, low altitude; HN, high nutrition level; LN, low nutrition level; NS, not significant.

[*] Differ significantly $P<0.05$. Means within rows with different superscripts differ ($P<0.05$).

Table 5 Effect of altitudes and nutrition levels on energy metabolism in Chinese Holstein cows

Items	HA		LA		SEM	Significance		
	HN	LN	HN	LN		Nutrition	Altitude	Interaction
Live weight (kg)	529.1	499.7	539.3	515.5	1.77	NS	NS	NS
Live weight gain[+] (kg/day)	0.398	0.266	−0.029	0.091	0.05	NS	NS	NS
BCS	2.69	3.00	2.88	2.70	0.66	NS	NS	NS
ME intake (MJ/kg)	122.5^{a}	84.6^{b}	110.2^{c}	76.9^{bd}	0.37	*	NS	NS
Milk energy output (MJ/kg)	39.9^{a}	29.2^{b}	43.2^{a}	27.3^{bc}	0.40	*	*	*
Milk energy/ME intake	0.331^{a}	0.349^{a}	0.396^{b}	0.365^{b}	0.01	NS	*	NS

[*] HA, high latitude; LA, low altitude; HN, high nutrition level; LN, low nutrition level; NS, not significant ($P>0.05$); BCS, body condition score; ME, metabolizable energy.

* Significantly differ $P<0.05$. Superscripts with different letters in the same row differ ($P<1.05$).

+ Mean difference in live weight between current and previous week and then divided by 7.

Effects on energy metabolism

The data on the effects of altitudes and nutrition levels on energy metabolism in Chinese Holstein cows are shown in Table 5. Live body weight（LW）, live weight gain and BCS were not affected by either altitudes or nutrition levels（$P > 0.05$）. The ME intake of the diet with a high nutrition level was higher than that of the diet with a low level of nutrition（116.35vs. 80.75MJ/day）. Milk energy output was affected by either altitudes or nutrition level（$P > 0.05$）. Ratios of milk energy to ME intake were significantly decreased by high altitude（$P > 0.05$）, irrespective of the nutrition levels.

Effects on rumen fermentation

The data on the effect of altitudes and nutrition levels on rumen fermentation in Chinese Holstein cows are shown in Table 6. Ruminal pH values were not affected by either altitudes or nutrition levels（$P > 0.05$）. Irrespective of altitudes, the ruminal ammonia nitrogen concentration was significantly increased by high nutrition levels（$P > 0.05$）. Total volatile fatty acid concentrations and propionate concentrations in ruminal fluid were significantly increased by the nutrition level, irrespective of altitudes（$P > 0.05$）. Acetate proportions in ruminal fluid remained unchanged for all of the treatments（$P > 0.05$）. The ratio of acetate to propionate in ruminal fluid was significantly decreased by the diet with a high nutritional level, irrespective of altitudes（$P > 0.05$）.

Table 6　Effects of altitudes and nutrition levels on rumen fermentation and blood glucose in Chinese Holstein cows

Items	HA		LA		SEM	Significance		
	HN	LN	HN	LN		Nutrition	Altitude	Interaction
pH	6.63	6.74	6.67	6.60	0.06	NS	NS	NS
NH_3-N（mg/100ml）	15.93^a	12.77^b	14.89^a	12.31^b	0.08	*	NS	NS
Total VFA（m_M）	131.04^a	120.97^b	129.11^a	117.38^b	0.49	*	NS	NS
VFA（%）								
Acetate（A）	63.05	65.09	64.77	65.13	0.11	NS	NS	NS
Propionate（P）	24.56^a	19.27^b	26.01^a	19.86^b	0.29	*	NS	NS
Butyrate	12.06	11.04	9.96	11.80	0.19	NS	NS	NS
A：P	2.61^b	3.19^a	2.54^b	3.12^a	0.02	*	NS	NS
Blood glucose（mM）	3.93^b	3.51^a	3.80^b	3.50ia	0.06	NS	*	NS

*HA, high latitude; LA, low altitude; HN, high nutrition level; LN, low nutrition level; NS, not significant.

*Significantly differ $P > 0.05$.

Means within a row with different superscripts differ significantly（$P > 0.05$）.

Discussion

Effects on DM intake and nutrient digestibility

To our knowledge, no *in vivo* digestibility trials in Chinese Holstein cows were carried out in the Tibetan plateau. In the present study, both the DM intake and the nutrient digestibility of the DM, OM, NDF, ADF and CP diets in Chinese Holstein cows were not affected by either altitude or nutrition levels. These results are in agreement with observations in dairy cattle by Bovolenta *et al.* (2002b), yaks by Han *et al.* (2002, 2003), sheep by Blaxter (1978) and goats by Forster *et al.* (1981), who all investigated the effect of high altitude on DM intake and nutrient digestibility in these mammals and found no effect of an increase in altitude on DM intake and nutrient digestibility. Leiber *et al.* (2004) investigated the contribution of altitude and the alpine origin of forage to DM intake in dairy cows, and a decrease in DM intake was found because of the low quality of alpine - origin forage. This result did not agree with the present study, in which a low - quality diet was also used, and no effect was observed on DM intake. The possible explanation is that the maize stover was well processed (cut short and rubbed into small pieces) and totally mixed with the concentrate in the current study. According to the reports of Berry *et al.* (2001b) and Weiss *et al.* (2009), the digestibility of NDF and ADF was decreased with an increased level of concentrate in the diet. In the present study, the digestibility of NDF and ADF was also decreased in Chinese Holstein cows fed diets with high nutrition levels, but it did not reach a significant level.

Effects on rumen fermentation

In some studies, ruminal ammonia nitrogen concentration was found to have a positive relationship with the CP level of the diet (Frank and Swensson, 2002; Hojman *et al.*, 2004). This phenomenon was also observed in this study. An increase in ruminal ammonia concentration was observed in Chinese Holstein cows fed diets with high CP concentrations (high nutrition, 19.1% vs. low nutrition, 9.7%). Rumen total volatile fatty acid concentrations and propionate concentrations were increased in Chinese Holstein cows fed a diet with high levels of nonstructural carbohydrates (high nutrition, 46.1% vs. low nutrition, 23.6%). These results are in agreement with studies by Zanton and Heinrichs (2009) and Zebeli *et al.* (2007), who fed cows diets with high levels of non - fibre carbohydrate and found that higher levels of total volatile fatty acids and propionate were produced in ruminal fluid. These effects were found only due to the nutrition level, irrespective of altitudes. It is reasonable to suggest therefore that high altitude has no effect on rumen fermentation in Chinese Holstein cows.

Effects on milk yield and milk composition

On average, throughout the entire experiment, milk yields at high altitudes were lower than at low altitudes in Chinese Holstein cows. However, in Chinese Holstein cows, there were no differences observed for DM intake and the apparent nutrient digestibility of each diet between high altitude and low altitude. One possible explanation for why the milk yield of cows at high altitudes was lower than that of the cows at low altitudes is that Chinese Holstein cows at higher altitudes need more nutrients to meet the requirement for maintenance, especially with regard to energy. The results of lower milk energy outputs in energy metabolism trials at high altitudes compared with those at low altitudes in the present study support this explanation.

Several studies were conducted in alpine dairy cows by Bovolenta *et al.* (1998, 2002a, b), to investigate the effect of energy concentrate supplementation (starchy or fibrous) on milk yield and milk compositions under grazing conditions. These groups found that energy concentrate supplementation increased milk yield and did not change milk compositions. The milk yield of cows fed diets with high levels of energy at high altitude was higher than that of cows fed diets with low levels of energy in the present study (17.0 vs. 32.9kg/day), which agrees well with these observations.

In the present study, the reduction in milk protein percentages was found attributable to the diet with lower CP levels (9.7%) rather than the altitude. This result correlates well with the observation in dairy cows by Leiber *et al.* (2004), who evaluated the effect of altitude and alpine origin forage on milk synthesis and found that low milk protein percentages were attributable to the lower CP content and digestibility of alpine - origin hay.

It was reported that blood glucose concentrations or milk lactose percentages have a strongly positive relationship with milk yield (Amamcharla and Metzger, 2011; Lemosquet *et al.*, 2009a, b; Law *et al.*, 2011). On average, over the entire experiment, the milk yield of cows at high altitudes was lower than that of those at low altitudes, irrespective of nutrition levels, but the blood glucose concentrations at high altitudes were similar to those at low altitudes. The partitioning pathway of energy metabolism may be changed when the Chinese Holstein cows are exposed to high - altitude areas. Less energy was partitioned to milk production. The result of constant milk fat concentrations at high or low altitudes with high or low nutrition levels of diets in the present study did not agree with observations in dairy cows by Leiber *et al.* (2003) and Kreuzer *et al.* (1998), both of whom investigated the response of milk constituents to high - altitude alpine grazing and found that the milk fat concentration was increased. The possible explanation of this result is that Chinese Holstein cow breeds are not able to synthesis more milk fat as a result of a genetic issue.

Effects on energy metabolism

It was reported (Leiber *et al.*, 2004) that the effects of high altitude on animals can comprise hypoxia, climatic conditions (ambient temperature and humidity), solar radiation and topographic challenges under grazing conditions. In the present study, cows at both experimental sites were kept in barns. Ambient temperatures of two experimental sites were close. The average daily ambient temperature at experimental sites during the experimental period ranged from 12.8 to 15.3℃ and 15.6 to19.4℃ in Lhasa and Lanzhou respectively. Thus, the contribution of hypoxia to the influence of cows on digestion, metabolism and productive performance was focused on in this present experiment, and the other factors (radiation, ambient temperature etc.) could be excluded. The effects of high altitude on energy metabolism in mammals have been well documented in the past 50years. Results from published studies indicate that the energy required for the maintenance in small to medium - sized species is usually elevated when they are exposed to high - altitude areas for 3 - 4weeks (Butterfield *et al.*, 1992). Increases in energy requirements for maintenance (0.72 times greater than lowland) were also observed in dairy cows at alpine altitudes (2 000m) when compared with those at lowland altitudes (400m; Berry *et al.*, 2001b). In the present study, an increase in altitude also increased the energy requirement for maintenance in Chinese Holstein cows. However, the opposite results were also found in yaks (Han *et al.*, 2002, 2003), sheep (Blaxter, 1978) and goats (Forster *et al.*, 1981), as it was found that the energy required for the maintenance in these mammals was not affected by the altitude ranging from 2 270 to 4 260m. These species are adapted to Tibet plateau ranging from 2 260 to 4 750m, while Chinese Holstein cows are a new breed imported from other low - altitude areas of China. It is therefore reasonable to demonstrate that the adaptability of Chinese Holstein cows is lower than that of the other species that are adapted to the Tibetan plateau.

Conclusions

DM intake, apparent nutrient digestibility and rumen fermentation parameters of Chinese Holstein cows remained unchanged when they were exposed to high altitudes above 3 600m. High altitudes increased the energy required for maintenance and decreased the ME used for milk energy output in Chinese Holstein cows. In conclusion, cows require extra energy for their adaptation to Tibet plateau, and diet with high density of energy is recommended. An improvement in forage quality could also be considered as an effective strategy to improve energy status of cows.

Acknowledgements

This research was funded by a key project of Tibet and Basal research foundation (BRF1610322010006). Authors also want to thank the staff of the Tibetan College of Agriculture, Institute of Animal & Veterinarian for their assistance with animal feeding and sample collection.

References

[1] Amamcharla J K, Metzger L E, Development of a rapid method for the measurement of lactose in milk using a blood glucose biosensor. *Journal of Dairy Science* 94, 4 800 – 4 809. 2011

[2] Association of Official Analytical Chemists, *Official Methods of Analysis*, 15th edn. AOAC, Arlington, VA, USA. 1990

[3] Bartl K, Gomez C A, Garcia M, Aufdermauer T, Kreuzer M, Hess HD, Wettstein HR, Milk fatty acid profile of Peruvian Criollo and Brown Swiss cows in response to different diet qualities fed at low and high altitude. *Archives of Animal Nutrition* 62, 468 – 484. 2008

[4] Berry N R, Jewell P L, Sutter F, Edward P J, Kreuzer M, Effect of concentrate on nitrogen turnover and excretion of P, K, Na, Ca and Mg in lactating cows rationally grazed at high altitude. *Livestock Production Science* 71, 261 – 275. 2001a

[5] Berry N R, Sutter F, Bruckmaizer R M, Blum J W, Kreuzer M, Limitations of high alpine grazing conditions for early – lactation cows: effect of energy and protein supplementation. *Animal Science* 73, 149 – 162. 2001b

[6] Blaxter K L, The effect of stimulated altitude on the heat increment of feed in sheep. *British Journal of Nutrition* 39, 659 – 661. 1978

[7] Bovolenta S, Ventura W, Piasentier E, Malossini, F, Supplementation of dairy cows grazing an alpine pasture: effect of concentrate level on milk production, body condition and rennet coagulation properties. *Annual Zootechnology* 47, 169 – 178. 1998

[8] Bovolenta S, Sacca E, Ventura W, Prasentier E, Effect of type and level of supplement on performance of dairy cows grazing on alpine pasture. *Italian Journal of Animal Science* 1, 255 – 263. 2002a

[9] Bovolenta S, Ventura W, Malosini W, Dairy cows grazing on alpine pasture: effect of pattern of supplement allocation on herbage intake, body condition, milk yield and coagulation properties. *Animal Research* 51, 15 – 23. 2002b

[10] Butterfield G E, Gates J, Fleming S, Brooks G A, Sutton J R, Reeves J T, Increased energy intake minimizes weight loss in men at high altitude. *Journal of Applied Physiology* 72, 1 741 – 1 748. 1992

[11] Dong Q M, Zhao X Q, Ma Y S, Li Q Y, Xu S X, Shi J J, Wang L Y, Effects of different dietaries on di-

gestion and live weight gain of feedlotting yaks in areas of Yangtze and Yellow river sources. Proceedings of the International Congress on Yak, Chengdu, Sichuan, P. R. China 2004. Section III nutrition and feeds. 2004

[12] Forster H V, Bisgard G E, Klein J P, Effect of peripheral chemorecepter denervation on acclimatization of goats during hypoxia. *Journal of Applied Physiology* 50, 392 – 398. 1981

[13] Frank B, Swensson C, Relationship between content of crude protein in rations for dairy cows and milk yield, concentration of urea in milk and ammonia emissions. *Journal of Dairy Science* 85, 1 829 – 1 838. 2002

[14] Han X T, Xie A Y, Bi C X, Liu S J, Hu L H, Effects of high altitude and season on fasting heat production in the yak *Bos grunniens or Poephagus grunniens*. *British journal of nutrition* 88, 189 – 197. 2002

[15] Han X T, Xie A Y, Bi X C, Liu S J, Hu L H, Effects of altitude, ambient temperature and solar radiation on fasting heat production in yellow cattle (*Bos taurus*). *British journal of nutrition* 89, 399 – 408. 2003

[16] Hojman D, Kroll O, Adin G, Gips M, Hanochi B, Ezra E, Relationships between milk urea and production, nutrition, and fertility traits in Israeli dairy herds. *Journal of Dairy Science* 87, 1 001 – 1 011. 2004

[17] Holter J B, Young A J, Methane prediction in dry and lactating Holstein cows. *Journal of Dairy Science* 75, 2 165. 1992

[18] Kreuzer M, Langhans W, Sutter F, Christen R E, Leuenberger H, Kunz P L, Metabolic response of early – lactating cows exposed to transport and high altitude grazing conditions. *Animal Science* 67, 237 – 248. 1998

[19] Law R A, Young F J, Patterson D C, Kilpatrick D J, Wylie A R G, Mayne C S, Effect of precalving and postcalving dietary energy level on performance and blood metabolite concentrations of dairy cows throughout lactation. *Journal of Dairy Science* 94, 808 – 823. 2011

[20] Leiber F, Nigg D, Kreuzer M, Wettstein H R, Influence of sea level and vegetative stage of the pasture on cheese – making properties of cow milk. *Proceedings of the Society of Nutrition Physiology* 12, 87. 2003

[21] Leiber F, Kreuzer M, Jorg B, Leuenberger H, Wettstein H R, Contribution of altitude and Alpine origin of forage to the influence of Alpine sojourn of cows on intake, nitrogen conversion, metabolic stress and milk synthesis. *Animal Science* 78, 451 – 466. 2004

[22] Lemosquet S, Raggio G, Lobley G E, Rulquin H, Guinard – Flament J, Lapierre H, Whole – body glucose metabolism and mammary energetic nutrient metabolism in lactating dairy cows receiving digestive infusions of casein and propionic acid. *Journal of Dairy Science* 92, 6 068 – 6 082. 2009a

[23] Lemosquet S, Delamaire E, Lapierre H, Blum J W, Peyraud J L, Effects of glucose, propionic acid, and nonessential amino acids on glucose metabolism and milk yield in Holstein dairy cows. *Journal of Dairy Science* 92, 3 244 – 3 257. 2009b

[24] Mulvanny P M, *Dairy Cow Condition Scoring*. Paper 4468. Natl. Inst. Res. Dairying, Reading, UK. 1977

[25] NRC, *Nutrient Requirements of Dairy Cattle*, 7th revised edn. National Academy Press, Washington, DC, USA. 2001

[26] Qiao G H, Introduction of dairy industry of China. In: G H Qiao (ed.), *Principles of Dairy Nutritive Manipulation and Technologies Application*. Gansu Science & Technology Press, Lanzhou, China, pp. 169 – 174. 2011

[27] Qiao G H, Shan A S, Ma N, Ma Q Q, Sun Z W, Supplemental Bacillus culture on rumen fermentation and milk yield in Chinese Holstein cow. *Journal of Animal Physiology and Animal Nutrition* 94, 429 – 436. 2010

[28] Van Soest P J, Robertson J B, Lewis B A, Symposium: carbohydrate methodology, metabolism and nutritional implications in dairy cattle. Methods for dietary fiber, neutral detergent fiber, and nonstarch polysaccharides in relation to animal nutrition. *Journal of Dairy Science* 74, 3 583 – 3 597. 1991

[29] Weiss W P, St – Pierre N R, Willett L B, Varying type of forage, concentration of metabolizable protein, and source of carbohydrate affects nutrient digestibility and production by dairy cows. *Journal of Dairy Science*

92, 5 595 – 5 606. 2009

[30] Zanton G I, Heinrichs A J, Digestion and nitrogen utilization in dairy heifers limit – fed a low or high forage ration at four levels of nitrogen intake. *Journal of Dairy Science* 92, 2 078 – 2 094. 2009

[31] Zebeli Q, Tafaj M, Weber I, Dijkstra J, Steingass H, Drochner W, Effects of varying dietary forage particle size in two concentrate levels on chewing activity, ruminal mat characteristics, and passage in dairy cows. *Journal of Dairy Science* 90, 1 929 – 1 942. 2007

(Published the article in Effects of Altitudes on Dairy Cows Production, 2012. affect factor: 0. 855)

Effect of High Altitude on Nutrient Digestibility, Rumen Fermentation and Basal Metabolism Rate in Chinese Holstein Cows on the Tibetan Plateau

Qiao G. H. [1,3], Yu C. Q. [2], Li J. H. [1], Yang X. [1], Zhu X. Q. [1] and Zhou X. H. [1]

(1. Lanzhou Institute of Animal and Veterinarian Pharmaceutics Science, Chinese Academy of Agricultural Sciences, Lanzhou, 730050, China; 2. Institute of Geographic Science and Natural Resources Research, Chinese Academy of Sciences, Beijing, 100102, China; 3. Corresponding author. E-mail: qiaoguohua_1980@hotmail.com)

Abstract: Two experiments were conducted to investigate the effect of two altitudes (3 600 and 1 600m) on nutrient digestibility, rumen fermentation and basal metabolism rate in Chinese Holstein cows. Experiment 1was conducted to investigate the effect of high altitude (3 600m)on rumen fermentation, *in vitro* dry matter degradability, and nitrogen metabolism compared with low altitude (1 600m) in Chinese Holstein cows. Results indicated that total volatile fatty acids concentration, total gas production and gas coefficient a, b and c, efficiency of microbial protein synthesis and *in vitro* dry matter degradability were lower at 3 600m than 1 600m ($P<0.05$). The number of protozoa and ammonia nitrogen concentration was higher at 3 600 mthan 1 600m ($P<0.05$). We concluded that carbohydrate fermentation in the rumen was impaired, and ammonia nitrogen used for microbial nitrogen synthesis was also decreased at high altitude of 3 600m compared with 1 600m. Experiment 2was conducted *in vivo* to investigate the effect of high altitude on apparent nutrient digestibility and basal metabolism rate in Chinese Holstein cows. Results indicated that the apparent digestibility of the diet's dry matter, organic matter, neutral detergent fibre and acid detergent fibre was lower at 3 600m than those at 1 600m, respectively ($P<0.05$). In Chinese Holstein cows, basal metabolism rate was increased with increasing level of altitude ($P<0.05$). The results indicated that the high altitude of the Tibetan plateau impaired rumen fermentation and elevated the basal metabolism rate of Chinese Holstein cows.

Received 26March 2012, accepted 24August 2012, published online 8January 2013

Introduction

The number of dairy cattle on the Tibetan plateau has been growing rapidly rising from 1.0 to 1.5 million since 1991. However, dairy productive performance in Tibet is lower than areas with lower altitudes in China (Qiao *et al.* 2012). Because there is no high performing local dairy breed in Tibet, Chinese Holstein cows (♂ Holstein origin × ♀, local yellow cattle) with average 305 day's milk production of 3 500kg excluding milk consumed by her calf has become the dominant breed of dairy cattle.

The conditions of the Tibetan plateau with altitude ranging from 2 900 to 3 900m are characterised by hypoxia, lower air pressure and lower ambient temperature compared with the other lower altitude areas of China. The effects of high altitude on dairy production including milk yield, main milk constituents and milk fat profile have been well documented in the past decade. To our knowledge, not many *in vivo* studies were carried out to investigate the effect of high altitude on rumen fermentation and BMR in dairy cows. Leiber *et al.* (2004) reported that in dairy cows, the energy requirement for maintenance was elevated under Alpine grazing conditions. However, different results have been observed by other scholars who carried out experiments on the effects of high altitude on yak and sheep that are native to the Tibet plateau, and found no effect on BMR or O_2consumption in yak (Han *et al.* 2002, 2003) and sheep (Blaxter1978), respectively. Qiao *et al.* (2012) conducted a comparative study on the effects of two altitudes (3 600vs 1 600m) on nutrient digestibility, dietary energy partitioning principles in Chinese Holstein cows, and found that the estimated metabolisable energy (ME) used for milk energy output was reduced in high - altitude areas. They concluded that the energy requirement for maintenance may be elevated at high altitude in Chinese Holstein cows. Little is currently known on the effect of high altitude on rumen microbial fermentation, digestion and BMR in Chinese Holstein cows. The authors hypothesised that high altitude of Tibet plateau has no effect on dry matter intake (DMI), digestion and rumen microbial fermentation, and has no effect on BMR in Chinese Holstein cows. Therefore, this study was conducted to investigate the effect of high altitude on nutrient digestibility, rumen microbial fermentation and BMR in Chinese Holstein cows and provide practical recommendations for dairy production on Tibetan plateau.

Materials and methods

Sites, cows, management and diets

The experimental sites selected were on the dairy research farm at the Tibetan Institute of Animal and Veterinary Science in Lhasa (29°40′37′N, 91°09′19′E) at an altitude of 3 600m and the Lanzhou Institute of Animal and Veterinarian Pharmaceutics Science (36°02′40′N, 103°45′22′E) at an altitude of 1 600m. A total of 12 Chinese Holstein dry cows (♂ Holstein origin × ♀, local yellow cattle), 6 cows ((525 ± 16) kg BW, age (25 ± 1.5) months) at altitude of 1 600m and 6 cows (531kg ± 15kg BW, age (27 ± 2.1) months) at altitude of 3 600m were selected and fed in barns at each experimental site. These 12 cows were selected from the Yongdeng dairy farm in Lanzhou city. They were halfsibs. Six of them were randomly selected and transferred to the dairy research farm of the Tibetan Institute of Animal and Veterinarian Science in Tibet when they were ~ 12months of age. The average daily ambient temperature at the experimental sites during the experimental period (1September - 25 October 2011) ranged from 12.8 to 15.3℃ and from 15.6 to 19.4℃ in Lhasa and Lanzhou, respectively. Cows at each experimental site were fed the same diet as presented below.

Each cow was fed mixed ration with 20% alfalfa hay, 40% corn silage and 40% concentrate (per kg dry matter of concentrate contains: 516g cracked corn, 290g soybean meal, 179g wheat bran and 15g premix per kg of premix contains: 40g Na, 28.5g Zn, 15g Mg, 28.5g Mn, 10g Fe, 4g Cu, 0.1g Co, 0.06g I, 3 600 000 IU VA, 630 000IU VD_3, 3.8g VE, 5g nicotinic acid). In order to keep concentrations of chemical components in the diet consistent, the concentrate was directly transferred from Lanzhou city to the dairy research farm in Tibet. The same breed of alfalfa (Zhonglan #1) was planted and harvested in both Tibet and Lanzhou when they were in flower stage (first cut). The same breed of corn silage (Gongzhuling #3) was planted and ensiled in both Tibet and Lanzhou when they were in milk ripe stage. All the chemical components (Table 1) of forages and diets were analysed in the laboratory according to the methods of AOAC (1984) with the exception of net energy for lactation, which was estimated by NRC (2001). No significant differences in chemical components were detected between the two diets at the experimental sites. Cows were individually fed *adlibitum* twice daily at 07:00 and 17:30hours. Cows at both altitudes were kept in barns. Clean water was available at any time throughout a day. The *in vivo* digestibility trial was conducted after a 20 - day adaptation period in both Lanzhou (1 600 m) and Tibet (3 600m). Subsequently, faeces from each cow were totally collected on 3 consecutive days. DM, organic matter (OM), acid detergent fibre (ADF), crude protein (CP) of diet and faeces were determined using methods of AOAC (1984). Neutral detergent fibre (NDF) was determined using the method of Van Soest *et al.* (1991).

In vitro *batch culture protocols*

Medicago sativa L. (alfalfa) was planted and harvested to ground level when they were in flower stage (first cut) at the dairy research farm of the Lanzhou Institute of Animal and Veterinarian Pharmaceutics Science, then put into an air force oven at 65℃ for 48h. Dry alfalfa hay was grounded through a 2mm holescreen and used as only substrate for the *in vitro* batch culture at each experimental site. Chemical components of alfalfa were shown in Table 1.

Table 1 Chemical components (±s. d.) of corn silage, alfalfa and diets used at the two experimental sites

Measurements	Corn silage		Alfalfa		Diet	
	Tibet	Lanzhou	Tibet	Lanzhou	Tibet	Lanzhou
Dry matter (%)	38.1 (1.1)	38.3 (1.4)	41.9 (0.7)	41.6 (1.3)	57.30 (1.3)	57.50 (1.1)
Crude protein (%)	7.5 (0.2)	7.4 (0.5)	17.0 (0.4)	17.3 (0.4)	18.02 (0.5)	18.11 (0.2)
Neutral detergent fibre (%)	43.5 (1.7)	41.3 (1.9)	48.5 (2.2)	48.3 (1.8)	33.19 (2.0)	32.77 (1.7)
Acid detergent fibre (%)	29.1 (1.9)	32.0 (2.3)	33.3 (2.5)	33.8 (1.6)	20.03 (2.1)	21.30 (2.3)
Ether extract (%)	2.5 (0.3)	2.6 (0.1)	2.1 (0.2)	2.2 (0.2)	3.20 (0.3)	3.31 (0.1)
NFC[A] (%)	41.7 (1.0)	39.9 (0.9)	25.9 (1.1)	26.3 (1.3)	44.33 (1.0)	43.26 (0.9)
NE_L (MJ/kg DM)[B]	—	—	—	—	6.06 (0.07)	5.91 (0.10)

[*] A. NFC: Non - fiber carbohydrates = 100 - (NDF + CP + EE + ash);

B. NE_L: net energy for lactation, calculated according to NRC (2001).

Identical *in vitro* batch culture experimental protocols were followed at Lanzhou Institute of Animal and Veterinarian Pharmaceutics Science and the Tibetan Institute of Animal and Veterinary Science. The fresh ruminal fluid of three cows at each experimental site was removed via oral tube before morning feeding into a vacuum flask that was flushed with CO_2. The ruminal fluid of each cow was squeezed through four layers of cheesecloth into a flask full of CO_2, and was kept in a water bath at the temperature of 38 ~ 39℃ for 72h. The ruminal fluid of three cows was mixed by equal volume. The fluid was then mixed with buffer (pH 6.9) (per L contains: 292mg of K_2HPO_4, 240mg of KH_2PO_4, 480mg of $(NH_4)_2SO_4$, 480mg of NaCl, 100 mg of $MgSO_4$, 64mg of $CaCl_2 \cdot 2H_2O$, 4 000mg of Na_2CO_3, and 600mg of cysteine hydrochloride) in a ratio of 1 : 2 (Russell and Martin 1984). After total mixing, 30ml of diluted rumen fluid was anaerobically transferred into a 100ml-calibrated glass syringe containing 200mg of dried and ground substrate. There were 30replicates at each experimental site. The syringes were kept in a water bath at 39℃ with a gentle shaking. The gas production (GP) and blank incubation (without substrate) were recorded after 0, 2, 4, 8, 12, 24, 48and 72h of incubation, and net GP was calculated accordingly. Incubation was terminated after recording the 72h-gas volume. The abovementioned *in vitro* batch culture experiment at each experimental site was repeated three times.

In vitro dry matter degradability (IVDMD) was calculated as the original weight of DM of alfalfa hay minus the dry residue weight (after 72h of incubation) divided by the original sample weight. These values were then multiplied by 100 to derive IVDMD percentage. Residues of alfalfa were obtained by centrifuging medium at 500g for 10min. The supernatant of each syringe was gently transferred to another tube. Residues of alfalfa were dried to a constant weight. Subsequently, nitrogen (N) contents of residues were determined using the method of AOAC (1984) and available N was calculated accordingly using the original alfalfa N minus residue N. The content of each tube was centrifuged at 12 000g for 10min at 4℃. Supernatant (0.5ml) in each tube was collected for further analysis of volatile fatty acid (VFA) concentration using gas chromatograph (GC – 2010, Shimadzu Co. Ltd, molecular sieve 5A column, 1.6m × 3.0mm I. D., 60 ~ 80mesh, Kyoto, Japan). Supernatant (0.5ml) was measured for ammonia – N concentration according to the method of Qiao *et al.* (2010). Ruminal fluid pH value of each syringe medium was measured immediately when 72h of incubation was completed using a portable pH meter (Hanna 30100 – 0221, Milford, MA, USA). The pellet from each tube was lyophilised and measured for microbial N production using the method of Zinn and Owens (1986) in which purines were treated as an internal maker with some modifications to improve the purine recovery of ruminal microorganisms. The hydrolysis conditions of 12M $HClO_4$, 121℃ for 2h were substituted by 1M $HClO_4$, 90℃ for 1 h (Makkar and Becker 1999). Ruminal fluid obtained via oral tube before the morning meal was lyophilised and then the contents of purine and N were determined.

Protozoa enumeration

Protozoa were counted using a counting chamber (Shanghai Biotech, Co. Ltd 1022B) on 3 consecutive days of the experimental period at each experimental site. Three ml of ruminal fluid obtained via oral tube was strained and 3ml methyl green formalin – saline solution was added. Protozoa were counted as described by Hungate (1966). Each sample of each cow at each experimental site was repeatedly counted for three times.

Basal metabolic rate determination

The procedure of Han *et al.* (2002) for determination of BMR [fasting heat production (FHP)] was strictly followed. Closedcircuit respiratory mask technique and identical gas collection protocols were used to determine BMR of Chinese Holstein cows at each experimental site after digestibility trial and *in vitro* batch culture study. Each dry cow at each experimental site was trained to adapt to the respiratory mask and were fed the diet at a maintenance level for 15 days. Ambient temperatures were recorded hourly every day and then 24h of a day were divided into four time zones based on the principle that there are no dramatic ambient temperature changes within each time zone. Then the cows at both experimental sites were fasted for 7days. No feed was fed to cows during this period. There was no ambient temperature difference between the two experimental sites during this period. Then air collection was performed on 3 consecutive days (Day5 ~ Day7) at 09:30, 15:30, 22:30 and 04:30 hours representing time zones of 08:01 ~ 12:00, 02:01 ~ 15:00, 15:01 ~ 20:00 and 20:01 ~ 08:00 hours, respectively. In brief, two snake - shaped pipes from the masks were connected to an air collection box with a volume of 2 m^3 and an exit for air sample collection. Body temperature (rectum temperature), respiratory rate and heart rate were recorded twice daily in the morning and afternoon from Day 1to Day7. Bodyweights of cows were recorded on Day5 ~ Day7. Air collection was performed for 15min in each time zone after the air in the box was homogenised by an air pump. Meanwhile, ambient air pressure was recorded and air samples were collected. FHP production in each time zone was obtained by multiplying 1min of heat production (FHP/min) and total minutes in each time zone. Daily FHP of each cow was obtained as the sum of heat production from all the time zones in the day. Air samples were analysed for concentrations of CH_4, O_2 and CO_2 using gas chromatograph (GC - 2010, Shimadzu Co. Ltd, Kyoto, Japan) with a stainless steel column (Polepack type Q, 80mesh, Waters Associates Inc., Milford, MA, USA). N_2 was used as a carrier gas. Daily FHP of each cow was calculated according to the equation $FHP\ (MJ) = [16.175 \times O_2n\ (L) + 5.021 \times CO_2\ (L) - 2.167 \times CH_4\ (L)] \times 4.18 \times 10^{-3}$ (Brouwer 1965).

Statistical analyses

In Experiment 1, the effect of different altitudes was tested by using the MIXED Procedure of SAS 8.1 (SAS Institute, Cary, NC, USA). The statistical model was $y = A_i + D_j + e_{ij}$, where y was the dependent variable, A_i was the fixed effect of altitude, D_j was the random effect of ruminal fluid sampling by day, and e_{ij} was the error term. Significance was declared at $P < 0.05$ or $P < 0.01$. Gas production coefficients of a, b and c were regressed using equation $GP = a + b\ [1 - \exp^{(-ct)}]$, where GP was the cumulative GP at time point t, a was the rapid gas fraction, b was the potential gas fraction, c was the constant rate of b (Menke *et al.* 1979). BMR data of cows were pooled and subjected to the equation $y = ax^b$, where y was FHP and x was fasted bodyweight.

In Experiment 2, each cow at each experimental site was treated as an experimental unit. Procedure t - test of SAS 8.1 was used to compare the differences between two altitudes. Significance was declared at $P < 0.05$ or $P < 0.01$.

Results

Effect of different altitudes on GP and its coefficients

Results of effects of different altitudes on GP coefficients are shown in Table 2. Total GP *in vitro* at high altitude was lower than low altitude ($P < 0.05$). The rapid GP coefficient a, potential GP coefficient b and GP rate c were also lower at high altitude compared with low altitude ($P < 0.001$).

Table 2 Effect of different altitudes on *in vitro* gas production and its coefficients

Measurements	3 600m	1 600m	s. e. m.	P - value
Total gas production (ml)	33.67a	43.33b	0.059	0.023
a (%)	0.09a	0.13b	0.005	<0.001
b (%)	0.61a	0.83b	0.010	<0.001
$a+b$ (%)	0.70a	0.96b	0.006	0.001
c (%/h)	0.013a	0.021b	0.009	<0.001

* Within rows, values followed by different letters are significantly different at $P = 0.05$.

Effect of high altitude on rumen fermentation parameters

Data on the effect of high altitude on rumen fermentation parameters were shown in Table 3. The pH value of the medium was lower at 1 600m than at 3 600m ($P < 0.05$). Total VFA concentration was significantly lower at 3 600m than at 1 600m ($P < 0.05$). However, altitude had no effect on acetate to propionate ratio ($P > 0.05$), percentage of acetate, propionate and butyrate ($P > 0.05$). The number of protozoa was higher at 3 600m than at 1 600m ($P < 0.05$).

Table 3 Effect of different altitudes on rumen fermentation parameters in Chinese Holstein cows

Measurements	High altitude (3 600m)	Low altitude (1 600m)	s. e. m.	P - value
pH value	6.63	6.11	0.11	<0.001
Ammonia - N (mg/dL)	24.1a	19.3b	0.67	0.01
Total volatile fatty acid (mmol/L)	73.2a	93.6b	1.01	0.008
Acetate (%)	64.1	64.5	0.98	n. s.
Propionate (%)	17.0	17.3	0.95	n. s.
Butyrate (%)	13.7	13.0	1.03	n. s.
Acetate : Propionate	3.76	3.74	0.83	n. s.
Protozoa ($\times 10^4$/ml)	42.1a	27.6b	0.96	0.002
In vitro dry matter degradability (%)	44.9a	53.9b	0.81	0.005

* Within rows, values followed by different letters are significantly different at $P = 0.05$

Effect of different altitudes on microbial protein synthesis

Data on the effect of different altitudes on microbial protein synthesis is shown in Table 4. Efficiency of microbial protein synthesis at 3 600m was lower than at 1 600m ($P < 0.05$). Microbial N production after 72h of incubation was lower at 3 600m than at 1 600m ($P < 0.001$). Alfalfa

N use for microbial protein synthesis was lower at 3 600m than at 1 600m （$P<0.01$）, available N of alfalfa use for bacterial protein was also lower at 3 600m than 1 600m （$P<0.01$）.

Table 4　Effect of different altitudes on microbial protein synthesis *in vitro*

Measurements	3 600m	1 600m	s. e. m.	P – value
EMPS[A]	17. 7a	20. 9b	1. 01	0. 03
MN[B] （mg N/72h）	0. 72a	1. 09b	0. 11	<0. 001
Bacterial – N/alfalfa N （%）	24. 1a	32. 5b	0. 22	0. 005
Bacterial – N/available N[C] （%）	30. 9a	40. 7b	0. 29	0. 008

[*] Within rows, values followed by different letters are significantly different at $P=0.05$.

a: EMPS = efficiency of microbial protein synthesis (grams of bacterial – N per kilogram of OM truly digested in the medium of batch culture);

b: MN = microbial nitrogen;

c: Grams of N per 200mg alfalfa-grams of N undegradable N per 200mg.

Effect of different altitudes on nutrient digestibility

As presented in Table 1, there were no significant differences between the chemical components of diets at each altitude. The results of effect of altitude on DMI and apparent total tract digestibility in Chinese Holstein cows were shown in Table 5. Digestibility of DM, OM, NDF and ADF was lower at 3 600m than at 1 600m （$P<0.05$）. No difference was observed for digestibility of CP between altitudes of 3 600and 1 600m （$P>0.05$）. DMI was not affected by altitude in Chinese Holstein cows in this experiment （$P>0.05$）.

Table 5　Effect of different altitudes on dry matter intake and apparent total tract digestibility in Chinese Holstein cows

Measurements	3 600m	1 600m	s. e. m.	P – value
Dry matter intake (kg/day)	13. 9	15. 5	0. 39	n. s.
Digestibility （%）				
Dry matter	58. 3	62. 1	0. 43	0. 03
Organic matter	60. 6	65. 2	0. 31	<0. 001
Crude protein	61. 6	60. 1	0. 99	n. s.
Neutral detergent fibre	49. 1	54. 7	0. 65	0. 01
Acid detergent fibre	42. 9	47. 7	0. 88	0. 02

[*] Within rows, values followed by different letters are significantly different at $P=0.05$.

Effect of different altitudes on physical characteristics and BMR

Responses of Chinese Holstein cows during fasting to different altitudes are shown in Table 6. The heart rate and respiratory rate of Chinese Holstein cows at an altitude of 1 600m were lower than those at altitudes of 3 600m （$P<0.05$）. No difference was observed in body temperature between altitudes of 1 600 and 3 600 m （$P>0.05$）. BMR of Chinese Holstein cows at altitudes of 1 600m is lower than that at altitudes of 3 600m （314. 13v. 399. 40kJ/kg $BW^{0.75}$; $P<0.05$）.

Table 6　Effect of different altitudes on heart rate, respiratory rate, body temperature and basal metabolism rate in Chinese Holstein cows

Measurements	Altitudes		s. e. m.	P – value
	3 600m	1 600m		
Heart rate (beats/min)	54.11	51.00	1.21	0.0647
Respiratory rate (times/min)	12.16a	10.34b	1.23	0.0355
Body temperature (℃)	38.60	38.64	0.33	0.7255
Basal metabolism rate (kJ/kg $BW^{0.75}$)	399.40a	314.13b	3.76	0.0088

* Within rows, values followed by different letters are significantly different at $P = 0.05$

Discussion

High altitude effects on livestock production should include several main aspects, namely hypoxia, low ambient temperature, diet type and radiation (Kreuzer *et al.* 1998). Many studies have been carried out to investigate this combination effect on animal production, from which the high altitude with hypoxia effect cannot be successfully differentiated. As presented in Materials and methods, average ambient temperatures were very close between the two experimental sites. Half – sib cows at both sites were kept in barns, and were fed the diet with similar chemical composition (Table 1). The contribution of hypoxia to the effect of altitude on dairy cows was likely to be the main influence in this study. VFA and microbial CP deriving from the rumen fermentation can meet ~ 70% of a cow's energy requirement and 80% of metabolisable protein requirement. The results of decreases of total VFA concentration, microbial crude protein production and efficiency, and an increase of BMR at high altitude in the present study did not support our hypothesis.

Effect of different altitudes on GP and nutrient digestibility

In vitro GP and its coefficients have been using in several studies to evaluate the capability of whole rumen fermentation and shift of microbial structure (Lila *et al.* 2003; Busquet *et al.* 2006; Castillejos *et al.* 2006). In this present study, decrease of total GP and IVDMD *in vitro* at 3 600m compared with 1 600m indicated that the capability of whole rumen fermentation was impaired at an altitude of 3 600m. Gas is mainly produced by the fermentation of carbohydrates, and is a result of interaction between the microorganisms present and the manner in which they digest the particular feed within the *in vitro* system (Mauricio *et al.* 2001). In this research, ground alfalfa was the only substrate in the medium. The total GP, rapid GP coefficient a, potential GP b and GP rate c were decreased at 3 600m compared with those at 1 600m. These results strongly indicate that the structure of rumen microbial communities was shifted when Chinese Holstein cows were exposed to an altitude of 3 600m. It is reasonable to assume that soluble carbohydrate – degrading bacteria and total carbohydrate degrading bacteria (soluble carbohydrate – fermenting bacteria, cellulose – degrading bacteria and hemicellulose – degrading bacteria) were suppressed at an altitude of 3 600m. Our result also agrees well with the other studies in which relationship between GP and IVDMD was investigated (Busquet *et al.* 2006; Castillejos *et al.* 2006; Coblentz and Hoffman 2009), and concluded that GP has a good positive relationship with IVDMD.

To our knowledge, in Chinese Holstein cows, almost no *in vivo* digestibility trials have been conducted on the Tibetan plateau. Responses of apparent nutrient digestibility in livestock to high altitude have been inconsistent. According to results of others who carried out experiments under Alpine (2 400m) grazing conditions, no effect of Alpine on nutrient digestibility was detected. In addition, no differences in *in vivo* apparent nutrient digestibility in yak (Han *et al*. 2003), sheep (Blaxter 1978) and goat (Forster *et al*. 1981) was found at the altitude ranging from 2 200 to 5 000m, respectively. A possible explanation is these species are native mammals living on the 'roof of world', and have an excellent adaptation to high altitude. In contrast, Chinese Holstein cows are not native to the Tibetan plateau. Our results are similar to observations in yellow cattle by Han *et al*. (2002), who conducted experiments on the Tibet plateau, and found a decrease of nutrient digestibility because of the suppression of rumen fermentation.

Effect of altitude on rumen fermentation parameters

Little is known on the effect of high altitude on rumen fermentation in Chinese Holstein cows. The purpose of this part of the experiment was to investigate whether low productive performance of Chinese Holstein cows is due to the suppression of rumen microbial fermentation at high altitude. An *in vitro* technique was used to investigate the response of main rumen fermentation parameters to high altitude. Ruminal pH is mainly affected by the saliva buffering system and the concentration of VFA. In the *in vitro* study here, saliva was simulated according to the method of Russell and Martin (1984), and the effect of the buffering system should be fixed and constant in the syringes. Therefore, fluctuation of pH value in the medium was mainly affected by accumulation of VFA. Hence, the decrease in pH value of ruminal fluid at 1 600m compared with that at 3 600m could be due to the accumulation of VFA produced by microbial fermentation in the batch culture. A decrease in total VFA concentration at 3 600m compared with 1 600m in the *in vitro* system in our experiment (Table 3) did support that the capacity of rumen fermentation was impaired. The present study's result of decrease of total VFA concentration at 3 600m agreed well with observations on the effect of high altitude on rumen fermentation in yak (Han *et al*. 1998; Xue and Han 1999), sheep (Blaxter 1978) and goat (Forster *et al*. 1981), where suppression effects were found at altitudes ranging from 2 200 to 5 000m.

Effects on rumen N metabolism and protozoa

Ammonia – N in ruminal fluid is an N source for some bacterial protein synthesis (Aguilera *et al*. 1992; González and Andres 2003). Ammonia – N concentration of ruminal fluid has been used to evaluate the usage status of ammonia – N (Busquet *et al*. 2006; Castillejos *et al*. 2006; Benchaar *et al*. 2008). In this *in vitro* study, decrease of microbial protein production at 3 600m compared with 1 600m (0.72vs 1.09mg N/72h) indicated that total bacteria number could also be decreased at 3 600m. Accumulation of ammonia – N could be due to the decrease of bacteria in the medium, or the capacity of bacteria to use ammonia – N, or both.

It has been reported that protozoa are a net consumer of N resources (Benchaar *et al*. 2008). Ruminal fluid ammonia – N concentration has a positive relationship with the number of protozoa. Protozoa cannot convert ruminal ammonia – N into true protein. They meet their requirement of protein from engulfed bacteria (Owen and Coleman 1976; Coleman and Laurie 1977; Coleman and Sandford 1979). Therefore, the ability of rumen fermentation may be impaired by increase of protozoa number. According to the results of this study, the number of protozoa was increased in the ruminal fluid of cows at high altitude compared with that at low altitude (42.1 vs 27.6×10^4/ml) after 72h of incubation. As a result, efficiency of microbial protein synthesis and bacterial – N to alfalfa N were also decreased in the in vitro system. The increase of ammonia – N concentration in the medium at altitude of 3 600m could be due to either the decrease of bacteria, or the increase of protozoa, or both.

Effect of high altitude on O_2 or BMR

The response of sheep (Blaxter 1978) and goats (Forster *et al.* 1981) to high altitude with low O_2 has been documented over the past 34 years. It has been found that O_2consumption or BMR was reduced when these mammals were exposed to the Tibetan plateau. But responses of O_2consumption in large – size mammals to high altitude, such as yaks, dairy cattle, and yellow cattle are variable. It has been reported that yak *Bos grunniens* and yak *Poephagus grunniens* (Han *et al.* 2002) and deer mice (Hayes 1989) that are native to high altitude areas are better adapted than the other species. The BMR or O_2consumption of these animals remained unchanged when they were exposed to different altitudes ranging from 2 260 to 4 270m. The possible explanation is these animals are native to the Tibetan plateau, and they have relatively large hearts and lungs and higher percentages of haemoglobin and blood red cells which can compensate for the deficiency in lower O_2supply at high altitude compared with other animals (Cai and Wiener 1995). In local yellow cattle, an increase of altitude increased BMR on the Tibetan plateau (Han *et al.* 2003). Leiber *et al.* (2004) also found that in dairy cattle, the energy requirement for maintenance was elevated under Alpine grazing conditions. In the present study, the BMR of Chinese Holstein cows was also increased at an altitude of 3 600m compared with 1 600m, which indicated that more energy was needed to meet the energy requirement for maintenance. It seems that the effect of high altitude with hypoxia on BMR is species – dependent.

It has been reported that observations or measurement of high altitude effects on animals should include hypoxia, humidity, radiation and cold temperature (Leiber *et al.* 2006). In our present study, hypoxia was likely to be the main aspect as experimental cows were kept in barns during the experimental period. The effect of a combination of complex environmental factors on dairy production under grazing conditions has been well documented in the past decade (Kreuzer *et al.* 1998; Han *et al.* 2003; Leiber *et al.* 2006). In these experiments, concentrations of blood glucose and blood – free amino acid were decreased, and blood β – hydroxy butyrate was increased, which strongly indicated deficiency in dietary energy provided (Kreuzer *et al.* 1998; Zemp *et al.* 1989a, 1989b; Leiber *et al.* 2006). Little is known on the effects of these aspects on Chinese Holstein cows. According to the results of this study, the BMR of Chinese Holstein cows determined using closed – circuit respiratory masks at an altitude of 3 600 m was higher than that at an altitude of 1 600m. The heart rate and respiratory rate of Chinese Holstein cows were also higher at an altitude of 3 600m than at an altitude of 1 600m (Table 6). These two physiological parameters have always been used to monitor an animal's energy metabolism status. These changed main body physiological parameters at an altitude of 3 600m indicated that Chinese Holstein cows underwent dramatic modulation when they were exposed to an altitude of 3 600m. They need more O_2 supply and energy for maintenance at high altitude than at low altitude. It has been thought that hypoxia induces a decrease in the set point for body temperature regulation in the brain, which in turn feeds back a decrease in O_2demand (Gautier 1996). The results of the present study, where there was unchanged body temperature between an altitude of 3 600m and an altitude of 1 600m did not agree with this hypothesis and O_2consumption was not reduced.

Conclusions

The rumen fermentation capacity of Chinese Holstein cow was impaired at an altitude of 3 600m compared with an altitude of 1 600m. The increase in altitude from 1 600 to 3 600m increased the number of protozoa in the rumen and decreased total VFA concentration in ruminal fluid and decreased the efficiency of microbial protein synthesis. The apparent nutrient digestibility of dietary DM, OM, NDF and ADF was lower at 3 600m than those at 1 600m, respectively. DMI remained unchanged between altitudes of 3 600 and 1 600m. In Chinese Holstein cows, the BMR increased with increasing level of altitude. Based on the results of the present study, a diet with a higher amount of energy is recommended for use in lactating dairy cows on the Tibetan plateau. An improvement of forage quality could also be considered as an effective strategy to improve energy status and production response of cows.

Acknowledgements

Authors want to thank staff of the Tibetan Institute of Animal and Veterinary Science for their generous work over this experiment. This research was funded by key scientific project of Tibet (XZ2012BAD－05) and Basal Research Foundation (BRF100203).

References

[1] Aguilera JF, Bustos M, Molina E, The degradability of legume seed meals in the rumen: effect of heat treatment. *Animal Feed Science and Technology* (1992) 36, 101 - 112. doi: 11. 1016/0377 - 8401 (92) 90090 - S

[2] AOAC 'Official methods of analysis. ' 14th edn. (Association of Official Analytical Chemists: Washington, DC) (1984)

[3] Benchaar C, McAllister TA, Chouinard PY Digestion, ruminal fermentation, ciliate protozoal populations, and milk production from dairy cows fed cinnamaldehyde, quebracho condensed tannin, or yucca schidigera saponin extracts. *Journal of Dairy Science* 91, 4 765 - 4 777. doi: 10. 3 168/jds. 2 008 - 1 338 (2008)

[4] Blaxter KL The effect of stimulated altitude on the heat increment of feed in sheep. *The British Journal* of *Nutrition* 39, 659 -661. doi: 10. 1079/BJN19780081 (1978)

[5] Brouwer E Report of sub - committee on constants and factors. In 'Energy metabolism'. EAAP Publication No. 11. pp. 441 -443. (Academic Press: London) (1965)

[6] Busquet M, Calsamiglia S, Ferret A, Kamel C Plant extracts affect in vitro rumen microbial fermentation. *Journal of Dairy Science* 89, 761 -771. doi: 10. 3168/jds. S0022 -0302 (06) 72137 -3. (2006)

[7] Cai L, Wiener G 'The yak. ' (FAO Regional Office for Asia and Pacific: Bangkok) (1995)

[8] Castillejos L, Calsamiglia S, Ferret A Effect of essential oil active compounds on rumen microbial fermentation and nutrient flow in in vitro systems. *Journal of Dairy Science* 89, 2 649 – 2 658. doi: 10. 3 168/ jds. S0022 – 0302 (06) 72341 – 4. (2006)

[9] Coblentz WK, Hoffman PC Effects of bale moisture and bale diameter on spontaneous heating, dry matter recovery, in vitro true digestibility, and in situ disappearance kinetics of alfalfa – orchardgrass hays. *Journal of Dairy Science* 92, 2 853 – 2 874. doi: 10. 3168/jds. 2008 – 1920. (2009)

[10] Coleman GS, Laurie JI The metabolism of starch, glucose, amino acids, purines, pyrimidines and bacteria by the rumen ciliate polyplastron multivesiculatum. *Journal of General Microbiology* 98, 29 – 37. doi: 10. 1099/00221287 – 98 – 1 – 29. (1977)

[11] Coleman GS, Sandford DS The uptake and utilization of bacteria, amino acids and nucleic acid components by the rumen ciliate *Eudiplodinium maggii. The Journal of Applied Bacteriology* 47, 409 – 419. doi: 10. 1111/j. 1 365 – 2 672. 1979. tb01201. x (1979)

[12] Forster HV, Bisgard GE, Klein JP Effect of peripheral chemorecepter denervation on acclimatization of goats during hypoxia. *Journal of Applied Physiology* 50, 392 – 398. (1981)

[13] Gautier H Interactions among metabolic rate, hypoxia, and control of breathing. *Journal of Applied Physiology* 81, 521 – 527. (1996)

[14] Gonzάlez J, Andres SRumen degradability of some legume seeds. *Animal Research* 52, 17 – 25. doi: 10. 1051/animres: 2003003 (2003)

[15] Han XT, Chen J, Han ZK Ruminal nitrogen metabolism and the flows of nitrogen fractions reaching the duodenum of growing yaks fed diets containing different levels of crude protein. *Acta Zoonutrimenta Sinica* 10, 34 – 43. (1998)

[16] Han XT, Xie AY, Bi CX, Liu SJ, Hu LH Effects of high altitude and season on fasting heat production in the yak *Bos grunniens or Poephagus grunniens. The British Journal of Nutrition* 88, 189 – 197. doi: 10. 1079/BJN2002610 (2002)

[17] Han XT, Xie AY, Bi XC, Liu SJ, Hu LH Effects of altitude, ambient temperature and solar radiation on fasting heat production in yellow cattle (*Bos taurus*). *The British Journal of Nutrition* 89, 399 – 408. doi: 10. 1079/BJN2003783 (2003)

[18] Hayes JP Field and maximal metabolic rates of deer mice (*Peromyscus maniculatus*) at low and high altitudes. *Physiological Zoology* 62, 732 – 744. (1989)

[19] Hungate RE 'The tureen and its microbes.' (Academic Press, Inc.: New York) (1966)

[20] Kreuzer M, Langhans W, Sutter F, Christen RE, Leuenberger H, Kunz PL Metabolic response of early – lactating cows exposed to transport and high altitude grazing conditions. *Animal Science (Penicuik, Scotland)* 67, 237 – 248. doi: 10. 1017/S1357729800009991 (1998)

[21] Leiber F, Kreuzer M, Jorg B, Leuenberger H, Wettstein HR Contribution of altitude and Alpine origin of forage to the influence of Alpine sojourn of cows on intake, nitrogen conversion, metabolic stress and milk synthesis. *Animal Science (Penicuik, Scotland)* 78, 451 – 466. (2004)

[22] Leiber F, Kreuzer M, Leuenberger H, Wettstein HR Contribution of diet type and pasture conditions to the influence of high latitude grazing on intake, performance and composition and renneting properties of the milk of cows. *Animal Research* 55, 37 – 53. doi: 10. 1051/animres: 2005041 (2006)

[23] Lila ZA, Mohammed N, Kanda S, Kamada T, Itabashi H Effect of sarsaponin on ruminal fermentation with particular reference to methane production in vitro. *Journal of Dairy Science* 86, 3 330 – 3 336. doi: 10. 3168/ jds. S0022 – 0302 (03) 73935 – 6. (2003)

[24] Makkar HP, Becker K Purine quantification in digesta from ruminants by pectrophotometric and HPLC methods. *The British Journal of Nutrition* 81, 107 – 120. (1999)

[25] Mauricio R, Owen E, Mould FL, Givens I, Teodorou MK, France J, Davies DR, Dhanoa MS Comparison of bovine rumen liquor and bovine faeces as inoculums for an in vitro gas production technique for evaluating forages. *Animal Feed Science and Technology* 89, 33 - 48. doi: 10.1016/S0377 - 8401 (00) 00234 - 0. (2001)

[26] Menke KH, Raab L, Salewski A, Steingass H, Fritz D, Schneider W The estimation of the digestibility and metabolizable energy content of ruminant feedingstuffs from the gas production when they are incubated with rumen liquor in vitro. *The Journal of Agricultural Science* 93, 217 - 222. doi: 10.1017/S0021859600086305 (1979)

[27] NRC 'Nutrient requirements of dairy cattle.' 7th revised edn. (National Academy Press: Washington, DC) (2001)

[28] Owen RW, Coleman GS The uptake and utilization of bacteria, amino acids and carbohydrates by the rumen ciliate *Entodinium longinucleatum* in relation to the sources of amino acids for protein synthesis. *The Journal of Applied Bacteriology* 41, 341 - 344. doi: 10.1111/j.1365 - 2672.1976.tb00641.x (1976)

[29] Qiao G H, Shao T, Yu C Q, Wang X L, Zhu X Q, Yang X, Lu Y A comparative study at two different altitudes with two dietary nutrition levels on rumen fermentation and energy metabolism in Chinese Holstein cows. *Journal of Animal Physiology and Animal Nutrition*. doi: 10.1111/j. 1439 - 0396.2012.01339.x (2012)

[30] Qiao G H, Shan A S, Ma N, Ma Q Q, Sun Z W Effect of supplemental Bacillus cultures on rumen fermentation and milk yield in Chinese Holstein cows. *Journal of Animal Physiology and Animal Nutrition* 94, 429 - 436. (2010)

[31] Russell JB, Martin SA Effect of various methane inhibitors on the fermentation of amino acids by mixed rumen microorganisms in vitro. *Journal of Animal Science* 59, 1 329 - 1 338. (1984)

[32] Van Soest PJ, Robertson JB, Lewis BA Methods for dietary fiber, neutral detergent fiber, and non - starch polysaccharides in relation to animal nutrition. *Journal of Dairy Science* 74, 3 583 - 3 597. doi: 10.3168/jds.S0022 - 0302 (91) 78551 - 2. (1991)

[33] Xue B, Han X T A comparative study on the protein degradability of foodstuffs in the rumen of growing yaks and growing Holsteins. *Chinese Journal of Herbivore Science* 1, 3 - 7. (1999)

[34] Zemp M, Blum J W, Leuenberger H, Kunzi N Influence of high altitude grazing on productive on physiological traits of dairy cows. II. Influence on hormones, metabolites ad haematological parameters. *Journal of Animal Breeding and Genetics* 106, 289 - 299. doi: 10.1111/j. 1439 - 0388.1989.tb00243.x. (1989a)

[35] Zemp M, Leuenberger H, Kunzi N, Blum JW Influence of high altitude grazing on productive on physiological traits of dairy cows. I. Influence on milk production and body weight. *Journal of Animal Breeding and Genetics* 106, 278 - 288. doi: 10.1111/j. 1439 - 0388.1989.tb00242.x. (1989b)

[36] Zinn RA, Owens FN Arapid procedure for purine measurement and its use for estimating net ruminal protein synthesis. *Canadian Journal of Animal Science* 66, 157 - 166. doi: 10.4141/cjas86 - 017. (1986)

(Published the article in Effect of High Aftitude on Dairy Cows, 2013 (1) affect factor: 0.986)

Efficiency of in Vitro Embryo Production of Yak (*Bos grunniens*) Cultured in Different Maturation and Culture Conditions

Guo Xian[1,2], Ding Xuezhi[1,2], Pei Jie[1,2], Bao Pengjia[1,2], Liang Chunnian[1,2], Chu Min[1,2], Yan Ping[1,2]

(1. Lanzhou Institute of Animal and Veterinarian Pharmaceutical Science, Chinese Academy of Agricultural Sciences, Lanzhou 730050, China; 2. Key Laboratory of Yak Breeding Engineering of Gansu Province, Lanzhou, China)

Abstract: A study was conducted to evaluate the potential of immature oocytes from the ovaries of yak. The effects of maturation times (23 ~ 30h), hormones (FSH, LH, E_2), sera (fetal calf serum, FCS, or oestrous cow serum, ECS) and culture system (co - culture with granulosa cell, GC, or with bovine oviduct epithelia cells, BOEC, tissue culture medium, TCM, 199absence co - culture cell, and synthetic oviduct fluid with amino acids, SOFaa) on the in vitro development of in vitro matured and fertilised yak oocytes were examined. Immature oocytes surrounded with compacted cumulus cell were cultured for 23 ~ 30h in TCM 199supplemented with 10% FCS and hormones. In vitro fertilisation (IVF) was performed with frozen - thawed, caffeine and heparin treated spermatozoa from Datong yak. Oocytes were incubated with 1×10^6/ml spermatozoa for 16 ~ 18h and then cultured in co - culture system with GC or BOEC and/or SOFaa for 7 ~ 8 days, respectively. Cleavage and development to blastocysts were recorded on days 2 and 8, respectively, after the start of culture. 27 ~ 28h was superior to other culture times and addition hormones (FSH, LH, E_2) were superior to absence either of them for oocyte maturation. The best maturation of oocyte in TCM 199with 5.0mg/L LH, 0.5mg/L FSH, 1mg/L E_2. FCS was superior to ECS for development to blastocysts. Co - culture with GC was superior to SOFaa and TCM 199absence co - culture cell for cleavage, development to blastocysts. The results show that choice of culture conditions has marked effect on the development of in vitro matured and fertilised yak oocytes. The present results indicate that the co - culture with GCs is the most important factor for IVF to development into blastocysts of yak oocytes matured in vitro.

Key words: Oocyte; In vitro maturation; In vitro fertilisation; Granulosa cell; Embryo; Yak

Introduction

Yaks (*Poephagus grunniens or Bos grunniens*) are regarded as one of the world's most interesting domestic animals since they not only thrive in conditions of extreme harshness and deprivation but also provide respectable amounts of meat, milk, wool and draft power for people. A herbivore, the yak lives predominantly on the 'roof of the world' from 2 500 to 6 000m above msl, as the Qinghai Tibetan Plateau is often called. The world's total yak population is estimated to be approximately 14 million of which more than 90% is in China (Wiener *et al.* 2003). Although yaks are multi – purpose bovids and have large population, the animal suffers from certain inherent reproductive problems, such as (1) late maturity, (2) seasonality of oestrus, (3) long post partum calving intervals and (4) low reproduction, which limits its reproductive efficiency (Sarkar *et al.* 2008a, 2008b). So, how to faster multiplication and conservation of yak germplasm have got worldwide attention.

In vitro production (IVP) is a well – established embryonic biotechnology with a variety of applications in basic and applied sciences. The technology supports the production of embryos used for research investigations, for treating human infertility, for enhancing the productivity of food animals, and for conservation of endangered mammals (Bavister 2002). IVP is generally referred to as a three – step procedure, namely oocyte in vitro maturation (IVM), in vitro fertilisation (IVF) and in vitro culture (IVC) of the zygote (Balasubramanian *et al.* 2007). Although great progress has been made in IVP of cattle, which is becoming one of the most exciting and progressive procedures available for today's producers, the efficiency of yak IVP is still low (Li *et al.* 2007a, 2007b; Yan *et al.* 2007; Zi *et al.* 2008). There are still many problems need to be solved or improved, such as optimum culture time and medium for cytoplasm maturation of oocytes, unknown factors affecting embryo developmental potential, and so on.

In the present study, we first tested the effect of IVM time (23 ~ 30h) of oocyte on the cleavage rate after fertilisation. We then compared the outcome of addition of different hormone combinations (FSH, LH, E_2) in maturation medium to the cleavage rate after *in vitro* fertilisation. The effects of different sera (fetal calf serum, FCS, or oestrous cow serum, ECS) on the efficiency of IVP in yak were also done. Finally, we evaluated the efficiency of different embryo culture system (co – culture with granulosa cell, GC, or with bovine oviduct epithelia cells, BOEC, tissue culture medium, TCM, 199 absence co – culture cell, and synthetic oviduct fluid with amino acids, SOFaa).

Materials and methods

Collection of oocytes

The ovaries were obtained from yaks killed at a local slaughter house and were transported in physiological saline 0.9% (w/v, NaCl) at 32 ~ 38℃ to the laboratory within 3h during October – December. The cumulus – oocyte complexes (COCs) (4 ~ 8/ovary) were collected from follicles of 2 ~ 8mm in diameter with an 18 – gauge needle attached to a 10ml disposable syringe. Only oocytes with an unexpanded cumulus oophorus and evenly granulated cytoplasm were cultured in a polystyrene cell culture dish (35mm × 10mm) containing maturation medium of 50μl microdrop covered with mineral oil.

IVM of oocytes

After washed three times in modified Dulbecco's phosphate buffered saline supplemented with 1% (w/v) polyvinyl pyrrolidone (Sigma) and once with the maturation medium (TCM 199 with Earle's salts and L – glutamine, Gibco, Cat. 31100 – 035, NY, USA) supplemented with 10% sera (FCS or ECS). The medium was also supplemented with FSH (Follicle stimulation hormone, Institute of zoology, Chinese academy of sciences), LH (Luteinizing hormone, Sigma, L7134 – 5mg Lot 066 K1569,) and E_2 (β – Estradiol, Sigma, Lot# 110M0138V). The oocytes were introduced into 50μl the maturation medium covered with mineral oil in a polystyrene cell culture dish (35mm × 10mm) and cultured for 27 ~ 28h at 38.5℃ in a CO_2 incubator under 5% CO_2 in air and high humidity.

The evaluation of oocyte maturation was referred by Choi *et al.* (1998). After culture, oocytes were fixed in acetic acid – ethanol (1 : 3), stained with 1% orcein and examined under a phase contrast microscopy (×400). Maturation has been assumed due to the presence of Metaphase II chromosome with the first polar body.

Experiment 1, 4 different culture time (23 ~ 24h, 25 ~ 26h, 27 ~ 28h and 29 ~ 30h) for oocyte maturation were designed in TCM 199 with Earle's salts and L – glutamine supplemented with 10% FCS, FSH, LH and β – Estradiol.

Experiment 2, 4different content hormons (0.5mg/L FSH + 1mg/L E_2, 5.0mg/L LH + 0.5mg/L FSH, 5.0mg/L LH + 1mg/L E_2, 5.0mg/L LH + 0.5mg/L FSH + 1mg/L E_2) for oocyte maturation were defined in TCM 199 with Earle's salts and L – glutamine supplemented with 10% FCS.

Experiment 3, 2 different serum (10% FBS or 10% ECS) for oocyte maturation were prepared in TCM 199 with Earle's salts and L – glutamine supplemented with hormones (FSH, LH and E_2) or absence.

Sperm preparation

Two 0.25ml frozen straws of semen from Datong male yak were thawed at 38.5℃ and were prepared for sperm capacitation. The thawed semen was layered under 1ml BO medium (Brackett and Oliphant 1975) (pH 7.4, capacitation medium) with 10mM caffeine (Sigma, C – 4144Lot 69H1232) and 3mg/ml BSA (Sigma, A6003 – 5G Lot# 010M7400V) in conical tubes for a swim – up procedure. The top 0.8ml medium was then collected after incubator for 1h at 38.5℃. The pooled medium containing spermatozoa was washed twice (400g, 10min) with capacitation medium. The final pellet of semen was re – suspended with 0.2ml BO medium (pH 7.4, fertilisation medium) with 6mg/ml BSA and 20μg/ml Heparin (Sigma – ALDRICH, H4784 – 250MG, Lot 050M1153). The spermatozoa were incubated for 30min at 38.5℃.

Fertilisation

The microdrops of 50μl fertilisation medium were prepared in 35mm × 10mm polystyrene cell culture dish using sterile tips. The microdrops were covered with mineral oil. The dishes put in incubator to equilibrate at least 2h. The COCs were treated by 2% Hyaluronidase (Sigma H4272). After 3min, oocytes were washed three times with fertilisation medium. Then maturation oocytes were transferred into the microdrop (about 20oocytes/microdrop). And 20 ~ 50μl of the fertilisation medium treated spermatozoa were also add to the microdrop to give the final sperm concentration of 1×10^6cells/ml. After in vitro insemination, oocytes and spermatozoa were incubated for 16 ~ 18h at 38.5℃ under 5% CO_2 and high humidity.

Subsequent culture

For examining the developmental capacity to blastocyts, the same procedures for in maturation and fertilisation were carried out. After 16 ~ 18h of fertilisation, all embryos in the four different groups were transferred to a co - culture medium (TCM 199 supplemented with 10% FCS) with BOEC or GC. BOEC was isolated by opening the oviduct longitudinally and scraping the mucosal epithelial layer with a sterile glass slide, and was further processed as described by Reischl *et al.* (1999). In brief, cells were collected in 2.5ml Hepes - buffered TCM 199 with 10% FCS, and were poured from three oviducts before being washed twice by centrifugation at 200g for 5min each. The cell pellet was incubated in 2ml 0.25% (w/v) trypsin 0.02% (v/v) EDTA solution (GIBCO, Canada Lot 939422) for 8min at 38.5℃. Finally, the cells were washed in TCM 199with 10% FCS, centrifuged at 170g for 5min and counted before plating (Rief *et al.* 2002). The GCs were collected from antral follicles of about 10mm in diameter after dissection and washed (500g, 5min) 2 times with TCM 199 addition 10% FCS with 5×10^6cells/ml. After 48h the GC layer was formed and attached to the bottom of culture dish. Then 20 embryos in each well were cultured for 8 days at 38.5℃ under 5% CO_2 in air and high humidity. The incubation medium was replaced half with new medium every 48h.

All of the culture media used were supplemented with antibiotics (100iu penicillin/ml and 100μg streptomycin/ml).

Observation and statistics

Some of oocytes used were randomly picked up for the calculation for maturation and fertilisation rates. Maturation rate was examined 27 ~ 28h after maturation and fertilisation rate was examined 47 ~ 48h after fertilisation. The early embryos were examined under the microscope every 24h after cultured. The data obtained were subjected to SPSS 16.0 analysis.

Results

Effects of culture time on IVM of yak oocyte

The medium of oocyte maturation was made of TCM 199 with Earle's salts and L - glutamine supplemented with 10% sera, FSH, LH and β - Estradiol. The maturation rate of oocytes examined after 28h of incubation was 84.3%. There were significant differences compared with 23 ~ 24h and 29 ~ 30h. The fertilisation rate examined after 18h of insemination was 46.5%. The cleavage rate was significantly higher than those of 23 ~ 24h and 29 ~ 30h. At the same time, the rates of maturation and fertilisation of 27 ~ 28h is higher than those of 25 ~ 26h. 25% of matured oocytes used developed to at least the eight cell stage by 3 ~ 4days after IVF in TCM 199 at 38.5℃ under 5% CO_2in air and high humidity. The results of the maturation rate and cleavage rate in different culture time are shown in Table 1.

Table 1 Effects of culture time on yak oocytes IVM

Culture time (h)	Number of oocytes	Percentage (mean ± s. e) of total examined	
		Matured	Cleaved (2 – cell)
23 ~ 24	276	51.3 ± 7.1[c]	25.6 ± 3.9[c]
25 ~ 26	293	78.4 ± 5.3[a]	43.7 ± 5.1[a]
27 ~ 28	284	84.3 ± 6.4[a]	46.5 ± 4.7[a]
29 ~ 30	269	72.5 ± 4.9[b]	36.4 ± 3.5[b]

* Values with different superscripts are significantly different ($P < 0.05$).

Effects of hormone on IVM and development of yak oocyte

Different hormones added to culture media, the maturation and cleavage rates of yak oocytes were different. The rate of oocyte maturation was lower than other groups in culture medium with LH and FSH except E_2. Adding FSH, LH and E_2 conditions, the rates of matured and cleaved was highest ($P < 0.05$) than FSH and E_2, LH and E_2, and FSH and LH, respectively. The results showed in Table 2 about effects of FSH, LH and E_2 to oocytes maturation and fertilisation.

Table 2 Effects of different hormons (FSH, LH and E_2) added to culture media on IVM and development into 2 – cell of yak oocytes

Hormones	Number of oocytes	Percentage (mean ± s. e) of total examined	
		Matured	Cleaved (2 – cell)
0.5mg/L FSH + 1mg/L E_2	531	76.0 ± 3.0[b]	32.5 ± 1.8[b]
5.0mg/L LH + 0.5mg /L FSH	498	13.5 ± 2.7[d]	0[d]
5.0mg/L LH + 1mg/L E_2	501	65.3 ± 4.2[c]	21.8 ± 4.0[c]
5.0mg/L LH + 0.5mg/L FSH + 1mg/L E_2	525	82.4 ± 1.9[a]	43.5 ± 3.5[a]

* Values with different superscripts are significantly different ($P < 0.05$).

Effects of serum on IVM and development of yak oocyte

Analysis of variance showed that the serum type did not affect cleavage or development into blasotocysts. The addition of hormones in both groups tended to reduce cleavage rates and there was no significant ($P < 0.05$) (Table 3). But absence of hormones in both groups to increase cleavage rates and there was significant ($P < 0.05$). The rates of maturation, fertilisation and blastocyst with FCS were higher than ones with ECS, but no significantly. Under absence of hormone conditions, the rates of maturation, fertilisation and blastocyst with FCS were lower than those with ECS ($P < 0.05$), and there were significantly, respectively.

Effects of culture system on in vitro development of yak blastocyst

Development into blastocyst was significantly ($P < 0.05$) affected by the addition of GCs and oviduct epithelial cells, but not serum or hormones. At 5 ~ 6 and 7 ~ 8days after IVF development to morulae and blastocysts were 28.7% and 10.6% in co – culture system with GC (Table 4), respectively. The rate of blastocysts in co – culture system with GC is higher than those in co – culture with BOEC, but not significantly ($P<0.05$). The rates of morulae and blastocysts in co - culture system with GC or BOEC were higher than the culture system of TCM199 with 10% FCS absence GC or BOEC and the culture system of SOFaa with 3mg/ml BSA, respectively.

The results showed in Figure 1about the photo of different development stage of oocyte and early embryo. From immature oocyte to matured oocyte, fertilised ovum to different stage of early embryo development, it needs 9 ~ 10days in TCM 199 at 38.5℃ under 5% CO_2 in air and high humidity.

Table 3 Effects of different serum on yak oocytes IVM, fertilisation and development

Serum	Hormone	Number of oocytes	Percentage (mean ± s. e) of total examined		
			Matured	Cleaved (2 - cell)	Morulae
FCS	+	379	87.2 ± 4.3^a	63.4 ± 4.7^a	27.1 ± 1.8^a
	-	342	21.3 ± 4.0^c	7.1 ± 3.8^c	3.6 ± 1.3^c
ECS	+	338	81.5 ± 3.4^a	61.7 ± 5.3^a	25.3 ± 4.2^a
	-	357	69.8 ± 2.2^b	39.6 ± 3.1^b	10.5 ± 3.6^b

*: Values with different superscripts are significantly different ($P<0.05$).

Table 4 Effects of different culture medium on yak early embryo in vitro development

Culture system	Number of Cleavage (2 - cell)	Percentage (mean ± s. e) of total examined		
		Morulae	Blastocysts	Hatched blastocysts
Co - culture with GC	242	28.7 ± 2.1^a	10.6 ± 2.5^a	4.3 ± 0.5^a
Co - culture with BOEC	239	$27.6K \pm 3.5^a$	8.7 ± 3.2^a	2.6 ± 0.3^a
TCM199 + 10% FCS absence GC and BOEC	229	0^c	0^c	0^c
SOFaa with 3 mg/ml BSA	217	11.0 ± 2.7^b	3.0 ± 2.6^b	0.8 ± 0.2^b

*: Values with different superscripts are significantly different ($P<0.05$).

Discussion

In vitro fertilisation of cow ova began in the 1970s (Brackett *et al.* 1978), and the first calf was born after IVF in 1981 (Brackett *et al.* 1982). Production in vitro offers a complementary strategy for enhancing and protecting the genetic diversity of rare or excellent populations. Although other calves have since been born (Brackett *et al.* 1984), the technology used has been considered unsuitable for field application. Few studies have scientific reported IVF of yak. In the present study we have obtained yak embryos that can be used of the introduction of new genetic information and a variety of other studies.

There are many factors effecting oocyte maturation and embryo cleavage in vitro in bovine. Some of the key investigations have focused on culture time for oocyte maturation. It is reported that 24h was the idea culture time for oocyte IVM of bovine (Picco *et al.* 2010). But in this experiment, it is improper time for oocyte maturation of yak; 27 ~ 28h was superior to 23 ~ 24h, 25 ~ 26h, 29 ~ 30h for oocyte maturation of yak, respectively. In vitro maturation time of yak oocyte is longer than other cattle, it is possible to relation to living environment and physiological character of yak.

We have obtained a high maturation rate (84. 3%) in our culture system. This is partly due to the fact that cumulus cells surrounding the oocyte expanded well during the maturation process. The importance of cumulus expansion during the maturation process has been suggested by Ball *et al.* (1984).

Culture of yak zygotes to the blastocyst stage in the most often recommended option for testing new batches of medium, its components, sera, hormones, BSA, etc. The results of such tests should be interpreted with caution because they depend on multiple factors such as the genetic background of the yak and the type of the medium used. Under conventional conditions for IVM culture, which is supplemented with FCS and hormones such as gonadotrophins and steroids, bovine oocytes with cumulus cells matured after 23 ~ 24h. It appears that hormone plays a significant role during IVC, and that hormone is mainly a problem in the presence of concentration in medium. In the experiment, FSH, LH and E_2 play important roles in oocyte maturation. Hormones are necessary during IVM of yak oocytes for subsequent development. Serum is a common constituent of culture media. Significant research has been done regarding the role of sera as growth factors in culture medium for mammalian preimplantation embryos. Although fetal formation is the ultimate assessment of embryo viability after IVC, embryo transfer is not always practical and is affected by additional factors involved in the procedure.

Successful embryo culture depends on multiple parameters, with the geneticbac kground of the embryos and the medium composition being the most important. It is well known that the culture of embryos in reduced volume of medium and/or in groups increases blastocyst development, blastocyst cell number and, most importantly, viability after transfer. The choice of the medium also depends on the actual purpose for which it is going to be used. Early formulations of embryo culture media based on balanced salt solutions, lacking co – culture cell and supplemented only with carbohydrates and BSA, supported development to term, but embryos exhibited a delayed cleavage rate and reduced viability after transfer. It had reported that the development of a two cell bovine embryo to the blastocyst stage had a block of development.

Historically, in vitro cultured bovine embryos struggled to pass the so – called '8 – to 16 – cell developmental block', a culture – induced event during the maternal – to – zygotic transition (Eyestone and First 1991). In vitro culture systems resembling the reproductive tract environment were improved to support preimplantation embryo development to pass this block. BOEC, GC and SOF (synthetic oviduct fluid) and other simple media were capable of supplying the embryonic requirement in vitro. The general lack of understanding about the developmental block in vitro, as well as early developmental processed and metabolism of bovine embryos, led to the advent of co – culture systems using somatic cells of the reproductive tract, typically BOEC (Thibodeaux *et al.* 1992).

In summary, mammalian preimplantation embryo culture is a constantly developing field. The use of specific media for IVC of preimplantation stageembryos allowed studies of the early development and manipulation of the mammalian genome. These media were designed for somatic cell culture (TCM199 and Ham's F10) and embryo culture media (SOF and modified Parker's medium). First, the effect of common cell culture media ((TCM) 199 and Han's F10) and embryo culture

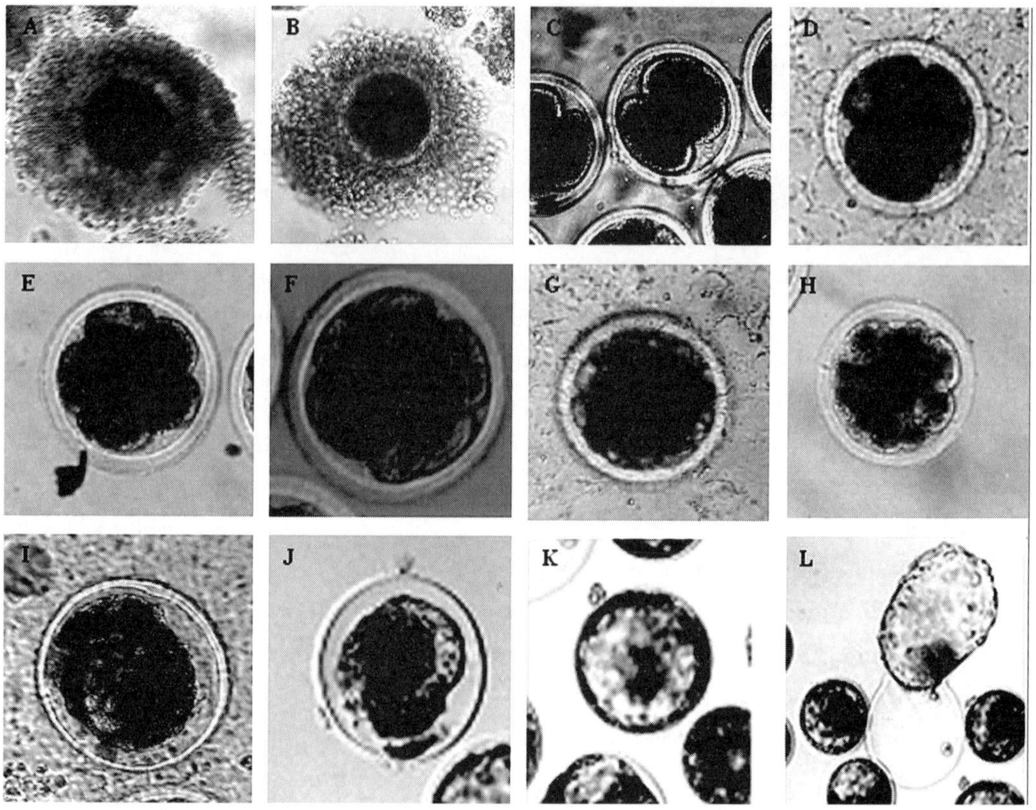

Figure The photo of oocyte and embryo in different development stage for yak

A: Immature oocyte (d - 1); B: Matured oocyte (d 0); C: 2 - cell embryo (d 1 - 2); D: 4 - cell embryo (d 2); E: 8 - cell embryo (d 2 - 3); F: 8 - 16cell embryo (d 3); G: 16 cell embryo (d 3 - 4); H: Compact Morula (d 4 - 5); I: Compact morula - early blastocyst (d 5 - 6); J: Early blastocyst (d 6 - 7); K: Blastocyst (d 7 - 8); L: Hatching blastocyst (d 8 - 9).

media (SOF and minimum essential medium (MEM)) on differentiated BOEC or GC growth was compared. TCM199 has been traditionally used in successful IVF for many years. Because hypoxanthine content is low in M199, it is beneficial to oocyte development in livestock. So, TCM199 is the basicmedium in the study. There are four different media to apply to study embryo development. SOFaa was unsuitable for culture early embryo of yak. In the present study the effects of BOEC or GC co - culture on yak IVP embryo development were evaluated to assess its suitability as a model for studying embryo - maternal communication. The beneficial effects on blastocyst development are thought to be due to embryotrophic factors provided by epithelial cells (Gandolfi *et al.* 1992) and glucose concentration by these cells (Bavister 1995). The effects of co - culture on embryo development are typically evaluated by rates of cleavage and development to the blastoyst stage. But as the co - culture system with GC is easier to establish than a co - culture system with BOEC. And in vitro development of yak embryo could also obtained beneficial factor from coculture system with GC.

Taking into consideration of all these factors, the zygotes are cultured in defined conditions. Blastocyst formation, assessed at a specific time, should reach more than 80%. Yak are considered to be seasonally polyestrous breeding occurs from July to November. Oocytes collected at the end of breeding season or in anestrous season of yak is the main reason lead to low cleavage rates and blastocyst rates in the study. Although the percentage of embryos developed to blastocysts was low (10.6%), the results of this study have verified the viability of embryos (blastocysts) developed in a co - culture system with yak cumulus cells.

In conclusion, the present study shows that the success of a particular culture system depends on the correct combination of a number of parameters. One of the most important findings of this study was that the IVM/IVF oocytes from yak could be matured in vitro and inseminated with frozen - thawed semen. In our laboratory, we prefer to culture oocyte or embryo with GC, 10% FCS, and incubations are most often carried out in a 38.5℃, humidified atmosphere of 5% CO_2, 95% air, regulated automatically.

Acknowledgements

The work was supported or partly supported by grants from Central Public - interest Scientific Institution Basal Research Fund (Contract No. 1610322011002), China Agriculture Research System (Contract No. CARS - 38), and Special Fund for Agro - Scientific Research in the Public Interest (Contract No. 201003061).

References

[1] Balasubramanian S, Son WJ, Mohana Kumar B, Ock SA, Yoo JG, Im GS, Choe SY, Rho GJ. 2007. Expression pattern of oxygen and stress - responsive gene transcripts at various developmental stages of in vitro and in vivo preimplantation bovine embryos. Theriogenology 68: 265 - 275.

[2] Ball GD, Leibfried ML, Ax RL, First NL. 1984. Maturation and fertilization of bovine oocyte in vitro. Journal of Dairy Science 67: 2 775 - 2 785.

[3] Bavister BD. 1995. Culture of preimplantation embryos: facts and artefacts. Human Reproduction Update 1/2: 91 - 148.

[4] Bavister BD. 2002. Early history of in vitro fertilization. Reproduction 124: 181 - 196.

[5] Brackett BG, Bousquet D, Boice ML, Donawick WJ, Evans JF, Dressel MA. 1982. Normal development following in vitro fertilization in the cow. Biology of Reproduction 27: 147 - 158.

[6] Brackett BG, Keefer LL, Troop LG, Donawick WJ, Bennett KA. 1984. Bovine twins resulting from in vitro fertilization. Theriogenology 21: Abstract 224.

[7] Brackett BG, Oh YK, Evans JF, Donawick WJ. 1978. In vitro fertilization of cow ova. Theriogenology 9: Abstract 89.

[8] Brackett BG, Oliphant G. 1975. Capacitation of rabbit spermatozoa in vitro. Biology of Reproduction 12: 260 - 274.

[9] Choi YH, Takagi M, Kamishita H, Wijayagunawardane MPB, Acosta TJ, Miyazawa K, Sato K. 1998. Developmental capacity of bovine oocytes matured in two kinds of follicular fluid and fertilized in vitro. Animal Reproduction Science 50: 27 - 33.

[10] Eyestone WH, First NL. 1991. Characterization of developmental arrest in early bovine embryos cultured in vitro. Theriogenology 35: 613 - 625.

[11] Gandolfi F, Brevini TAL, Modina S, Passoin L. 1992. Early embryonic signals: embryo maternal interactions before implantation. Animal Reproduction Science 28: 269 - 276.

[12] Li Y, Dai Y, Du W, Zhao C, Wang L, Wang H, Liu Y, Li R, Li N. 2007a. In vitro development of yak (*Bos grunniens*) embryos generated by interspecies nuclear transfer. Animal Reproduction Science 101: 45 – 59.

[13] Li Y, Li S, Dai Y, Du W, Zhao C, Wang L, Wang H, Li R, Liu Y, Wan R, Li N. 2007b. Nuclear reprogramming in embryos generated by the transfer of yak (*Bos grunniens*) nuclei into bovine oocytes and comparison with bovine – bovine SCNT and bovine IVF embryos. Theriogenology 67: 1 331 – 1 338.

[14] Picco SJ, Anchordoquy JM, de Matos DG, Anchordoquy JP, Seoane A, Mattioli GA, Errecalde AL, Furnus CC. 2011. Effect of increasing zinc sulphate concentration during in vitro maturation of bovine oocytes. Theriogenology 74: 1 141 – 1 148.

[15] Reischl J, Prelle K, Schol H, Neumuller C, Einspanier R, Sinowatz F, Wolf E. 1999. Factors affecting proliferation and dedifferentiation of primary bovine oviduct epithelial cells in vitro. Cell and Tissue Research 296: 371 – 383.

[16] Rief S, Sinowatz F, Stojkovic M, Einspanier R, Wolf E, Prelle K. 2002. Effects of a novel co – culture system on development, metabolism and gene expression of bovine embryos produced in vitro. Reproduction 124: 543 – 556.

[17] Sarkar M, Chakraborty P, Sharma BC, Deka BC, Duttaborah BK, Mohanty TK, Prakash BS. 2008a. Assessment of superovulatory responses in terms of palpable corpora lutea and embryo recovery using plasma progesterone in yaks (*Poephagus grunniens* L.). Research in Veterinary Science 85: 233 – 237.

[18] Sarkar M, Sengupta DH, Dutta Bora B, Rajkhoa J, Bora S, Bandopadhaya S, Ghosh M, Ahmed FA, Saikia P, Mohan K, *et al.* 2008b. Efficacy of Heatsynch protocol for induction of estrus, synchronization of ovulation and timed artificial insemination in yaks (*Poephagus grunniens* L.). Animal Reproduction Science 104: 299 – 305.

[19] Thibodeaux JK, Menezo Y, Roussel JD, Hansel W, Goodeaux LL, Thompson DL, Jr Godke RA. 1992. Coculture of in vitro fertilized bovine embryos with oviductal epithelial cells originating from different stages of the estrous cycle. Journal of Dairy Science 75: 1 448 – 1 455.

[20] Wiener G, Jianlin H, Ruijun L. 2003. The Yak. 2nd ed. Bangkok, Thailand: RAP publication 2003/06, FAO. Yan P, Xu BZ, Guo X, Pan HP, Yang BH. 2007. In vitro maturation of yak oocytes. Chinese Journal of Veterinary Science 27: 130 – 133.

[21] Zi X D, Lu H, Yin R H, Chen S W. 2008. Development of embryos after in vitro fertilization of bovine oocytes with sperm from either yaks (*Bos grunniens*) or cattle (*Bos taurus*). Animal Reproduction Science 108: 208 – 215.

(Published the article in Journal of Applied Animal Research affect factor: 0.4)

Genetic Diversity Analysis of DRB3. 2 in Domestic Yak (*Bos grunniens*) in Qinghai-Tibetan Plateau

Bao Pengjia[1,2], Yan Ping[1,2], Liang Chunnian[1,2], Guo Xian[1,2], Pei Jie[2], Chu Min[2] and Zhu Xinshu[2]

(1. Lanzhou Institute of Animal Science and Veterinary Pharmaceutics, Chinese Academy of Agricultural Science, Lanzhou, Gansu 730050, China. 2. Key Laboratory for Yak Genetics, Breeding and Reproduction Engineering of Gansu Province, Lanzhou 730050, China.)

Abstract: DRB3 gene has been extensively evaluated as a candidate marker for association with many bovine disease and immunological traits. A hemi-nested polymerase chain reaction – sequencing method was used to investigate the polymorphisms of DRB3. 2 gene from 209 individuals in three different domestic yak (*Bos grunniens*) populations (62 Tianzhu white yaks, 78 Gannan yaks and 69 Datong yaks) from the Qinghai-Tibetan Plateau. Sixty-three polymorphic sites and 143 haplotypes were detected. The percentage of polymorphic sites in Gannan Yak (GNY), Tianzhu white Yak (TWY) and Datong Yak (DTY) were 21. 80, 29. 95 and 12. 95%, while the haplotype diversity were 0. 9987, 0. 9984 and 0. 9855, respectively. At the amino acid level, Glu had the highest content; the percentage was 12. 326%, followed by Arg (10. 315%), Phe (10. 804%), Val (8. 346%), Gly (8. 315%), Leu (6. 606%) and Ala (5. 851%), whereas Met and Ile were below than 1%. Only 19 amino acids were found in DTY, Met was lost. Among the synonymous codons, whose third base was G and/or C had a higher usage frequency. Most variability were found in amino acid residues 11, 13, 26, 28, 30, 32, 37, 56, 57, 59, 60, 61, 67, 70, 71, 72, 73 and 74. In GNY, the residues at positions 71, 11and 72 were highly polymorphic with 8, 7 and 7, at 50, 58, 70, 74 and 78, the residues were selectively polymorphic than other yak populations; the other polymorphic sites were common in the populations. The results of this study indicated that the Chinese domestic yak populations in the Qinghai – Tibetan Plateau have abundant polymorphism in DRB3. 2, and the GNY was the highest, followed by TWY and DTY.

Key words: Domestic Yak; Hemi – nested PCR; BoLA – DRB3. 2; Polymorphism.

Introduction

The bovine lymphocyte antigen (BoLA) system is the major histocompatibility complex (MHC) of cattle. The genes located in the MHC class II region encode glycoproteins that are composed of α – and β – chains that bind exogenous peptides within the cell and present them to CD4 – positive T helper cells (Banchereau and Steinman, 1998). The immunological importance of the MHC genes and their possible role in disease resistance have been a major impetus for research on the MHC system in cattle denoted BoLA. The MHC class II genes in cattle have been shown to be similar to those of humans in structure (Bensaid *et al.*, 1991). So far, a single DRA locus and three DRB loci, two DQA and DQB loci, and single DOB, DNA, DYA, DYB and DIB loci have been characterized (Andersson *et al.*, 1986a, b, 1988a; Andersson and Rask, 1988b; Stone and Muggli – Cockett, 1990). DR and DQ have been identified as the twoprincipal class II molecule in ruminants. In DR sub – region of cattle, at least three different DRB loci have been described along with pseudogene and gene fragments (Ellis and Ballingall, 1999). However, DRA and DRB3 have been found as major expressed gene pair (Lewin *et al.*, 1999). Furthermore, DRB3 has been found to be highly polymorphic and it is responsible for the difference in the susceptibility to infectious disease. Polymorphism of BoLA – DRB3 is confined mainly to the second exon that encodes for β1 domain, responsible for peptide-binding sites (Sachinandan De *et al.*, 2011).

Yak species (*Bos grunniens*) is the most important grazing livestock for beef and milk productions on the Qinghai – Tibetan Plateau, as it represents a unique bovine species adapted to the Tibetan Plateau of China at altitudes of 3 000m above sea level, where oxygen content is only 33% of that at sea level and the intensity of ultraviolet radiation is 3 to 4 times that in lowland areas (Storz *et al.*, 2010). Consequently, due to natural selection adapted to such environment, yak likely have special physiological mechanisms of resistance to infectious diseases because they are usually not artificially immunized. Recent studies have demonstrated that the MHC allele diversity is associated with the ability to recognize a large number of antigens, resulting in a more efficient immune response (Behl *et al.*, 2007; Fernández *et al.*, 2008). Analysis of the BoLA-DRB3 gene is of special interest at least for two reasons: a high functional importance of the gene (one of the key genes control bacterial infections) and a high level of polymorphism (Ali and Abbas, 2011). So it may be interesting to assess the level of allele diversity in the BoLA-DRB3 gene in a population that is under great pressure for survival under tough conditions, such as the domesticated yak populations in the Qinghai – Tibetan Plateau. In addition, this polymorphism can be used to study the genetic relationships between populations and to assess their levels of genetic differentiation.

Presently, extensive information is available on the levels of genetic diversity of exon 2of the DRB3 gene in different populations of cattle obtained by amplification of this segment by PCR, and subsequent digestion with endonucleases (Gilliespie *et al.*, 1999; Mota *et al.*, 2004; Behl *et al.*, 2007). However, information on DRB3. 2 gene and its polymorphisms in domestic yak still remains very scarce. The objective of this study was to investigate the level of genetic diversity present in the BoLA – DRB3. 2 locus in three domestic yak populations from Qinghai-Tibetan Plateau in Northwest China.

Materials and Methods

Sample collection and genomic DNA isolation

Blood samples were obtained from 209 yaks belonging to three Chinese domestic yak populations: Tianzhu White Yak (TWY, n = 62), Gannan Yak (GNY, n = 78) and Datong Yak (DTY, n = 69). Approximately 10ml of blood was collected from each animal via the jugular vein. The whole blood was preserved in acid citrate dextrose solution and stored at −70℃. Genomic DNA was isolation from the blood by the Relax Gene Blood DNA System (TIANGEN Biotech, China).

Amplification of DRB3. 2

A hemi-nested PCR method was used to amplify the DRB3. 2 gene by using the primers published by Van Eijk *et al.* (1992). The oligonucleotide primers HL030 (5′ − ATCCTCTCTCTGCAGCACATTTCC − 3′), HL031 (5′ − TTTAAATTCGCGCTCACCTCGCCGCT − 3′) and HL032 (5′ − TCGCCGCTGCACAGTGAAACTCTC − 3′) were used in the polymerase chain reaction (PCR). The first round PCR was carried out in a final volume of 25μL containing: 12. 5μL 2 × Hotstart Taq PCR Master Mix, 0. 5mM of each HL030 and HL031 primers, 100ng of DNA, and ddH_2O up to 25μL. The cycling conditions were as follows: an initial denaturation step of 4min at 95℃ followed by 10cycles of 1min at 94℃, 2min at 60℃, and 1min at 72℃. The last polymerization step was extended for 10min at 72℃.

Briefly, 3μL of first − round reaction product was transferred to a new tube with 50μL of PCR buffer containing primers HL030 and HL032. Primer HL032 is internal to the sequence of the amplified product of the first − round PCR and has eight bases that overlap with primer HL031 (underlined in the text above) and 25μL of 2 × Hotstart Taq PCR Master Mix at the same concentrations as described above. The cycling conditions for the second round of PCR were as follows: 30 cycles of 45s at 94℃ and 90s at 65℃ as the annealing extension step, followed by a final extension step of 5min at 72℃ and conserved at 4℃.

Sequence preparation and genetic analysis

All of the DNA samples were purified by using a TIANgel Midi Purification Kit (TIANGEN Biotech, China), then tested and entrusted to Sangon Biotech Co., Ltd. (Shanghai, China) for sequencing. The results were corrected by comparing with the sequencing peak map; all the sequences were blasted with the DRB3. 2 gene of yak sequence, and the inaccurate and non − coding fragments of the gene were deleted. The BioEdit program v7 (Hall TA, 1999.) was used to ClustalW multiple alignment. The DnaSP program v5 (Librado and Rozas, 2009) was used to analyze DNA polymorphism and haplotype. The MEGA program v5 (Kumar *et al.*, 1993) was used to calculate the relative frequencies of nonsynonymous (dy) and synonymous substitutions (ds) according to Nei and Gojobori (1986), Jukes and Cantor's (1969) correction was applied for multiple hits. The peptide binding groove was according to the model of Brown *et al.* (1993).

Results

The result of polymerase chain reaction

Using the primers HL030, HL031and HL032 to amplify yak DRB3. 2, we got a 284bp segment (Figure), blast with the DRB3. 2 gene of yak (download from the GenBank) found that it is the DRB3. 2 gene of yak.

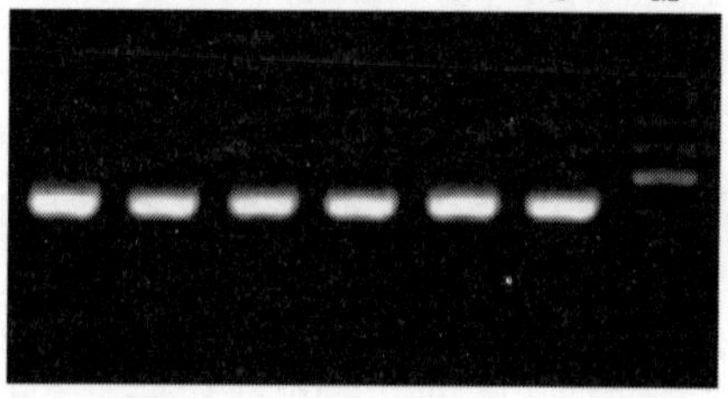

Figure The result of polymerase chain reaction. Lanes 1 to 6, PCR product; lane M, 100bp DNA ladder marker

The nucleotide diversity of DRB3. 2 in yak

Polymorphic loci variation

In the 209 animals tested, 63 polymorphic sites in BoLADBR3. 2 were indentified (Table 1). The percentage of polymorphic sites is 31. 50%, singleton variable sites are 17 (26. 98%), and parsimony informative sites are 46 (73. 02%). The transition and transversion of DRB3. 2 in the three yak populations are shown in Table 2. In DTY and TWY, transversion ratio was higher than transition, in GNY this ratio is contrary, and T/C variance is the major variation in transition.

Table 1 Polymorphic information of DRB3. 2 in yak

Populations	N	P (%)	SP	PIP
TWY	62	29. 95	21	41
DTY	29	12. 95	14	15
GNY	46	21. 80	2	44
Yak	63	31. 50	17	46

* N, Number of polymorphic sites; P, percentage of polymorphic sites; SP, single polymorphic sits; PIP, parsimony informative polymorphic sites.

Table 2 Transition and transversion of DRB3. 2 in three yak populations

Population		A	T	C	G	N	C*	N/C*
DTY	A	—	5. 94	7. 37	7. 69	38. 85	61. 14	0. 64
	T	6. 7	—	14. 55	10. 56			
	C	6. 7	11. 73	—	10. 56			
	G	4. 88	5. 94	7. 37	—			
GNY	A	—	4. 42	6. 02	11. 81	52. 81	47. 19	1. 12
	T	5. 12	—	19. 32	8. 04			
	C	5. 12	14. 16	—	8. 04			
	G	7. 52	4. 42	6. 02	—			
TWY	A	—	5. 18	7. 05	9. 38	43. 24	56. 76	0. 76
	T	6. 45	—	15. 92	9. 7			
	C	6. 45	11. 71	—	9. 7			
	G	6. 23	5. 18	7. 05	—			

* Each entry shows the probability of substitution (r) from one base (row) to another base (column). For simplicity, the sum of r values is made equal to 100. Rates of different transitional substitutions are shown in bold and those of transversionsal substitutions are shown in italics. The transition/transversion rate ratios are k1 = 0. 906 (purines) and k2 = 2. 175 (pyrimidines). The overall transition/transversion bias is R = 0. 689, where R = [A * G * k1 + T * C * k2] / [(A + G) * (T + C)]. N is transition and C* is transversion.

Haplotype distribution

For the 209 sequences, 143 haplotypes were detected (Table 3), and GNY had the highest haplotype diversity, while DTY had the lowest.

Table 3 Haplotype diversity of DRB3.2 in yak

Population	N	Np	Pi (%)	K	HD
DTY	69	54	2.155	4.806	0.9855
GNY	78	74	5.530	11.667	0.9987
TWY	62	59	4.117	8.481	0.9984
Yak	209	143	3.594	7.188	0.9843

* N, Sample size; Np, no. of haplotype; Pi, nucleotide diversity; K, average number of nucleotide differences; HD, haplotype diversity.

Nucleotide divergence and net genetic distance

Among the three yak populations, GNY and DTY have the biggest net genetic distance, whereas GNY and TWY is the smallest (Table 4). The biggest nucleotide divergence was between TWY and GNY, while the smallest was between TWY and DTY.

Table 4 The net genetic distance (Da) and nucleotide divergence (Dxy) among three yak populations

Population	DTY	GNY	TWY
DTY	—	0.0057	0.0029
GNY	0.0406	—	0.0013
TWY	0.0302	0.0446	—

* Above diagonal is the net genetic distance (Da) and below diagonal is the nucleotide divergence (Dxy).

The amino acid polymorphism of DRB3.2 in yak

Amino acid constitute

The amino acid constituents in the three yak populations are very different. There are 20 amino acids in GNY and TWY, but DTY only have 19 ones (Met is lost). Glu has the highest percentage (12.326%), followed by Arg, Phe, Val, Gly, Leu and Ala, with percentage of 10.3151%, 10.804%, 8.3463%, 8.3151%, 6.606% and 5.8512%, respectively. Met and Ile were below 1%.

Amino acid variation

The frequencies of codon usage in DRB3.2 among the three yak populations are different. For synonymous codons, the ones whose third base is G and/or C have a higher usage frequency (Table 5). The sharing of DRB3.2 polymorphism at the amino acid level found in yak populations is presented in Table 6. Most variability were found in amino acid residues at sites 11, 13, 26, 28, 30, 32, 37, 56, 57, 59, 60, 61, 67, 70, 71, 72, 73 and 74. In GNY, amino acid residues at positions11, 71and 72 were highly polymorphic with 7, 8 and 7 amino acids. However, residues at sites 50, 58, 70, 74, 78 showed more selectively polymorphic than other yak populations. The amino acids for other polymorphic sites were common in the experimental populations. The level of polymorphism was the highest in GNY, followed by TWY and DTY.

Table 5 The codon usage frequency of yak

Codon	Count	RSCU	Codon	Count	RSCU	Codon	Count	RSCU	Codon	Count	RSCU
UUU (F)	0	0.01	UCU (S)	0.1	0.23	UAU (Y)	2	0.96	UGU (C)	1.5	1.19
UUC (F)	8.4	1.99	UCC (S)	0.1	0.23	UAC (Y)	2.1	1.04	UGC (C)	1	0.81
UUA (L)	0	0	UCA (S)	0	0	UAA (*)	0	0	UGA (*)	0	0
UUG (L)	0.1	0.18	UCG (S)	0	0.03	UAG (*)	0	3	UGG (W)	2	1
CUU (L)	0	0	CCU (P)	0	0	CAU (H)	2.3	1.18	CGU (R)	0	0.03
CUC (L)	0	0.02	CCC (P)	0	0	CAC (H)	1.6	0.82	CGC (R)	1	0.67
CUA (L)	0.1	0.15	CCA (P)	0	0	CAA (Q)	0	0.02	CGA (R)	0	0
CUG (L)	3.9	5.65	CCG (P)	0.9	4	CAG (Q)	1.6	1.98	CGG (R)	5.6	3.83
AUU (I)	0	0	ACU (T)	0	0	AAU (N)	1.3	0.67	AGU (S)	0	0
AUC (I)	0.3	3	ACC (T)	2	2.71	AAC (N)	2.6	1.33	AGC (S)	2.9	5.52
AUA (I)	0	0	ACA (T)	0	0	AAA (K)	0	0.01	AGA (R)	2	1.38
AUG (M)	0	0	ACG (T)	1	1.29	AAG (K)	3.9	1.99	AGG (R)	0.1	0.09
GUU (V)	0	0	GCU (A)	0	0.04	GAU (D)	0	0	GGU (G)	0	0
GUC (V)	0.7	0.57	GCC (A)	2.1	2.7	GAC (D)	5.1	2	GGC (G)	1	1
GUA (V)	0	0	GCA (A)	0	0	GAA (E)	1	0.21	GGA (G)	1	0.99
GUG (V)	4	3.43	GCG (A)	1	1.26	GAG (E)	8.3	1.79	GGG (G)	2	2.01

* All frequencies are averages over all taxa. The relative synonymous codon usage (RSCU) is given in parentheses following the codon frequency; the numbers in the box means the usage frequency of this synonymous codon is high than others.

Discussion

Information about bovine MHC polymorphisms is important in the beef and dairy industry since MHC contributes substantially to fitness and resistance/susceptibility to disease (Ripoli *et al.*, 2004). Therefore, we used a hemi – nested PCR – sequencing method to detect polymorphisms in Chinese domestic yak populations from Qinghai – Tibetan Plateau. PCR products were represented by a 284bp fragment that was expected on the basis of the nucleotide sequence of the gene (Figure 1).

Polymorphic loci variations analyses demonstrated that the yak DRB3. 2 locus had a high degree of polymorphism. Li *et al.* (2005) indicated that the polymorphism of yak DRB3 exon 2 in 24 Chinese domestic yaks was rich, and they also observed 115 polymorphic sites in sequence of 234 bp segment and the percentage of polymorphism loci was 49. 15%. The results were similar to our findings. Moreover, comparing to other bovine breeds, the DRB3. 2 gene of yak was almost similar in the Holstein herds studied by other researchers (Sofia and Leif, 1995; Gelhaus *et al.*, 1995).

The haplotype analysis of the three Chinese yak populations showed that GNY has the highest haplotype diversity, nucleotide divergence and net genetic distance, followed by TWY and DTY. The reason for this result may be that as GNY is an original breed and artificial selection is focus on the older, obvious physical defects and frail individuals, so the selective pressure is very small. The DTY is a bred variety, and in the process of breeding formation, artificial selection plays a leading role, not only concerned about body appearance, but also on high adaptability, disease resistance and production performance. Hence, they faced a greater pressure in artificial selection.

Mammalian genomes are highly heterogeneous in base composition (Laurent *et al.*, 2002). A strong synonymous codon usage biased towards the codons ending at C or G was observed in DRB3.2 genes (Table 5). Codon analysis of a variety of biology confirmed that the composition limits (Bulmer, 1991), transcription selection (Sharp *et al.*, 1988), tRNA abundance (Pan and Fu, 2001), mutation pressure (Pan and Fu, 2001), gene function (Ma *et al.*, 2002; Comeron, 2004), protein secondary structure (Gu *et al*, 2004; Kahali *et al.*, 2007), gene length (Marais and Duret, 2001; Miyasaka, 2002) and CpG islands (Scaiewicz *et al.*, 2006; Woo *et al.*, 2007) are factors that form codon bias. In our study, high GC3s content was mainly due to an important factor in keeping the structure and function of gene under selection constraint.

Extensive polymorphism was also revealed in the peptide – binding amino acid region in yak populations. Out of all peptide – binding sites, in position 71, eight different amino acids were encountered followed by seven amino acids in the position 11. In the non peptide – binding region position 72, 74 and 57 were found to be highly variable, containing seven, six and six amino acid substitutions. Sachianandan *et al.* (2011) analyzed the variation of allelic forms of MHC – DRB3.2 of cattle and buffalo and compared the variation with sheep, goats and other ruminant species. They found that in peptide – binding region (PBR), positions 37 and 11 encountered seven and six amino acids. Furthermore, in the non peptidebinding region, positions 57 and 67 were found containing four amino acid substitutions. In this study, we found a high ratio of non – synonymous substitution to synonymous substitution in PBR, and this high ratio indicates that non – synonymous sites evolved faster than synonymous sites and implies balancing selection favored new variants and increased allelic polymorphism (Bergstrom and Gyllensten, 1995). The pattern and level of DRB3.2 polymorphism revealed in our study could be a consequence of adaption to the cold and oxygendeficient climate of Qinghai – Tibetan Plateau, and lack or abundance of the forage that relatively changes with the seasons.

High polymorphism of DRB3 gene promotes its use as a highly informative marker in molecular genetics and phylogenetic studies. The functional roles of DRB3 gene are different. There were many works have been done in other mammals, such as the relationship between BoLADRB3 gene and resistance/susceptibility to persistent lymphocytosis caused by bovine leukemia virus (Xu *et al.*, 1993; Udina *et al.*, 2003), mastitis caused by Staphylococcus sp. (Rupp *et al.*, 2007; Zahra *et al.*, 2003), and other diseases (Maillard *et al.*, 1996; Lewin *et al.*, 1999). Previous studies showed that many diseases observed in cattle were also reported in yak, and it appears that the incidences of some diseases may be high and this is attributed to lack of economic incentive for prevention and treatment in many cases (Gerald *et al.*, 2003). But in our previous study, we found that the prevalence of yak is obviously smaller than cattle (unpublished data).

Up to now, the relationship between the polymorphism of MHC and the disease resistance in yak is still unknown. Therefore, the understanding of MHC diversity could be very useful, and our future work is to discover the relationship between them.

Table 6 Comparison of polymorphic amino acid substitutions for DRB3. 2 molecules in yak populations

Code position	Amino acid of TWY	Amino acid of GNY	Amino acid of DTY	Code position	Amino acid of TWY	Amino acid of GNY	Amino acid of DTY
6	H	H	H	*28	DHE	DHE	DH
7	F	F	F	29	RS	R	R
8	LF	L	L	*30	YCHSQ	YCSHP	Y
*9	EQ	EQ	E	31	FY	FY	F
10	Y	Y	Y	*32	YHN	YHN	YHN
*11	SYCHR	SYCHRFA	SYCHR	33	N	N	N
12	KT	KT	KT	34	G	G	G
*13	SRGK	SRGK	SR	35	E	E	E
14	E	E	E	36	E	E	E
15	C	C	C	*37	FYVT	FYSNTL	F
16	H	H	H	*38	V	V	V
17	F	F	F	39	R	R	R
18	F	F	F	40	F	F	F
19	N	N	N	41	D	D	D
20	G	G	G	42	S	S	S
21	T	T	T	43	D	D	D
22	E	E	E	44	W	W	W
23	R	R	R	45	GD	GD	G
24	VL	VL	V	46	E	E	E
25	R	R	R	*47	FY	FY	F
26	FLY	FLY	FY	48	R	R	R
27	L	L	L	49	AP	A	A
50	V	VL	V	67	FIL	FILTS	FILT
51	T	T	T	68	L	L	L
52	E	E	E	69	E	E	E
53	LE	L	L	*70	REQ	REQG	REQ
54	G	G	G	*71	EGRK	EGRKANQT	ERK
55	RQ	RQ	R	72	ARGES	ARGESDT	ARS
*56	PRQ	PRQ	PRQ	73	ARG	ARG	AG
57	ADV	ADVSFP	ADVF	74	ENK	ENKATS	ENK
58	A	APT	A	75	V	V	V
59	EK	EKSVD	EK	76	D	D	D
*60	HYSQ	HYSQTA	HYSQ	77	TR	TR	T
*61	WCLR	WCL	W	78	Y	YV	Y
62	N	N	N	79	C	C	C
63	S	S	S	80	R	R	R
64	Q	Q	Q	81	H	H	H
65	K	K	K	82	KN	KN	KN
66	DE	DE	D				

Conclusion

The results obtained from the current study suggest that the Chinese domestic yak populations also have abundant polymorphism in DRB3. 2, and GNY was the highest, followed by TWY and DTY, thus suggesting an unfavorable state of the DTY population that is probably caused by inbreeding depression due to a long-term isolation and a small population size.

Acknowledgements

The study was supported by the Central Public-interest Scientific Institution Basal Research Fund (1610322009002), Program of National Beef Cattle and Yak Industrial Technology System (CARS-38), Publicinterest Industry Agricultural Science and Technology Special Program (201003061), Gansu Provincial Key Laboratory of Yak Breeding Project.

References

[1] Ali S, Abbas D. Polymorphism of BoLA – DRB 3. 2 gene in Iranian native cattle by using PCR – RFLP method. J. Cell Anim. Biol. (1986a) 5: 47 – 52.

[2] Andersson L, Bohme J, Peterson PA, Rask L. Genomic hybridization of bovine class II major histocompatibility complex genes: 2. Polymorphism of DR genes and linkage disequilibrium in the DQ – DR region. Anim. Genet. (1986a) 17: 295 – 304.

[3] Andersson L, Bohme J, Rask L, Peterson PA. Genomic hybridization of bovine major histocompatibility genes: 1. Extensive polymorphism of DQA and DQB genes. Anim. Genet. (1986b) 17: 95 – 112.

[4] Andersson L, Lunden A, Sigurdardottir S, Davies CJ, Rask L. Linkage relationships in the bovine MHC region. High recombination frequency between class II subregions. Immunogenetics (1988a) 27: 273 – 280.

[5] Andersson L, Rask L. Characterization of the MHC class II region in cattle. The number of DQ genes varies between haplotypes. Immunogenetics (1988b) 27: 110 – 120.

[6] Banchereau J, Steinman RM. Dendritic cells and the control of immunity (review). Nature (1998) 392: 245 – 252.

[7] Behl JD, Verma NK, Behl R, Mukesh M, Ahlawat SPS. Characterization of genetic polymorphism of the bovine lymphocyte antigen DRB3. 2 locus in Kankrej cattle (*Bos indicus*). J. Dairy Sci. (2007) 90 : 2 997 – 3 001.

[8] Bensaid A, Kaushal A, Baldwin CL, Clevers H, Young JR, Kemp SJ, MacHugh ND, Toye PG, Teale AJ. Identification of expressed bovine class I MHC genes at two loci and demonstration of physical linkage. Immunogenetics (1991) 33: 247 – 254.

[9] Bergstrom T, Gyllensten U. Evolution of the Mhc class II polymorphism: the rise and fall of class II gene function in primates. Immunol. Rev. (1995) 143: 14 – 31.

[10] Brown JH, Jardetzky TS, Gorgaetal JC. Three – dimensional structure of the human class II histocompatibility antigen HLA – DR1. Nature (1993) 364: 33 – 39.

[11] Bulmer M. The selection – mutation – drift theory of synonymous codon usage. Genetics (1991) 129: 897 – 907.

[12] Comeron JM. Selective and mutational patterns associated with gene expression in humans: influences on synonymous composition and intron presence. Genetics (2004) 167 : 1 293 – 1 304.

[13] Ellis SA, Ballingall KT. Cattle MHC: evolution in action Immunol. Rev. (1999) 167: 159 – 168.

[14] Fernάndez IG, José GRR, Amanda GV, Raúl UA, Rogelio AAM. Polymorphism of locus DRB3. 2 in popu-

lations of Creole Cattle from Northern Mexico. Genet. Mol. Biol. (2008) 31: 880 – 886.

[15] Gelhaus A, Schnittger L, Mehlitz D, Horstmann RD, Meyer CG. Sequence and PCR RFLP analysis of 14 novel BoLA – DRB3 alleles. Anim. Genet. (1995) 26: 147 – 153.

[16] Gerald W, Han J, Long R. The yak, second ed. The regional office for Asia and the Pacific Food and Agriculture Organization of the United Nations Bangkok, Thailand. (2003)

[17] Gilliespie BE, Jayarao BM, Dowlen HH, Oliver SP. Analysis and frequency of bovine lymphocyte antigen DRB3. 2 alleles in Jersey cows. J. Dairy Sci. (1999) 82 : 2 049 – 2 053.

[18] Gu W, Zhou T, Ma J, Sun X, Lu Z. The relationship between synonymous codon usage and protein structure in *Escherichia coli and Homo sapiens*. Biosystems (2004) 73: 89 – 97.

[19] Hall TA. BioEdit: a user – friendly biological sequence alignment editor and analysis program for Windows 95/98/NT. Nucl. Acids Symp. Ser. (1999) 41: 95 – 98.

[20] Jukes TH, Cantor CR. Evolution of protein molecules, in: Munro HN, editor, Mammalian Protein Metabolism. Academic Press, New York, pp. (1969) 21 – 132.

[21] Kahali B, Basak S, Ghosh TC. Reinvestigating the codon and amino acid usage of *S. Cerevisiae* genome: a new insight from protein secondary structure analysis. Biochem. Biophys. Res. Commun. (2007) 354: 693 – 699.

[22] Kumar S, Tamura K, Nei M. MEGA: Molecular Evolutionary Genetics Analysis User Manual. Pennsylvania State University, University Park. (1993)

[23] Laurent D, Marie S, Gwenae P, Dominique M, Nicolas G. Vanishing GC – Rich isochores in Mammalian Genomes. Genetics (2002) 162 : 1 837 – 1 847.

[24] Lewin HA, Russel GC, Glass EJ. Comparative organization and function of the major histocompatibility complex of domesticated cattle. Immunol. Rev. (1999) 167: 145 – 158.

[25] Li Q F, Li Y H, Zhao X B, Li X B, Pan Z X, Xie Z, Li N. Sequence Variation at exon2 of MHC DRB3 Locus in Bovinae. J. Agric. Biotechnol. (2005) 13: 441 – 446.

[26] Librado P, Rozas J. DnaSP v5: A software for comprehensive analysis of DNA polymorphism data. Bioinformatics (2009) 25 : 1 451 – 1 452.

[27] Ma J, Campbell A, Karlin S. Correlations between Shine-Dalgarno sequences and gene features such as predicted expression levels and operon structures. J. Bacteriol. (2002) 184 : 5 733 – 5 745.

[28] Maillard JC, Martinez D, Bensaid A. An amino acid sequence coded by exon 2 of the BoLA – DRB3 gene associated with a BoLA class I specificity constitutes a likely genetic marker of resistance to dermatophiloses in Brahman Zebu cattle of Martinique (FWI), Ann. N. Y. Acad. Sci. (1996) 791: 185 – 197.

[29] Marais G, Duret L. Synonymous codon usage, accuracy of translation and gene length in Caenorhabditis elegans. J. Mol. Evol. (2001) 52: 275 – 280.

[30] Miyasaka H. Translation initiation AUG context varies with codon usage bias and gene length in Drosophila melanogaster. J. Mol. Evol. (2002) 55: 52 – 64.

[31] Mota AF, Martinez ML, Coutinho LL. Genotyping BoLA – DRB3 alleles in Brazilian Dairy Gir cattle (*Bos indicus*) by temperaturegradient gel electrophoresis (TGGE) and direct sequencing. Eur. J. Immunogen. (2004) 31: 31 – 35.

[32] Nei M, Gojobori T. Simple methods for estimating the numbers of synonymous and nonsynonymous nucleotide substitutions. Mol. Biol. Evol. (1986) 3: 418 – 426.

[33] Pan XH, Fu JL. Molecular evolution of MHC – DOA genes I. The maintenance of interallelic divergence and the influence of GC content on gene structure. Acta Genet. Sinica, (2001) 24: 195 – 205.

[34] Ripoli MV, Liŕon JP, De JC, Luca, Rojas F, Dulout FN, Giovambattista G. Gene Frequency Distribution of the BoLA – DRB3 Locus in Saavedre ñ o Creole Dairy Cattle. Biochem. Genet. (2004) 42: 231 – 240.

[35] Rupp R, Hernandez A, Mallard BA. Association of Bovine Leukocyte Antigen (BoLA) DRB3.2 with immune response, mastitis, and production and type traits in Canadian Holsteins. J. Dairy Sci. (2007) 90: 1 029 – 1 038.

[36] Sachianandan D, Raj KS, Biswajit B. Allelic Diversity of Major Histocompatibility Complex Class II DRB Gene in Indian Cattle and Buffalo. Mol. Biol. Int. (2011) 1: 1 – 7.

[37] Scaiewicz V, Sabbia V, Piovani R, Musto H. CpG islands are the second main factor shaping codon usage in human genes. Biochem. Biophys. Res. Commun. (2006) 343: 1 257 – 1 261.

[38] Sharp PM, Cowe E, Higgins DG, Shields DC, Wolfe KH, Wright F. Codon usage patterns in *Escherichia coli*, Bacillus subtilis, *Saccharomyces cerevisiae*, *Schizosaccharomyces pombe*, *Drosophila melanogaster* and *Homo sapiens*: a review of theconsiderable within-species diversity. Nucleic Acids Res. (1988) 16 (17): 8 207 – 8 211.

[39] Sofia M, Leif A. Extensive MHC class II DRB3 diversity in African and European cattle. Immunogenetics (1995) 42: 408 – 413.

[40] Stone RT, Muggli – Cockett NE. Partial nucleotide sequence of a novel bovine MHC class II b – chain gene, BoLA – DIB. Anim. Genet. (1990) 21: 353 – 360.

[41] Storz JF, Scott GR, Cheviron ZA. Phenotypic plasticity and genetic adaptation to high – altitude hypoxia in vertebrates. J. Exp. Biol. (2010) 213: 4 125 – 4 136.

[42] Udina IG, Karamysheva EE, Turkova SO, Orlova AR, Sulimova GE. Genetic mechanisms of resistance and susceptibility to leukemia in Ayrshire and black pied cattle breeds determined by allelic distribution of gene Bola-DRB3. Russ. J. Genet. (2003) 39: 306 – 317.

[43] Van Eijk MJT, Stewart – Haynes JA, Lewin HA. Extensive polymorphism of the BoLA – DRB3gene distinguished by PCR – RFLP. Anim. Genet. (1992) 23: 483 – 496.

[44] Woo PC, Wong BH, Huang Y, Lau SK, Yuen KY. Cytosine deamination and selection of CpG suppressed clones are the two major independent biological forces that shape codon usage bias in corona viruses. Virology (2007) 369: 431 – 442.

[45] Xu A, van Eijk MJT, Park C, Lewin H. Polymorphism in BoLADRB3 exon 2 correlates with resistance to persistent lymphocytosis caused by bovine leukemia virus. J. ImmunoI. (1993) 151: 6 977 – 6 985.

[46] Zahra A, Niel K, Bonnie AM. Biological effect of varying peptide binding affinity to the BoLA – DRB3*2703 allele. Genet. Sel. (2003) 35 (1): 51 – 65.

(Published the article in Africa. Journal of Biotechnolgy, 2012
affect factor: 0.573)

Reducing Methane Emissions and the Methanogen Population in the Rumen of Tibetan Sheep by Dietary Supplementation with Coconut Oil

Ding Xuezhi, Long Ruijun, Zhang Qian, Huang Xiaodan,
Guo Xusheng, Mi Jiandui

(Lanzhou Institute of Animal and Veterinaian Pharma ceutice, CAAS, Lanzhou730050, China)

Abstract: The objective was to evaluate the effect of dietary coconut oil on methane (CH_4) emissions and the microbial community in Tibetan sheep. Twelve animals were assigned to receive either a control diet (oaten hay) or a mixture diet containing concentrate (maize meal), in which coconut oil was supplemented at 12g/day or not for a period of 4 weeks. CH_4 emissions were measured by using the 'tunnel' technique, and microbial communities were examined using quantitative real – time PCR. Daily CH_4 production for the control and forage – to – concentrate ratio of 6 : 4 was 17.8 and 15.3g, respectively. Coconut oil was particularly effective at reducing CH_4 emissions from Tibetan sheep. The inclusion of coconut oil for the control decreased CH_4 production (in grams per day) by 61.2%. In addition, there was a positive correlation between the number of methanogens and the daily CH4production (R = 0.95, $P < 0.001$). Oaten hay diet containing maize meal (6 : 4) plus coconut oil supplemented at 12 g/day decreases the number of methanogens by 77% and a decreases in the ruminal fungal population (85% ~95%) and *Fibrobacter succinogenes* (50% ~98%) but an increase in *Ruminococcus flavefaciens* (25% ~ 70%). The results from our experiment suggest that adding coconut oil to the diet can reduce CH_4 emissions in Tibetan sheep and that these reductions persist for at least the 4 – week feeding period.

Key words: Methane; Coconut oil; Tibetan sheep; Microbial communities

Introduction

Global warming is currently viewed as one of the most important environmental impacts arising from greenhouse gases. Animal husbandry, particularly of ruminants, has been identified as a significant contributor to global anthropogenic CH_4 emissions. For environmental and economic considerations, renewed efforts are being made internationally to find sustainable strategies for reducing the CH_4 production from domestic ruminants (Goel *et al.* 2009; Morgavi *et al.* 2010). The range in emissions depends greatly on the level of feed intake and composition of the diet (Hegarty *et al.* 2007; Eckard *et al.* 2010).

Certain plant oils and dietary lipids have been identified as potential methane – suppressing feed ingredients for ruminants (Beauchemin *et al.* 2007; Chilliard *et al.* 2009). Medium – chain fatty acids have been demonstrated to defaunate the rumen in vivo (Machmüller and Kreuzer 1999; Beauchemin *et al.* 2008) and significantly reduce methanogen numbers in vitro (Jordan *et al.* 2006), with coconut oil identified as being very effective. In vivo CH_4 emissions in sheep have been reduced significantly through the incorporation of coconut oil (Liu *et al.* 2011). Dietary interactions with coconut oil have been reported both in vitro (Jordan *et al.* 2006; Kongmun *et al.* 2010) and in vivo (Machmüller and Kreuzer 1999) with the greatest reduction in CH_4 occurring on diets with a low forage – to – concentrate ratio. Although several studies have highlighted the effectiveness of coconut oil as an anti – methanogenic dietary supplement in vitro (Lovett *et al.* 2003; Beauchemin *et al.* 2008) and at low planes of nutrition in vivo (Jordan *et al.* 2006), we argue that as an indigenous sheep breed which might have developed a different rumen microbial population during isolation on the Tibetan plateau. In this experiment, we tested the hypothesis that coconut oil and diets containing a mixture of forage and concentrate would reduce CH_4 emissions and numbers of methanogens in Tibetan sheep when compared to forage diets without coconut oil.

Materials and methods

Animals and treatments

Twelve wether Tibetan sheep (11 months old, with an average live weight of 25 ± 5kg) were randomly allocated into four groups; each group was fed one of four diets: (1) oaten hay as a control diet (OH); (2) control diet including frozen solid coconut oil which was directly fed at 12g/day (OHC); (3) oaten hay diet containing maize meal with a forage – to – concentrate ratio of 6 : 4 (FC); and (4) oaten hay diet containing maize meal (6 : 4) plus coconut oil supplemented at 12g/day (FCC). The sheep were adapted to each treatment for 3weeks prior to a 7 – day CH_4 measurement. The diets were offered in two equal portions at 0800 and 1600 hours and water was available at all times. Feed intake was recorded daily and adjusted to maintain 5% to 10% orts.

Methane measurement system

A 'tunnel' approach developed by Lockyer and Jarvis (1995) with some modifications was employed to monitor CH_4 emission. Briefly, this system consisted of a polythene clad tunnel of approximately 9 × 4 × 2.5m. Two small wind tunnels were used to blow air into and draw air out of the large tunnel. There was also a Hewlett Packard 5 890 gas chromatograph (Gardena, CA, USA) to measure and record the concentration of CH_4 in air entering and leaving the large tunnel, as well as apparatus to monitor and record airspeeds and temperatures.

Sampling and analytical methods

Samples of each diet were taken weekly and pooled by period for chemical analysis. The samples were analysed weekly for dry matter (DM) content to calculate daily DM intake for each sheep. Analytical DM content of the samples was determined by drying at 135℃ for 3 h (ID 930.15, AOAC 1990). Additionally, approximately 20ml of rumen fluid was obtained directly before the start of the treatment feeding (day 0), and then rumen fluid was collected every 7 days of the successive experimental periods. These samples were collected before morning feeding with a flexible stomach tube inserted into the rumen through the oesophagus. The rumen samples were filtered through four layers of cheesecloth and used for ciliate count analysis and for real – time PCR to quantify the relative numbers of bacteria, fungi, methanogens, *Ruminococcus flavefaciens* and *Fibrobacter succinogenes*.

Rumen microbial population and ciliate protoza

Rumen microbial DNA isolation was done according to the 'CTAB' protocol described by Zhou *et al.* (1996) and Murray and Thompson (1980). Species – specific PCR primers such as total bacteria, fungi, methanogen, *R. flavefaciens* and *F. succinogenes* that were used to amplify partial 16S rDNA regions (target DNA) were obtained from the International Atomic Energy Agency (Table 1). The numbers of different microbial groups were estimated in the samples using SYBR green qPCR assay. Real – time PCR amplification and detection were performed using an iQ5 detection system (Bio – Rad). The PCR conditions for different communities were from Denman and McSweeney (2006) and Denman *et al.* (2007). The population size of different microbial groups was determined relative to the total bacterial population. Relative population sizes of *R. flavefaciens*, *F. succinogenes*, total rumen fungi and methanogens were expressed as a proportion of total rumen bacterial 16S rDNA. The delta Ct values were calculated by subtracting the Ct of the target gene from the Ct value of the reference gene (16S rDNA of total bacteria). The relative expression of different groups was calculated according to the formula as $2^{-\Delta Ct}$ (Denman and McSweeney 2006). The shifts in microbial communities owing to supplementation of coconut oil and concentrate were determined by using the microbial population in the oaten hay (control) as 100. The PCR conditions used to amplify DNA from ciliate protozoa was designed as previously described by Sylvester *et al.* (2004).

Table 1 Primers designed to target species – specific regions of the different microbial groups

Target species	F/R primer	Primer sequence	Amplicon length (bp)
Bacteria	Forward	5′-CGGCAACGAGCGCAACCC-3′	200
	Reverse	5′-CCATTGTAGCAACTFTFTAFCC-3′	
Fungi	Forward	5′-GAGGAAGTAAAAGTCACAAGGTTTC-3′	200
	Reverse	5′-CAAATTCACAAAGGGTAGGATGATT-3′	
Methanogen	Forward	5′-TTCGGTGGATCDCARAGRGC-3′	200
	Reverse	5′-GBARGTCGWAWCCGTAGAATC-3′	
F. succinogenes	Forward	5′-GTTCGGAATTACTGGGCGTAAA-3′	200
	Reverse	5′-CGCCTGCCCCTGAACTATC-3′	
R. flavefaciens	Forward	5′-CGAACGGAGATAATTTGAGTTTACTTAGG-3′	200
	Reverse	5′-CGGTCTCTGTATGTTATGAGGTATTACC-3′	
Ciliate protozoa	Forward	5′-GCTTTCGWTGGTAGTGTATT-3′	200
	Reverse	5′-CTTGCCCTCYAATCGTWCT-3′	

Statistical analysis

All the measures [dry matter intake (DMI), CH_4 production and different microbial populations] were performed in triplicate. Data were subjected to one – way ANOVA followed by Duncan's multiple range test for comparison of the means among different treatments ($P < 0.05$). Analyses were conducted using the SPSS 12.0 (SPSS Inc., Chicago, IL, USA). Significant differences are indicated by different superscript letters in the tables.

Results

DMI and methane emissions

The effect of different diets on DMI and CH_4 production are presented in Table 2. The average DMI was significantly ($P < 0.05$) reduced by inclusion of either dietary coconut oil or concentrate. A 20.2% decline in DMI was observed when coconut oil and concentrates were both present in the diet.

Table 2 Effect of coconut oil supplementation on methane output and protozoa numbers

	OH	OHC	FC	FCC	SEM
DMI (kg/day)	0.84a	0.72b	0.79c	0.67d	0.019
CH_4 production					
g/day	17.8a	6.9c	15.3b	8.4cd	0.42
l/kg DMI	24.9	9.6	20.1	11.8	3.57
g/kg DMI	21.2	10.3	19.4	11.7	2.72
Protozoa numbers ($10^5 ml^{-1}$)					
Day 0	6.67	6.03	5.26	5.74	0.295
Day 7	7.42a	3.52b	1.71c	2.21d	1.293
Day 14	5.70	4.95	3.3	4.70	0.501
Day 28	5.42	4.73	3.13	4.45	0.478

* Means with different letters (a, b, c) within rows are significantly different ($P < 1.05$);
SEM standard error of the mean

Daily CH_4 emissions of the Tibetan sheep with different dietary treatments were significantly different ($P < 0.05$) (Table 2). CH_4 production from the animals fed the oaten hay alone average 17.8g/day, and adding concentrate to the diet led to a reduction in methane production of 19.3% ($P < 0.05$). However, the diet of oaten hay including coconut oil decreased the CH_4 production (in grams per day) by 61.2%, and methane emissions were lower in the animals consuming the diet contained both coconut oil and concentrate when compared to the diet containing oaten hay and coconut oil.

Microbial populations

The shifts in microbial communities are presented in Fig. 1. Similar changes in microbial groups were observed when coconut oil was included in the diet. The oaten hay with coconut oil was most inhibitory (91%) to methanogens followed by forage – to – concentrate plus coconut oil (77%). Using the microbial population in the control diet (oat hay) as 100%, the total anaerobic fungal population decreased in the treatments. The FCC treatment was most inhibitory to fungi (94%), followed by OHC and FC diet (90% and 86%). The population of fibre - degrading bacteria, *R. falvefaciens* increased with concentrate or coconut oil addition. However, *F. succinogenes* was decreased by 89% and 98% for OHC and FCC, respectively.

As shown in Table 2, coconut oil reduced rumen protozoa numbers in all diets. At the start of the treatment period, ciliate numbers in rumen fluid did not differ among the animals, which were subsequently allocated to the different treatments (day 0). However, protozoa counts were reduced through the dietary inclusion of coconut oil after 7days of application and were reduced most in the FCC diet, but this was not significantly more than adding coconut oil to the control diet ($P > 0.05$).

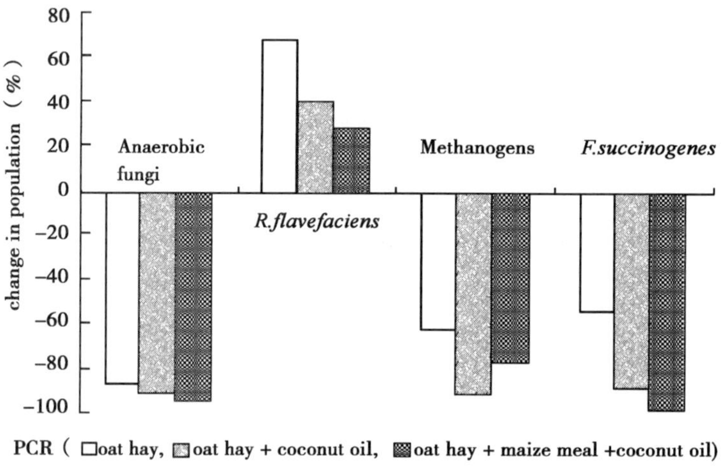

Fig. Shifts in microbial population as determined by quantitative

Discussion

Methane emissions

Adjusting the ingredients of the diet is an effective means to manipulate ruminal fermentation pathways and reduce enteric CH_4 production (Lovett *et al.* 2003; Grainger and Beauchemin 2011). In our experiment, concentrate mixtures led to a reduction in daily CH_4 production, which confirms the earlier findings of Lovett *et al.* (2003). The reduction in CH_4 production observed with supplementation of coconut oil supports the in vitro results reported by Machmüller *et al.* (1998) and Kongmun *et al.* (2010), in addtion in vivo (Machmüller and Kreuzer 1999; Liu *et al.* 2011). These reductions in methane emissions were attributed to a decline in methanogen numbers and also for the reason of a might decline in the metabolic rate of methanogens, ruminal defaunation (Lovett *et al.* 2003; Liu *et al.* 2011). Jordan *et al.* (2006) reported a linear decline in CH_4production of 0.36 g/day per gram of coconut oil added to a diet containing a 0.5 : 0.5 forage - to - concentrate ratio in vivo in beef heifers. However, in this study, when 12 g/day of coconut oil was used as a supplement with the roughage diet (control), the CH_4 production was decreased by 0.78 g/day; this size reduction has not been previously reported in sheep. Our results indicate that coconut oil can be used as an effective feed supplement for reducing methane emissions in Tibetan sheep.

Effect on different microbial communities

It has been reported that the numbers of methanogens in the rumen can be significantly reduced when exposed to dietary lipids, with coconut oil being particularly effective (Morgavi *et al.* 2010). In our experiment, the greatest reduction in methanogen numbers with supplementation of coconut oil occurred in consuming the diet containing both concentrate and coconut oil (FCC). The increased Ct values for bacteria in the FCC diet indicate a decrease in the total bacterial population, which is consistent with the findings of Jordan *et al.* (2006).

The greater decrease in the fungal population mediated by coconut oil was associated with an increase in population of different fibre – degrading bacteria. An antibiosis between bacteria and fungi has been suggested by Liu *et al.* (2011) and there are reports of the inhibitory effects of *R. flavefaciens*, *R. albus* and *Butyrivibrio fibrisolvens* on ruminal fungi, when grown in co – culture techniques (Goel *et al.* 2009).

Decreased protozoal counts with supplementation of coconut oil have been reported in many studies. Ruminal protozoa counts were decreased in the present study with dietary inclusion of coconut oil. This agrees with earlier in vivo (Machmüller and Kreuzer 1999; Lovett *et al.* 2003) and in vitro data. The rapidity of the anti-protozoal effects of coconut oil as identified by Machmüller *et al.* (1998) was also observed in this experiment, where protozoal counts were within 7 days. However, despite the magnitude of the response being similar, the difference measured in our experiment was not statistically significant. Our results demonstrate the long – term efficacy of coconut oil as a defaunating agent for group – housed Tibetan sheep, although Sutton *et al.* (1983) reported that 5 weeks after finishing feeding coconut oil, protozoa levels were still low for sheep maintained in isolation.

Correlation between microbe interactions and methane production

Regarding the effects of coconut oil fractions on methanogens and methane levels, the present data revealed unexpected results. The relationship between rumen protozoa numbers and CH_4 production was weak in the present study. Differences in the proportion of the diet being digested post – ruminally could be caused by various fed dietary treatments (Lovett *et al.* 2003). As CH_4 production is lower in the hindgut than the rumen (Ellis *et al.* 2008), this would further weaken any relationship between rumen protozoa numbers and in vivo CH_4 production (Jordan *et al.* 2006). Similarly, it has been demonstrated in vitro that the effects of coconut oil on methane production are not dependent on its protozoainhibiting effect (Jordan *et al.* 2006), and Machmüller (2006) confirmed this using sheep in vivo. A significantly positive correlation between the methanogenic population and the daily CH_4 production ($R = 0.95$, $P < 0.001$) was observed in our experiment which supports the idea that coconut oil inhibits rumen methanogens directly and may change their metabolic activity, as well as the composition of the rumen methanogenic population (Machmüller 2006; Liu *et al.* 2011).

References

[1] AOAC (Association of Official Analytical Chemists), Official Methods of Analysis, vol. I, 15th ed. AOAC,

Arlington, VA, USA. 1991.

[2] Beauchemin KA, Kreuzer M, O'Mara F, McAllister T A. Nutritional management for enteric methane abatement: a review, Australian Journal of Experimental Agriculture, 48, 21 -27. 2008.

[3] Beauchemin KA, McGinn SM, Petit, HV., Methane abatement strategies for cattle: lipid supplementation of diets, Canadian Journal of Animal Science, 87, 431 -440. 2007.

[4] Chilliard Y, Martin C, Roue J, Doreau, M., Milk fatty acids in dairy cows fed whole crude linseed, extruded linseed, or linseed oil, and their relationship with methane output, Journal of Dairy Science, 92, 5 199 -5 211. 2009.

[5] Denman SE, McSweeney CS., Development of a realtime PCR assay for monitoring anaerobic fungal and cellulolytic bacterial populations within the rumen, FEMS Microbiology Ecology, 58, 572 -582. 2006.

[6] Denman SE, Tomkins NW, McSweeney CS., Quantification and diversity analysis of ruminal methanogenic populations in response to the anti methanogenic compound bromochloromethane, FEMS Microbiology Ecology, 62, 313 -322. 2007.

[7] Eckard RJ, Grainger C, de Klein CAM., Options for the abatement of methane and nitrous oxide from ruminant production: A review, Livestock Sciences, 130, 47 -56. 2011.

[8] Ellis JL, Dijkstra J, Kebreab E, Bannink A, Odongo NE, McBride BW, France, J., Aspects of rumen microbiology central to mechanistic modelling of methane production in cattle, The Journal of Agricultural Science, 146, 213 -233. 2008.

[9] Goel H, Makaar HPS, Becker K., Changes in microbial community structure, methanogenesis and rumen fermentation in response to saponin - rich fractions from different plant materials, Journal of Applied Microbiology, 105, 770 -777. 2009.

[10] Grainger C, Beauchemin KA. Can enteric methane emissions from ruminants be lowered without lowering their production? Animal feed science and technology, 166 - 167, 308 -320. 2011.

[11] Hegarty RS, Goopy JP, Herd RM, McCorkell B. Cattle selected for lower residual feed intake have reduced daily methane production. Journal of Animal Science, 85, 1 479 -1 486. 2007.

[12] Jordan E, Lovettl DK, Hawkins M, Callan JJ, O'Mara FP. The effect of varying levels of coconut oil on intake, digestibility and methane output from continental cross beef heifers, Animal Science, 82, 859 -865. 2006.

[13] Kongmun P, Wanapat M, Pakdee P, Navanukraw C., Effect of coconut oil and garlic powder on in vitro fermentation using gas production technique, Livestock Sciences, 127, 38 -44. 2010.

[14] Liu H, Vaddella V, Zhou D., Effects of chestnut tannins and coconut oil on growth performance, methane emission, ruminal fermentation, and microbial populations in sheep, Journal of Dairy Science, 94, 6 069 -6 077. 2011.

[15] Lockyer DR, Jarvis SC., The measurement of methane losses from grazing animals, Environment Pollution, 90, 383 -390. 1995.

[16] Lovett D, Lovell S, Stack L, Callan J, Finlay M, Connolly J, O'Mara FP., Effect of forage: concentrate ratio and dietary coconut oil level on methane output and performance of finishing beef heifers, Livestock Production Science, 84, 135 -146. 2003.

[17] Machmüller A, Kreuzer M., Methane suppression by coconut oil and associated effects on nutrient and energy balance in sheep, Canadian Journal of Animal Science, 79, 65 -72. 1999.

[18] Machmüller A., Medium - chain fatty acids and their potential to reduce methanogenesis in domestic ruminants, Agriculture Ecosystems and Environment, 112, 107 -114. 2006.

[19] Machmüller A, Ossowski DA, Wanner M, Kreuzer M., Potential of various fatty feeds to reduce methane release from rumen fermentation in vitro (Rusitec), Animal feed science and technology, 71,

117 - 130. 1998.

[20] Morgavi DP, Forano E, Martin C, Newbold CJ., Microbial ecosystem and methanogenesis in ruminants, Animal, 4, 1 024 - 1 036. 2011.

[21] Murray HG, Thompson WF., Rapid extraction of high molecular weight DNA. Nuclear Physics A, 8, 4 321 - 4 326. 1981.

[22] Sutton JD, Knight R, McAllan AB, Smith RH., Digestion and synthesis in the rumen of sheep given diets supplemented with free and protected oils, British Journal of Nutrition, 49, 419 - 432. 1983.

[23] Sylvester JT, Karnati KR, Yu, Z., Development of an assay to quantify rumen ciliate protozoal biomass in cows using rela - time PCR, The Journal of Nutrition, 134, 3 378 - 3 384. 2004.

[24] Zhou J, Bruns MA, Tiedje JM., DNA recovery from soils of diverse composition, Applied and environmental microbiology, 62, 316 - 322. 1996.

(Published the article in Trop Anim Health Prod, 2012
affect factor: 1. 115)

Sex Determination in Ovine Embryos Using Amelogenin (*AMEL*) Gene by High Resolution Melting Curve AnabIsis

Yue Yaojing, Liu Jianbin, Guo Tingting, Feng Ruilin, Guo Jian, Sun Xiaoping, Niu Chun E. and Yang B. H.

(Chinese Academy of Agricultural Sciences, Lanzhou Institute of Animal and Veterinary Pharmaceutics Sciences, Lanzhou, 730050, China)

Abstract: In the study, researchers have established and tested the reliability of a method for sex determination of ovine embryos using the *AMEL* gene by melting curve analysis of PCR amplification. It can be carried out in a regular laboratory or under farm conditions within 1. 5 h for 96 samples. The PCR amplicons of 99/99 and 99/54 base pairs produced from female and male sheep, respectively are easily distinguished by both melting curve analysis and gel electrophoresis. The specificity of the method was earlier demonstrated by testing 9 blood samples from small – tailed sheep (5 males and 4 females). No amplification failures and very high agreement between genotypic and phenotypic sex was found (9/9). The sensitivity of the AMEL sexing assay was established for values >10 pg ovine genomic DNA. Forty five biopsied embryos were transferred into 22 recipient sheep on the same day that the embryos were collected and sex of the kid was confirmed after parturition. About 17 kids of predicted sex were born. The sex, as determined by PCR corresponded to the anatomical sex in all cases. To the knowledge, this was the first time that sex determination using the amelogenin gene was performed in ovine embryos by melting curve analysis.

Key words: Ovine embryos; Amelogenin gene; Sexing; High – resolution melting, blood; China

Tntroduction

The embryo transfer technology represents a powerful tool for the acceleration of various breeding programs in sheep. Known sex of embryos produced for use in ET programs can more effectively help to manage producer resources because more heifer calves per ET can be produced. This approach can improve the genetic potential of sheep breeds in shorter time intervals.

Several protocols have been established for sexing embryos in farm animals such as karyotyping (King, 1984), H – Y antigen detection (Andersib, 1987), X – linked enzymatic determination (Monk and Handyside, 1988) and based on the identification of the Y chromosome such as *SRY*,

ZFY and *TSPY* genes include *in situ* hybridization, Southern dot blotting, Polymerase Chain Reaction (PCR), Loop - Mediated Isothermal Amplification (LAMP) (Miller, 1991; Bredbacka and Peippo, 1992; Gutierrez - Adan *et al.*, 1996; Ng *et al.*, 1996; Sohn *et al.*, 2002; Jinming *et al.*, 2007). Among of these methods, PCR - based sexing assays are generally favored because of the advantages of being relatively simple, rapid and inexpensive. Some of the existing protocols are only based on the PCR - detection of Y chromosome specific sequences such as genes: SRY (Takahashi *et al.*, 1998; Mara *et al.*, 2004) and TSPY (Lemon *et al.*, 2005) or repeated sequences (Schroder *et al.*, 1990; Bredbacka and Peippo, 1995; Kageyama *et al.*, 2004). The presence of no signal does not necessarily mean that the sample has a female origin because experimental errors can also lead to negative results. Then, sexing protocols also need a PCR product being out of the Y chromosome as a positive control of template (DNA) in the sample or an X chromosome specific fragment. The amelogenin (*AMEL*) gene which exists on both X (AMELX) and Y (AMELY) chromosomes has been used to determine the sex in humans (Sullivan *et al.*, 1993), cattle (Chen *et al.*, 1999; Nicolai Z. Ballin), sheep and deer (Pfeiffer andBrenig, 2005), goats (Chang *et al.*, 2006; Weikard *et al.*, 2006) as well as in the related species (Weikard *et al.*, 2006). The complexity of these methods and the need for multiple steps to perform them greatly increase the risk of cross - contamination and thereby misdiagnosis. Recently, a novel technique, High - Resolution Melting (HRM) has been investigated for the detection of point mutations, Single - Nucleotide Polymorphism (SNP), internal tandem duplications, simultaneous mutation scanning and genotyping in bacteriology, cancer research and human platelet antigens (Wittwer *et al.*, 2003). The sample preparation consists of a standard PCR reaction with a dsDNA intercalation fluorescent dye and does not require any post - PCR handling. Products can be analyzed directly after PCR amplification using specially designed instruments for High - Resolution Melting (HRM) analysis. HRM is a mutation detection and scanning technique that has high reliability. It has been reported to have near 100% sensitivity and specificity when the analyzed PCR products were up to 400 by in length. In this research, researchers tried to design a protocol of sex determination in ovine embryos using Amelogenin gene (*AMEL*) by melting curve analysis of PCR amplifications.

Materials and Methods

Sheep blood samples and DNA extraction: In order to test the specificity of the technique, samples with known sex (male and female) from 9 Small Tailed Han (4 female, 5 male). Blood samples were obtained from. Genomic DNA was obtained from blood samples following the manufactured instruction of the Tiangen Biotech Commercial kit.

The DNA was quantified using NanoDrop ND - 2000 Spectrophotometer (NanoDrop Technologies, Inc.). The sensitivity of the AMEL sexing assay was proven using dilution series of ovine genomic DNA, ranging from 1 ~ 10 pg (1 ng, 100 pg, 10 pg). DNA purity was evaluated by comparing the absorbance ratios A260/280 and A260/230.

Collection of sheep embryos: About 10 donors (Poll Dorset x Small Tailed Han intercross F1) were treated with CIDR (Fluorogestone acetate 300 mg) (New Zealand) sponges for 12 days and were super ovulated with oFSH (ovine Follicle Stimulating Hormone 20 mg/ml) (Canada) in 4 decreasing doses (2days ×0.5ml and 2days ×1.0ml) at 12h intervals from 10 ~ 13th days. On the 12th day, the sponges were removed and PMSG (300 IU) (Canada) was injected intramuscular (i. m.) and oestrus was detected by rams at 12h intervals after sponge withdrawal (Wang *et al.*, 2006). Intrauterine insemination with fresh semen (Poll Dorset x Small Tailed Han intercross F1) was carried out twice at 12h intervals 48h after sponge withdrawal. Embryo recovery took place 7 ~ 8days after sponge removal.

Embryo manipulation and DNA extraction: Four to ten cells from 67 compact morulae were aspirated through the zona pellucida using micro – manipulation systems equipment (TransferTip (ES), Eppendor f). After biopsy, embryos were incubated inl X PBS (Phosphate Buffered Saline, PBS (Gibico) pH = 7.4) supplemented with 0.4% BSA (Bovine Serum Albumin, BSA) at 10℃ during 10 h (Hong *et al.*, 2005). In this period sexing was performed from the aspirated cells by high – rsesolution melting analysis. The following day, embryos were transferred to FGA + eCG (equine Chorionic Gonadotropin) treated recipients (2 embryos of the same predetermined sexlewe) (Dervishi *et al.*, 2008).

The DNA was isolated from the embryos by the single step method described by Saravanan *et al.* (2003) in 1 × PCR buffer (10 mM Tris – Cl, pH 8.3, 50 mM KCI, 1.5 mM $MgCl_2$) containing proteinase K (150 mg/ml) and incubated for 30 min at 37℃. Then proteinase K was inactivated by incubating at 97℃ for 10 min. The tubes were kept frozen at –20℃ until sexing was carried out.

High – resolution melting analysis of *AMEL* gene: Primers were designed using Primer 5.0 (Rozen and Skaletsky, 2000). The primers AMEL – SF and AMEL – SR hadthe sequences 5′ – ATCCAGCCRCAGCCTCACC – 3′ and 5′ – GATGGGGTGCACGGGTGG – 3′, respectively.

All DNA was amplified in a 10 μL final volume containing 1 × Tiangen Biotech taq PCR Master Mix, 0.1μmol/L forward primer CAMEL – SF), 0.1μmol/L reverse primer (AMEL – SR) and 1 × LCGreenPlus +. The PCR program consists of an initial preheating at 95℃ for 5 min to activate the Taq DNA polymerise, followed by 30 amplification cycles. Each cycle is comprised of an annealing step at 62.5℃ for 15 sec, an elongation step at 72℃ for 15 sec and denaturation at 94℃ for 45 sec. The final melting program consists of three main steps beginning with a denaturation at 95℃ for 1 min, renaturation at 25℃ for 1 min.

The PCR products were separated in 2% agarose gel for 1 h and then photographed. The Lightscanner (Idaho technology) is an instrument that measures high – resolution DNA melting curves from samples in a 96 well PCR plates. This is achieved by monitoring the fluorescence change of the fluorescent DNA intercalating dye, LCGreenPlus + as the sample is melted. Turnaround time per sample is approximately 1 ~ 2 min, depending on how broad the temperature range is required to be. The Lightscanner was heated at 0.3℃/sec. *AMEL* gene was simultaneously analyzed between 40 and 98℃ with a turnaround time of approximately, 7 min/96 samples. The Light – Emitting Diode (LED) power was auto adjusted to 90% fluorescence.

Sequencing of *AMEL* gene PCR products: The samples were directly sequenced from the Lightscanner PCR amplification. The PCR products were mixed with Nucleic Acid Purification kit (Tiangen Biotech) to remove the remaining primers and bidirectional sequenced with forward and reverse primers using ABI PRISM terminator cycle sequencing kit Version 1.1 (Takra Biotechnology (DaLian) Co., Ltd.) on the ABI PRISM 3730 genetic analyzer (Takra Biotechnology (DaLian) Co., Ltd.).

Results and Discussion

Sexing using sheep blood sample: As expected from the sheep sequence, a 99 by product, representing amplification from the X chromosome amelogenin was detected for the ewe. For the ram, researchers detected the 99 bp, representing X chromosome and a new 54 bp band, representing specific Y chromosome amelogenin amplification. As shown in Fig. 1, in total, 9 animals (4 females and 5 males) were analyzed, showing no amplification failures and a high agreement between genotypic and phenotypic sex (10/10) indicating that the sexing method based PCR amplified amelogenin gene was 100% reproducible and reliable. Both amplified fragments were sequenced and analyzed showing that the 54 bp (AMELY) and 99 bp (AMELX) fragments had a 100% of identity with variant 2 of AMELY (GenBank acc DQ469593) and *AMELX* genes (GenBank acc DQ469591), respectively.

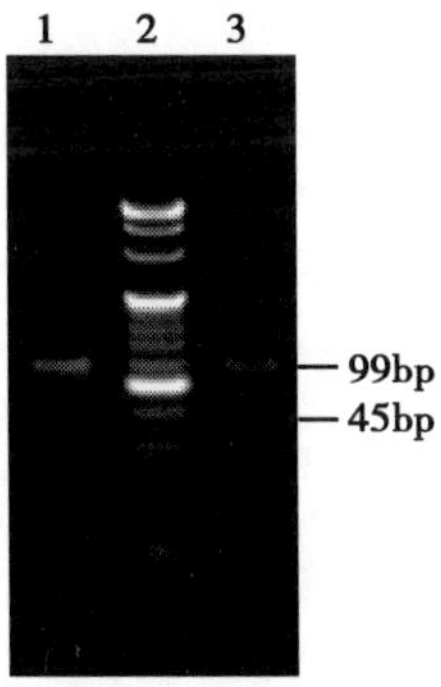

Fig. 1 Total 2% gel electrophoresis of PCR amplicons; Lane 1, a 20 bp ladder; Lane 2 and 3, PCR amplicons from beef male and female DNA, respectively

Amplicons from PCR analysis of ram (AMELX and AMELY) DNA with the primer pair AMEL - SFIR showed two distinct peaks in the high - resolution melting curve analysis (Fig. 2).

Gel electrophoresis of the PCR amplicons confirmed the melting curve results by showing two distinct fragments (Fig. 1) around the expected 54 and 99 base pairs (Fig. 1). Ewe (2 X AMELX) DNA subject to PCR analysis with the primer pair AMEL—SF/R showed as expected a single peak in the melting curve analysis (Fig. 2) and a single band on the gel electrophoresis (Fig. 1). The sensitivity of the AMEL sexing assay was established for values >10 pg ovine genomic DNA (Fig. 3).

Sexing of sheep embryos: A total of 51 embryos were used to test the efficiency and accuracy of the method. After biopsy, all embryos were incubated during 24 h and their developmental ability was evaluated. Five embryos were discarded because they showed abnormal development for their age. Forty five were transferred to twenty two recipient ewes. About 5 weeks after the transfer of the

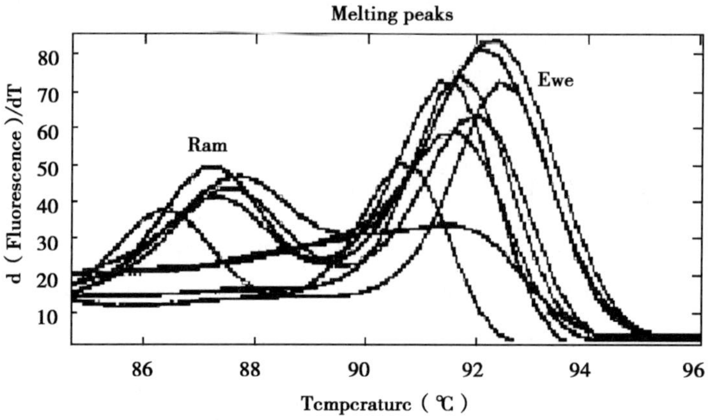

Fig. 2 Melting curve analysis of PCR amplicons. The different male and female curves represent DNA products from 9 Small Tailed Han

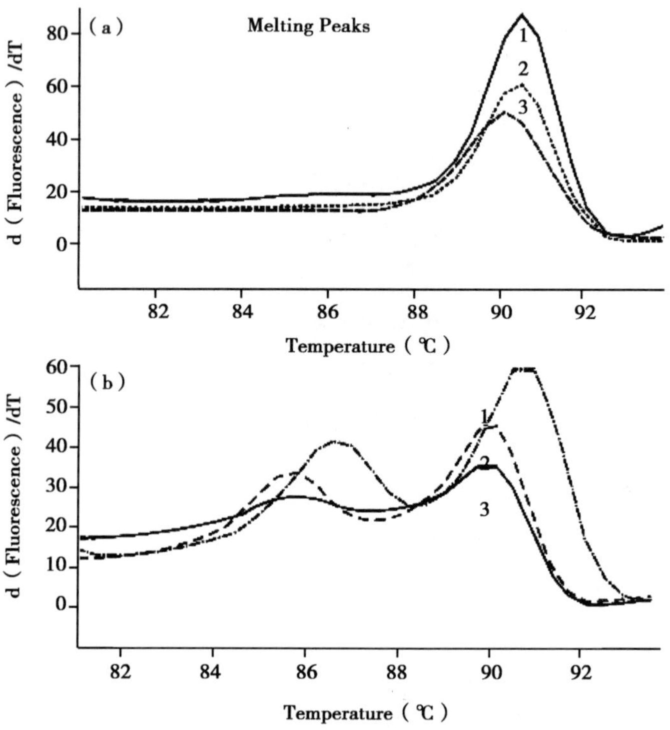

Fig. 3 Dilution series of the amelogenin assay; (a) Lines 1 ~ 3, male DNA dilution series, (1) 1 ng; (2) 100 pg and (3) 10pg; (b) Lines 1 ~ 3, male DNA dilution series; (1) 1ng; (2) 100pg and (3) 10pg

sexed blastocysts, pregnancy rate of the recipient of does was confirmed. About 17 lambs were born, 6 of which were males and 11 were females. The sex as determined by PCR corresponded to the anatomical sex in all cases. The sex determination took < 3 h including DNA extraction and PCR amplification.

The amelogenin gene encodes an important protein in the developing mammalian tooth enamel matrix that has been conserved during the evolution of vertebrates. The amelogenin (*AMEL*) gene which exists on both X (AMELX) and Y (AMELY) chromosomes has been used to determine the sex in humans (Sullivan *et al.*, 1993), cattle (Chen *et al.*, 1999), sheep and deer (Pfeiffer and Brenig, 2005), goats (Chang *et al.*, 2006; Weikard *et al.*, 2006) as well as in the related species (Weikard *et al.*, 2006). Pfeiffer and Brenig described a 45 bp deletion in the amelogenin gene at Y chromosome. In this way females amplify only a 263 bp band while males produce two bands of 263 and 218 bp. In this study, researchers also fund that females amplify only a 99 bp band while males produce two bands of 99 and 54 bp. The major advantage of this method is the co-amplification, in a single tube, of two specific fragments, one from Y-chromosome and one for the X chromosome using a single primer pair and making the use of a PCR control unnecessary (Gerardo *et al.*, 2007). But all analyses of amplifications must be done by gel electrophoresis and therefore a potential future alternative. In the study, researchers have established and tested the reliability of a method for sex determination of ovine embryos using the *AMEL* gene by melting curve analysis of PCR amplification. This assay provides a rapid and sensitive method for sexing. It can be carried out in a regular laboratory or under farm conditions within 2 ~ 3 h for 96 samples. This is especially important for the future application of the protocol to sheep embryos sexing. The protocol was showing that the efficiencies in sex determination were 100% by evaluating genomic DNA from 4 females and 5 males. This result was comparable to those reported by Nicolai Z. Ballin with the same primers. Assuming a DNA content of about 6 pg per cell, this is in agreement with the result in the dilution series assay displaying a detection limit of 10 pg genomic DNA (Fig. 3). The results showed that the threshold of the amelogenin assay using DNA samples extracted from the embryos seems to be 2 cells.

The method of biopsy and sex determination of goat embryos that we used can accurately and efficiently predict the sex of embryos before transfer. The microblade used to biopsy the embryo is easy to use and has proven to be effective with bovine and sheep embryos (Herr and Reed, 1991; Kochhar *et al.*, 2000). Quality of the embryo and the size of the biopsy can influence the outcome of the procedure (Ju *et al.*, 2001). Researchers used morphologically normal blastocysts and obtained a 37.3% kidding rate after transfer. These results are comparable with *in vitro* produced ovine embryos, biopsied and sexed by similar technique (Mara *et al.*, 2004).

Conclusion

In conclusion, researchers have established and tested the reliability of a method for sex determination of ovine embryos using the *AMEL* gene by melting curve analysis of PCR amplification. The advantage of this assay is that neither additional control amplicons with a second locus specific autosomal primer pair nor the gel electrophoresis is necessary for sex determination and control of the PCR reaction. This assay provides a rapid and sensitive method for sexing. It can be carried out in a regular laboratory or under farm conditions within 2 ~ 3 h for 96 samples. The rapid sex determination using amelogenin gene allows transferring sexed fresh embryos in MOET and IVF (*In Vlitro* Fertilization) programmers to make them more efficient (Dervishi *et al.*, 2008).

Acknowledgements

This research was supported by the Central Level, Scientific Research Institutes for Basic R&D Special Fund Business (No.: BRF100102) by the Earmarked Fund for Modern China Wool and Cashmere Technology Research System (No.: nycytx – 40 – 2) by the National High Technology Research and Development Program of China (863 Program) (No.: 2008AA101011 – 2).

References

[1] Andersib G B. Identification of embryonic sex by detection of H – Y antigen. Theriogenol., 1987. 27: 87 – 97.

[2] Bredbacka P and J Peippo. Sex diagnosis of ovine and bovine embryos by enzymatic amplification and digestion of DNA from the ZFYIZFX locus. Agric. Sci. Fin., 1992. 2: 233 – 238.

[3] Bredbacka P and J Peippo. PCR – sexing of bovine embryos: A simplified protocol. Theriogenol., 1995. 44: 167 – 176.

[4] Chang Z, Fan X, Luo M, Wu Z and Tan J. Factors affecting super – ovulation and embryo transfer in Boer goats. Asian – Aust. J. Anim. Sci., 2006. 19: 341 – 346.

[5] Chen C M, Hu C L, Wang C H, Hung C M, Wu H K, Choo K B and Cheng W T. Gender determination in single bovine blastomeres by polymerise chain reaction amplification of sex – specific polymorphic fragments in the amelogenin gene. Mol. Reprod. Dev., 1999. 54: 209 – 214.

[6] Dervishi E, A M artinez – Royo P Sanchez, J L Alabart, M J Cocero, J Folch and J H Calvo. Reliability of sex determination in ovine embryos using amelogenin gene (*CAMEL*). Theriogenol., 2008. 70: 241 – 247.

[7] Gerardo P. L, A. Ivan, F. P. Lucia, G. Felix and J. R. Luis, 2007. A sexing protocol for wild ruminants based on PCR amplification of amelogenin genes AMELX and AMELY. Arch. Tierz. Dummerstorf, 50: 442 – 446.

[8] Gutierrez – Adan, A., E. Behboodi, J. F. Medrano, J. D. Murray and G. B. Anderson, 1996. Nested PCR primers for sex determination across a range of mammalian orders. Theriogenology, 45: 189 – 189.

[9] Herr, C. M. and K. C. Reed, 1991. Micromanipulation of bovine embryos for sex determination. Theriogenol., 35: 45 – 54.

[10] Hong Q H, Z Y Zhao, Q Y Shao and W H Jin, 2005. Industrialization of embryo transfer in sheep and cattle. Yurrnan J. Anim. Husbandry Vet., 1: 65 – 66.

[11] Jinming, H., Y. Wei, W. Naike and T. Xiuwen. Use of the Non – electrophoretic method to detect testis specific protein gene for sexing in preimplantation bovine embryos. Asian – aust. J. Anim. Sci., 2007. 20: 866 – 871.

[12] Ju, J. C., Y. C. Chang, W. T. Huang, P. C. Tang and S. P. Cheng, 2001. Super ovulation and transplantation of demi – and aggregated embryos in rabbits. Asian – Aust. J. Anim. Sci., 14: 455 – 461.

[13] Kageyama, S., I. Yoshida, A. K. Kawakura, K. Chikun, 2004. A novel repeated sequence located on the bovine Y chromosome: Its application to rapid and precise embryo sexing by PCR. J. Vet. Med. Sci., 665: 509 – 514.

[14] King, W. A., 1984. Sexing embryos by cytological methods. Theriogenol., 21: 7 – 17.

[15] Kochhar, H. S., B. C. Buckrell, J. W. Pollard and W. A. King, 2000. Production of sexed lambs after biopsy of ovine blastocyts produced *in vitro*. Can. Vet. J., 41: 398 – 400.

[16] Lemos, D. C., A. F. Lopesrios, L. C. Caetano, R. B. Lobo and R. A. Vila *et al.*, 2005. Use of TSPY gene for sexing cattle. Genet. Mol. Biol., 28: 117 – 119.

[17] Mara, L., S. Pilichi, A. Sanna, C. Accardo and B. Chessa *et al.*, 2004. Sexing of *in vitro* produced ovine embryos by duplex PCR. Mol. Reprod. Dev., 69: 35 – 42.

[18] Miller, J. R., 1991.. Isolation of Y chromosome - specific sequences and their use in embryo sexing. Reprod. Dourest Anim., 26: 58 -65.

[19] Monk, M. and A. H. Handyside. Sexing of preimplantation mouse embryos by measurement of X - linked gene dosage in a single blastomere. J. Reprod. Fertil, 1988. 82: 365 -368.

[20] Ng, A., K. Sathasivam, S. Laurie and E. Notaria. Determination of sex and chimaerism in the domestic sheep by DNA amplification using HMG - box and icrosatellite sequences. Anim. Reprod. Sci., 1996. 41: 131 -139.

[21] Pfeiffer, I. and B. Brenig, 2005. X - and Y - chromosome specific variants of the amelogenin gene allow sex determination in sheep (*Ovis cries*) and European red deer (*Cervus elaphus*). BMC. Genetics, Vol. 6. 10. 1186/1471 -2156 -6 -16.

[22] Rozen, S. and H. Skaletsky. Primer3 on the WWW for general users and for biologist programmers. Methods, Mol. Boil., 2000. 132: 365 -386.

[23] Saravanan, T., A. M. Nainar and A. Kumarewsan, 2003. Sexing of sheep embryos produced in vitro by polymerise chain reaction and sex - specific polymorphism. Asian - Aust. J. Anim. Sci., 16: 650 -654.

[24] Schroder, A., J. R. Miller, P. D. Thomsen and B. Avery, 1990. Sex determination of bovine embryos using the polymerise chain reaction. Anim. Biotechnol., 1: 121 -133.

[25] Sohn, S. H., C. Y. Lee, E. K. Ryu, J. Y. Han, S. Multan and S. Pathak, 2002. Rapid sex identification of chicken by fluorescence in situ hybridization using a W chromosome - specific DNA probe. Asian - Aust. J. Anim. Sci., 15: 1 531 -1 535.

[26] Sullivan, K. M, A. Marmucci, C. P. Kimpton and P. Gill, 1993. A rapid and quantitative DNA sex test fluorescence based PCR analysis of X Y homologous gene amelogenin. BioTechnique, 15: 636 -641.

[27] Takahashi, M., R. Masuda, H. Uno, M. Yokoyama, M. Suzuki, M. C. Yoshida and N. Ohtaishi, 1998. Sexing of carcass remains of the Sika deer (*Cervus nippon*) using PCR amplification of the Sry gene. J. Vet. Med. Sci., 60: 713 -716.

[28] Wang, Y. Q., Y. Z. Zhao and F. D. Li, 2006. Application of Superovulation Technique to Embryo Transfer Practice in Poll Dorset Sheep. China Herbivores, 26: 3 -5.

[29] Weikard, R., C. Pitra and C. Kuhn, 2006. Amelogenin cross - amplification in the family Bovidae and its application for sex determination. Mol. Reprod. Dev., 73: 1 333 -1 337.

[30] Wittwer, C. T., G. H. Reed, C. N. Gundry, J. G. Vandersteen and R. J. Pryor, 2003. Highresolution genotyping by amplicon melting analysis using LCGreen. Clin. Chem., 49: 853 -860.

(Published the article in Journal of Animal and Veterinary Advances, 2012 affect factor: 0. 39)

Microscopic Structure of Rabbit Hair

GUO Tianfen[1,2], WANG Xinrong[3], LI Weihong[1,2], NIU Chune[1,2]

(1. Lanzhou Institute of Animal Science and Veterinary Pharmaceutics, CAAS, Lanzhou 730050, China; 2. Quality supervising, Inspecting and Testing Center for Animal Fiber, Fur, Leather and Products MOA, Lanzhou 730050, China; 3. Gansu Agricultural University, Lanzhou 730050, China)

Abstract: The paper was to explore the microscopic structure of rabbit hair. Single rabbit hair with typical features was selected to observe its microscopic structure from tip to root, and its fiber diameter was also measured. The rabbit hair tip was constituted by scale layer and cortical layer, without medullary layer; the middle part was generally constituted by scale layer, cortical layer and medullary layer; the root had no medullary layer, and the scale layer was wheatear – shaped. This was the property of rabbit hair, which could be used for comparative studies with other animal fiber and species identification. Rabbit hair had developed medullary layer, and fiber diameter was positively related to column number of medullary cavity. The hair generally was single column, and coarse hair was multi – column. Single rabbit hair was the finest in the tip, coarse in the middle and tapering in the root. The diameter difference of various parts was large, and the external growth characteristics was spindle – shaped. Using biological microscope method to identify different animal fur and product species is more objective and simple.

Key words: Rabbit; Hair; Microscopic structure; Average diameter

(Published the article in Agricultural Science and Technology)

大通牦牛提纯复壮当地牦牛效果的研究

王宏博，郎　侠，梁春年，丁学智，郭　宪，包鹏甲，阎　萍

（中国农业科学院兰州畜牧与兽药研究所，甘肃省牦牛繁育
工程重点实验室，兰州 730050）

摘　要：为改良青海当地地方牦牛的生产性能和品种退化，利用大通牦牛冻精与当地母牛进行二元杂交，进行提纯复壮当地牦牛。结果表明：改良 F_1 代母牦牛的受胎率、产犊率、犊牛成活率、繁殖成活率依次分别为 75.2%、95.1%、90.15%、46.5%（n = 562），比当地牦牛的上述繁殖指标依次分别提高 0.04%、0.06%、0.02%、0.086%。因此，引入大通牦牛改良当地牦牛，其繁殖性能略有提高，如果再加强牦牛饲养管理，改善饲养条件，可提高当地牦牛的繁殖性能。

关键词：改良；效果；大通牦牛；当地牦牛

大通种牛场从 20 世纪 80 年代起与中国农业科学院兰州畜牧与兽药研究所、青海牧业科学院等科研单位合作，在普通家牦牛中导入野牦牛血液来提纯家牦牛，通过对“野牦牛驯化”、“复壮犊牦牛”等课题的研究，于 2002 年培育出牦牛新品种“大通牦牛”。大通牦牛以野牦牛为父本，野牦牛常年生长在海拔 4 500 ~ 6 000m 高寒荒漠地带，由于生存环境严酷，其身高、体重、生长速度、抗逆性和生活力等性状的平均遗传力远高于家牦牛[1~5]。而青海省祁连县野牛沟乡和海晏县等地的地方牦牛品种长期受自然环境、社会经济及科学文化等多种因素的制约，牦牛缺乏人为的定向培育，其生产性能很低，成年牦牛公牛平均体重约 320kg，母牦牛约 200kg，适龄母牦牛年头均挤奶量不足 180kg，成年牦牛屠宰率 48%。特别是近年来，由于掠夺式经营和草场的退化及超载过牧，地方品种的牦牛处于低水平的饲养状态。同时严重的近亲繁殖和品种不断退化，致使牦牛体重逐年下降，据资料报道，在 50 年间下降率平均为 40% 以上[6~10]。因此，利用大通牦牛对家牦牛的提纯复壮显得尤为重要，本研究比较了大通牦牛提纯复壮当地家牦牛的效果。

1　材料与方法

1.1　测定基点与方法的选择

测定基点选择在祁连县野牛沟乡、海晏县、默勒镇、刚察、大通牛场。

采用测杖测定牦牛体高、体斜长；采用卷尺测定其管围；采用卡尺（圆形测定器）测定其胸围，采用磅秤测定其活重，大通牦牛的体尺、活重、产乳性能和屠宰性能等数据均来自青海大通种牛场，改良后代和当地牦牛的相关数据是实地测定所得。

1.2　大通牦牛改良当地牦牛的方法

利用大通牦牛冻精与当地母牛进行二元杂交。向当地母牛群发放冻精的办法，严格控制

并登记各头公牛冻精配种及产犊情况；在犊牛断奶分群时，佩戴不同颜色的耳号以便辨认。

1.3 统计分析

所测得的牦牛相关数据的处理和分析采用 SPSS（11.0）进行，差异显著性分析用 t 检验进行，计算结果均以平均值 ± 标准差表示。

2 结果与分析

2.1 大通牦牛与当地牦牛生长发育比较

由表 1 可知，大通牦牛 6 月龄、1 岁、1.5 岁、3 岁和 6 岁龄体重、体尺均显著高于当地牦牛（$P<0.01$）。

表 1 大通牦牛、当地牦牛活重和体尺测定表

项目	性别	样本数	体高（cm）	体斜长（cm）	胸围（cm）	管围（cm）	活重（kg）
大通牦牛							
六月龄	♂	171	89.42±5.75^{A}	91.12±5.83^{A}	114.05±8.52	12.23±0.96^{A}	82.56±9.76^{A}
	♀	137	88.09±5.12^{A}	87.75±4.75^{A}	113.74±6.53	12.89±0.98^{A}	79.92±7.63^{A}
1 岁	♂	157	91.71±6.15^{A}	91.64±5.76^{A}	118.79±7.21	13.83±0.83^{A}	92.66±10.26^{A}
	♀	178	88.65±5.97^{A}	90.92±5.25^{A}	115.57±6.77	12.96±0.85^{A}	83.15±11.35^{A}
1.5 岁	♂	161	96.98±5.43^{A}	108.51±8.14^{A}	138.35±7.91	11.69±0.85^{A}	143.90±21.32^{A}
	♀	147	91.93±5.03^{A}	98.12±7.49^{A}	118.47±7.58	12.31±0.75^{A}	117.53±6.24^{A}
3 岁	♂	124	101.88±6.55^{A}	114.14±6.68^{A}	172.45±9.29	15.05±0.98^{A}	237.31±15.31^{A}
	♀	139	95.13±4.36^{A}	108.21±5.10^{A}	131.06±7.94	13.83±0.83^{A}	169.57±11.96^{A}
6 岁	♂	57	121.32±6.67^{A}	142.53±9.78^{A}	195.6±11.53	19.20±1.80^{A}	381.7±29.61^{A}
	♀	63	106.81±5.72^{A}	121.19±6.57^{A}	153.46±8.43	15.44±1.59^{A}	220.3±27.19^{A}
当地牦牛							
六月龄	♂	62	75.34±4.26^{B}	74.66±3.87^{B}	91.32±6.5	29.7±0.37^{B}	46.78±5.73^{B}
	♀	54	69.56±3.45^{B}	71.78±2.98^{B}	87.45±7.43	9.4±0.57^{B}	43.65±4.56^{B}
1 岁	♂	34	77.47±8.12^{B}	79.0±8.37^{B}	83.21±9.68	9.7±0.37^{B}	52.6±6.5^{B}
	♀	27	76.21±9.02^{B}	78.11±9.54^{B}	79.34±8.14	9.4±0.57^{B}	50.6±7.44^{B}
1.5 岁	♂	70	83.68±4.87^{B}	86.77±4.77^{B}	103.45±5.23	9.9±0.28^{B}	93.67±16.48^{B}
	♀	68	79.56±3.78^{B}	80.58±4.25^{B}	94.35±4.25	9.6±0.48^{B}	84.17±15.09^{B}
3 岁	♂	89	98.77±3.67^{B}	98.68±3.98^{B}	114.34±6.54	11.2±0.54^{B}	159.10±10.64^{B}
	♀	95	89.87±5.45^{B}	90.78±3.98^{B}	109.45±5.35	10.6±0.68^{B}	151.50±18.58^{B}
6 岁	♂	101	109.18±4.78^{B}	119.10±6.68^{B}	151.35±6.18	15.70±1.05^{B}	190.97±26.08^{B}
	♀	75	107.78±6.43^{B}	117.16±11.39^{B}	149.47±9.52	15.70±1.19^{B}	184.96±35.69^{B}

注：同列数据肩标不同大写字母者表示差异极显著（$P<0.01$），同列大通牦牛和当地牦牛相同年龄间进行多重比较，其不同年龄间不进行多重比较。下表同

2.2 大通牦牛改良家牦牛效果比较

表2结果表明，改良 F_1 代牦牛初生重、6 月龄和一岁半龄体重分别为 12.89、50.94 和 90.04kg，比当地家牦牛各年龄段体重（初生重、6 月龄和 18 月龄体重分别 11.55、43.98、81.87kg）体重增加 1.34、6.69、8.07kg。体重比同龄家牦牛分别提高 11.6%、15.8%、9.85%。F_2 代牦牛 18 月龄平均体高、体斜长、胸围分别为 91、93.9、118.58cm。比同龄家牦牛上 3 项指标分别高 6.22、6.03、5.15 cm。提高幅度分别为 7.34%、6.86%、4.54%，体格大于家牦牛。经

生物统计方法检验，F_1 代牦牛和家牦牛体重、体尺差异极显著（$P<0.01$）。

表2　改良 F_1 代牦牛、当地牦牛生长发育状况

组别	年龄	样本数	体高（cm）	体斜长（cm）	胸围（cm）	体重（kg）
改良 F1 代牦牛	初生	22	53.36 ± 1.64^A	51.03 ± 2.39^A	58.35 ± 2.00^A	12.89 ± 0.81^A
	六月龄	23	77.95 ± 3.33^A	80.65 ± 6.15^A	99.29 ± 7.37^A	50.94 ± 7.47^A
	十八月龄	54	91.00 ± 5.20^A	93.90 ± 6.29^A	118.58 ± 6.07^A	90.04 ± 14.03^A
当地牦牛	初生	42	51.22 ± 1.63^B	48.28 ± 2.41^B	56.30 ± 2.09^B	11.55 ± 0.89^B
	六月龄	33	73.54 ± 4.16^B	75.76 ± 3.97^B	93.02 ± 5.62^B	43.98 ± 5.83^B
	十八月龄	48	84.78 ± 3.97^B	87.87 ± 5.07^B	113.43 ± 4.21^B	81.97 ± 10.10^B

2.3　改良 F_1 代的产肉性能

引进大通牦牛（♂）改良母牦牛，其 F_1 代和当地牦牛的产肉性能见表3。

表3　改良 F_1 代与当地牦牛产肉性能比较

品种	性别	年龄（岁）	样本数	体重（kg）	胴体重（kg）	屠宰率（%）
改良 F_1 代	♂	2	3	108.5	50.10	46.18
当地牦牛	♂	2	3	97.5	44.02	45.15

由表3可见，在相同地区、相同环境及相同饲养管理条件下，改良后代的产肉性能优于当地牦牛，胴体重增加6.08kg，屠宰率提高1.03个百分点。

2.4　改良 F_1 代的产乳性能

经测定牦牛120d的产奶量，结果表明，改良 F_1 代牦牛产奶量较家牦牛略有提高（表4），但差异不显著。

表4　改良 F_1 代与当地母牦牛产奶性能对比

品种	胎次	样本数	120d 产奶量（kg）	日均产奶量（kg）	平均乳脂率（%）
改良 F_1 代	1	20	192.0 ± 20.18	1.60 ± 0.16	5.30 ± 0.29
当地牦牛	1	10	184.59 ± 10.54	1.53 ± 0.10	5.35 ± 0.41

2.5　改良 F_1 代母牦牛的繁殖性能

改良 F_1 代母牦牛的受胎率、产犊率、犊牛成活率、繁殖成活率依次分别为75.2%、95.1%、90.15%、46.5%（n=562），比当地牦牛的上述繁殖指标依次分别提高0.04%、0.06%、0.02%、0.086%。

3　讨论

大通牦牛具有明显的野牦牛特征，嘴、鼻、眼睑为灰白色；具有清晰可见的灰色背线；公牛均有角，母牛多数有角，体形外貌符合规定：体型结构紧凑，偏向肉用，体质结实，发育良好，体重、体尺符合育种指标，毛色全黑色或夹有棕色纤维，背腰平直，前胸开阔，肢高而结实[1~2]。

本研究观察发现，经大通牦牛改良的当地牦牛 F_1 后代具有明显的大通牦牛特征，且改

良 F_1 代的体重、体尺指标表现出与大通牦牛高度的一致性。改良 F_1 代的繁殖性能均比本地牦牛有不同程度的提高，这与李生存[4]和马俊贵等[5]报道的结果一致。由此可见，引入大通牦牛改良本地牦牛其繁殖性能略有提高，如果再加强牦牛饲养管理，改善饲养条件，可提高家牦牛繁殖性能。

经对祁连县野牛沟乡、默勒镇、海晏、刚察引进的大通牦牛后裔适应性观察和生长发育的抽样测定结果表明，种公牦牛引进后适应性良好，生长发育正常。公牦牛情期配种期，性欲旺盛，雄性特征明显，表现凶悍；其后裔体形外貌大多似母本，但脊背部灰白色背线，鼻吻部、眼睑处的灰白色较突出。

大通牦牛生长发育速度较快，初生、6 月龄、18 月龄体重比家牦牛平均提高 15% ~ 27%；具有较强的抗逆性和适应性，连续 5 年的统计表明牦牛越冬死亡率小于 1%，比同龄家牦牛群体的 5% 越冬死亡率降低 4 个百分点。繁殖率较高，初产年龄由原来的 4.5 岁提前到 3.5 岁，经产牦牛为 3 年 2 胎，产犊率为 75%。突出表现在越冬死亡率明显降低，觅食能力强，采食范围广[1~2]。

培育成功的杂交后代具有明显的野牦牛特征，改良 F_1 代生长发育速度较快，初生、6 月龄、18 月龄体重比家牦牛平均提高 15% ~27%；具有较强的抗逆性和适应性，连续 5 年的越冬死亡率小于 1%，比同龄家牦牛降低 4 个百分点；繁殖率较高，初产年龄由原来的 4.5 岁提前到 3.5 岁，经产牛为 3 年产 2 胎，产犊率为 75%；体魄强健，四肢有力，觅食能力强，采食范围广，可充分利用高山草场。

跟踪调查发现，改良牦牛不但表现出很好的适应性，而且对改良当地的家牦牛发挥了重大作用。改良 F_1 代各年龄段体重比当地家牦牛提高幅度都在 15% 以上，并表现出很强的高山放牧能力和很显著的耐寒、耐饥和抗病能力，体型外貌都具有“大通牦牛”之特征，这也从另一角度证明了“ 大通牦牛”稳定的遗传力。因此，利用大通牦牛提纯复壮当地牦牛效果明显。

参考文献

[1] 阎萍，潘和平．野牦牛种质特性的研究与利用［J］．中国畜牧兽医，2004，40（12）：31－33.

[2] 陆仲璘，何晓玲，阎萍．世界上第一个牦牛培育新品种：“大通牦牛”简介［J］．中国草食动物，2005（专刊）：12－14.

[3] 常顺兰．海晏县牦牛提纯复壮的现状及对策［J］．中国畜牧兽医，2007，34（8）：136－137.

[4] 李生存．青海高原野血牦牛生产性能及利用［J］．黑龙江畜牧兽医，2003（9）：21.

[5] 马俊贵，马朝银．海北州引进含 1/2 野血牦公牛提纯复壮家牦牛效果观测［J］．草业与畜牧，2008（4）：45－46.

[6] 保善科，孔占林．海北州牦牛提纯复壮技术推广调查报告［A］．北京：中国畜牧兽医学会，2009.

[7] 袁明忠．大通牦牛品种特性及其提纯复壮的研究现状［J］．养殖与饲料，2011（7）：73－74.

[8] 包永清．青海大通牦牛与甘南牦牛杂交效果初报［J］．畜牧兽医杂志，2008，27（1）：73－74.

[9] 索南多杰，马超龙，麻文林．祁连县畜种改良工作情况调查［J］．青海畜牧兽医杂志，2009，39（3）：25－27.

[10] 马忠．牦牛提纯复壮工作现状及发展对策［J］．养殖与饲料，2010（6）：93－94.

（发表于《中国畜牧》）

EGF、KGF在不同细度甘肃高山细毛羊皮肤中的表达规律

郭婷婷，杨博辉，岳耀敬，郭　健，孙晓萍，冯瑞林，刘建斌

（中国农业科学院兰州畜牧与兽药研究所，兰州730050）

摘　要：为了研究角质细胞生长因子（KGF）、表皮生长因子（EGF）在不同细度甘肃高山细毛羊皮肤中的表达规律，探索其与羊毛细度的关系，试验以不同细度的成年甘肃高山细毛羊为研究对象，采用组织芯片技术与免疫组织化学技术相结合的方法对EGF和KGF在不同细度的甘肃高山细毛羊皮肤中蛋白表达水平和定位进行研究。结果表明：EGF和KGF在不同细度的甘肃高山细毛羊皮肤组织中均有表达，且不同部位的表达存在差异。说明EGF和KGF参与毛囊的发育，对羊毛细度有一定影响。

关键词：角质细胞生长因子（KGF）；表皮生长因子（EGF）；甘肃高山细毛羊；免疫组织化学法

（发表于《黑龙江畜牧兽医》）

Agouti 与 MITF 在不同颜色被毛藏羊皮肤组织中 mRNA 表达量研究

韩吉龙，岳耀敬，郭　健，刘建斌，牛春娥，郭婷婷，
冯瑞林，孙晓萍，吴瑜瑜，杨博辉

（中国农业科学院兰州畜牧与兽药研究所，兰州 730050）

摘　要：为了研究决定藏羊不同颜色被毛形成的分子机制，采用实时荧光定量 PCR 检测皮肤组织中决定色素形成信号通路中上游调控刺鼠信号蛋白基因（*agouti signaling protein*（ASIP）gene，Agouti）和下游调控小眼畸形相关转录因子基因（*microphthalmia - associated transcription factor*，MITF）mRNA 的相对表达量，分析同一基因在黑色眼圈、棕黑色颈部和白色体侧被毛皮肤组织的相对表达量差异，对两基因在同一个体同一颜色被毛皮肤组织表达量进行相关性分析。结果表明：Agouti 基因在同一个体 mRNA 表达水平为棕黑色皮肤组织显著性高于黑色和白色（$P<0.05$），MITF 基因 mRNA 表达量黑色皮肤组织显著性高于棕黑色和白色（$P<0.05$）。通过相关性分析得知，两基因在同一个体同一毛色皮肤组织 mRNA 相对表达量无显著相关性。推论得知，MITF 基因 mRNA 高表达与藏羊眼圈周围黑色被毛显著相关，Agouti 基因 mRNA 高表达、MITF 基因 mRNA 低表达与其颈部棕黑色被毛可能为其原因，体侧部的白色毛与 Agouti 和 MITF 下基因 mRNA 表达无明显相关。

关键词：欧拉型藏羊；实时荧光定量 PCR；Agouti 基因；MITF 基因；毛色

（发表于《中国草食动物科学》）

不同地点、胎次对牦牛乳中脂肪酶及淀粉酶活性的影响研究

席　斌[1]，褚　敏[1]，辛蕊华[1]，李维红[1,2]

（1. 中国农业科学院兰州畜牧与兽药研究所，兰州 730050；
2. 农业部动物毛皮及制品质量监督检验测试中心，兰州 730050）

摘　要： 为了比较分析青海天峻、甘肃甘南、甘肃天祝牦牛乳中脂肪酶（LPL）及淀粉酶（AMS）活性，试验采用紫外分光光度法对采自青海天峻、甘肃甘南各 30 份牦牛乳以及天祝抓喜秀龙乡红疙瘩村、岱乾村和碳山岭镇四台沟村的 90 份牦牛乳中的 LPL 及 AMS 活性进行检测和分析。结果表明：甘肃天祝牦牛乳中 LPL 活性与甘南地区相比差异不显著（$P>0.05$），但与青海天峻地区相比差异极显著（$P<0.01$）；甘肃天祝牦牛乳中 AMS 活性与青海天峻相比无明显差异（$P>0.05$），但与甘南地区相比差异极显著（$P<0.01$）；而天祝 3 个地区之间牦牛乳中 LPL、AMS 活性均无显著差异（$P>0.05$）。说明不同胎次对牦牛乳中 LPL 及 AMS 活性均无显著影响（$P>0.05$）；不同地点对牦牛乳中 LPL、AMS 活性有一定影响。

关键词： 地点；胎次；牦牛乳；脂肪酶（LPL）；淀粉酶（AMS）；活性

（发表于《黑龙江畜牧兽医》）

大通牦牛 DRB3.2 基因遗传多样性

包鹏甲，梁春年，郭　宪，丁学智，阎　萍
（中国农业科学院兰州畜牧与兽药研究所；甘肃省牦牛繁育工程
重点实验室　兰州 730050）

关键词：主要组织相容性复合物；遗传多样性；巢式 PCR；大通牦牛

主要组织相容性复合物（maior histocompatibility complex，MHC）是由染色体上紧密连锁、高度多态的基因位点组成的一个区域，在机体免疫系统中发挥着非常重要的抗原递呈作用，与家畜的抗病性和易感性密切相关，已成为近年来动物疾病诊断和家畜抗病育种研究的热点。大通牦牛是中国农业科院兰州畜牧与兽药研究所与青海省大通种牛场历经 20 多年培育成的牦牛新品种。由于其稳定的遗传性，较高的产肉性能，优良的抗逆性和对高山高寒草场的利用能力，深受牦牛饲养地区群众的欢迎，青海、西藏自治区（全书称西藏）、甘肃、川西北及毗邻地区的高寒牧区每年都要购买数千头大通牦牛种牛，改良当地牦牛，获得了显著成绩。DRB3 是牛 MHC 中功能最重要、多态性最丰富的区域，参与机体的免疫应答，已被越来越多地用于研究抗病力遗传、调查种群遗传结构、确定近缘物种间的关系和进化历史，本研究拟对其 DRB3.2 基因遗传多样性进行研究。

（发表于《Proceeding of the 13th National Symposium on Animal Genetic Markers》）

甘南藏羊不同群体遗传多样性的微卫星分析

郎　侠[1,2]，王彩莲[3]

（1. 中国农业科学院兰州畜牧与兽药研究所，兰州 730050；2. 甘肃省牦牛繁育工程重点实验室；3. 甘肃省农业科学院草畜研究所）

摘　要：文章旨在分析甘南藏羊不同类群的微卫星 DNA 多态性，以期了解该资源的遗传多样性，为该羊种保种、选育和品质的进一步提高提供一些有益资料。利用 8 对微卫星引物，以欧拉型藏羊、甘加型藏羊和乔科型藏羊为研究对象，通过计算基因频率、多态性信息含量、有效等位基因数和杂合度，评估其品种内的遗传变异。结果表明：在 8 个座位中，共检测到 96 个等位基因，每个座位平均为 12 个等位基因；座位平均杂合度 0.876；平均有效等位基因数 9.902；平均多态性信息含量 0.896。结果提示，甘南藏羊群体存在丰富的遗传多样性，所选微卫星标记可用于绵羊遗传多样性评估。

关键词：欧拉型藏羊；甘加型藏羊；乔科型藏羊；微卫星；遗传多样性

（发表于《中国草食动物科学》）

甘肃高山细毛羊数量性状遗传相关和通径分析

孙晓萍，刘建斌，郎　侠，郭　健

（中国农业科学院兰州畜牧与兽药研究所甘肃省牦牛繁育工程重点实验室，兰州 730050）

摘　要：为了探讨甘肃高山细毛羊数量性状的遗传参数，研究应用 2006—2007 年甘肃皇城绵羊繁育技术推广站采集的甘肃高山细毛羊 6 个核心群育种资料，对初生重，出生等级，断奶毛长，断奶重，1.5 岁毛长、细度、产毛量及体重的遗传力以半同胞组内相关法进行估测，对遗传相关以公羊的方差组分和协方差组分进行估测，并对计算出的相关系数进行显著性检验。结果表明：以上性状对产毛性状的影响存在很大差异，通径分析进一步揭示污毛量、净毛率及与其互作的性状（断奶重、剪毛后体重、毛丛长度、束强）是影响产毛量最主要的因素。

关键词：甘肃高山细毛羊；数量性状；遗传力；遗传相关；通径分析

（发表于《黑龙江畜牧兽医》）

河西绒山羊绒毛生长规律的研究

魏云霞[1]，肖玉萍[1]，蒲万霞[1]，杨正谦[2]

（1. 中国农业科学院兰州畜牧与兽药研究所，兰州 730050；
2. 临夏（州）农业学校，甘肃临夏 731100）

河西绒山羊原始品种是一个受自然环境和人类影响的地方品种，来源无确切考证，有学者把它划归为西藏山羊类群，毛色以黑、白为主，是国家最早命名的优良地方品种，属绒、肉、皮兼用型。经过几十年的改良选育，河西绒山羊不论是个体品质或群体数量均达到了培育目标，主要生产性能与辽宁绒山羊、内蒙古绒山羊和柴达木绒山羊基本接近，已跨入全国高产绒山羊的行列。但因各地的改良基础、草场、饲养条件差异较大，整体水平不够一致，产绒量还较低。为进一步提高河西绒山羊的产绒量，开展了河西绒山羊绒毛生长规律研究，现报道如下。

1 材料

1.1 试验时间与试验动物

试验时间为 2008 年 6 月 5 日至 2009 年 5 月 5 日（河西绒山羊绒毛开始萌发至脱落）。于 2008 年 5 月底从甘肃永昌河西绒山羊养殖农户中购买 20 只（公母各半）6 月龄、体重相近［（15.21 ± 0.61）kg］的河西绒山羊，饲养在离靖远县 5km 的河金坪村（海拔 1 700m，年均气温为 8 ~ 10℃）。所有试验羊维持自然光照（月均日照时数为 220h 左右）。试验期各月份日照和平均气温见表 1。

表 1　2006 年 6 月至 2007 年 5 月各月日照和平均气温

项目	2008 年							2009 年				
	6 月	7 月	8 月	9 月	10 月	11 月	12 月	1 月	2 月	3 月	4 月	5 月
日照（h）	277.1	238.8	251.8	193.1	175.7	193.0	214.4	158.5	177.9	248.8	237.6	267.7
平均气温（℃）	23.6	24.1	21.8	18.3	9.2	3.3	-6.3	-4.4	-0.6	5.8	13.0	17.9

试验羊在自然草场放牧。采用放牧加人工补饲相结合的方式饲养，冬春季除放牧外，每日给每只试验羊补饲精料 150g。补饲精料参照山羊饲养标准配制，精料组成及营养水平见表 2。

1.2 绒毛样品的采集

在试验开始前，在试验羊体侧中部区域（大约 10cm × 10cm）用染发剂将毛被贴根染成黑色作为标记。每月 5 日，在标记部位紧贴皮肤剪取约 2g 被毛作为分析毛样，备测绒毛长度。同时每月用直尺测量未被染色（即新长出）的绒毛长度。

表2 精料组成和营养水平

精料组成	含量	营养水平	含量
玉米（%）	66.6	干物质（%）	89.0
向日葵仁饼（%）	9.0	代谢能（MJ/kg）	10.4
麦麸（%）	10.0	粗蛋白（%）	10.6
米糠（%）	14.0	粗脂肪（%）	4.6
添加剂（%）	0.5	粗纤维（%）	7.4
食盐（%）	0.5	无氮浸出物（%）	61.9
		灰分（%）	4.6

注：代谢能为理论计算值。

1.3 试样的制备

将试验样品充分混合，用多点法从正、反两面随机抽取纤维（不少于40个点）约150mg，充分混合，平分成3份，其中，2份用于平行试验，1份备用。

1.4 排图

将抽取的试样用手反复整理成一端接近平齐且纤维自然顺直的小绒束，右手握住小绒束平齐的一端，将另一端贴于绒板并用左手的大拇指摁住该端，将纤维由长至短从绒束中缓慢拔出，使逐次被拔出的纤维沿绒板左上端自上而下、自左而右、一端平齐地贴覆在绒板上，当手中的纤维全部拔完后用镊子将试样取出，再理成小绒束。如此操作数遍（不多于5遍），直至将试样均匀地排成底边长度为（250±10）mm，且纤维分布均匀的长度分布图。

1.5 作图

将手排长度标准板置于已排好的长度分布图上，目光直视图形的每个观测点，按照手排长度标准板上的刻度，将相关的数值记录下来。长度分布图的底边为横坐标、纤维长度为纵坐标，从原点自左向右每间隔10mm标出横坐标x_1，x_2，……，x_i，……，x_{n-1}，如果末组组距小于10mm，标出终点坐标点x_n，测量每组中点对应的纤维长度H_1，H_2，……H_i，……,H_n。长度分布图底边总长度为x_n。

1.6 绒毛手排长度计算

用以下公式计算：

$$L=\frac{10\sum_{i=1}^{n-1}H_i-(x_n-x_{n-1})H_n}{x_n}$$

式中：L为平均长度（mm），H_i为第i组中点坐标对应的纤维长度（mm），x_n为长度分布图底边总长度（mm），H_n为末组中点坐标对应的纤维长度（mm）。

2 结果与分析

试验期间测得的羊绒长度见表3、表4、图1。

表3　河西绒山羊公羊绒毛生长长度

测定日期	绒毛实际长度（mm）	绝对增长量（mm）	相对增长量（%）
2008-09-05	13.8 ±1.8	13.8	32.5
2008-10-05	25.4 ±3.7	11.6	27.4
2008-11-05	35.2 ±3.3	9.8	23.1
2008-12-05	42.1 ±4.4	6.9	16.3
2009-01-05	42.4 ±4.3	0.3	0.7
合计	42.4	42.4	100

表4　河西绒山羊母羊绒毛生长规律

测定日期	绒毛实际长度（mm）	绝对增长量（mm）	相对增长量（%）
2008-09-05	11.2 ±2.2	11.2	30.4
2008-10-05	21.6 ±3.7	10.4	28.2
2008-11-05	31.3 ±3.3	9.5	25.8
2008-12-05	36.6 ±2.4	5.5	14.9
2009-01-05	36.8 ±4.3	0.2	0.5
合计	36.8	36.8	100

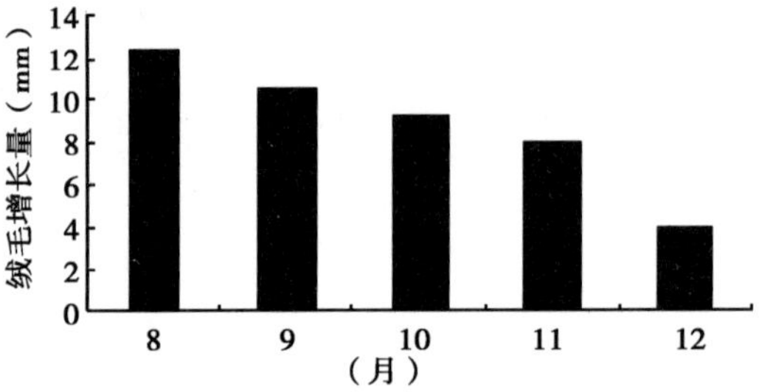

图1　河西绒山羊公母羊总体绒毛生长情况

在自然情况下，试验开始（2008 年 6 月 5 日）时所有试验羊均没有绒纤维长出。7 月下旬绒山羊绒毛开始萌发，7 月进入生长阶段，翌年 1 月份基本停止生长。公、母羊绒毛均以 8 月生长最快，分别占全年总生长量的 32.5% 和 30.4%；8～11 月的生长量分别占全年生长量的 83.0% 和 84.4%。1 月时羊绒基本停止生长，羊绒生长强度均呈直线下降趋势，直至终止。从总体上讲，公羊各月份的生长强度均高于母羊，方差分析结果表明，9，12 和翌年 1 月公羊绒毛实际长度极显著高于母羊（$P < 0.01$），10 月和 1 月显著高于母羊（$P < 0.05$）。据观察，公羊绒毛萌发的时间比母羊早 15d 左右。

3　讨论

绒山羊绒毛生长周期和繁殖周期一致，都与光周期变化有关，光照由长变短时绒毛开始生长，光照由短变长时逐渐停止生长。而低纬度地区由于光周期变化不明显，绒山羊绒毛生长很少或不长绒[1]。B. J. MCDonald 等[2]、贾志海[3]、达文政[4]、曲永年[5]、高文厚[6]的研究表明，光周期是影响绒山羊绒毛季节性生长的主要因素，绒毛的主要生长期在 8 ~ 11 月，高峰期为 9 月，光周期缩短可显著增加绒纤维长度。本试验结果表明，绒山羊公、母羊绒毛的最大生长量发生于夏季，最小生长量发生于冬季。表明当日照逐渐变短时绒毛萌发且迅速增长，日照逐渐增加时羊绒停止生长；短光照可以刺激绒毛生长。本试验得出的绒毛生长模式和前人的结果基本一致，但绒毛快速生长期和高峰期均提前 1 ~ 2 个月，可能是营养因素和光照以外的因素所致。

在放牧条件下，7 ~ 12 月的牧草总体营养价值逐渐降低，绒山羊采食量也逐渐下降，摄入的有效养分量也相应减少。研究表明，内蒙古白绒山羊 7 ~ 12 月摄入的粗蛋白越来越少，养分摄入量不足抑制和延迟了绒毛生长[7~8]。而牛一兵等[9]的研究表明，在放牧条件下，随牧草营养水平的下降和光照递减，自然光照组的内蒙古白绒山羊绒毛的生长速度在8 ~ 11 月逐渐上升，12 月下降，呈慢—快—慢的增长模式，主要生长期在 9 月下旬 ~ 12 月上旬，高峰期为 11 月。本试验在放牧条件下，随牧草营养水平的下降和光照递减，7 月下旬绒毛才开始萌发，8 月进入生长阶段，翌年 1 月基本停止生长，2 月完全停止生长，5 月脱落，也呈慢—快—慢的增长模式。

4 结论

河西绒山羊绒毛在 7 月开始萌发生长，生长期为 8 ~ 12 月；8 月为生长高峰期，生长量从 8 ~ 12 月呈线性下降趋势，至翌年 1 月绒毛基本停止生长，2 月完全停止生长，5 月绒毛脱落，呈慢—快—慢的增长模式。公羊的生长强度均大于母羊，绒毛萌发的时间也比母羊早 15d 左右。

参考文献

[1] 中国农业年鉴编辑委员会 . 中国农业年鉴 [M]. 北京：中国农业出版社，2007：202 – 208.

[2] MCDONNELD B J, HOEY W A, HOPKINS P S. CyCliCal fleece growth in cashmere goats [J]. Austr J AgriC Res, 1987, 38: 597 – 609.

[3] 贾志海 . 不同光周期和褪黑激素对绒山羊生产性能影响 [J]. 中国畜牧杂志，1998 (4)：8 – 10.

[4] 达文政 . 山羊绒生长期的测定 [J]. 中国养羊，1991 (2)：45 – 46.

[5] 曲永年 . 辽宁山羊绒生长规律的研究 [J]. 中国养羊，1995 (4)：37.

[6] 高文厚 . 对二狼山白绒山羊被毛脱换的探讨 [J]. 中国养羊，1996 (2)：42 – 43.

[7] 孙海洲，侯先志，于志红，等 . 日粮蛋白和能量水平对内蒙古阿尔巴斯白绒山羊产绒性能的影响[J]. 内蒙古畜牧科学，1998，19 (3)：5 – 7.

[8] 彭玉麟，贾志海，卢德勋，等 . 不同蛋白质水平的日粮对内蒙古白绒山羊消化代谢的影响 [J]. 畜牧兽医学报，2002，33 (4)：321 – 326.

[9] 牛一兵，贾志海，卢德勋，等 . 放牧条件下绒山羊绒毛季节性生长变化规律的研究 [J]. 动物营养学报，2005，17 (6)：33 – 36.

（发表于《黑龙江畜牧兽医》）

家兔绒毛的显微结构研究

郭天芬[1,2]，王欣荣[3]，李维红[1,2]，牛春娥[1,2]

（1. 中国农业科学院兰州畜牧与兽药研究所，兰州 730050；2. 农业部动物毛皮及制品质量监督检验测试中心，兰州 730050；3. 甘肃农业大学，兰州 730050）

摘　要：［目的］了解家兔绒毛的显微形态结构，选择具有典型特征的单根家兔绒毛，从尖部到根部对其进行显微结构观察，并检测其纤维直径。家兔绒毛的毛尖由鳞片层和皮质层组成，无髓质层；中部一般由鳞片层、皮质层和髓质层组成；根部无髓质层，鳞片层呈麦穗状。这是家兔绒毛的特性，可用于与其他动物纤维的比较研究及种类鉴别。兔毛具有发达的髓质层，纤维直径与髓腔列数成正相关。绒毛一般为单列，粗毛为多列。单根兔绒的尖部最细，中部变粗，根部又变细，且各部分直径差异较大，外形生长特性呈纺锤形。利用生物显微镜法鉴别不同动物毛皮及其产品种类是较为客观、简便的方法。

关键词：家兔；绒毛；显微结构；平均直径

（发表于《安徽农业科学》）

利用高分辨率溶解曲线分析法检测藏羊 *Agouti* 基因突变

韩吉龙，杨博辉，岳耀敬，郭婷婷

（中国农业科学院兰州畜牧与兽药研究所，兰州 730050）

关键词：藏羊；*Agouti* 基因；基因突变；毛色

刺鼠信号蛋白基因（*Agouti*），由 4 个外显子和 3 个内含子构成，编码一段约 20 个氨基酸的信号肽和一段功能区。绵羊的 *Agouti* 基因编码区大小为 5 353bp，位于第 13 号染色体。研究表明 *Agouti* 基因对于毛色的调控起着决定作用，*Agouti* 基因的表达会引起褐黑素的产生，而 *Agouti* 不表达时则会引起真黑素的表达，从而调节被毛颜色的形成。J Gratten 等（2010）研究发现 *Agouti* 基因 g. 100 – 104 缺失 AGGAA 和 g. 5172T→A 的非同义突变与毛色相关。藏羊作为原始绵羊品种，其毛色是否与 *Agouti* 基因相关目前尚没有相关报告。本研究拟通过高分辨率溶解曲线分析法来进行 *Agouti* 基因的多态性与毛色的关联分析，以期望找到调控藏羊毛色的 QTN。

（Published the article in Proceeding of the 13th National Symposium on Animal Genetic Markers）

毛囊形态发生中信号转导通路研究进展

吴瑜琦，岳耀敬，杨博辉，郭　健，刘建斌，牛春娥，
郭婷婷，冯瑞林，孙晓萍，韩吉龙

（中国农业科学院兰州畜牧与兽药研究所，兰州 730050）

羊毛是人类在纺织史上最早利用的天然纤维之一，羊毛纤维柔软而富有弹性，羊毛制品具有低碳环保、手感丰满、吸湿性强、保暖性好、穿着舒适等优点。66%的羊毛用于服装生产，30%用于室内纺织品生产，如地毯、块毯及其他室内装饰品，4%用于产业用纺织品的生产。影响羊毛价格的性状主要包括羊毛细度、密度、强度、长度和净毛率，这些性状除净毛率外都是受多基因或多基因座联合调控的数量性状，且性状间还存在负相关。如毛囊密度与羊毛产量呈正相关（相关系数 0.35 ± 0.19），与羊毛直径呈负相关（相关系数 -0.65 ± 0.12），选择降低羊毛细度将增加毛囊密度，反之亦然。同样选择增加羊毛产量将导致羊毛细度、毛囊密度和羊毛强度增加。羊毛性状间的遗传相关表明基因通过对羊毛皮肤毛囊不同发育阶段的复杂调控网络来影响羊毛生长，因此，开展影响我国羊毛产量、质量的羊毛毛囊发育调控的分子发育机制的应用基础研究，是提高羊毛质量，不断改进和发展羊毛生产根本途径。现将毛囊形态发生中信号转导通路的研究进展进行综述，以期为开展毛囊形态发生的分子机制研究提供理论基础。

（发表于《中国草食动物科学》）

牦牛 HORMAD1 基因的克隆及生物信息学分析

金　帅[1,2,3]，齐社宁[1]，阎　萍[2,3]，梁春年[2,3]，郭　宪[2,3]，包鹏甲[2,3]，
裴　杰[2,3]，褚　敏[2,3]，丁学智[2,3]，刘文博[2,3]，吴晓云[2,3]，刘　建[2,3]

（1. 兰州大学基础医学院，兰州 730030；2. 中国农业科学院兰州畜牧与兽药研究所，
兰州 730050；3. 甘肃省牦牛繁育工程重点实验室，兰州 730050）

摘　要： 利用分子克隆技术获得了牦牛 HORMAD1 基因编码区序列，并采用生物信息学方法对该基因及其编码蛋白的基本理化性质、疏水性、信号肽、二级结构等方面进行了预测和分析。结果表明，牦牛 HORMAD1 基因包含一个长度为 1 182bp 的开放阅读框，编码 393 个氨基酸；其编码蛋白属于亲水性蛋白，无明显的信号肽，含有很多个磷酸化位点。二级结构主要以无规则卷曲和 α 螺旋为主。HORMAD1 基因编码产物氨基酸邻接系统树表明，牦牛 HORMAD1 与黄牛、马和猪等物种的 HORMAD1 氨基酸遗传距离较近，具有高度相似性。

关键词： 牦牛；HORMAD1 基因；克隆；生物信息学分析

（发表于《生物技术通报》）

牦牛 *MSTN* 基因分子克隆及序列分析

梁春年，丁学智，包鹏甲，郭　宪，裴　杰，王宏博，
褚　敏，刘文博，吴晓云，阎　萍

（中国农业科学院兰州畜牧与兽药研究所，甘肃兰州 730070）

摘　要：设计特定引物对牦牛 *MSTN* 基因 PCR 分段扩增并克隆和测序，利用分子生物学软件进行序列拼接，获得牦牛 *MSTN* 基因序列（GenBank 登录号 EU926670）。该基因由 3 个外显子和 2 个内含子组成，CDS 序列全长为 1 128 bp（GenBank 登录号 EU926671），由 375 个氨基酸组成。外显子大小分别为 373，374，381bp，内含子大小分别为 1 843，2 028bp。牦牛与普通牛的 *MSTN* 基因编码区中，在 417 位发生一次碱基转换（C→T），但未造成氨基酸改变。不同物种间在该基因编码区核苷酸序列和氨基酸序列上有较高的相似性，牦牛与普通牛、绵羊、猪、人、狗、小鼠、马、兔子、鸡、猩猩各物种间核苷酸相似性大小分别为 99.9%，96.5%，94.3%，89.1%，91.9%，91.3%，93.6%，91.7%，82.0%，92.0%。牦牛与普通牛 *MSTN* 氨基酸序列相似性最高，为 100%；而与绵羊、猪、小鼠、人、狗、马、兔子、鸡、猩猩各物种间相似性大小分别为 93.3%，95.5%，92.5%，94.1%，93.3%，94.9%，94.4%，88.0%，94.4%。生物信息学软件分析发现：牦牛 *MSTN* 基因编码蛋白的理论分子量约为 42.6kDa，PI 值为 6.14，Leu 的含量最高（9.9%），其次是 Lys（7.2%）。牦牛 *MSTN* 基因编码蛋白二级结构以 β 折叠为主，属于跨膜蛋白，跨膜区位于 $AA_6 \sim AA_{23}$；具有一个分泌信号肽结构，其氨基酸序列 MQKLQICVYIYLFMLIVA 具有 TGF－β 家族的特征。

关键词：牦牛；*MSTN* 基因；克隆；序列分析

（发表于《华北农学报》）

牦牛 *PRKAG3* 基因部分序列多态性分析

焦　斐[1,2]，杨富民[1]，阎　萍[2]，梁春年[2]，郭　宪[2]，包鹏甲[2]，裴　杰[2]，
丁学智[2]，褚　敏[2]，刘文博[2]，吴晓云[2]，刘　建[2]

（1. 甘肃农业大学食品科学与工程学院，兰州 730070；2. 中国农业科学院兰州畜牧与兽药研究所/甘肃省牦牛繁育工程重点实验室，兰州 730050）

摘　要：采用 PCR – SSCP 技术，对甘南牦牛、大通牦牛和天祝白牦牛的 *PRKAG3* 基因第 3、第 9 内含子的多态性进行检测。结果表明，牦牛 *PRKAG3* 基因在第 3、第 9 内含子中均存在多态位点，2 个位点分别存在 AA、AB 和 EE、EF 基因型。其中，A 和 E 为优势等位基因，2 个位点在 3 个牦牛群体中均处于 Hardy – Weinberg 平衡状态。独立性检验结果表明，在 1 806bp 处天祝白牦牛与甘南牦牛之间差异显著（$P<0.05$），大通牦牛与甘南牦牛、天祝白牦牛之间差异不显著（$P>0.05$）；在 4 487bp 处大通牦牛和天祝白牦牛、甘南牦牛之间差异极显著（$P<0.01$），天祝白牦牛与甘南牦牛之间差异不显著（$P>0.05$）。

关键词：牦牛；*PRKAG3* 基因；PCR – SSCP；多态性

（发表于《江苏农业科学》）

牦牛 *VEGF-A* 基因的克隆及生物信息学分析

吴晓云[1,2,3]，阎　萍[2,3]，梁春年[2,3]，郭　宪[2,3]，包鹏甲[2,3]，裴　杰[2,3]，
褚　敏[2,3]，丁学智[2,3]，刘文博[2,3]，焦　斐[2,3]，刘　建[2,3]

（1. 甘肃农业大学动物科学技术学院，兰州 730070；2. 中国农业科学院兰州畜牧与兽药研究所，兰州 730050；3. 甘肃省牦牛繁育工程重点实验室，兰州 730050）

摘　要：利用分子克隆技术获得了牦牛 *VEGF-A* 基因编码区序列，并采用生物信息学方法对该基因及其编码蛋白的基本理化性质、疏水性、信号肽、二级结构等方面进行了预测和分析。结果表明，牦牛 *VEGF-A* 基因包含一个 573bp 的开放阅读框，编码 190 个氨基酸；其编码蛋白属于亲水性蛋白，具有明显的信号肽。二级结构主要以无规则卷曲和 α 螺旋为主。*VEGF-A* 基因编码产物氨基酸邻接系统树表明，牦牛 *VEGF-A* 与黄牛、绵羊和成都麻羊等物种的 *VEGF-A* 氨基酸遗传距离较近，具有高度同源性。

关键词：牦牛；*VEGF-A* 基因；克隆；生物信息学分析

（发表于《华北农学报》）

牦牛功能基因的研究进展

肖玉萍[1]，魏云霞[1]，张百炼[2]，吴晓睿[1]，师　音[1]，周　磊[1]，李维红[1]

（1. 中国农业科学院兰州畜牧与兽药研究所，兰州 730050；2. 甘肃省农村饮用水安全管理办公室，兰州 730030）

牦牛是分布于海拔 3 000m以上，以我国青藏为中心的高山、亚高山地区的牛种之一，是唯一能够充分利用青藏高原草地资源进行动物性生产的优势牛种和特有的遗传资源。牦牛对高海拔地区严寒、缺氧、缺草等恶劣条件具有极强的适应能力，可提供奶、肉、毛、绒、皮革、役力、燃料等生产、生活必需品，在高寒牧区具有不可替代的生态、社会、经济地位。20 世纪 80 年代后期，国际上的动物育种开始进入分子水平，这给牦牛育种带来了曙光。牦牛分子育种是根据其个体所有性状的基因和基因型的组织结构及功能效应进行整体或全基因组选择、选配、保种，要开展分子育种，首先需要知道牦牛的基因组情况；因此，牦牛功能基因的研究成为近年来国内外学者研究的热点。本文就牦牛功能基因的研究情况作一综述，以期为牦牛选种、选育提供理论依据。

（发表于《黑龙江畜牧兽医》）

牦牛无角性状相关基因 SNP 标记筛选

刘　建，刘文博，阎　萍，梁春年，郭　宪，包鹏甲，裴　杰，王宏博，褚　敏

（中国农业科学院兰州畜牧与兽药研究所，甘肃省牦牛繁育工程重点实验室，兰州 730050）

角是牦牛重要的外貌性状，目前，随着家养牦牛数量的逐渐提高，避免集中运输和屠宰过程中牛角对牛只和饲养人员造成的伤害损失显得十分重要。除了传统的人工去角技术外，培育牦牛无角新品系是一种更为有效的途径。根据经典遗传学理论，角是由常染色体上的单个基因座所控制，在普通牛上该基因座称为无角基因座（*POLL* locus），且无角 P 对有角 p 呈显性。因此，在培育无角新品系的过程中，利用分子遗传标记剔除杂合子个体，可有效地保证后代无角性状的遗传稳定性。目前，普通牛的无角基因座已精细定位于 1 号染色体长臂近着丝粒区一个 1Mb 大小的染色体区间内，该区域包括多个已知的编码蛋白的基因。本研究将通过候选基因法，利用 DNA 测序和高分辨率溶解曲线 SNP 检测技术筛选与牦牛无角性状呈显著性关联的 SNP 标记。

（Published the article in Proceeding of the 13th National Symposium on Animal Genetic Markers）

牛 RHOQ 基因的电子克隆与生物信息学分析

吴晓云，阎　萍，梁春年，郭　宪，包鹏甲，裴　杰，丁学智，
褚　敏，刘文博，焦　斐，刘　建
（中国农业科学院兰州畜牧与兽药研究所，甘肃省牦牛繁育工程重点实验室，兰州 730050）

摘　要：利用电子克隆技术获得牛 RHOQ 基因 cDNA 序列，采用生物信息学方法对该基因及其编码蛋白的基本理化性质、疏水性、信号肽、二级结构和亚细胞定位等方面进行预测和分析。结果表明，牛 RHOQ 基因的 cDNA 序列全长 1 517bp，包含 1 个 618bp 开放阅读框，编码 205 个氨基酸；其编码蛋白属疏水性蛋白，不存在信号肽及跨膜结构，定位于分泌系统囊泡；二级结构主要以无规则卷曲和 α－螺旋为主。牛 RHOQ 基因编码蛋白可能具有生长因子功能，可能在神经再生和突触延伸过程中起重要作用。

关键词：牛；RHOQ 基因；电子克隆；生物信息学分析

（发表于《生物技术通报》）

三个牦牛群体 DRB3.2 基因 PCR－RFLP 多态性研究

包鹏甲，阎　萍，梁春年，郭　宪，丁学智，裴　杰，褚　敏，朱新书
（中国农业科学院兰州畜牧与兽药研究所甘肃省牦牛繁育工程重点实验室，兰州 730050）

摘　要：为了研究牦牛 DRB3.2 基因 exon 2 遗传多样性，试验采用 PCR－RFLP 对天祝白牦牛、甘南牦牛、大通牦牛 3 个类群 757 头个体的 MHC－DRB3.2 基因进行 PCR－RFLP 分析。结果表明：共检测出 8 个 Hae Ⅲ酶切位点、11 种基因型；在 3 个牦牛类群中，Hae Ⅲ C 基因型（225bp /175bp / 85bp /35bp）在大通牦牛、天祝白牦牛中是优势基因型，基因型频率分别为 0.387 和 0.366；而在甘南牦牛中，Hae Ⅲ A 基因型（175bp /85bp /35bp）为优势基因型，基因型频率为 0.306。3 个群体的多态信息含量分别为 0.739，0.754，0.743，均达到了高度多态（PIC >0.50）。说明牦牛 DRB3.2 基因具有高度多态性，在研究牦牛抗病育种和提高牦牛生产性能方面具有独特的效力和广泛的应用前景。

关键词：牦牛；DRB3.2 基因；PCR－ RFLP

（发表于《黑龙江畜牧兽医》）

微卫星标记 BMS1714 和 INRA61 在甘肃高山细毛羊中的遗传多样性研究

郭婷婷，杨博辉，郭　健，牛春娥，岳耀敬，孙晓萍，冯瑞林，刘建斌

（中国农业科学院兰州畜牧与兽药研究所，兰州 73050）

摘　要：为了研究甘肃高山细毛羊的遗传多样性，为地方山羊品种的选种、选育及保种工作奠定基础。本研究选取位于绵羊第 25 号染色体上的微卫星标记 BMS1714 和 INRA61，对其进行遗传多样性研究。结果表明：微卫星标记 BMS1714 和 INRA61 在甘肃高山细毛羊上呈现多态性。BMS1714 基因座有 10 个等位基因，其片段大小在 121～138bp；BMS1714 基因座多态信息含量（PIC）、有效等位基因数（Ne）、平均杂合度（He）分别为 0.714，4.029，0.731。INRA61 基因座有 11 个等位基因，其片段大小在 279～292bp；INRA61 基因座的 *PIC*、*Ne*、*He* 分别为 0.76，5.127，0.809。由此表明，微卫星位点 BMS1714 和 INRA61 在甘肃高山细毛羊上均为高度多态位点，可用于甘肃高山细毛羊的遗传多样性分析。

关键词：BMS1714；INRA61；微卫星标记；遗传多样性；甘肃高山细毛羊

（发表于《中国草食动物科学》）

*MyoD*1 基因 3′UTR SNPs 多态性与甘南牦牛生长性状相关性的研究

褚　敏[1,2]，阎　萍[1,2]，梁春年[1,2]，郭　宪[1,2]，裴　杰[1,2]，丁学智[1,2]，包鹏甲[1,2]，朱新书[1,2]

（1. 中国农业科学院兰州畜牧与兽药研究所，兰州 730050；2. 甘肃省牦牛繁育工程重点实验室，兰州 730050）

摘　要：利用 PCR－SSCP 的方法对 180 头成年甘南牦牛 *Myo*D1 基因（GeneID：NC_007313）外显子 3 全序列及 3′UTR 部分序列进行 SNPs 筛选，并将筛选到的突变位点与牦牛五项体尺指标进行相关性分析。结果在 3′UTR1976 位检测到一处新突变，突变由 C→T 的转换造成，由此产生 3 种基因型。利用最小二乘分析研究了该位点不同基因型对甘南牦牛体尺性状的影响。结果表明，CC 型个体的胸围、体重显著高于 CT、TT 型个体（$P<0.05$），CC 型个体的体斜长显著高于 CT、TT 型个体（$P<0.05$），3 种基因型对体高及管围的影响差异不显著。

关键词：Myod1 基因；SNPs；牦牛；生长性状

（发表于《畜牧与兽医》）

大通牦牛 *MyoD*1 基因 3′UTR SNPs 多态性及其与生长性状相关性的研究

褚　敏，阎　萍，梁春年，郭　宪，裴　杰，丁学智，包鹏甲，朱新书

（中国农业科学院兰州畜牧与兽药研究所，兰州 730050）

摘　要：利用 PCR－SSCP 对 296 头成年大通牦牛 *MyoD*1 基因（GenBank 登录号：NC_007313）外显子 3 全序列及 3′UTR 部分序列进行 SNPs 筛选，并将筛选到的突变位点与牦牛五项体尺指标进行相关性分析。结果发现，在 3′UTR 1976bp 处检测到 1 个新突变，突变由 C→T 的转换造成，由此产生 3 种基因型：CC、CT、TT。利用最小二乘分析研究了该位点不同基因型对大通牦牛体尺性状的影响。结果表明，CC 基因型个体的胸围、体重显著高于 CT、TT 基因型个体（$P<0.05$），CC、CT 基因型个体的体斜长显著高于 TT 基因型个体（$P<0.05$），3 种基因型对体高及管围的影响差异不显著（$P>0.05$）。

关键词：*MyoD*1 基因；SNPs；牦牛；生长性状

（发表于《中国畜牧科学》）

大通牦牛与甘南当地牦牛杂交改良效果分析

郭　宪[1,2]，包鹏甲[1,2]，裴　杰[1,2]，梁春年[1,2]，丁学智[1,2]，阎　萍[1,2]

（1. 中国农业科学院兰州畜牧与兽药研究所，甘肃省牦牛繁育工程重点实验室，兰州 730050；2. 国家肉牛牦牛产业技术体系遗传育种与繁殖研究室，牦牛选育岗位）

牦牛（*Bos grunniens*）主要分布于青藏高原及其毗邻地区，是青藏高原特有的遗传资源。由于对高海拔地带严寒、缺氧、缺草等恶劣条件的良好适应能力而成为高寒牧区最基本的生产生活资料，可提供肉、奶、毛、绒、皮革、役力、燃料等，在高寒牧区具有不可替代的生态、社会、经济地位[1]。但牦牛产区自然生态条件严酷，加之生产方式落后、饲养管理水平低、技术转化率低等原因，导致牦牛生产性能低、产品附加值低，从而影响牦牛养殖业经济效益。为提高牦牛生产性能、增加农牧民收入、促进区域经济发展，科研工作者在提高牦牛生产性能方面开展了大量的科学研究，其品种改良是主要研究手段和方法之一。

（发表于《中国草食动物科学》）

甘南玛曲夏季牧场欧拉型藏羊牧食行为的研究

王宏博[1,2]，郎　侠[1,2]，丁学智[1,2]，梁春年[1,2]，刘文博[1,2]，阎　萍[1,2]

（1. 中国农业科学院兰州畜牧与兽药研究所，兰州 730050；2. 甘肃省牦牛繁育工程重点实验室，兰州 730050）

摘　要：为了研究夏季牧场草地状况对欧拉型藏羊牧食行为的影响，本试验选取甘肃甘南藏族自治州玛曲县作为研究地点，研究夏季牧场状况对欧拉型藏羊的牧食行为，从理论上探讨欧拉型藏羊牧食行为的变化规律。根据草地植被特征，通过野外调查，调查了解了甘肃甘南藏族自治州玛曲县夏季牧场的草地状况，并选取1牧户，采用跟踪观测的方法对欧拉型藏羊的牧食行为进行了观察。结果表明：夏季牧场位于祁连山谷地，属于高寒草甸和沼泽地，以高寒草甸地为主。夏季牧场植物物种丰富，植被以莎草科甘肃嵩草（*Kobresia kansuensis*）为优势种（优势度：0.9759）、毛茛科唐松草（*Thalietrum aquilegifolium var. sibiricum*）为亚优势种（0.5170），植被总盖度达9%，甘肃嵩草的盖度为28.3%，毛茛科小花草玉梅（*Anemone rivularis* var. *flore-minore*）的盖度为3.3%。甘肃嵩草和小花草玉梅的鲜重分别为77.3g/m^{2}和21.3g/m^2，干物质量分别为34.8g/m^{2}和4.2g/m^2，频度分别为100%和73.3%。在夏季草场，欧拉型藏羊的昼采食时间最长（450min），游走时间较长（80min）；反刍和站立时间基本相近，分别为32min和28min；夏季牧场白天的卧息时间非常少，8min。采食时间与地上生物量和牧草盖度成正相关（$r=0.782$，0.902；$P>0.05$）；昼反刍时间与牧草高度成极显著正相关（$r=0.995$，0.9996；$P<0.01$）；昼卧息时间与地上生物量成极显著负相关（$r=-0.9994$；$P<0.01$）；昼站立时间与牧草高度成显著正相关（$r=0.989$；$P<0.05$），与牧草盖度成极显著负相关（$r=-0.995$；$P<0.01$）；昼游走时间与牧草盖度成负相关（$r=-0.905$；$P>0.05$）。随牧草高度的增加，欧拉型藏羊的昼采食时间呈下降趋势，但采食时间占放牧时间的71.4%；由于夏季牧场牧草幼嫩多汁，欧拉型藏羊采食牧草后的反刍和卧息行为较少。

关键词：欧拉型藏羊；牧食行为；夏季草场

（发表于《动物营养与食品安全》）

河西绒山羊绒毛生长的季节性变化规律及其与生长激素关系的研究

魏云霞，肖玉萍，杨保平，程胜利，杨博辉

（中国农业科学院兰州畜牧与兽药研究所，兰州 730050）

摘　要：随机选择六月龄、平均体重 15.21kg ± 0.61kg 的河西白绒山羊 20 只（公母各半），从 5 月脱绒后开始对试验羊按月连续 12 个月采集血样和绒毛样品。用放射免疫法测定绒山羊血清生长激素（GH）浓度，用手排长度法测定各月绒毛生长长度。结果：①河西绒山羊绒毛在 6 月和 7 月开始生长，生长期自 8、9、10、11 至 12 月共 5 个月；8 月为生长高峰期，占其全年总生长量的 31.5%；生长量从 8～12 月呈线性下降趋势，至 1 月停止生长，呈慢—快—慢增长模式。②河西绒山羊血清 GH 浓度呈现明显的季节性变化规律；经相关性分析，绒毛生长速度与血清 GH 浓度水平呈极显著正相关（$r = 0.81$，$P < 0.01$）；性别对血清 GH 浓度没有影响。GH 浓度对绒毛生长可能有促进作用。

关键词：河西绒山羊；绒毛生长；生长激素

（发表于《中国草食动物科学》）

河西走廊放牧利用退化荒漠草原等级划分研究

周学辉，常根柱，杨红善

（中国农业科学院兰州畜牧与兽药研究所，兰州 730050）

摘　要：在对我国草地退化现状分析的基础上，根据全球环境基金（GEF）项目的研究结果，研究提出了河西走廊放牧利用荒漠草原退化草地分级的原则、指标、判定规则、利用状况指标和草原质量综合评价方法。以植物群落特征、群落组成、指示植物、草地上部生物量、土壤养分、土壤氮素含量、载畜量、家畜采食率和土壤含水量为主要监测指标；以地表特征、土壤容重和土层全氮含量为辅助检测指标；以未退化草地为对照，将荒漠草原退化草地分为轻度退化、中度退化和重度退化三级。该划分方案可作为荒漠类退化草地分级的主要参考依据，同时为该类草原的合理利用与综合治理提供了理论依据。

关键词：放牧利用；荒漠草原；退化草地；分级；指标

（发表于《中国畜牧业协会草业分会》）

角蛋白关联蛋白基因与羊毛性状关系的研究进展

郭婷婷，杨博辉，岳耀敬，牛春娥，孙晓萍，郭　健，冯瑞林，刘建斌
（中国农业科学院兰州畜牧与兽药研究所，兰州 730050）

摘　要：综述了与羊毛性状有关的角蛋白关联蛋白基因的定位、生物学功能及其与羊毛性状之间的关系，旨在为绒毛用羊品种改良和分子标记辅助选择育种提供理论基础。
关键词：*KAP* 基因；羊毛性状；基因定位

（发表于《安徽农业科学》）

利用羊穿衣技术提高绵羊生产性能的研究

孙晓萍，刘建斌，郎　侠，郭　健
（中国农业科学院兰州畜牧与兽药研究所，甘肃省牦牛繁育工程
重点实验室，兰州 730050）

摘　要：探讨细毛羊罩衣对绵羊羊毛纤维产量和质量的影响。以聚氯乙稀为主要原料，制成不同型号的羊罩衣，在绵羊剪毛后给绵羊穿上，来年剪毛时脱下。研究细毛羊罩衣对羊毛被毛结构、羊毛纤维产量、羊毛纤维品质、羊毛油汗、羊毛净毛率和杂质、羊毛毛纺加工的影响。绵羊穿上罩衣 1 年后，个体羊毛纤维长度平均增长 0.33cm（$P<0.05$），净毛率提高 12.44 个百分点（$P<0.01$），净毛产量提高 0.44kg（$P<0.01$），羊毛中油脂含量提高 6.44 个百分点（$P<0.01$），毛丛污染深度下降 3.10cm（$P<0.01$），灰尘含量下降 21.25%（$P<0.01$）。绵羊穿上罩衣后，明显改善了羊毛纤维的质量，提高了绵羊的生产性能。
关键词：羊衣；提高；羊毛；产量；质量

（发表于《安徽农业科学》）

牦牛的繁殖特性

阎　萍

（中国农业科学院兰州畜牧与兽药研究所，甘肃省
牦牛繁育工程重点实验室，兰州 730050）

牦牛（*Bos grunniens*）是青藏高原牧区的主体畜种和重要的生产生活资料，长期的自然选择造就了牦牛适合高寒特殊环境的体质构造和生理特性，形成了有别于其他牛种的繁殖特点。妊振期为 250～260d，具有一年产一胎的繁殖能力。但是牦牛的实际繁殖水平一般仅两年一胎或三年两胎。母牦牛一般是 2～4 岁时才发情受配。初配年龄主要取决于当地的草场和饲养管理条件，营养状况好，个体发育正常，初配年龄就早，营养状况差，发育受阻，初配年龄就推迟。因此，了解和掌握牦牛生殖生理和繁殖性能，是提高牦牛生产力的重要途径。

（发表于《中国畜牧科学》）

牦牛肉品质研究现状及展望

阎　萍[1,2]，王宏博[1,2]

（1. 中国农业科学院兰州畜牧与兽药研究所，兰州 730050；
2. 甘肃省牦牛繁育工程重点实验室，兰州 730050）

摘　要：本文通过论述牦牛肉的食用品质、营养品质、脂肪酸、风味物质以及重金属等，并与其他牛肉进行比较，阐述牦牛肉的营养价值和食用价值，为消费者提供参考。
关键词：牦牛肉；肉品质；食用品质；营养品质

（发表于《2012 中国牛业进展》）

曲霉发酵豆粕对育成猪磷代谢的影响

刘锦民[1]，王晓力[2]

（1. 甘肃省粮食局，兰州 730000；2. 中国农业科学院兰州畜牧与兽药研究所，兰州 730000）

摘　要： 选用 8 周龄的二元杂交育成猪 15 头（平均体重 23.6kg），随机分为 3 组，每组 5 头，采用全收粪法比较了饲喂曲霉（*Aspergillus usami*）发酵豆粕和普通豆粕对育成猪磷及其他营养成分的消化率。结果表明，发酵豆粕组育成猪磷消化率显著高于普通豆粕组，同时，粗蛋白质消化率也得到明显改善。试验结果提示，饲喂曲霉（*Aspergillus usami*）发酵的豆粕不仅可以提高育成猪磷和蛋白质的消化率，而且可以降低磷和氮的排泄量。

关键词： 发酵豆粕；育成猪；磷；消化率

（发表于《中兽医医药杂志》）

乳品及饲料中三聚氰胺及三聚氰酸的检测方法研究进展

王华东[1]，冯晓春[2]

（1. 中国农业科学院兰州畜牧与兽药研究所，兰州 730050；2. 兰州理工大学技术工程学院）

2008 年 9 月，甘肃、江苏等多个省市集中出现了“三鹿”奶粉污染致婴儿结石事件，以及随后在液态奶、奶糖、雪糕等制品中相继检测出三聚氰胺，引发了一系列的乳制品与食品安全事件，导致奶牛养殖与乳品市场萎缩。为促进奶业健康、有序、可持续发展，保障人类自身健康，必须加大对乳制品中三聚氰胺及三聚氰酸的监测力度。早期使用苦味酸法、升华法和电位滴定法对三聚氰胺进行检测，其前处理方法和检测限达不到目前对食品中三聚氰胺及三聚氰酸残留的检测要求。近年来，随着免疫学方法以及色谱/质谱等高端检测技术的发展与广泛应用，对乳制品与饲料中三聚氰胺及三聚氰酸残留实现了更准确、快速、灵敏的检测。现就其进展情况综述如下。

（发表于《中兽医医药杂志》）

饲草青贮系统中乳酸菌及其添加剂研究进展

王晓力[1]，张慧杰[2]，孙启忠[2]，玉　柱[3]，郭艳萍[3]

（1. 中国农业科学院兰州畜牧与兽药研究所，兰州 730050；2. 中国农业科学院草原研究所；3. 中国农业大学草地研究所）

青贮主要是微生物利用植物组织中的糖类（贮藏在植株细胞中）作为底物，并把它们转化为有机酸类，从而实现对饲草进行保鲜贮藏的过程。成功的青贮发酵取决于3个基本的必要条件：适合的微生物菌群（乳酸菌）结构、充足的可溶性碳水化合物含量，以及适当的物理－化学环境（厌氧、酸性）。青贮系统中参与活动的微生物种类繁多而复杂，为防止青贮饲料发生霉变，使得其营养成分得到长久并完整的保存，了解其中的微生物活动规律尤其是有利于青贮的菌群极为重要，其中，乳酸菌的质量和数量更是青贮成功与否的关键。笔者等着重介绍了饲草青贮系统中乳酸菌的种类、作用及乳酸菌青贮添加剂，并对近些年国内外的研究现状进行分析。

（发表于《中兽医医药杂志》）

西藏达孜混播油菜对箭筈豌豆种子产量影响的研究

李锦华[1]，乔国华[1]，张小甫[1]，杨　晓[1]，田福平[1]，余成群[2]

（1. 中国农业科学院兰州畜牧与兽药研究所，兰州 730050；2. 中国科学院地理科学与自然资源研究所，北京 100101）

摘　要：雨季造成的光照不足是西藏箸豌豆种子高产的制约因素之一。其种子田混播支撑作物，改善草层的通风透光性，是有效的增产措施。试验品种选择西牧324，支撑作物选择油菜，进行混播试验。结果表明，箭豆与油菜“间行播”处理的箭豆种子产量最高，达1 844.00kg/hm^2，与箭豆单播处理的差异达到显著水平（$P<0.05$）。在油菜混播量为0～8.3kg/hm^{2}的范围内设5个水平的试验中，随油菜播量增加，西牧324种子产量随之增加。在西牧324不同行距条件下混播油菜的试验中，箭豆播种行距为20cm的种子产量（2 386.39 kg/hm^2）显著高于行距为33.3cm的种子产量（1 431.67kg/hm^2）（$P<0.05$）。分析西牧324种子产量与其秸秆产量和株高的相关性，结果表明，种子产量与秸秆产量有显著正相关关系（$P<0.05$），而与株高没有显著相关性。说明与支撑作物混播时，较高的箭筈豌豆光合器官量对于保证其种子高产有较大作用，而株高的影响较小。

关键词：箭筈豌豆；种子；油菜；混播

（发表于《中国草食动物科学》）

野牦牛的抗逆性与牦牛的抗逆育种研究

朱新书，阎　萍，梁春年，郭　宪，裴　杰，包鹏甲，褚　敏，丁学智

（中国农业科学院兰州畜牧与兽药研究所甘肃省牦牛繁育工程重点实验室，兰州 730050）

摘　要：为了阐述牦牛抗逆育种的途径和方法，文章对野牦牛的强抗逆性进行介绍，经过残酷的自然选择和特殊的闭锁繁育，野牦牛在体格大小、生长速度、生活力等方面远优于家牦牛，对青藏高原的各种自然条件具有极强的抗逆性，是改良、复壮家养牦牛的天然优良基因库，导入野牦牛血液，通过抗逆育种，对提高家牦牛的生活力和生产力具有重要意义。

关键词：野牦牛；牦牛；抗逆性；抗逆育种

（发表于《黑龙江畜牧兽医》）

中草药饲料添加剂对河西肉牛血液生化指标影响的研究

周学辉[1]，杨世柱[1]，李　伟[1]，郭兆斌[2]

（1. 中国农业科学院兰州畜牧与兽药研究所，兰州 730050；
2. 甘肃农业大学食品科学与工程学院）

摘　要：为了提升河西肉牛产品特别是血液制品的质量及产业效益，利用饲料中添加中草药添加剂的方法对河西肉牛架子牛进行了育肥、屠宰试验。采取低剂量中草药组、高剂量中草药组和对照组肉牛血样，用专门仪器和方法测定了血液中 5 种血清矿物质离子、7 种血清代谢产物及 5 种血清酶等生化指标的含量。结果表明：高剂量中草药组肉牛血清中的 Na、Mg 离子及乳酸脱氢酶含量均极显著高于低剂量组和对照组（$P<0.01$），Cl 离子含量极显著高于对照组（$P<0.01$），其葡萄糖、肌酐、甘油三酯含量均显著高于对照组（$P<0.05$），谷丙转氨酶显著低于其他两组肉牛（$P<0.05$），而谷草转氨酶（AST）却显著高于其他两组肉牛（$P<0.05$）；低剂量组和高剂量组肉牛血清中的高密度脂蛋白极显著高于对照组（$P<0.01$），碱性磷酸酶极显著低于对照组肉牛（$P<0.01$），而谷丙/谷酸则显著高于对照组肉牛（$P<0.05$）。结论：试验配制的中草药饲料添加剂对肉牛血液生化指标有良好的影响，对高剂量组肉牛的影响更为明显；饲喂中草药饲料添加剂的肉牛生长发育、营养状况均显著好于对照组，但高剂量组更优；利用高剂量中草药饲料添加剂饲喂的肉牛，其血液在制造保健用品方面有广阔开发前景。

关键词：中草药饲料添加剂；河西肉牛；血清矿物质离子；血清代谢产物；血清酶

（发表于《中国草食动物科学》）

中草药添加剂对河西牛肉挥发性化合物的影响

周学辉[1]，杨世柱[1]，李 伟[1]，杨濯羽[2]，郭兆斌[2]，余群力[2]
（1. 中国农业科学院兰州畜牧与兽药研究所，兰州 730050；
2. 甘肃农业大学食品科学与工程学院）

摘 要：为了研究中草药添加剂对牛肉挥发性化合物的影响，试验选取成年河西肉牛，饲养过程分为中草药组和对照组，屠宰后运用气相色谱－质谱联用技术（GC－MS）进行牛肉挥发性化合物含量的测定。结果表明，中草药组和对照组牛肉挥发性物质在种类和含量上都有差异，对照组烃类、酮类、芳香族化合物、含硫类含量均高于中草药组，而中草药组醛类物质的含量比对照组高 30.11%，同时认为中草药的添加与 α－蒎烯、β－丁香烯、柠檬烯等特征挥发性化合物的产生有关。说明中草药添加剂能够增加牛肉中醛类、醇类以及杂环类化合物的含量，并能产生一些中草药添加剂所特有的挥发性化合物。

关键词：牛肉；中草药饲料添加剂；挥发性化合物

（发表于《中兽医医药杂志》）

中草药添加剂育肥河西肉牛脏器中生物活性物质研究

周学辉[1]，李 伟[1]，杨世柱[1]，郭兆斌[2]
（1. 中国农业科学院兰州畜牧与兽药研究所，兰州 730050；
2. 甘肃农业大学食品科学与工程学院）

摘 要：为了提升河西肉牛产品质量和产业效益，采用不同方法对用低、高剂量中草药饲料添加剂育肥河西西杂肉牛屠宰后不同脏器中的 L－肉碱、谷胱甘肽、肝素钠、牛磺酸等 4 种生物活性物质的含量进行测定。结果表明，中草药添加剂低剂量组肉牛肝脏、心脏、肾脏中的 L－肉碱含量最高，且均极显著高于高剂量组和对照组（$P<0.01$）。3 组肉牛之间肾脏中谷胱甘肽含量差异极显著（$P<0.01$），以低剂量组最高，谷胱甘肽在肉牛肝脏中的含量较为丰富。肝素钠在肺脏中的含量最高，低剂量组肉牛心脏、肺脏中肝素钠的含量均高于高剂量组，3 个试验组肉牛之间心脏肝素钠含量均差异极显著（$P<0.01$）。高剂量组肝脏中牛磺酸含量极显著高于低剂量组和对照组（$P<0.01$），肺脏中牛磺酸含量最高。结果提示，低剂量组肉牛肝脏、心脏、肾脏中的 L－肉碱，肾脏及肺脏中的牛磺酸以及高剂量组肉牛肝脏中的牛磺酸具有开发利用的潜力和价值；从试验牛肝脏中分离提取谷胱甘肽以及从肺脏中提取肝素钠切实可行。

关键词：中草药饲料添加剂；河西肉牛；脏器；L－肉碱；谷胱甘肽；肝素钠；牛磺酸

（发表于《中兽医医药杂志》）

5种动物毛皮种类的鉴别

李维红[1,2]，席　斌[1,2]，郭天芬[1,2]，王宏博[1,2]

（1. 中国农业科学院兰州畜牧与兽药研究所，兰州 730050；2. 农业部动物毛皮及制品质量监督检验测试中心（兰州），兰州 730050）

摘　要：对狐、貉、貂、狸、麝鼠5种毛皮动物的毛绒进行了研究，利用扫描电镜法比较其鳞片层及其横截面超微结构特征，以达到5种毛皮动物种类鉴别的目的。并测定了5种毛皮动物针毛及绒毛的鳞片翘角、鳞片高度、鳞片厚度等指标。结果显示：5种毛皮动物的毛纤维超微结构形态特征各异，其鳞片翘角、鳞片高度、鳞片厚度等指标的数据各不相同，同种动物针毛与绒毛之间的鳞片翘角、鳞片高度、鳞片厚度等指标也有差异。

关键词：毛皮动物；扫描电镜；微观结构

（发表于《毛纺科技》）

标记辅助选择技术在山羊育种中的应用

肖玉萍，周　磊，程胜利，杨保平，魏云霞

（中国农业科学院兰州畜牧与兽药研究所，兰州 730050）

摘　要：标记辅助选择技术与传统育种方式相结合，可大大提高动物育种效率，缩短育种周期。因此，加强标记辅助选择技术在山羊育种中的应用研究具有重要的实践意义。文章综述了标记辅助选择在动物育种应用中的基本策略、影响选择效率的因素及其在山羊育种中的应用，同时探讨了该技术在动物实际育种中应用较少的原因，并对其应用前景进行了展望。

关键词：标记辅助选择；山羊；育种

（发表于《中国草食动物科学》）

低碳经济与绒毛用羊业发展之路

岳耀敬[1]，杨博辉[1]，王天翔[2]，牛春娥[1]，郭婷婷[1]，孙晓萍[1]，
郎　侠[1]，刘建斌[1]，冯瑞林[1]，郭　健[1]
(1. 中国农业科学院兰州畜牧与兽药研究所，兰州 730050；2. 甘肃省绵羊繁育技术推广站)

气候变暖已引起了国际社会的强烈关注。人类活动产生的温室气体累积导致的全球气候变暖已成为科学界的共识。备受世界瞩目的哥本哈根世界气候大会已经落幕，来自 192 个国家的环境部长和政府官员们郑重承诺在 2012 年前共同削减温室气体排放，并帮助脆弱地区应对变暖带来的灾害。发展低碳经济是应对全球气候变化、减少温室气体排放的紧迫要求。世界观察研究所（WWI）在《世界观察》杂志（2009 年 11/12 月刊）中刊登的一篇题为《牲畜与气候变化》的报告中指出："牲畜和它们的副产品实际上至少排放了 325.64 亿 tCO_2 当量的温室气体，占世界总排放的 51%。" 远远超过联合国粮农组织先前估计的数值 18%。我国现有绒毛用羊约 1.2 亿只（细毛羊约 3 000万只，半细毛羊约 3 000万只，绒山羊约 6 000万只），至 2008 年年底，全国羊毛总产量 36.77 万 t，其中，细羊毛产量 12.39 万 t，半细羊毛产量 10.48 万 t，山羊绒产量达 1.72 万 t，同时生产羊肉约 126t（《2009 年中国统计年鉴》）。据新西兰 Andrew Bathe and Glenys Pellow 研究表明，羊毛在农牧场生产过程中温室气体排放的排放量为 CO_2 985kg/t，羊肉温室气体排放量为 1 520kg/t，依次估计我国绒毛用羊生产过程中温室气体排放量将达 2 157 415t/年，可见，减少绒毛用羊业生产的温室气体排放，发展低碳绒毛用羊生产经济应对气候变化，是实现我国绒毛用羊业可持续发展的根本出路。

（发表于《中国草食动物科学》）

河西肉牛最优杂交组合筛选试验

周学辉[1]，杨世柱[1]，李　伟[1]，戴德荣[2]，王东辉[2]

（1. 中国农业科学院兰州畜牧与兽药研究所，兰州 730050；
2. 张掖市甘州区畜牧技术管理站，张掖 734000）

摘　要：为了确定河西肉牛最优杂交组合，试验采用单因素方差分析，将河西肉牛 5 个杂交组合的 F_1 代公牛架子牛分为 5 个试验组（夏♂×西杂♀组、西♂×西杂♀组、皮♂×西杂♀组、安♂×西杂♀组、德♂×西杂♀组），用自行设计的配方日粮进行育肥及屠宰试验。结果表明：5 个杂交组合 F_1 代育肥牛的始重、末重（屠宰前活重）、胴体重及屠宰率差异均不显著（$P>0.05$）；安♂×西杂♀F_1 代肉牛日增重最高，2.18±0.06kg，且与其他 4 个组合相比差异显著（$P<0.05$），说明安♂×西杂♀为最优杂交组合。

关键词：河西肉牛；杂交组合；筛选；育肥；屠宰试验

（发表于《黑龙江畜牧兽医》）

牦牛的繁殖技术

阎　萍[1,2]，郭　宪[1,2]，梁春年[1,2]

（1. 国家肉牛牦牛产业技术体系遗传育种与繁殖研究室，牦牛选育岗位；2. 中国农业科学院兰州畜牧与兽药研究所，甘肃省牦牛繁育工程重点实验室，兰州 730050）

牦牛（*Bos grunniens*）是青藏高原牧区的主体畜种和重要的生产生活资料，长期的自然选择造就了牦牛适合高寒特殊环境的体质构造和生理特性，形成了有别于其他牛种的繁殖特点。牦牛妊振期 250～260d，具有一年产一胎的繁殖能力。但是牦牛的实际繁殖水平一般仅两年一胎或三年两胎。母牦牛一般是 2～4 岁龄才发情受配。初配年龄主要取决于当地的草场和饲养管理条件，营养状况好，个体发育正常，初配年龄就早；营养状况差，发育受阻，初配年龄就推迟。因此，了解和掌握牦牛生殖生理和繁殖性能，是提高牦牛生产力的重要途径。

（发表于《中国草食动物科学》）

母牦牛的繁殖特性与人工授精

郭　宪，裴　杰，包鹏甲，梁春年，丁学智，褚　敏，阎　萍
（中国农业科学院兰州畜牧与兽药研究所，甘肃省牦牛繁育工程重点实验室，兰州 730050）

摘　要：牦牛是青藏高原特有的遗传资源，是高寒牧区的基本生产生活资料。牦牛繁殖是牦牛生产中的关键环节，深入了解母牦牛的繁殖特性是实施牦牛繁殖调控的基础。本文从初情期、性成熟、发情、妊娠期、分娩、繁殖力等方面综述了牦牛的繁殖特性，并探析了提高牦牛人工授精效率的技术措施，旨在为牦牛品种改良提供技术参考。
关键词：牦牛；繁殖特性；人工授精

（发表于《中国牛业科学》）

奶牛真胃扭转的手术治疗

高昭辉[1,3]，董书伟[1,3]，荔　霞[1,3]，朱新荣[2]，施福明[2]，
刘学成[2]，杨天鹏[2]，严作廷[1,3]，刘永明[1,3]
（1. 中国农业科学院兰州畜牧与兽药研究所农业部兽用药物创制重点实验室，兰州 730050；2. 甘肃省荷斯坦奶牛繁育示范中心，兰州 730080；3. 甘肃省中兽药工程技术中心，兰州 730050）

摘　要：近年来奶牛真胃扭转发生率不断升高，特别是初产牛最易发生，占患病牛的80%，给奶牛养殖带来一定的经济损失。笔者工作中也常遇真胃扭转病例，手术治疗可获得较好的疗效，现将手术治疗一例初产荷斯坦牛真胃扭转的过程作一介绍。
关键词：奶牛；真胃扭转；手术治疗

（发表于《中国奶牛》）

食品中农药残留检测前处理技术进展

熊　琳[1,2]，杨博辉[1]，牛春娥[1]，郭婷婷[1]

（1. 中国农业科学院兰州畜牧与兽药研究所，兰州 730050；2. 农业部动物毛皮及制品质量监督检验测试中心，兰州 730050）

摘　要：样品前处理是检测食品中农药残留的关键部分，为提高食品中农药残留检测的准确度和精确度，各种农药残留检测前处理技术不断地出现，极大地提高了食品中农药残留检测的效率，综述了食品中农药残留检测的样品前处理技术。

关键词：食品安全；农药残留检测；前处理方法；萃取；净化

（发表于《江苏农业大学学报》）

中药饲料添加剂“速肥绿药”对架子牛育肥试验

周学辉，杨世柱，李　伟

（中国农业科学院兰州畜牧与兽药研究所，兰州 730050）

摘　要：采用单因素方差分析设计，设 3 个处理（空白对照组、添加 50g 组、100g 组），每个处理设 6 个重复。用自行研制的纯中草药饲料添加剂“速肥绿药”对西♂×西杂♀1 代肉牛架子牛进行了育肥及屠宰试验。结果表明，3 组间初始重差异不显著（$P>0.05$）；添加 50g 组和 100g 组末重（宰前重）均显著高于对照组（$P<0.05$）；100g 组胴体重、屠宰率、日增重分别为 402.67kg、59.28%、1.75kg/d，后 2 项指标均较 50g 组及对照组差异显著（$P<0.05$）。

关键词：中草药饲料添加剂；速肥绿药；架子牛；育肥；屠宰试验

（发表于《中兽医医药杂志》）

兽药学科

A15-day oral dose toxicity study of aspirin eugenol ester in Wistar rats

Li Jianyong[1], Yu Yuanguang[2], Yang Yajun[1], Liu Xiwang[1], Zhang Jiyu[1], Li Bing[1], Zhou Xuzheng[1], Niu Jianrong[1], Wei Xiaojuan[1], Liu Zhiqi[3]

(1. Key Lab of New Animal Drug Project, Gansu Province, Key Lab of Veterinary Pharmaceutical Development, Ministry of Agriculture, Lanzhou Institute of Animal and Veterinarian Pharmaceutical Science of CAAS, Lanzhou 730050, China; 2. Lanzhou Institute of Veterinary Sciences of CAAS, Lanzhou 730030, China; 3. The Academy of Life Science and Engineering, Lanzhou University of Technology, Lanzhou 730050, China)

Abstract: The subchronic toxicity of aspirin eugenol ester (AEE) was evaluated after 15 - day intragastrically administration in rats at daily doses of 50, 1 000, and 2 000mg/kg. AEE at low - dose showed no toxicity to the tested rats. Following repeated exposure to medium - or high - dose of AEE, apparent changes were observed in the levels of blood glucose, AST, ALP, ALT and TB in both male and female rats, and appeared to be dose - independent. There were no significant gender differences in most indexes of subchronic toxicity throughout the experimental period with the exception of food consumption and body weight. The no - observed - adverse - effect level (NOAEL) of AEE was considered to be 50mg/kg/day under the present study conditions.

Key words: Aspirin eugenol ester (AEE); Subchronic toxicity; Rats

1 Introduction

Aspirin has been used as a drug for treatment of inflammation and fever for more than a century. The biochemical mechanism of action of aspirin has been described previously (Vane, 1971; Flower *et al.*, 1972; Vane and Botting, 1987). Aspirin produces its therapeutic (e. g. anti - inflammatory and analgesic) and side effect (e. g. gastrointestinal ulcers) via inhibition of cyclooxygenase (COX) which is a key enzyme to catalyze prostaglandin formation (Vane, 1971). Recently, its use has been extended to prevention and treatment of cardiovascular diseases based on its anti - thrombotic action in platelets since inhibition of COX by aspirin blocks thromboxane A2 production which is crucial for blood clotting (Patrono, 1989). The anti - platelet effect of aspirin has been tested in various forms of coronary artery disease, pregnancy - induced hypertension and pre-eclampsia in angiotensin - sensitive primigravida at low dosage and showed positive results in most of the reports (Schoemaker *et al.*, 1998; Wallenburg *et al.*, 1986).

Eugenol is the main component of volatile oil extracted from dry alabastrum of Eugenia caryophyllata Thumb. Various therapeutic effects of eugenol have been demonstrated, including antivirus, antibacterial, antipyretic, analgesia, anti - inflammation, anti - platelet aggregation, anticoagulation, antioxidation, antidiarrhea, anti - hypoxia and, antiulcer, and inhibition of intestinal movement and arachidonic acid metabolism (Tragoolpua and Jatisatienr, 2007; Chami *et al.*, 2005; Gill and Holley, 2004; Nagababu and Lakshmaiah, 1997; Hashimoto *et al.*, 1988; Feng and Lipton, 1987; Raghavendra and Naidu, 2009). It is used to treat toothache, hepatopathy and gastrointestinal diseases.

However, the side effect of aspirin, such as gastrointestinal damage, is very serious and eugenol is irritative and vulnerable to oxidation. Chemically carboxyl group of aspirin and hydroxyl group of eugenol are responsible for these side effects and structural instability. Therefore, based on prodrug principal, aspirin and eugenol can be combined into aspirin eugenol ester (AEE) with reduced side effect and increased therapeutic effect and stabilization. AEE is supposed to be decomposed into aspirin and eugenol by the enzymes after absorption, which would show their original activities again and might act synergistically. The results of the acute toxicity and activities of AEE showed that the acute toxicity of AEE was less than the controls, which was 0.02 times of aspirin and 0.27 times of eugenol. Moreover, its anti - inflammatory, analgesic and antipyretic effects were similar as aspirin and eugenol, but lasted for a longer period (Li *et al.*, 2011a, b; Ye *et al.*, 2011; Li *et al.*, 2010). Thus, AEE is a promising drug candidate for treatment of inflammation, pain and fever and prevention of cardiovascular diseases with few side effects.

Since AEE is being developed for the treatment and prevention of many diseases, it is important to characterize its subchronic toxicity after 15days' oral administration in animals. The present study was performed to assess the subchronic toxicity of AEE, specifically after repeated oral administrations for 15 - days in rats at daily dose levels of 50, 1 000 and 2 000mg/kg. Blood or tissue samples were collected at various time points for the analysis of hematology, clinical chemistry and pathology. As an evaluation of preclinical safety, this study will provide guidance for the design of further preclinical toxicity studies and clinical trials of AEE.

2 Materials and methods

2.1 Chemicals and reagents

Aspirin eugenol ester (AEE), transparent crystal (purity: 99.5% with RE - HPLC), was prepared in Key Lab of New Animal Drug Project of Gansu Province, Key Lab of Veterinary Pharmaceutical Development of Agricultural Ministry, Lanzhou Institute of Husbandry and Pharmaceutical Sciences of CAAS. CMC - Na (carboxyl methyl cellulose sodium) was supplied by Tianjin Chemical Reagent Company (Tianjin, China).

2.2 Animals

Eighty wistar rats of both sexes with clean grade (Certificate No.: SCXK (Gan) 2008 - 0075), aged 7weeks and weighing 150 ~ 160g, were purchased from the animal breeding facilities of Lanzhou University (Lanzhou, China) and housed individually to allow recording of individual feed consumption, and to avoid bias from hierarchical stress. They were housed in plastic Macrolon cages of appropriate size with stainless steel wire cover and chopped bedding. Light/dark regime was 12/12h and living temperature is (22 ±2)℃ with relative humidity of 55% ±10%. Standard compressed rat feed from the animal breeding facilities of Lanzhou University and drinking water were supplied ad libitum. The study was performed in compliance with the Guidelines for the care and use of laboratory animals as described in the US National Institutes of Health. Animals were allowed a 2 – week quarantine and acclimation period prior to start of the study.

2.3 Dosing

The high –, medium –, and low – doses were selected as 2 000, 1 000 and 50mg/kg BW, respectively, based on the results of acute toxicity testing (which identified the approximate lethal dose to be 10, 937mg/kg BW) and preliminary studies in rats. AEE dose suspension liquids were prepared in 0.5% of CMC – Na. Rats were randomized into four groups: three test groups (n = 20rats each, male : female = 1 : 1) and a vehicle group as control (n = 20rats, male: female = 1 : 1). AEE was administered intragastrically in each rat based on individual daily body weights once daily for 15days.

2.4 Study design

During the administration and recovery period, clinical symptoms and mortality were observed daily. Body weight and food consumption for each rat was measured daily and subjected to statistical analysis once a week.

Hematology, the examinations of clinical biochemistry, visceral index and pathology were carried out at the end of drug administration. A portion of each blood sample was treated with EDTA – Na_2 and analyzed for hematological indexes such as the leukocyte count (WBC), erythrocyte count (RBC), hemoglobin concentration (HGB), hematocrit, mean corpuscular volume (MCV), mean corpuscular hemoglobin (MCH), mean corpuscular hemoglobin concentration (MCHC), platelet (PLT), neutrophil (MID), and lymphocyte (LYM) count with a hematology analyzer (Cell Dy900, Baker, USA). Moreover, an automatic biochemistry analyzer (7060, Hitachi, Japan) was used to examine the serum obtained from another portion of the blood sample for the content of glutamic pyruvic transaminase (GPT), aspartate aminotransferase (AST), alkaline phosphatase (ALP), creatinkinase (CK), lactate dehydrogenase (LDH) glucose (GLU), total protein (TP), albumin (ALB), blood urea nitrogen (BUN), creatinine (CREA), total bilirubin (TBIL), triglycerides (TG) and total cholesterol (TC).

Necropsy was carried out 24h after the final administration. All rats from each group were euthanized by exsanguinations in femoral artery. Skin/mammary gland, liver, pancreas, spleen, thymus, heart, lungs, stomach, duodenum, ileum, jejunum, colon, ovaries, uterus, urine bladder, kidneys, adrenals, trachea/thyroid gland, brain, pituitary gland, testes, epididymis, seminal vesicles (including fluid) with coagulation gland, prostate, popliteal lymph nodes, sciatic nerve, fat, muscle and sternum were dissected. All organs and tissues were macroscopically examined for gross pathology. Following necropsy, the brain, heart, liver, spleen, lung, kidneys, adrenal glands, testes and uterus were individually isolated and weighed to calculate the ratios of organ weight to body weight. These organs and ovaries, stomachs were then fixed in 10% neutral buffered formalin for histological processing based on the methods reported previously (Van der Ven *et al.*, 2006).

2.5 Histology

After fixation (and subsequent weighing, vide supra), organs sampled for histological examination were dehydrated and paraffinized and embedded according to standard sampling and trimming procedures. Sections of 4μm were stained with hematoxylin and eosin (HE) in an automated way. Microscopic observations were done by initial unblinded comparison of control and top dose samples. Blind and/or semi-quantitative scoring was applied when changes were detected by the initial inspection.

2.6 Statistics

All data are expressed as mean ± standard deviation (SD). The differences of ratios of organ weight to body weight were analyzed using ANOVA with LSD or Dunette's test (SPSS 12.0 software, USA). Other data were tested using ANOVA with Repeated Measures built in General Linear Model (SPSS 12.0 software, USA), and inter-group comparisons were made using the Multivariate of General Linear Model. P-values of <0.05 were considered statistically significant.

3 Results

3.1 Mortality and clinical observations

No deaths were observed in any group over the administration periods. Compared with the control group, the test groups exhibited no treatment-related changes in clinical signs such as external appearance, behavior, mental state, and daily activities.

With regard to food consumption, all rats in medium- and highdose groups showed a significant decrease compared with the control rats. The food consumption of all rats in low-dose groups was almost identical to that in the control group throughout the administration period. In all groups, the male rats consumed more foods than the female rats.

As a result of the changes in food consumption, all rats with same gender in the medium- and high-dose group showed a significant reduction in body weight gain compared with control animals ($P<0.01$) (Table 1). All rats in the medium- and highdose groups had significantly lower body weights during administration. All rats in the low-dose group tested on the other hand showed a tendency towards increased body weights that were not significantly different from those in the control group.

Table 1 Body weights changes of male rats and female rats before and after the administration of aspirin eugenol ester ($\bar{x} \pm s$)

Group		Body weight prior to administration (g)	Body weight after administration (g)	Changes in body weight (g)
control	Male	151.25 ± 6.99	249.25 ± 8.84	98.00 ± 6.22
	Female	153.00 ± 9.76	195.00 ± 12.08	46.25 ± 9.67

(Continued)

Group		Body weight prior to administration (g)	Body weight after administration (g)	Changes in body weight (g)
50 (mg/kg)	Male	156.60 ±8.66	248.20 ±10.91	93.60 ±7.38
	Female	150.11 ±13.58	180.44 ±11.64	43.33 ±5.83
1 000 (mg/kg)	Male	153.40 ±13.29	197.90 ±13.44	44.50 ±6.42**
	Female	150.44 ±13.15	167.77 ±16.70	17.33 ±12.79△△
2 000 (mg/kg)	Male	156.40 ±11.93	184.30 ±13.76	27.90 ±15.97**
	Female	155.14 ±12.19	149.85 ±15.35	−4.41 ±12.17△△

Note: Significantly difference compared with male control, ** $P<0.01$; Significantly difference compared with female control, △△ $P<0.01$.

3.2 Hematology

In the hematological analysis (Table 2), no treatment changes were observed in rats. However, the platelet count of rats following 15 days administration of all AEE doses was significantly decreased compared with that of control rats ($P<0.01$). This change was transient as there was no longer any significant difference between the three tested groups and the control group at the end of a 1-week recovery period (data not shown, $P>0.05$). In addition, there were no apparent changes in all indexes of hematology except platelet at each dose of AEE for both male and female rats after 15-day repeated dosing compared to the values obtained before administration (data not shown, $P>0.05$), and no significant differences were observed between the male and female rats during the experimental period.

Table 2 Hematological findings in rats administrated intragastrically with AEE for 15 days ($\bar{x}\pm s$)

Dose	Control	50 (mg/kg)	1 000 (mg/kg)	2 000 (mg/kg)
Male				
No. of animals examined	10	10	10	10
Leukocytes (10^9/L)	6.31 ±0.91	6.95 ±1.02	7.41 ±1.33	6.35 ±1.15
Erythrocytes (10^{12}/L)	7.58 ±0.85	8.79 ±0.82	8.31 ±0.81	8.06 ±1.61
Lymphocyte (10^9/L)	6.10 ±0.84	6.19 ±0.86	6.58 ±0.49	6.90 ±0.83
Hemoglobin (g/L)	214.52 ±17.89	218.00 ±15.59	211.75 ±17.79	210.05 ±18.76
Neutrophil (10^9/L)	0.97 ±0.11	0.84 ±0.16	0.82 ±0.19	0.95 ±0.25
MCHC (g/L)	380.11 ±9.78	376.01 ±6.68	381.29 ±5.31	392.65 ±15.76
MCH (pg)	26.36 ±0.75	25.85 ±0.53	25.51 ±0.36	26.24 ±1.09
MCV (fl%)	67.57 ±1.14	66.42 ±0.43	66.92 ±0.21	66.83 ±0.99
Platelets (10^9/L)	962 ±65.08	566.5 ±72.17**	698.5 ±28.33**	690.25 ±58.49**
Female				
No. of animals examined	10	10	10	10
Leukocytes (10^9/L)	6.83 ±0.59	7.97 ±1.85	7.67 ±1.31	6.87 ±1.23
Erythrocytes (10^{12}/L)	8.12 ±0.31	8.21 ±0.36	8.63 ±0.41	8.09 ±0.20
Lymphocyte (10^9/L)	6.18 ±0.84	6.49 ±0.38	6.35 ±0.13	6.51 ±0.37

(Continued)

Dose	Control	50 (mg/kg)	1 000 (mg/kg)	2 000 (mg/kg)
Hemoglobin (g/L)	210. 5 ± 15. 8	203. 3 ± 22. 9	227. 4 ± 35. 7	206. 3 ± 28. 5
Neutrophil (10^9/L)	1. 09 ± 0. 09	1. 09 ± 0. 18	1. 05 ± 0. 13	1. 07 ± 0. 16
MCHC (g/L)	390. 1 ± 9. 7	403. 7 ± 23. 3	407. 4 ± 10. 9	391. 14 ± 17. 5
MCH (pg)	26. 86 ± 0. 75	27. 33 ± 1. 65	27. 55 ± 0. 57	25. 56 ± 0. 42
MCV (fl%)	67. 05 ± 1. 21	66. 83 ± 0. 53	67. 62 ± 0. 55	67. 09 ± 0. 27
Platelets (10^9/L)	849 ± 65. 08	513. 75 ± 50. 02**	599 ± 132. 84**	486. 5 ± 174. 9**

Note: Significantly difference compared with control, ** $P < 0.01$.

3. 3 Blood biochemistry

The results of blood chemical tests are given in Table 3. This showed both female and male rats had a trend of decreased blood glucose in medium – and high – dose groups compared with the control animals, and levels of blood glucose in medium – and high – dose groups were significantly reduced in both sexes of rats after 15 – day repeated dosing compared to the pre – dosing levels ($P <$ 0. 05 and < 0.01). In addition, levels of ALT, ALP, AST and total bilirubin in all rats were significantly elevated in the medium – and high – dose groups ($P < 0.01$) in comparison with the control animals following 15 – day repeated dosing. With exception of TC and TG, no treatment – related changes in low – dose groups were observed in the other indexes examined. In each dose group, there was a trend of decreased levels of TC and TG compared to control animals following 15 – day repeated dosing. However, the decreased levels of TC and TG were countered with dosage, which was difficult to be explained. After a 1 – week recovery period, all of the above indexes were restored and difference between the three test groups and the control group was no longer statistically significant. No treatment – related changes were observed in the other indexes examined and no significant gender differences were detected in all indexes of blood chemistry during the experimental period.

Table 3 Biochemical parameters in rats administrated intragastrically with AEE for 15days ($\bar{x} \pm s$)

Dose	control	50 (mg/kg)	1 000 (mg/kg)	2 000 (mg/kg)
Male				
No. of animals examined	10	10	10	10
Creatinine (μmol/L)	80. 56 ± 5. 42	85. 00 ± 7. 68	85. 25 ± 2. 22	91. 25 ± 4. 31
LDH (U/L)	2003 ± 153	2172 ± 123	2041 ± 59	2016 ± 134
ALT (U/L)	70. 15 ± 13. 23	81. 57 ± 8. 93	117 ± 4. 55**	133. 3 ± 15. 39**
ALP (U/L)	210. 5 ± 23. 1	196. 8 ± 19. 06	269. 5 ± 21. 62**	325. 7 ± 20. 35**
AST (U/L)	286. 2 ± 17. 5	305. 0 ± 18. 96	394. 8 ± 26. 81**	392. 0 ± 14. 70**
Total protein (g/L)	65. 3 ± 5. 37	68. 3 ± 3. 13	65. 2 ± 3. 86	59. 6 ± 3. 71
Albumin (g/L)	38. 9 ± 1. 35	37. 5 ± 1. 02	37. 9 ± 0. 95	31. 10 ± 1. 07
Total bilirubin (μmol/L)	6. 50 ± 0. 17	6. 67 ± 0. 11	7. 75 ± 0. 14**	8. 01 ± 0. 16**

(Continued)

Dose	control	50 (mg/kg)	1 000 (mg/kg)	2 000 (mg/kg)
Blood urea nitrogen (mmol/L)	9.10 ±0.22	8.96 ±0.16	9.15 ±0.41	9.33 ±0.43
Glucose (mmol/L)	1.78 ±0.22	1.81 ±0.72	1.67 ±0.05*	1.16 ±0.91**
Total cholesterol (mmol/L)	2.20 ±0.15	1.36 ±0.16**	1.85 ±0.17**	1.82 ±0.17**
Triglycerides (mmol/L)	2.12 ±0.21	1.63 ±0.11**	1.84 ±0.14*	1.95 ±0.13
Creatinkinase (U/L)	4001 ±316	4098 ±326	3999 ±376	3802 ±171
Female				
No. of animals examined	10	10	10	10
creatinine (μmol/L)	85.12 ±6.98	86.25 ±10.05	88.33 ±1.53	89.50 ±5.96
LDH (U/L)	2109 ±153	2008 ±168	2086 ±56	2077 ±215
ALT (U/L)	101.01 ±10.23	98.26 ±22.26	113.33 ±14.01	136.50 ±15.05**
ALP (U/L)	205.69 ±8.16	215.25 ±10.78	244.66 ±8.02**	293.75 ±14.61**
AST (U/L)	290.45 ±11.91	299.50 ±21.12	319.66 ±16.70*	333.75 ±21.27**
Total protein (g/L)	66.32 ±1.36	66.75 ±2.38	65.46 ±1.40	65.35 ±3.54
Albumin (g/L)	38.15 ±0.42	37.80 ±0.67	38.30 ±0.69	35.12 ±0.36
Total bilirubin (μmol/L)	7.15 ±0.22	7.19 ±0.19	8.13 ±0.20**	8.20 ±0.23**
Blood urea nitrogen (mmol/L)	8.55 ±0.29	8.73 ±0.23	7.96 ±0.50	8.57 ±0.46
Glucose (mmol/L)	1.75 ±0.15	1.82 ±0.42	1.57 ±0.36*	1.27 ±0.31**
Total cholesterol (mmol/L)	2.13 ±0.12	1.72 ±0.05**	1.83 ±0.12**	1.90 ±0.08**
Triglycerides (mmol/L)	2.23 ±0.12	1.66 ±0.15**	1.86 ±0.13**	2.08 ±0.28
Creatinkinase (U/L)	4120 ±265	4240 ±361	4139 ±316	4127 ±218

Note: Significantly difference compared with control, * $P<0.05$, ** $P<0.01$.

3.4 Organ weights

Liver, spleen, epididymis and kidney weight were increased following AEE administration and statistical significance was achieved in the groups receiving 2 000mg/kg bw/day (mkd) in both females and males ($P<0.01$) (Table 4). There are dose dependent Changes. Only levels of liver and kidney in high – dose groups were significantly increased in both sexes of rats after 15 – day repeated dosing compared to the control and low – dosing levels ($P<0.01$).

Several other organ weights showed no dose dependent Changes. With regard to AEE dosage, no treatment – related changes were observed in medium – and low – dose groups while significant Changes in liver and kidney were found in high – dose group.

Table 4 Relative organ weights in rats treated intragastrically with AEE for 15days ($\bar{x}\pm s$)

Dose	control	50 (mg/kg)	1 000 (mg/kg)	2 000 (mg/kg)
Male				
No. of animals examined	10	10	10	10

(Continued)

Dose	control	50 (mg/kg)	1 000 (mg/kg)	2 000 (mg/kg)
Brain	6. 37 ±0. 38	6. 43 ±0. 56	6. 57 ±0. 78	6. 38 ±0. 71
Liver	3. 56 ±0. 20	3. 62 ±0. 19	3. 67 ±0. 13	4. 04 ±0. 50 $^{**,\#\#}$
Heart	3. 40 ±0. 18	3. 71 ±0. 42	3. 26 ±0. 09	3. 53 ±0. 35
Spleen	1. 66 ±0. 15	1. 74 ±0. 14	1. 75 ±0. 17	1. 78 ±0. 16
Kidney	5. 96 ±0. 19	6. 01 ±0. 33	6. 21 ±0. 11	7. 02 ±0. 61 $^{**,\#\#}$
Adrenal gland	1. 23 ±0. 12	1. 35 ±0. 26	1. 35 ±0. 35	1. 80 ±0. 52
Testis	1. 13 ±0. 07	1. 17 ±0. 05	1. 28 ±0. 06	1. 08 ±0. 11
Epididymis	9. 10 ±0. 52	9. 13 ±0. 59	9. 24 ±0. 33	9. 40 ±0. 22
Thymus	1. 15 ±0. 32	1. 28 ±0. 18	1. 17 ±0. 23	1. 19 ±0. 21
Female				
No. of animals examined	10	10	10	10
Brain	8. 21 ±0. 51	8. 31 ±0. 86	8. 25 ±0. 11	8. 12 ±0. 26
Liver	2. 58 ±0. 20	2. 42 ±0. 07	2. 74 ±0. 22	3. 57 ±0. 44 $^{**,\#\#}$
Heart	3. 36 ±0. 11	3. 25 ±0. 41	3. 24 ±0. 17	3. 15 ±0. 30
Spleen	1. 79 ±0. 26	1. 79 ±0. 32	1. 88 ±0. 26	1. 95 ±0. 35
Kidney	6. 08 ±0. 19	6. 08 ±0. 18	5. 78 ±0. 38	6. 44 ±0. 25 $^{**,\#\#}$
Adrenal gland	1. 80 ±0. 25	1. 83 ±0. 34	1. 77 ±0. 55	1. 91 ±0. 56
Testis	1. 65 ±0. 36	1. 57 ±0. 39	1. 72 ±0. 33	1. 81 ±0. 16
Epididymis	2. 55 ±0. 32	2. 52 ±0. 26	2. 62 ±0. 35	2. 71 ±0. 23
Thymus	1. 52 ±0. 16	1. 61 ±0. 36	1. 38 ±0. 34	1. 54 ±0. 21

Note: compared with control, $^{**}P<0.01$; compared with low – dose group, $^{\#\#}<0.01$.

3.5 Histopathology

There were no histopathological changes in any organ from the animals given with low – dose of AEE. In medium – dose groups, stomach mucous membrane epithelium became uneven. For duodenum, the evidences of degeneration, necrosis and ecclasis of intestinal villus epithelium, hyperemia of lamina propria, quantity increase of intestinal gland goblet cell and mucus excretion hyperfunction can be found. Other visceral organs assessed were normal. In high – dose group, the histopathological changes of stomach mucous membrane epithelium, duodenum and ileum were the similar as that in medium – dose group. Minor granular degeneration of kidney tubules epithelial cell and liver cell were also observed. We did not find any other histopathological changes in any of the other organs assessed.

The increased liver weights were associated with centrolobular hypertrophy (Fig. 1). In males, there was a trend that the rats receiving high dose had a higher ratio of apoptotic cells in their livers compared with control. Finally, an increased ratio of binucleated hepatocytes was observed in exposed livers.

The results demonstrated that in the kidney hyperemia happened in the glomerulus (Fig. 2) of male groups and the renal tubule interval of female groups following AEE administration at high

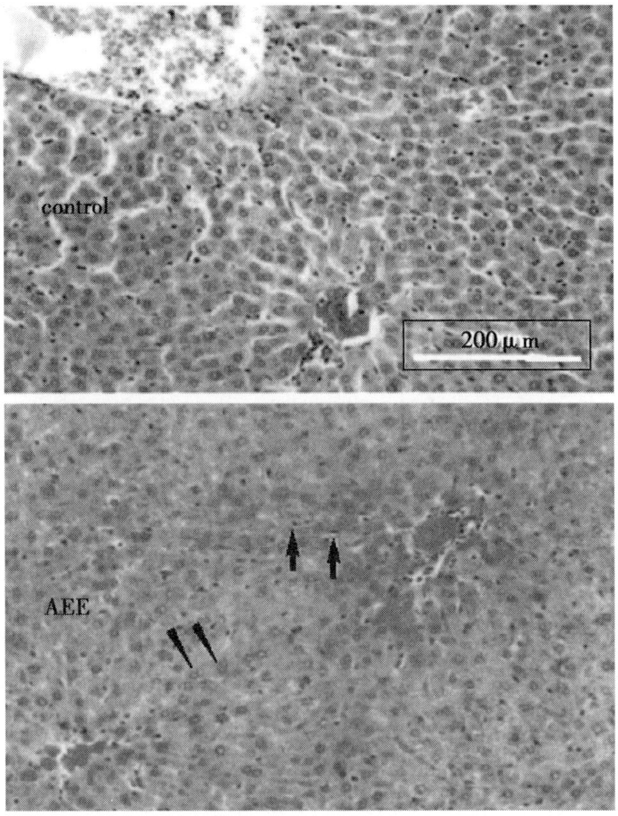

Fig. 1 Medium power microphotographs of rat liver, illustrating centrilobular hepatocellular hypertrophy as a result of exposure to AEE (2000 mg AEE/ (kg BW · day)). Furthermore, there is an increased ratio of binucleated cells (arrowheads) and apoptotic cells (arrows)

dose. Pathological changes such as degeneration and necrosis appeared in renal tubular epithelial cells in both high - dose and middle - dose groups, while more significant changes took place in higher dose groups.

The stomach, duodenum and ileum showed an increased mucosa height after high - dose AEE exposure, this was mainly due to increased height of the villus (Fig. 3).

4 Discussion

AEE is being developed for anti - inflammatory, analgesic, antipyretic, antiatheroscloresis and anti - thrombosis pharmaceutical. Our previous studies showed that this compound was effective against the symptoms of inflammation, fever, soreness (Li *et al.*, 2010, 2011a, b; Ye *et al.*, 2011). Because the duration of AEE dosage in clinical use as anti - inflammatory, analgesic and antipyretic will be about 3 days, 15 - day oral dose toxicity chosen as a preliminary to a long term study is reasonable. While long - termuse of AEE will be required in treatments of atherosclerosis and thrombosis, the chronic toxicity of AEE are important for better understand its safety after repeated administrations.

The food consumption by, and the body weight of, rats in the high - and medium - dose

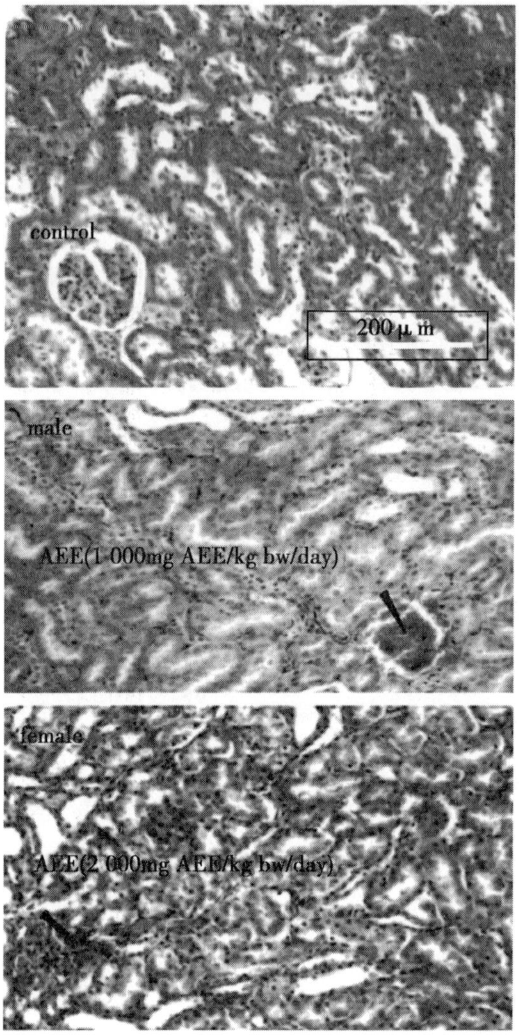

Fig. 2 Medium power microphotographs of rat kidney, illustrating hyperemia happened in the glomerulus of male groups and the renal tubule interval of female groups after high - and medium - dose AEE exposure

groups were decreased during the study. In addition, the body weights of male rats in all groups grew more than that of female rats. We speculate the reduction in food consumption and body weight observed in rats during the test may be due to hydrolysis of AEE into aspirin which can lead to gastrohelcosis. Quantities of AEE in low - , medium - and high - dose groups were 100, 2 000 and 4 000times of its clinic therapy dose, but AEE in low - dose showed no toxic to the tested rats. This indicated that AEE should be safe at clinical recommended dose. AEE should be decomposed in rare proportion at gastrointestinal tract. of course, further studies should be conducted to investigate the degree of AEE decomposition in stomach.

With regard to clinical chemistry and hematology, some in - dexes showed obvious changes after repeated dosing compared with the values in the control group or obtained prior to drug administration and were completely reversible after a 1 - week recovery period. The changes in TC, TG and

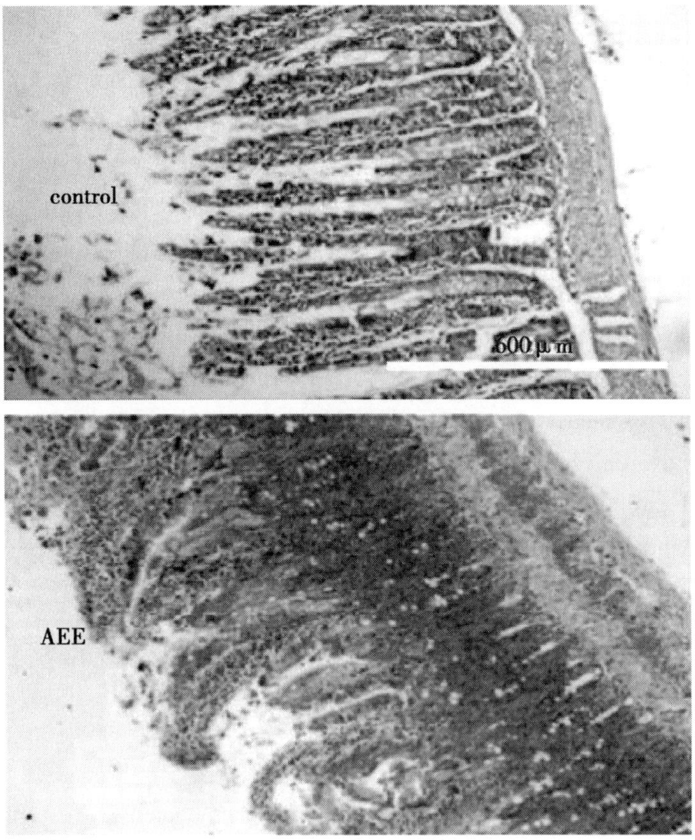

Fig. 3 Medium power microphotographs of rat duodenum, illustrating increased villus height after AEE (1 000 mg AEE/kg BW/day) exposure

platelet counts are be – lieved to be interesting since they appeared to be most significant in low – dose groups. This is very beneficial to the therapy of anti – atherescloresis and anti – thrombosis. Blood glucose levels in both sexes of rats showed a significantly decreasing trend at medium – and high – dose groups compared with those in the control group and the pre – dosing levels. The changes in AST, ALP, ALT and TB counts appeared to be dose – independent and these are relative to more AEE metabolism.

The test of organ weight and histopathology showed no changes in low – dose groups, this was in accordance with other indexes. Medium – dose AEE can make stomach, duodenum and ileum epithelium a slight degenerate and necrotic, while high – dose AEE can also make kidney tubules epithelial cell and liver cell a mild degenerate and necrotic. This may result from irritation of AEE and its metabolites.

The NOAEL values for aspirin and eugenol were 75 and 250 mg/ (kg · day), respectively (Marin Municipal Water District Vegetation Management Plan, 2008; Opinion of the scientific committee on cosmetic products and non – food products intended for consumers, 2001). The NOAEL value of aspirin was from teratogenic and embryotoxic test, whose results showed that following oral administration salicylic acid was neither teratogenic nor embryotoxic up to 75 mg/ (kg · day) in rodents and up to 100 mg/ (kg · day) in monkey. The NOAEL value of eugenol was from single

– dose studies in rats. The NOAEL value of AEE was considered to be 50 mg/（kg · day）under the present study conditions for dosage design. However, considering possible metabolism of AEE and comparing with the NOAEL values for aspirin and eugenol, we make sure the NOAEL value for AEE should be more than 50 mg/（kg · day）, which need to be determined.

In summary, there were no significant gender – specific differences in most indexes of subchronic toxicity throughout the experimental period regardless of dosage with the exception of food consumption and body weight. The NOAEL was considered to be 50 mg/（kg · day）, under the present study conditions. The results suggest it is important to closely monitor renal function, platelet and blood glucose levels when AEE is used as a long – term therapy. However, AEE should be used as such a long – term agent as antiatheroscloresis and anti – thrombosis under the dosage of 50 mg/（kg · day）, which also need to investigate and confirm by its chronic toxicity.

As part of the preclinical safety evaluation of AEE, these findings not only lay the groundwork for additional studies to further investigate preclinical toxicity associated with chronic AEE use, but also provide guidance in the design of clinical trials to ensure safety of prolonged AEE use in clinical therapy.

Conflict of Interest

The authors declare that there are no conflicts of interest.

Acknowledgements

The project was supported by special project of fundamental scientific research professional fund for central public welfare scientific research institute（2012ZL085）and the earmarked fund for China Agriculture Research System（cars – 38）.

References

[1] Chami N, Bennis S, Chami F, Aboussekhra A, Remmal A. Study of anticandidal activity of carvacrol and eugenol in vitro and in vivo. Oral Microbiol. Immunol. 2005.（20）106 – 111.

[2] Feng J, Lipton, J M, Eugenol: antipyretic activity in rabbits. Neuropharmacology 1987.（26）1 775 – 1 778. Flower, R, Gryglewshi, R, Herbaczynska – cedro, K, Vane, J. R, 1972. Effects of anti – inflammatory drugs on prostaglandin biosynthesis. Nat. New Biol. 238, 104 – 106.

[3] Gill, A. O, Holley, R. A, 2004. Mechanisms of bactericidal action of cinnamaldehyde against *Listeria monocytogenes* and of eugenol against L. *monocytogenes* and *Lactobacillus sakei*. Appl. Environ. Microbiol. 70, 5 750 – 5 755.

[4] Hashimoto, S, Uchiyama, K, Maeda, M, Ishitsuka, K, Furumoto, K, Nakamura, Y, 1988. In vivo and in vitro effects of zinc oxide – eugenol（ZOE）on biosynthesis of cyclo – oxygenase products in rat dental pulp. J. Dent. Res. 67, 1 092 – 1 096.

[5] Li, J, Yu, Y, Wang, Q, Yang, Y, Wei, X, Zhou, X, Niu, J, Li, B, Zhang, J, Ye, D, 2010. Analgesic roles of aspirin eugenol ester and its mechanisms. Anim. Husband. Vet. Med. 42（10）, 20 – 24.

[6] Li, J, Yu, Y, Wang, Q, Zhang, J, Yang, Y, Li, B, Zhou, X, Niu, J, Wei, X, Liu, X, Liu, Z, 2011a. Synthesis of aspirin eugenol ester and its biological activity. Med. Chem. Res. 20, 1 – 6. http://dx.doi.org/10.1007/s00044 – 011 – 9609 – 1.

[7] Li, J, Wang, Q, Yu, Y, Yang, Y, Niu, J, Zhou, X, Zhang, J, Wei, X, Li, B, 2011b. Anti – in-

flammatory effects of aspirin eugenol ester and the potential mechanisms. China J. Pharmacol. Toxicol. 25 (1), 57 -61.

[8] Marin Municipal Water District Vegetation Management Plan, 2008. Herbicide risk assessment. DRAFT - 8/28/08 (chapter 6 - Clove Oil (Eugenol)).

[9] Nagababu, E, Lakshmaiah, N, 1997. Inhibition of xanthine oxidase - xanthine - iron mediated lipid peroxidation by eugenol in liposomes. Mol. Cell. Biochem. 166, 65 - 71.

[10] Opinion of the scientific committee on cosmetic products and non - food products intended for consumers, 2001. Evaluation and opinion on: salicylic acid. SCCNFP/0522/01, final.

[11] Patrono, C, 1989. Aspirin and human platelets from clinical trials to acetylation of cyclooxygenase and back. TiPS 10, 453 -458.

[12] Raghavendra, R. H, Naidu, K. A, 2009. Spice active principles as the inhibitors of human platelet aggregation and thromboxane biosynthesis. Prostaglandins Leukot. Essent. Fatty Acids 81, 73 - 78.

[13] Schoemaker, R. G, Saxena, P. R, Kalkman, E. A, 1998. Low - dose aspirin improves in vivo hemodynamics in conscious, chronically infarcted rats. Cardiovasc. Res. 37, 108 - 114.

[14] Tragoolpua, Y, Jatisatienr, A, 2007. Anti - herpes simplex virus activities of *Eugenia caryophyllus* (Spreng.) Bullock & S. G. Harrison and essential oil, eugenol. Phytother. Res. 21, 1 153 - 1 158.

[15] Van der Ven, L. T. M, Verhoef, A, Van de Kuil, A, Slob, W, Leonards, P. E. G, Visser, T. J, Hamers, T, Herlin, M, Hakansson, H, Olausson, H, Piersma, A. H, Vos, J. G, 2006. A 28 - day oral dose toxicity study enhanced to detect endocrine effects of hexabromocyclododecane (HBCD) in wistar rats. Toxicol. Sci. 94 (2), 281 -292.

[16] Vane, J. R, 1971. Inhibition of prostaglandin synthesis as a mechanism of action for aspirin - like drugs. Nat. New Biol. 231, 232 -235.

[17] Vane, J. R, Botting, R, 1987. Inflammation and mechanism of anti - inflammatory drugs. FASEB J. 1, 89 -96.

[18] Wallenburg, H. C, Dekker, G. A, Makovitz, J. W, Rotmams, P, 1986. Low - dose aspirin prevents pregnancy - induced hypertension and pre - eclampsia in angiotensin - sensitive primigravide. Lancet 1, 1 -3.

[19] Ye, D, Yu, Y, Li, J, Yang, Y, Zhang, J, Zhou, X, Niu, J, Wei, X, Li, B, 2011. Antipyretic effects and its mechanisms of aspirin eugenol ester. China J. Pharmacol. Toxicol. 25 (2), 1 -5.

(Published the article in Food and Chemical Toxicology affect factor: 2.999)

Cloning and Prokaryotic Expression of cDNAs from Hepatitis E Virus Structural Gene of the SW189 Strain

Hao Baocheng[1], Lan Xi[2], Xing Xiaoyong[3], Xiang Haitao[3], Wen Fengqin[3], Hu Yuyao[4], Hu Yonghao[3], Liang Jianping[1,3], Liu Jixing[2,3]

(1. Key Laboratory of New Animal Drug Project, Gansu Province/Key Laboratory of Veterinary Plarmaceutics Discovery, Ministry of Agricultural/Lanzhou Institute of Husbandry and Pharmaceutical Sciences of CARS, Lanzhou 730050, China; 2. Lanzhou Veterinary Research Institute of CAAS, Lanzhou 730046, China; 3. College of Veterinary Medicine, Gansu Agricultural University, Lanzhou 730070, China; 4. College of Veterinay Medicine, South China Agricultural University, Guangzhou 510642, China)

Abstract: It is necessary to study the second open reading frame (open reading frame 2, ORF2) and antigenic peptides characteristic of genotype 4 of HEV, and create one kind of high sensitivity and specificity of anti－HEV IgM detection kit. The objective of this study was to obtain recombinant antigen for development of anti－HEV ELISA method and vaccine against hepatitis E virus infection. A 728 base cDNA was collected from 5′－terminus of open reading frame 2 (ORF2) among epidemic hepatitis E virus (HEV) isolated from Gansu, Western China. The fragment was digested with *Not* I and *Nco* I, and inserted into vector pET32a (+). The recombinant plasmid was transformed into E. *coli* Rosetta and the fusion protein expressed was confirmed by Western blot analysis. The recombinant plasmid was identified and confirmed with enzyme digestion, polymerase chain reaction (PCR) and sequencing, respectively. A protein band of about 45. 3 kDa was demonstrated by SDS－PAGE. The result of Westcrn blot analysis suggested that the fusion protein reacted with anti－HEV positive sera at a dilution of 1 : 1 500. The recombinant protein pORF2 may be useful in developing anti－HEV ELISA kit and vaccine against hepatitis E virus infection.

Key words: Cloning; Hepatitis E virus; Prokaryotic expression; Western blot

Introduction

Hepatitis E is caused by the hepatitis E virus (HEV), has two forms, epidemic and individual. Not only human can be infected with HEV, but also pig, boar, chicken and bird (Sun *et al.*, 2004), rat (Makoto *et al.*, 2003), cat (Choi *et al.*, 2004; Okamoto *et al.*, 2004), and deer. HEV genome consists of 5′and 3′non－genetic structure regions and three open reading frames (Engle *et al.*, 2002; Emerson *et al.*, 2003). It is approximately 7. 5 kb in length. The ORF2 is 2 kb in length, encoding the viral capsid protein, which is the major structural gene coding region, where the nucleotide sequences is the most conservative. HEV has more antigenic epitopes. The structure is more complex, in addition to the N – terminal 25 ~ 38aa segment, the majority antigenic epitopes located in the C – terminal 2/3. Many experiments are based on recombinant ORF2 antigens expressed in *E. coli* (Li *et al.*, 1997; Anderson *et al.*, 1999), insect (Li *et al.*, 2000; Sehgal *et al.*, 2003), or animal (Jameel *et al.*, 1996) cells. It is similar to other RNA viruses, RNA – dependent RNA polymerase of HEV does not copy the proofreading function, so it is prone to mutation and to produce the genetic and antigenic groups on the related virus, the existence is different HEV genotypes and subtypes. The study can lay the foundation for diagnostic antigens research with the protein of ORF2 of HEV, and provide the basis for the screening of the antigenic epitopes.

HEV is divided into four genotypes based on sequence homology in nucleotide, amino acid homology of the size and phylogenetic tree analysis. The four kinds of hepatitis E virus genotype are genotypes 1 and 2 which are limited to humans, and genotypes 3 and 4 strains which are found in both humans and pigs (Meng, 2010). In recent years, genotype 4 of hepatitis E is mainly found in China (Hao *et al.*, 2009; Li *et al.*, 2009). A pandemic of HE involving 119 280 patients took place from 1986 to 1988 in Xinjiang, Western China, killing of 707 people. It is the largest and most serious hepatitis E epidemic in the world (Aye *et al.*, 1992).

In recent years, the genotype 4 of ORF2 protein fragment of HEV has become a hot research topic (Li *et al.*, 2009; Ran *et al.*, 2010) because the viral proteins of ORF2 encoded can be used as a potential candidate of the hepatitis E virus vaccine. The objective of this study was to obtain recombinant antigen for development of anti – HEV ELISA method and vaccine against hepatitis E virus infection. Cloning and expression of ORF2 of swCH189 strains laid the foundation for studying the biological function of genes of HEV.

Materials and Methods

Virus strain: The swCH189 strains of HEV is positive by RT – PCR (Hao *et al.*, 2011) and were preserved by laboratory of animal infectious diseases which belongs to Lanzhou Veterinary Research Institute of CAAS, PR China. The swCH189 strains of HEV belong to genotype IV. The GenBank ID number is FJ6101232.

Bacterias and vector: *E. coli* strains, pMD18 – T carrier (Takara, Japan), Express bacterium Escherichia coli (*E. coli*) Rosetta and expression vector pET32a (+) were preserved by the laboratory of animal infectious diseases.

Primers: According to analysis of the translation product of ORF2 and the distribution of the antigen activity site, a pair of primer was designed. The primers were modified to contain restriction sites to facilitate cloning and ligating with expressed vector. The primers were synthesized (Takara, Japan). The nucleotide sequences of the primers were as follows:

U1: 5′CATGCCATGGGTATTGCGCTAACCTTGTTTAATCTTGCTGATA – 3′ (forward)

L1: 5′ATTTGCGGCCGCTCAATACTCCCGGGTTTTACCCACCTT – 3′ (reverse)

(*Note*: the underlined part of primers are enzymes digested sites, CCATGG is *Nco* I digested site, GCGGCCGC is *Not I* digested sites.)

RT-PCR: Viral RNAs were extracted from 200 μl of serum sample with RNA easy Mini Kit (Takara, Japan) according to the manufacturer's instructions, The main principles of the extraction are as follows: the main ingredient were guanidine thiocyanate and phenol in the TRIZOL. Guanidine thiocyanate could lysis cells, promote ribosome dissociation, and separate RNA and protein, then RNA was released into the solution. When adding chloroform, acidic phenol was extracted, which contributed to the RNA into the aqueous phase. The solution formed aqueous layer and organic layer after centrifugation, so RNA and the remaining organic phase proteins and DNA were separated. The extracted RNA was dissolved and precipitated with 60 μl no-RNase water. It was stored at -20℃. The PCR amplification was carried out in a 20 μl reaction mix containing 8 μl of extracted RNA, 100 pmol of each of forward and reverse primers and 8 μl of 2xTakara OneStep RT-PCR Enzyme Mix. Water was added to obtain a final volume of 20 μl. The cycle conditions were 42℃ for 60 min and 70℃ for 15 min for RT, 94℃ for 2 min (for hot start) and then 35 cycles at 94℃ for 30 sec, 50℃ for 30sec and 72℃ for 1 min (for PCR amplification), and a final extension at 72℃ for 15 min. Finally, PCR products (2 μl) were electrophoresed in 1.0% agarose gels in a standard TAE buffer and visualised by UV light after staining with ethidium bromide.

Digestion and plasmid construction: The 728 bp of ORF2 cDNA was digested with *Not I* and *Nco I*, and the pET32a (+) vector was digested with *Not I* and *Nco I*. The purpose fragments were recycled with MiniBEST Agarose Gel DNA Extraction Kit (Takara, Japan) according to the manufacturer's instructions. The restrict fragment was inserted to the pET32a (+) of prokaryotic expression vector and formed the recombinant plasmid pET32a (+) -CP239. The recombinant was then transformed into *E. coli* Rosetta and plated with agar containing ampicillin. For confirmation of target gene, the plasmid DNA was extracted by the alkaline lysis procedure, then examined for inserts of the expected sizes by enzymedigestion (*Not I* and *Nco I*), PCR and sequencing. The nucleotide sequences of 728bp from ORF2 cDNA were identified (Takara, Japan) (dideoxynucleotide method using a commercial kit).

Production of pORF2 fusion proteins: The recombinant product was transformed into *E. coli* Rosetta. The transformants of pET32a (+) -CP239 was incubated into L-broth medium containing 100 μg/ml of ampicillin for 4 hours at 37℃. The cultures were then induced with 0.3 mmol isopropyl-β-D-thiogalactopyranoside (IPTG) for 4 hours at 37℃ with constant shaking. The supernatant and precipitate were separated through centrifugation after the bacterial pellet was ultrasonically broken (300V, 3x5 sec). Then the fusion proteins were separated on 12% SDS-PAGE gels. Bacilli were harvested by centrifugation at 3 000 rpm for 10 min at 4℃ and kept frozen at -70℃. The pellet was then resuspended in a 1/50 volume of phosphate-buffered saline (PBS) and subjected to three cycles of freeze-thawing in liquid nitrogen and cold (4~10℃) water. The bacilli suspension was sonicated by five bursts of 30 sec each at 60% maximal power and centrifuged at 13 000 rpm for 15 min at 4℃. 20% Triton X-100 lysis buffer was added to the postsonic pellet. The suspension was incubated for 30 min at room temperature, and centrifuged at 12 000 rpm for 10 min. The pellet containing most of the fusion proteins was washed with PBS, and resuspended in 500 μl of 10 mmol Tris (pH 7.5) and stored in aliquots at 4℃.

SDS-PAGE and western blot: The fusion protein was separated by SDS-polyacrylamide gel electrophoresis (PAGE) gel, and transferred to nitrocellulose membrane for Western blot with anti - HEV positive sera confirmed by a commercial ELISA kit (WanTai, China). The membrane was blocked with 2% BSA/PBS for 2 hours at 37℃. After a rinse with wash buffer (PBS containing 0.05% Tween 20, PBS - T20), strips were cut out and incubated with anti - HEV positive sera at 1 : 1 500 dilution in PBS. After a 2 hours incubation with shaking at 37℃, the strips were rinsed three times for 5 min each time with wash buffer. This was followed by incubation with a 1 : 1 500 dilution (in wash buffer) of anti - sheep IgG horseradish peroxidase conjugate for 1 hour at 37℃. After a rinse with wash buffer as above, color development was carried out with 3, 3′ - diamino-benzidine as a substrate and photographed.

Results

RT - PCR: The detection of HEV RNA positive was obtained from the serum of swine in Gansu Slaughter plant. A 728 basepair of ORF2 cDNA was collected by RT - PCR (Fig. 1).

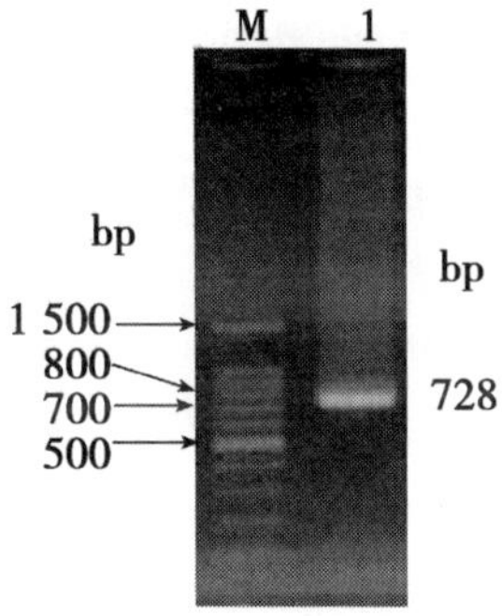

Fig. 1　A728 basepair of ORF2 cDNA was collected

Construction of recombinant plasmids: The 728 basepair of ORF2 cDNA was inserted to the pET32a (+) of prokaryotic expression vector and formed the recombinant plasmid pET32a (+) - CP239 (Fig. 2).

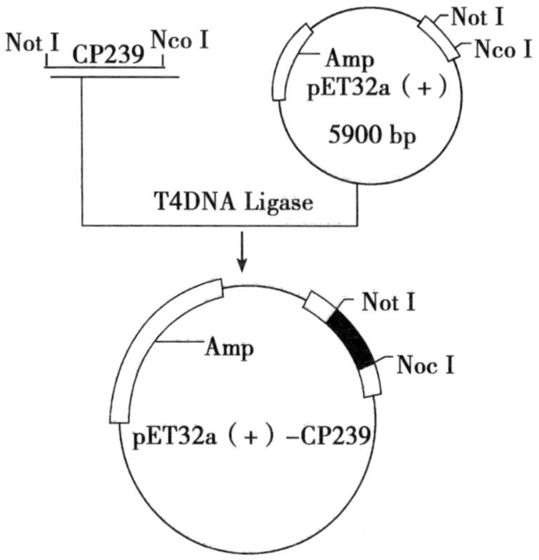

Fig. 2　The 728 basepair of ORF2 cDNA was inserted

The recombinant was identified with enzyme digestion (Fig. 3).

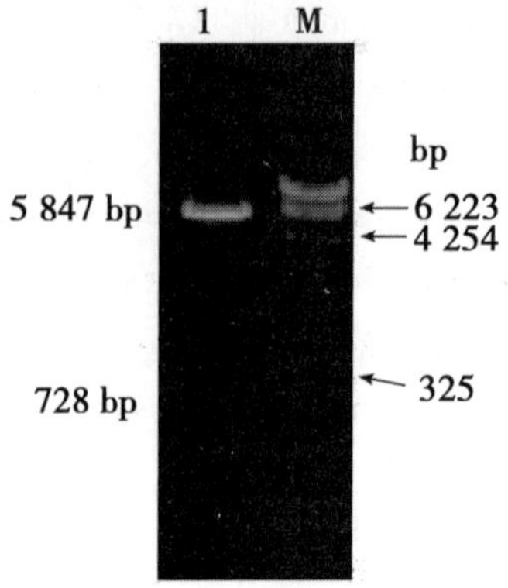

Fig. 3 The recombinant was identified

A nucleotide segment of about 728 bp was demonstrated and confirmed by sequencing. HEV nucleotide of recombinant was 100% homologous with that of the swCH189 strains of HEV.

Expression of HEV structural protein: The recombinant protein which contained 728 bp cDNA of ORF2 was expressed in *E. coli* Rosetta to produce the protein fused with GST (normal molecular mass, about 20 kDa). A clear protein band of about 45.3 kDa (Fig. 4), consistent with the prediction of the molecular mass of the translation of hepatitis E virus sequences, was demonstrated by SDS - PAGE. This protein was designated GST - pORF2. The result of Western blot analysis showed that pORF2 protein was reactive with anti - HEV - positive serum at a dilution of 1 : 1 000.

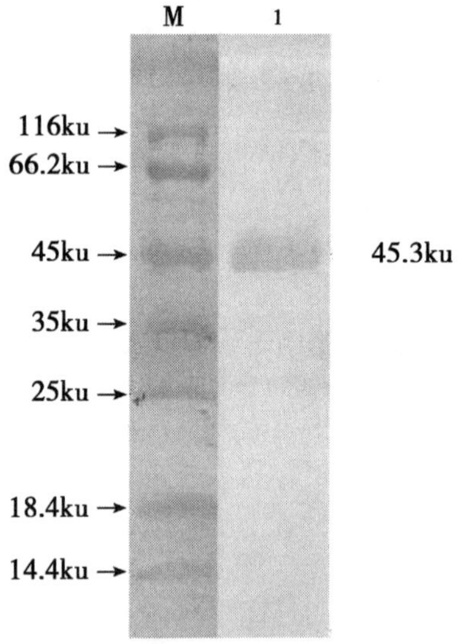

Fig. 4 A clear protein band of about 45.3 kDa

Discussion

The ORF2 is the major structural gene coding region. Kaur *et al.* (1992) positioned three antigen activity regions of pORF2: 25 ~ 38, 341 ~ 354, 517 ~ 530. Khudyakov *et al.* (1992) also positioned four antigen activity regions of pORF2: 319 ~ 340, 631 ~ 648, 641 ~ 660. Li *et al.*

(1997) found that the expression of ORF2 full - length protein C - terminal peptide chain folding was good because of gene epitopes and cut the N - terminal peptide chain ORF2 antigen.

In this study, because GC content of ORF2 gene region was higher, we added the GC x buffer to improve efficiency in the cloning process. In addition, annealing temperature is one important factor in this experiment. because the undeserved annealing temperature can cause non specific amplification. If we improve the annealing temperature, it may affect the efficiency of the amplification, resulting in products decreased. The last annealing temperature was optimized at 50℃.

In this cloning process, the RT - PCR method was used because it has higher sensitivity, quick and efficient than other methods. The index amplification of RT - PCR is a kind of very sensitive technology, it can detect very low RNA copy. RT - PCR amplification technique is widely used in the diagnosis of genetic diseases, and it may be used to quantitative monitoring the content of some RNA. The target fragment was successfully cloned, but its specificity was low, also some cloning errors occurred, We applied TaKaRa Taq™ Polymerase (Takara, Japan) to prevent its occurrence because it did not have 5′ - 3′exonuclease activity, but had the 3′ - 5′exonuclease activity. It could adjust the generated error of PCR amplification process, so the base mismatch rate of product was extremely low.

The recombinant protein pORF2 from 728 bp fragment of HEV ORF2 was expressed in E. coli Rosetta and confirmed by protein analysis. The molecular of the protein was 45.3 kDa, which are liable to form inclusion bodies (Kuang *et al.*, 2009), it may be related to higher expression temperature (37℃) and higher expression levels. It was reported that the expression protein of human HEV produced polymerization phenomenon (Meng *et al.*, 2001; Zhang *et al.*, 2001), but the expression product is not seen more emergence of the strip in this experiment.

The recombinant protein was expressed as fusion protein that facilitates the purification process. The pET - 32a (+) expression vector is a commonly used as prokaryotic expression vector. It is a higher level expression system. In this system the fusion protein is the most abundant cellular protein and is easily purified as a soluble protein. We used the method of gel purified and ultimately obtained the target protein. The western blot results showed that the fusion protein could be identified by positive serum of pig HEV and it had the main antigen epitope of natural strains of HEV. This showed that the pORF2 protein could be used in a Western blot format to detect anti - HEV in animal sera.

The next step of the research work will mainly focus on observation of whether immunization can be improved by adding or duplicating some certain antigenic epitopes and whether anti - pORF2 titers decline significantly over time. This part of the viral proteins can be used as a potential candidate of the hepatitis E virus vaccine, if we could demonstrate that it is immunologically reactive with a wide variety of human sera, not only in outbreak settings but also in sporadic cases.

Acknowledgements

This study was supported by "Eleventh Five - Year" National Technology Support Program Fund of China (Safe and Environment - Friendly Veterinary Drug Research and Development No. 2006BAD31B05).

References

[1] Anderson D A, Li F., Riddell M A. ELISA for IgG – class antibody to hepatitis E virus based on a highly conserved, conformational epitope expressed in *Eschericia coli*. J Virol Methods. 1999. 81: 131 – 142.

[2] Aye T T, Uchida T, Ma X Z, lida F, Shikata T, Zhuang H and Win K M. Complete nucleotide sequence of a hepatitis E virus isolated from the Xinjiang epidemic (1986 – 1988) of China. Nucleic Acids Res . 1992. 20 (13): 3 512.

[3] Choi C, Ha S K and Chae C. Development of nested RT – PCR hepatitis E virus in formalin – fixed, paraffin – embedded tissues hybridization. J Virol Methods. 2004. 115 (1): 67 – 71.

[4] Emerson S U and Purcell R H. Hepatitis E virus. Rev Med Virol. 2003. 13: 145 – 154.

[5] Engle R E, Yu C and Emerson S U. Hepatitis E virus (HEV) capsid antigens derived from viruses of human and swine origin are equally efficient for detecting anti – HEV by enzyme immunoassay. J Clin Microbiol. 2002. 40: 4 576 – 4 580.

[6] Hao B C, Lan X, Liu J X and Hu Y H. Cloning and sequence analysis of ORF2 gene of hepatitis E virus swCHGL189 strain. Biotechnol Bull. 2009. 5: 122 – 126.

[7] Hao B C , Lan X, Hu Y H, Liu J X and Liang J P. Establishment and application of duplex RT – PCR detection diagnostic methods of the swine hepatitis E virus. Chinese J Vet Med. 2011. 47 (5): 35 – 36.

[8] Jameel S, Zafrullah M and Ozdener M H. 1996. Expression in animal cells and characterization of the hepatitis E virus structural proteins. J Virol. 1996. 70: 207 – 216.

[9] Kaur M , Hymas, K C and Purdy M A. 1992. Human liner B – cell epitopes encoded by the hepatitis E virus include determinants in the RNA – dependent RNA polymerase. Proc Natl Acad Sci USA. 1992. 89 (9): 89 – 96.

[10] Khudyakov Y E, Khudyakova N S and Fields H A. Epitope mapping in proteins E virus. Virology. 1993. 194 (1): 89 – 96.

[11] Kuang A L, Chen Y Y, Peng Z F, Zhen L P, Fu R Y, Xu X, Guo L T and Wang C Q. The reasons of formation of inclusion body and its processing method. Shanghai J Anim Husb Vet Med. 2009. 1: 62 – 63.

[12] Li D D, Zhang X X, Yu J, Zhao L P and Xiang W H. Cloning and prokaryotic expression of structural gene of ORF2 of SHEV. Chinese J Vet Med. 2009. 45 (3): 17 – 19.

[13] Li F, Torresi J and Stephen A. Amino – terminal epitopes are exposed when full – length open reading frame 2 of hepatitis E virus is expressed in *Escherichia coli*, but carboxy – terminal epitopes are masked. J Med Virol . 1997. 52: 289 – 300.

[14] Li T C, Zhang J and Shinzawa H. Empty virus – like particle – based enzyme – linked immunosorbent assay for antibodies to hepatitis E virus. J Med Virol. 2000. 62: 327 – 333.

[15] Li W G, Li J H, Yang G S, Shu P and Ying G F. Swine hepatitis E serum epidemiological survey in Yunnan Province. J Yunnan Agr Univ. 2009. 24 (6): 917 – 919.

[16] Makoto H, Xin D, Li T C, Naokazu T, Hiroki K, Nobuo K, Teruki K, Ikuo G, Toshiyuki M, Masaji N, Katsuya T, Toshiro K, Tsutomu T, Haruo W and Kenji A. 2003. Evidence for widespread infection of hepatitis E virus among wild rats in Japan. Hepatol Res. 2003. 27 (1): 1 – 5.

[17] Meng J, Dai X, Chang J C, Lopareva E, Pillot J, Fields H A and Khudyakov Y E. Identification and characterization of the neutralization epitope (s) of the hepatitis E virus. Virology. 2001. 288 (2): 203 – 211.

[18] Meng X J. Hepatitis E virus: Animal reservoirs and zoonotic risk. Vet Microbiol. 2010. 140: 256 – 265.

[19] Okamoto H, Takahashi M and Nishizawa T. Presence of antibodies to hepatitis E virus in Japanese pet cats. Presence of an Japanese pet cats. Infection. 2004. 32 (1): 57 – 58.

[20] Ran X H, Wen X B, Wang M, Li D Y and Hou X L. Expression of C – terminal major epitope domain of

ORF2 of SHEV in *E. coli*. Progress Vet Med. 2010. 31 (1): 24 – 27.

[21] Sehgal D, Malik P S and Jameel S. Purification and diagnostic utility of a recombinant hepatitis E virus capsid protein expressed in insect larvae. Protein Expr Purif. 2003. 27: 27 – 34.

[22] Sun Z F, Larsen C T and Dunlop A. Genetic identification of from healthy chicken flocks and characterization of the capsid from chickens with hepatitis splenomegaly syndrome in different United States. J Gen Virol. 2004. 8 (Pt3): 693 – 700.

[23] Tei S, Kitajiama N and Takahashi K. Zoonotic transmission to human beings. Lancet. 2003. 362 (9381): 371 – 373.

[24] Zhang J Z, Ng M H, Xia N S, Lau S H, Che X Y, Chau T N, Lai S T and Im S W. Conformational antigenic determinants generated by interactions between a bacterially expressed recombinant peptide of the hepatitis E virus structural protein. J Med Virol. 2001. 64: 125 – 132.

(Published the article in Thai J Vet Med
affect factor: 0. 194)

Dietary Supplementation of Female Rats with Elk Velvet Antler Improves Physical and Neurological Development of Offspring

Chen Jiongran[1,2], Woodbury Murray R.[3], Alcorn Jane[4], Honaramooz Ali[2]

(1. Key Laboratory of New Animal Drug Project of Gansu Province, Key Laboratory of Research and Development for Veterinary Pharmaceutics, Ministry of Agriculture, Lanzhou Institute of Animal and Veterinary Pharmaceutics Science, Chinese Academy of Agricultural Sciences, Lanzhou, 730050, China; 2. Department of Veterinary Biomedical Sciences, Western College of Veterinary Medicine, University of Saskatchewan, Saskatoon, SK, Canada S7N 5B4; 3. Department of Large Animal Clinical Sciences, Western College of Veterinary Medicine, University of Saskatchewan, Saskatoon, SK, Canada S7N 5B4; 4. College of Pharmacy and Nutrition, University of Saskatchewan, Saskatoon, SK, Canada S7N 5C9.)

Elk velvet antler (EVA) has a traditional use for promotion of general health. However, evidence of EVA effects at different lifestages is generally lacking. This paper investigated the effects of long-term maternal dietary EVA supplementation on physical, reflexological and neurological development of rat offspring. Female Wistar rats were fed standard chow or chow containing 10% EVA for 90 days prior to mating and throughout pregnancy and lactation. In each dietary group, 56 male and 56 female pups were assessed for physical, neuromotor, and reflexologic development postnatally. Among the examined physical developmental parameters, incisor eruption occurred one day earlier in pups nursing dams receiving EVA. Among neuromotor developmental parameters, duration of supported and unsupported standing was longer for pups nursing EVA supplemented dams. Acquisition of neurological reflex parameters (righting reflex, negative geotaxis, cliff avoidance acoustic startle) occurred earlier in pups nursing dams receiving EVA. Longterm maternal EVA supplementation prior to and during pregnancy and lactation accelerated certain physical, reflexologic, and neuromotor developmental milestones and caused no discernible adverse effects on developing offspring. The potential benefits of maternal EVA supplementation on postnatal development warrants further investigation to determine whether EVA can be endorsed for the promotion of maternal and child health.

1 Introduction

Velvet antler is a well – known Chinese *materia medica* used clinically as an ancient oriental natural product for thou – sands of years in the treatment of various diseases and as a tonic[1]. In the 1 990s, North American elk farming for the purpose of velvet antler production was a growing industry bolstered by an increasing demand for elk velvet antler (EVA). However, chronic wasting disease, a prion disease, in farmed elk and deer closed many of the traditional markets to North American antler products. A demand for EVA for Traditional Chinese Medicine (TCM) has, nevertheless, remained, and dried antler is the most frequently used animal – derived ingredient in TCM prescriptions. Recently, the North American holistic health and natural product industry is rediscovering antler's beneficial effects, promot – ing its use as a food supplement and nonpharmaceutical therapeutic agent for both human and veterinary medicine applications.

According to the principles of TCM, velvet antler has a variety of beneficial health effects. It is prescribed by TCM doctors to enhance the sense of well – being and vitality, improve musculoskeletal function, elevate resistance to dis – ease, and modulate the immune system to decrease allergic response. It is believed also to improve blood circulation, modify blood pressure, and promote rapid healing of tissues and bones[2~4].

Elk velvet antler is harvested from live elk in late spring or early summer while still in its growth phase. Antlers are appendages that grow annually from antler generating structures called pedicles located on the frontal skull bone of male elk and deer. The antler in the growing stage is a soft, cartilaginous, blood – filled tissue covered by skin with a velvet – like texture, hence the name velvet antler. Antlers grow very rapidly reaching growth rates of approxi – mately 2 cm/day. Growth and mineralization is complete at approximately 120 days after it begins. Antlers reach peak commercial value at approximately 70 to 80 days of growth; after this time, the economic value and medical usefulness of antler is gradually lost due to progressive mineralization and concomitant loss of biological activity. The soft, blood – filled cartilage tissue in the distal portions of antler will be replaced by mineralized cartilage and bone tissue advancing up from the base within this period[4~6].

In North America, the processed antler is ground and encapsulated in varying milligram amounts and sold over the counter[7,8]. In the East, the antler is cut into wafer – thin slices and combined with other natural ingredients in prescriptions for various TCM purposes.

Despite a long history and widespread use, to our knowledge, little data is available about the effect of EVA on physical and neurological development of neonates particu – larly following maternal dietary supplementation with EVA. One study evaluated short – term pre – and postnatal exposure of rat pups to EVA (from the 18th day *in utero* to 86 days after birth) and reported no changes in normal physical, devel – opmental, or behavioral indices[2]. To add to the limited database of EVA effects during postnatal development, our study was designed to comprehensively evaluate the effects of long – term maternal EVA supplementation on physical, reflexological, and neurological development of offspring and whether gender differences existed in the offspring's response to maternal supplementation.

2 Materials and Methods

2.1 Animal Diets

The control diet consisted of regular rodent chow (Prolab RMH 3000, PMI Nutrition International, St. Louis, MO, USA) ground into powder using a laboratory mill. A 10% w/w EVA diet[2] was made using 1 : 9 EVA powder : powered regular rat chow and mixed to homogeneity. EVA powder was derived from the freeze – drying method and was supplied by Norelkco Nutraceuticals (Moosomin, SK, Canada). Briefly, the velvet antler was frozen immediately after harvesting, transported to the plant where the button end (about 2 cm) was cut off in order to help control bacteria and the hide (skin) was removed. The rest of the frozen stick was shredded onto stainless steel trays to expose maximum surface area. The shred was quickly (about 12 h) dried by a blast of cold (−1℃), dry air, which results in minimum loss of bioavailable nutrients and no pathogen buildup. The dried shred was then ground in a nitrogen – cooled grinder to a fine powder.

2.2 Animals

Sixteen Wistar adult female rats weighing 200 to 250 g and 16 Wistar adult male rats weighing 400 to 450 g were obtained from Charles Rivers (Montreal, PQ, Canada). Male and female rats were randomly assigned to the two treatment groups and housed in plexiglass cages lined with sawdust. Rats were provided with water ad *libitum* and fed regular powdered rat chow (female and male controls) or 10% EVA powdered chow (male and female treatment groups) *ad libitum* for 90 consecutive days prior to mating. All animals were kept under constant temperature at 21℃ and 60% humidity and on a 12 h : 12 h light : dark schedule. Female rats were bred to males fed the same chow as the females. Female and male rats were paired for 5 days or until a vaginal mucus plug (as evidence of mating) was confirmed. Females were checked twice daily until pups were born. The day of birth was designated postnatal day 0 (PND 0). During pregnancy and lactation, the female adult rats from either group continued to receive the same diet as before parturition. All females were allowed to deliver naturally and rear their young to weaning.

After parturition, pups were housed with their dam for a lactation period of 21 days (PND 21), after which they were weaned and provided *ad libitum* with the same chow as their mothers for the remainder of the experimental period of 31 days (PND 31). In order to prevent maternal responses from playing a role in litter manipulations, efforts were made to ensure the early living environment contributed by maternal behavior was the same across groups[9~11]. Pups born to dams fed EVA or regular chow were selected for study across all litters in each diet group to ensure equivalent nutritional intake and maternal care for pups. On PND 2, litters were sexed and 224 pups were assigned to one of 14 pup groups (7 control and 7 EVA groups, corresponding to the diet fed to their mothers), each group containing 8 males and 8 females. This work was approved by the University of Saskatchewan's Animal Research Ethics Board and adhered to the Canadian Council on Animal Care guidelines for humane animal use.

2.3 Examination of Physical Development

At parturition, litter size, total litter weight, and mean birth weight of the pups were recorded for each dam. Pups were weighed every day from PND 0 to PND 21. Physical development and maturation of the pups within each dietary group was evaluated by observing the dates of onset and completion of pinna unfolding, incisor eruption, eye opening, and growth of anogenital hair[12~19].

2.4 Examination of Neuromotor Development

Crawling, head waving, and supported and unsupported standing were examined as indicators of neuromotor development. Mean age of onset and completion of the indicators in each pup group, the number of animals in each group displaying the indicator activity from day of onset to day of completion, and the time spent in these activities were recorded. The times spent in grooming and total relative activity during a fixed test period were also assessed as parameters of spontaneous activity. The equipment and detailed procedures employed in making these observations were previously described[20].

2.5 Examination of Neurological Reflexes

The pups were given a battery of tests to evaluate the development of neu - rological reflexes. This occurred every morning beginning at 9 : 00 AM from PND 1 to PND 21 and by the same assessor. The pups were tested in the same room as the dam during separation, under soft white light.

The righting reflex was considered positive if the pup turned over to regain a normal posture within 2.5 s of being placed on its back on a hard flat surface. The acoustic startle test was positive if a loud, sharp, buzzer elicited a sudden extension of head and limbs. A positive cliff avoidance test was recorded if the pup turned and crawled away when placed on a bench top with its forepaws extending over the edge. A pup was judged to possess a negative geotaxis capability if, when placed with its head downward on a grid tilted 45° to horizontal, it could rotate a full 180° and climb the grid within a maximum time of 30 s. If the pup could flex its digits to grasp a paper clip used to stroke the forepaw, it was regarded as having achieved bar holding ability or a palmar grasp. A positive vibrissa placing response was recorded if the pup raised its head and extended its forelimbs to grasp the bench surface when lowered by the tail so that the vibrissae touched the bench. The visual placing test was positive if the pup attempted to grasp the bench surface when lowered by the tail to the bench top but without touching it.

Age (days) of onset and completion of the reflex responses described above for each pup were recorded. The number of animals in each group displaying the positive reflex from onset to completion was also recorded daily. Detailed descriptions of these responses have been published elsewhere[14,17~19,21].

2.6 Statistical Analysis

All data were analyzed using SPSS for windows (version 17.0). Body weight and the time spent standing supported and unsupported were subjected to repeated - measures ANOVA using a general linear model (GLM) considering age and treatment simultaneously. The mean time of onset and completion for acquisition of the activities was subject to a t - test for independent samples. The percentage of animals acquiring the physical, neuromotor, and reflex developmental parameters at various postnatal days were subject to a Chi - Square ANOVA using the Crosstab Procedure. The level of significance was set at $P \leqslant 0.05$. Relative EVA effects (i. e., the percent difference between EVA and control groups) and gender effects (i. e., the percent difference between male and female groups) were tested using analysis of covariance (ANCOVA).

3 Results

3.1 Growth and Physical Development

The diet supple – mented with 10% EVA was readily consumed by the adult rats. No obvious differences in the weight of food consumed by dams were detected between groups. Pups from EVA supplemented dams showed no apparent teratogenic effects and no postnatal deaths occurred during the experimental period. No significant differences between treatment groups were observed with regard to litter size, litter weight, or individual birth weight. Rate of body weight gain monitored until postnatal day 21 was not different between pups nursing mothers fed regular chow and EVA – supplemented chow (data not shown), and no gender differences were noted in the rate and pattern of body weight gain.

To determine the possible effects of maternal EVA sup – plementation on physical development, growth indices and achievement of physical development milestones such as pinna unfolding, eye opening, anogenital hair growth, and incisor eruption were assessed. Neither EVA supplementa – tion nor gender influenced the onset and completion of outer ear flap unfolding (postnatal day 3 and 4, resp.), eye opening (postnatal day 13 and 15, resp.), and anogenital hair growth (postnatal day 14 and 18, resp.) (data not shown). However, the onset and completion of incisor eruption occurred earlier in pups nursing dams supplemented with EVA (postnatal day 9 and 11 versus postnatal day 10 and 12, resp.) ($P < 0.05$) with no differences observed between genders within the respective dietary groups.

3.2 Neuromotor and Reflexologic Development

The effects of maternal dietary EVA supplementation on the postnatal day of onset and completion of group acquisition of neuromotor skill and neurological reflex are shown in Table 1. The average age of onset of the neuromotor skills for head – waving (2.0 days) and crawling (4.5 days) occurred at the same age for both groups and genders (data not shown). Pups nursing mothers supplemented with EVA demonstrated earlier onset and completion of several measures of reflex and neuromotor development. Although no statistically significant differences were observed between the diet groups in either age of onset or completion of righting reflex, negative geotaxis, cliff avoidance behaviour, and acoustic startle response, pups from EVA – supplemented mothers tended to develop these activities sooner and more quickly than the control pups (Tables 1 and 2). The mean day of complete group acquisition for both supported and unsupported standing was earlier ($P \leqslant 0.05$) in the EVA supplemented group. Few gender differences were noted except where the mean day of onset for unsupported standing in EVA males was earlier than that of the control group, but this difference was not seen with EVA females (Table 1).

When acquisition of reflex and neuromotor skills was assessed on the basis of the average total percentage of pups acquiring the specific skill on a particular postnatal day of development (Table

2), a significantly greater number of pups nursing EVA – supplemented dams showed earlier acquisition of the measured indices. Additionally, pups from EVA – supplemented dams demonstrated significantly greater ability to stand supported or unsupported, measured by time spent in the activity, than pups from control diet dams (Table 3). However, no significant difference between the genders within either group was noted.

No significant differences between groups were observed in the amount of time spent in grooming during a six – minute observation period or in the onset and completion of bar holding ability, vibrissa, and visual placing, and the proportion of pups showing these reflex activities over time was not different (data not shown).

Table 1 Effects of maternal dietary EVA supplementation on the onset and completion of neuromotor skill and neurological reflex acquisi – tion in rat pups

Signs[2]	Day of onset[1]				Day of completion[1]			
	Male		Female		Male		Female	
	Control	EVA	Control	EVA	Control	EVA	Control	EVA
Righting reflex	2.0 (0.01)	1.0 (0.05)	2.0 (0.02)	1.1 (0.07)	4.9 (0.25)	3.1 (0.12)	4.9 (0.26)	3.1 (0.12)
Negative geotaxis	4.1 (0.18)	4.0 (0.20)	4.0 (0.20)	4.0 (0.20)	6.9 (0.27)	6.8 (0.28)	6.5 (0.30)	6.5 (0.30)
Cliff avoidance	6.6 (0.30)	6.3 (0.26)	6.9 (0.20)	6.9 (0.33)	8.0 (0.35)	6.6 (0.28)	7.9 (0.32)	7.9 (0.35)
Acoustic startle	11.0 (0.45)	9.4 (0.40)	11.0 (0.45)	9.4 (0.40)	12.0 (0.50)	11.0 (0.45)	12.0 (0.50)	11.6 (0.51)
Supported standing	8.0 (0.34)	7.6 (0.33)	8.0 (0.34)	7.6 (0.33)	11.2 (0.65)	10.0 (0.00)*	10.9 (0.48)	9.1(0.34)*
Unsupported standing	15.9 (0.66)	14.9 (0.65)*	15.8 (0.56)	15.0 (0.70)	17.9 (0.66)	15.8 (0.30)*	17.9 (0.66)	16.1 (0.40)*

*1. Data are expressed as mean ± standard error of man (SEM).

2. The acquisition of various reflexes (yes/no) was assessed at specific postnatal ages, for example, Righting Reflex at PND 1 – 5, Negative Geotaxis at PND 4 – 6, Cliff Avoidance at PND 6 – 8, Acoustic Startle at PND 9 – 12, Supported standing at PND 7 – 11, and Unsupported standing at PND 15 – 18.

*Significant ($P \leq 0.05$) differences between control and EVA groups (within gender) in the onset or completion of physical and neurological signs of development.

Table 2 Percentage of rat pups possessing an acquired neuromotor and neurological reflex on specific postnatal days

Signs		% of pups				Diet effect (P value)	Gender effect (P value)
		Male		Female			
		Control	EVA	Control	EVA		
Righting reflex	PND 1	0	10.7	0	5.4	0.002	NS
	PND 2	28.8	80.4	39.3	71.4	<0.001	NS
	PND 3	53.6	100	76.6	100	<0.001	NS
	PND 4	96.4	100	94.6	100	0.024	NS
Negative geotaxis	PND 4	23	25	35	38	0.019	NS
	PND 5	56	54	70.2	71	0.035	NS
	PND 6	82	85	100	100	0.004	NS

(Continued)

Signs		% of pups				Diet effect (*P* value)	Gender effect (*P* value)
		Male		Female			
		Control	EVA	Control	EVA		
Cliff avoidance	PND 6	10. 7	30. 3	14. 3	26. 7	0. 003	NS
	PND 7	50	70	55	73. 2	0. 004	NS
Acoustic startle	PND 9	0	16. 1	0	19. 6	<0. 001	NS
	PND 10	5. 4	42. 9	14. 3	53. 6	<0. 001	NS
	PND 11	53. 6	100	41	93	<0. 001	NS
Supported standing	PND 7	0	26. 8	0	26. 8	<0. 001	NS
	PND 8	35. 7	7. 8	36	73. 2	<0. 001	NS
	PND 9	62. 5	100	59	100	<0. 001	NS
	PND 10	73. 2	100	89. 3	100	<0. 001	0. 039
Unsupported standing	PND 15	0	44. 6	0	21. 4	<0. 001	0. 019
	PND 16	33. 9	100	39. 2	75	<0. 001	NS
	PND 17	74. 9	100	66. 1	100	<0. 001	NS
	PND 18	100	100	89. 3	100	0. 013	0. 013

* NS: not significant.

PND: postnatal day, recorded as the first PND where a significant difference ($P \leqslant 0.05$) was noted.

Table 3 Effect of maternal dietary supplementation of elk velvet antler on neuromotor coordination (standing) of rat pups

Dietary group	Time spent standing[1]							
	Supported					Unsupported		
	PND 7	PND 8	PND 9	PND 10	PND 11	PND 16	PND 17	PND18
				Male				
Control	0. 0 (0. 00)	4. 4 (0. 21)	7. 4 (0. 31)	7. 2 (0. 26)	10. 2 (0. 65)	6. 5 (0. 33)	9. 6 (0. 46)	12. 3 (0. 55)
EVA	3. 4 (0. 11)**	4. 2 (0. 25)	8. 1 (0. 33)*	10 (0. 45)*	11. 1 (0. 67)	9. 8 (0. 44)*	13. 7 (0. 61)*	18. 7 (0. 77)*
				Female				
Control	0. 0 (0. 00)	4. 3 (0. 23)	6. 8 (0. 27)	7. 3 (0. 30)	10. 5 (0. 53)	6. 8 (0. 28)	8. 9 (0. 36)	13. 6 (0. 54)
EVA	3. 2 (0. 12)**	4. 5 (0. 24)	8. 4 (0. 40)*	9. 8 (0. 40)*	11. 3 (0. 56)	10. 2 (0. 57)*	13. 3 (0. 58)*	19. 1 (0. 70)*

* Time in seconds during a six – minute observation period. Each value represents the mean of 56 pups ± standard error of mean (SEM).

The level of significance was set at $^{*}P \leqslant 0.05$ or $^{**}P < 0.001$.

4 Discussion

The primary aim of our study was to investigate the effects of long – term maternal dietary EVA supplementation in the diet of female rats prior to mating, and for the duration of pregnancy and lactation, on the physical, reflex, and neuro – motor development of the progeny. We employed a battery of behavioral tests that reflect the developmental maturation of the central nervous system[22] and assessed various parameters of physical development. We used long – term maternal and paternal supplementation with EVA as several studies identify the importance of diet on fetal and postnatal development through diet – mediated epigenetic regulation[23,24]. Furthermore, we assessed potential gender differences as previous studies have noted differences according to gender in the *in utero* and/ or postnatal development of progeny from dams undergoing experimental dietary or drug interventions[25~30].

Assessment of EVA effects at this lifestage is relevant as EVA contains a number of bioactive substances including polyamines, chondroitin sulfate, minerals (calcium, phos-phorous, magnesium, iron, and potassium), glucosamine, glycosaminoglycans, and hyaluronic acid as well as all 24 amino acids[31,32]. Furthermore, growing velvet antler contains neutrophin-3[33] and a number of other growth factors and proto-oncogene mRNA[34]. Cultured antler cells respond to IGF-1 and IGF-2[35] and display receptors for estradiol and IGF-1[36]. Deer antler has been shown to contain biologically active molecules including antinar-cotic factor (s)[37], growth promoting polypeptides[38], diaphorase activity[39], and epidermal growth factor[40].

We observed no discernable adverse effects of long-term maternal EVA supplementation on developmental indices of rat offspring. Our results are consistent with Hemmings and Song[2] who reported that EVA supplemented chow had no long-term effect on the offspring of Fischer rats fed EVA chow from the 18th day of gestation to 88 days after birth. This particular study concentrated on the effect of maternal EVA supplementation from the telophase of gestation throughout the lactation period[2]. Our study suggests that maternal supplementation prior to and throughout pregnancy causes no teratogenic or adverse outcomes on postnatal physical and neurological developmental parame-ters of offspring.

In our study, we used typical physical and neurological development parameters[41~44] to assess the effects of long-term maternal EVA supplementation on postnatal development. The physical maturation parameters, pinna unfolding, eye opening, and anogenital hair are developmentally and hormonally controlled[45], but were not significantly affected in either gender by the dam's consumption of EVA. On the other hand, the onset and completion of incisor eruption occurred one day earlier in offspring from EVA-fed dams. Incisor eruption is influenced by epidermal growth factor and transforming growth factor, both of which are found in EVA[46,47]. Our data suggests that maternal EVA supplementation may modulate the factors leading to incisor eruption in offspring.

In the rat, the first 2 weeks after birth constitute a period of rapid brain growth, a critical phase in neurobehavioral maturation, which corresponds to the last months of human fetal brain growth[48]. Among the battery of tests used in the examination of neurological reflex development, righting reflex, negative geotaxis, cliff avoidance, and acoustic startle were positively influenced by EVA supplementation. Furthermore, the number of pups acquiring these skills in the EVA group showed significant improvement through the observation period compared to those in the control group. Although the earlier developing exploratory activities such as crawling, head waving[49], and grooming were not affected by EVA, pups of EVA-supplemented dams were significantly younger in completing more advanced skills and exhibiting more purposeful movements. For example, the righting reflex is thought to measure the development of dynamic postural adjustments that require the integrity of muscular and motor function and adequate acquisition of symmetrical coordination between left and right sides of the body[50]. Both male and female pups suckling dam receiving EVA diet showed earlier onset and completion of this reflex as compared with control pups. This result indicates that EVA maternal supplementation benefits the development of dynamic postural adjustments and acquisition of symmetrical coordination of the body and has a positive effect on the development of these skills. It has been reported that there is a gender difference in neuronal maturation during brain development[51], but we did not observe a gender difference in any of the tests of neurological reflex development. Furthermore, our data revealed no difference between control and EVA - supplemented pups in the onset and completion of a battery of reflexes that assess maturation of sensory motor cortex and cerebellar and vestibular systems (namely, bar holding ability, vibrissa placing, and visual placing)[52,53].

5 Conclusions

In summary, our study suggests that long - term maternal dietary consumption of EVA has beneficial effects on physical and neurological development of offspring rats. The text on Traditional Chinese Medicine, Pen Ts'ao Kang Mu, written in the 16th century in China, considers EVA a universal tonic and an important medicament for sexual virility. This text does not identify the bioactive component (s) of EVA contributing to these health benefits. Similarly, the bioactive constituents resulting in earlier development of physical, reflexological, and neuromotor skills in offspring of long - term EVA supplementation of dams are not known but EVA polypeptides have been associated with growth promoting activities[38]. Whether these contribute to the acceleration of various indices of physical and neurological development is not known but further research is warranted before the use of EVA as a general tonic for humans during period of pregnancy, parturition, and lactation can be promoted.

Acknowledgments

This paper was supported through grants from the Natural Sciences and Engineering Research Council (NSERC) to A. Honaramooz, and by funds from the Canada - Saskatchewan Agrifood Innovation Fund (AFIF) through the Research Chair, WCVM Specialized Livestock Health and Production, M. R. Woodbury. The authors thank M. Burmester, P. Mason, and their staff for their diligent animal care. Elk velvet antler powder was a kind gift from Norelkco Nutraceuticals, Moosomin, SK, Canada.

References

[1] Z. Q. Zhang, Y. Zhang, B. X. Wang, H. O. Zhou, Y. Wang, and H. Zhang, "Purification and partial characterization of anti - inflammatory peptide from pilose antler of *Cervus nippon Temminck*," *Acta Pharmaceutica Sinica*, Vol. 27, No. 5, pp. 321 - 324, 1992.

[2] S. J. Hemmings and X. Song, "The effects of elk velvet antler consumption on the rat: development, behavior, toxicity and the activity of liver γ - glutamyltranspeptidase," *Comparative Biochemistry and Physiology* C, Vol. 138, No. 1, pp. 105 - 112, 2004.

[3] J. R. Mikler, C. L. Theoret, and J. C. Haigh, "effects of topical elk velvet antler on cutaneous wound healing in strepto - zotocin - induced diabetic rats," *Journal of Alternative and Complementary Medicine*, Vol. 10, No. 5, pp. 835 - 840, 2004.

[4] M. Moreau, J. Dupuis, N. H. Bonneau, and M. Lécuyer, "Clinical evaluation of a powder of quality elk velvet

antler for the treatment of osteoarthrosis in dogs," *Canadian Veterinary Journal*, Vol. 45, No. 2, pp. 133 - 139, 2004.

[5] P. Fennessey, "Pharmacology of velvet," in *Proceedings of the Deer Course for Veterinarians*, Vol. 6, pp. 96 - 103, Deer Branch of the New Zealand Veterinary Association, 1989.

[6] M. T. Bagonluri, M. R. Woodbury, R. S. Reid, and J. O. Boison, "Analysis of lidocaine and its major metabolite, monoethyl - glycinexylidide, in elk velvet antler by liquid chromatography with UV detection and confirmation by electrospray ionization tandem mass spectrometry," *Journal of Agricultural and Food Chemistry*, Vol. 53, No. 7, pp. 2 386 - 2 391, 2005.

[7] M. Allen, K. Oberle, M. Grace, A. Russell, and A. J. Adewale, "A randomized clinical trial of elk velvet antler in rheumatoid arthritis," *Biological Research for Nursing*, Vol. 9, No. 3, pp. 254 - 261, 2008.

[8] A. Duarte and J. Abdo, "The 2000 year old medicine. Velvet deer antler," *Life Extension Magazine*, Vol. 14, pp. 99 - 103, 1994.

[9] C. Caldji, B. Tannenbaum, S. Sharma, D. Francis, P. M. Plotsky, and M. J. Meaney, "Maternal care during infancy regulates the development of neural systems mediating the expression of fearfulness in the rat," *Proceedings of the National Academy of Sciences of the United States of America*, Vol. 95, No. 9, pp. 5 335 - 5 340, 1998.

[10] D. Francis, J. Diorio, D. Liu, and M. J. Meaney, "Nongenomic transmission across generations of maternal behavior and stress responses in the rat," *Science*, Vol. 286, No. 5 442, pp. 1 155 - 1 158, 1999.

[11] D. Liu, J. Diorio, B. Tannenbaum *et al.*, "Maternal care, hippocampal glucocorticoid receptors, and hypothalamic - pituitary - adrenal responses to stress," *Science*, Vol. 277, No. 5 332, pp. 1 659 - 1 662, 1997.

[12] M. S. Lamptey and B. L. Walker, "Physical and neurological development of the progeny of female rats fed an essential fatty acid - deficient diet during pregnancy and/or lactation," *Journal of Nutrition*, Vol. 108, No. 3, pp. 351 - 357, 1978.

[13] D. Regloödi, P. Kiss, A. Tamás, and I. Lengvári, "The effects of PACAP and PACAP antagonist on the neurobehavioral development of newborn rats," *Behavioural Brain Research*, Vol. 140, No. 1 - 2, pp. 131 - 139, 2003.

[14] J. L. Smart and J. Dobbing, "Vulnerability of developing brain—II. effects of early nutritional deprivation on reflex ontogeny and development of behaviour in the rat," *Brain Research*, Vol. 28, No. 1, pp. 85 - 95, 1971.

[15] J. L. Smart and J. Dobbing, "Vulnerability of developing brain—VI. relative effects of foetal and early postnatal undernutrition on reflex ontogeny and development of behaviour in the rat," *Brain Research*, Vol. 33, No. 2, pp. 303 - 314, 1971.

[16] C. R. Neal Jr., G. Weidemann, M. Kabbaj, and D. M. Vázquez, "Effect of neonatal dexamethasone exposure on growth and neurological development in the adult rat," *American Journal of Physiology*, Vol. 287, pp. R375 - R385, 2004.

[17] J. Altman and B. McCrady, "The influence of nutrition on neural and behavioral development—IV. effects of infantile undernutrition on the growth of the cerebellum," *Develop - mental Psychobiology*, Vol. 5, No. 2, pp. 111 - 122, 1972.

[18] W. M. Fox, "Reflex - ontogeny and behavioural development of the mouse," *Animal Behaviour*, Vol. 13, No. 2, pp. 234 - 241, 1965.

[19] D. Wahlsten, "A development time scale for postnatal changes in brain and behavior of B6D2F2 mice," *Brain Research*, Vol. 72, No. 2, pp. 251 - 264, 1974.

[20] L. Svennerholm, C. Ailing, A. Bruce, I. Karlsson, and O. Sapia, "Effects on offspring of maternal malnutrition in the rat," in *Lipids, Malnutrition and the Developing Brain*, pp. 141 - 157, Associated Scientific Pub-

lishers, Amsterdam, The Nether lands, 1972.

[21] M. G. Alton – Mackey and B. L. Walker, "Graded levels of pyridoxine in the rat diet during gestation and the physical and neuromotor development of offspring," *American Journal of Clinical Nutrition*, Vol. 26, No. 4, pp. 420 – 428, 1973.

[22] P. E. Wainwright, "Dietary essential fatty acids and brain function: a developmental perspective on mechanisms," *Proceedings of the Nutrition Society*, Vol. 61, No. 1, pp. 61 – 69, 2002.

[23] R. A. Waterland and R. L. Jirtle, "Transposable elements: targets for early nutritional effects on epigenetic gene regulation," *Molecular and Cellular Biology*, Vol. 23, No. 15, pp. 5 293 – 5 300, 2003.

[24] R. A. Waterland and R. L. Jirtle, "Early nutrition, epigenetic changes at transposons and imprinted genes, and enhanced susceptibility to adult chronic diseases," *Nutrition*, Vol. 20, No. 1, pp. 63 – 68, 2004.

[25] P. W. Nathanielsz, "Animal models that elucidate basic princi – ples of the developmental origins of adult diseases," *Institute for Laboratory Animal Research Journal*, Vol. 47, No. 1, pp. 73 – 82, 2006.

[26] M. E. Symonds, T. Stephenson, D. S. Gardner, and H. Budge, "Tissue specific adaptations to nutrient supply: more than just epigeneticsff" *Advances in Experimental Medicine and Biology*, Vol. 646, pp. 113 – 118, 2009.

[27] A. J. Drake, B. R. Walker, and J. R. Seckl, "Intergenerational consequences of fetal programming by in utero exposure to glucocorticoids in rats," *American Journal of Physiology*, Vol. 288, No. 1, pp. R34 – R38, 2005.

[28] A. Garofano, P. Czernichow, and B. Bréant, "In utero undernu – trition impairs rat beta – cell development," *Diabetologia*, Vol. 40, No. 10, pp. 1 231 – 1 234, 1997.

[29] Z. Khan, J. Carey, H. J. Park, M. Lehar, D. Lasker, and H. A. Jinnah, "Abnormal motor behavior and vestibular dysfunction in the stargazer mouse mutant," *Neuroscience*, Vol. 127, No. 3, pp. 785 – 796, 2004.

[30] E. Zambrano, G. L. Rodríguez – González, C. Guzmán *et al.*, "A maternal low protein diet during pregnancy and lactation in the rat impairs male reproductive development," *Journal of Physiology*, Vol. 563, No. 1, pp. 275 – 284, 2005.

[31] H. – S˙. Chang and P. Pui – Hay, *Pharmacology and Applications of Chinese Material Medica*, World Scientific, Singapore, 1987.

[32] J. S. Sim, H. H. Sunwoo, R. J. Hudson, and B. T. Jeon, *Antler Science and Product Technology*, University of Alberta, Edmon – ton, Canada, 2001.

[33] R. L. Garcia, M. Sadighi, S. M. Francis, J. M. Suttie, and J. S. Fleming, "Expression of neurotrophin – 3 in the growing velvet antler of the red deer *Cervus elaphus*," *Journal of Molecular Endocrinology*, Vol. 19, No. 2, pp. 173 – 182, 1997.

[34] S. M. Francis and J. M. Suttie, "Detection of growth factors and proto – oncogene mRNA in the growing tip of red deer (*Cervus elaphus*) antler using reverse – transcriptase polymerase chain reaction (RT – PCR)," *Journal of Experimental Zoology*, Vol. 281, No. 1, pp. 36 – 42, 1998.

[35] M. Sadighi, S. R. Haines, A. Skottner, A. J. Harris, and J. M. Suttie, "Effects of insulin – like growth factor – I (IGF – I) and IGF – II on the growth of antler cells in vitro," *Journal of Endocrinology*, Vol. 143, No. 3, pp. 461 – 469, 1994.

[36] L. K. Lewis and G. K. Barell, "Regional distribution of estradiol receptors in growing antlers," *Steroids*, Vol. 59, No. 8, pp. 490 – 492, 1994.

[37] H. S. Kim, H. K. Lim, and W. K. Park, "Antinarcotic effects of the velvet antler water extract on morphine in mice," *Journal of Ethnopharmacology*, Vol. 66, No. 1, pp. 41 – 49, 1999.

[38] Q. L. Zhou, Y. Q. Liu, Y. Wang, Y. J. Guo, and B. X. Wang, "A comparison of chemical composition and

bioactivity of polypeptides from velvet antlers of *Cervus nippon Temminck* and *Cervus elaphus Linnaeus*," *China Journal of Chinese Materia Medica*, Vol. 26, No. 10, pp. 699 - 702, 2001.

[39] P. M. Barling and J. Shirley, "Diaphorase activity in sebaceous glands and related structures of the male red deer," *Comparative Biochemistry and Physiology B*, Vol. 123, No. 1, pp. 17 - 21, 1999.

[40] K. M. Ko, T. T. Yip, and S. W. Tsao, "Epidermal growth factor from deer (*Cervus elaphus*) submaxillary gland and velvet antler," *General and Comparative Endocrinology*, Vol. 63, No. 3, pp. 431 - 440, 1986.

[41] L. P. Spear, "Neurobehavioral assessment during the early postnatal period," *Neurotoxicology and Teratology*, Vol. 12, No. 5, pp. 489 - 495, 1990.

[42] N. Sousa, O. F. X. Almeida, and C. T. Wotjak, "A hitchhiker's guide to behavioral analysis in laboratory rodents," *Genes, Brain and Behavior*, Vol. 5, No. 2, pp. 5 - 24, 2006.

[43] J. Altman and K. Sudarshan, "Postnatal development of locomotion in the laboratory rat," *Animal Behaviour*, Vol. 23, No. 4, pp. 896 - 920, 1975.

[44] I. Khan, V. Dekou, M. Hanson, L. Poston, and P. Taylor, "Pre - dictive adaptive responses to maternal high - fat diet prevent endothelial dysfunction but not hypertension in adult rat offspring," *Circulation*, Vol. 110, No. 9, pp. 1 097 - 1 102, 2004.

[45] A. Minelli, *The Development of Animal Form: Ontogeny, Morphology, and Evolution*, Cambridge University Press, Cam - bridge, UK, 2003.

[46] A. Borella, M. Bindra, and P. M. Whitaker - Azmitia, "Role of the 5 - HT (1A) receptor in development of the neonatal rat brain: preliminary behavioral studies," *Neuropharmacology*, Vol. 36, No. 4 - 5, pp. 445 - 450, 1997.

[47] F. Cirulli and E. Alleva, "Effects of repeated administrations of EGF and TGF - α on mouse neurobehavioral development," *NeuroToxicology*, Vol. 15, No. 4, pp. 819 - 826, 1994.

[48] J. Dobbing and J. Sands, "Quantitative growth and development of human brain," *Archives of Disease in Childhood*, Vol. 48, No. 10, pp. 757 - 767, 1973.

[49] M. G. Alton - Mackey and B. L. Walker, "The physical and neuromotor development of progeny of female rats fed graded levels of pyridoxine during lactation," *American Journal of Clinical Nutrition*, Vol. 31, No. 1, pp. 76 - 81, 1978.

[50] M. Dierssen, V. Fotaki, M. Martínez de Lagrán *et al.*, "Neurobehavioral development of two mouse lines commonly used in transgenic studies," *Pharmacology Biochemistry and Behavior*, Vol. 73, No. 1, pp. 19 - 25, 2002.

[51] K. M. Olesen and A. P. Auger, "Sex differences in Fos protein expression in the neonatal rat brain," *Journal of Neuroen - docrinology*, Vol. 17, No. 4, pp. 255 - 261, 2005.

[52] P. M. E. Waile and D. J. Tracey, "Somatosensory system," in *The Rat Nervous System*, G. Paxinos, Ed., pp. 689 - 724, Academic, San Diego, Calif, USA, 1995.

[53] C. Welker, "Microelectrode delineation of fine grain somatotopic organization of SmI cerebral neocortex in albino rat," *Brain Research*, Vol. 26, No. 2, pp. 259 - 275, 1971.

(Published the article in Evidence - Based Complementary and Alternative Medicine affect factor: 4. 774)

Effects of Yeast Polysaccharide on Immune Enhancement and Production Performance of Rats

Wang Hui[1,2], Zhang Xia[3], Cheng Fusheng[1], Luo Yongjiang[1] and Dong Pengcheng[1]

(1. Key Laboratory of New Animal Drug Project, Gansu Province, Key Laboratory of Veterinary Pharmaceutics Discovery, Ministry of Agriculture, Lanzhou Institute of animal and Veterinarian Pharmaceutical Science, Chinese Academy of Agricultural Sciences, Lanzhou 730050, China; 2. College of Veterinary Medicine, 3. College of Life Sciences and Technology, Gansu Agricultural University, Lanzhou 730070, China)

Abstract: The aims of the present study were to investigate the immune enhancement effect and production performance of Yeast Polysaccharide (YPS) on rats. The results showed that each index in YPS groups was higher than that in blank control group. Any dose of YPS by orally administrated significantly raised spleen and thymus index, serum IgA, IgG, AKP and LZM level, phagocytic index and phagocytic activity of macrophages in the rats. Meanwhile it can increase production performance of rats. Results suggested that any dose of YPS can enhance immunologic function and production performance of rats and the dose of 100 mg/kg^{-1} has the most obvious efficacy.

Key words: Yeast polysaccharide; Immune enhancement; Production performance; Rats; cell; China

Introduction

One of the most promising recent alternatives to classical antihiotic treatment is the use of immunomodulators for enhancing host defense responses (Tzianabos, 2000). Several types of immunomodulators have been identified including mammalian proteins such as Interferongamma (IFN - γ) (Murray, 1996), granulocyte colony - stimulating factor (Nemunaitis, 1997) etc. In recent decades, polysaccharides isolated from natural sources (plants, fungi, bacteria, algae and animals) have also attracted a great deal of attention in the biomedical arena because of their broad spectrum of therapeutic properties (antitumor, immunostimulant, antiaging, antioxidant, antiviral, antidiabetic, antiatherosclorosis and antiinflammatory activities) and relatively low toxicity (Pang *et al.*, 1999; Smestad, 2001; Wasser, 2002; Li *et al.*, 2006; Chattopadhyay *et al.*, 2008; Dai *et al.*, 2009; Sun *et al.*, 2009; Khotimchenko *et al.*, 2006; De Barba *et al.*, 2011; Sohail *et al.*, 2011; Thetsrimuang *et al.*, 2011).

Fungi is an important source of materials in traditional Chinese medicine. Polysaccharide extracts from many species of fungi exhibit immunostimulating and/or antitumor activities (Ohno *et al.*, 2001; Wasser, 2002; Zhang *et al.*, 2003). Yeast is one of fungi, yeast cell wall is a thick envelope (100 – 200 nm) representing 15% ~ 25% of the dry mass of the cell (Farkas, 1979). The major components of cell wall are polysaccharides (up to 90%), mainly β – D – glucans and α – D – mannans. Yeast Polysaccharides (YPS), β – D – glucans and α – D – mannans have been previously demonstrated to antitumor (Khalikova *et al.*, 2005), antioxidant (Kogan *et al.*, 2005) and nutrition (Kogan and Kocher, 2007). In this study, rats were treated with different doses of YPS under the same condition. By observing spleen and thymus indexes, the levels of serum IgA, IgG, AKP and LZM, the phagocytic index and phagocytic activity of Macrophages, Average Daily Gain (ADG) and Average Daily Intake (ADI) effects of YPS on immunologic function and production performance of rats were analysed.

Materials and Methods

Preparation of polysaccharide: Yeast powder was independently researched and developed by Lanzhou Institute of Animal Science and Veterinary Pharmaceutics, Chinese Academy of Agricultural Sciences, China. Yeast powder was mixed with distilled water in the proportion of 1 : 3 (volume ratio) then break – walled sequentially by autoclaving (15 pounds, 35 min), freezing (– 40℃, 2h) thawing (100℃, 15 min) andultraphonic (500 W, 25 min), and then 3% of potassium hydroxide (KOH) was added. After a 1.5 h reaction at 100℃ and neutralized with 20% of acetic acid immediately then centrifuged at 3 500 r min^{-1} for 10 min. The supernatant was precipitated with three volumes of ethanol and stored overnight at 4℃. After centrifugation, the precipitates were collected and then washed sequentially with acetone (2 times) and ether. And after freeze drying, Yeast Polysaccharide (YPS) was obtained. YPS was stored at – 20℃ and freshly dissolved in distilled water immediately before use.

Experimental rats: About 120 male, healthy and unvaccinated rats (provided by laboratory animal center of Lanzhou University, China), each weighing around 220 g were randomly allotted into 12 cages. Every 3 cages formed a group, 10 in each cage. Rats were housed and maintained under pathogen – free conditions in microisolator cages.

Animal care was in compliance with recommendations of The Guide for Care and Use of Laboratory Animals (National Research Council). About 120 rats were randomly divided into 4 groups.

Groups division: Groups I – III were the YPS groups. Group IV was the blank control group. Polysaccharide contents: rats were administered with YPS at the doses of 200, 100 and 50 mg kg^{-1} body weight marked as I – III, respectively. Group IV contained no polysaccharide in its distilled water. YPS groups were orally administrated daily with YPS solution according to design dose and weighed in time every day; equivalent physiological saline was orally administrated at the same doses in the blank control group during 30 days. Rats were sacrificed by femoral bloodletting on day 31. Spleen and thymus were collected and weighed. The immunological parameters and relevant indexes were described in this study.

Effect of YPS on spleen and thymus indexes: After bloodstain of spleen and thymus without fat was washed with 4℃ equivalent physiological saline and sipped up with filter paper then weighted. The immune organs index was calculated as follows:

$$\text{Immune organs index (mg/g)} = \frac{\text{Immune organs weight (mg)}}{\text{Body weight (g)}}$$

Effect of YPS on serum IgA, IgG, AKP and LZM: On termination of the experiment, blood samples were collected; serums were separated and stored at −80℃ before assay. Two − site sandwich Enzyme − Linked Immunosorbent Assays (ELISA) were performed for quantify IgA and IgG (Sigma, USA) and colorimetry assays detected quantify Alkaline Phosphatase (AKP) and Lysozyme (LZM) (Nanjing Jiancheng Bioengineering Institute, Nanjing, China) with kit according to the manufacturer's instruction.

Macrophage phagocytosis assay (carbon clearance): The test of carbon clearance was carried out according to the method of Yang *et al.* (2007) and Han *et al.* (2010). Normal rats were divided into four groups (8 per group) and orally administrated with distilled water 2.0 ml or with YPS 2.0 ml (50, 100 and 200 mg/kg body weight, respectively) for 10 days. After 10 days of gastric feed, Indian ink according to 0.1 ml/10 g body weight was injected into the tail vein of the rats.

A total of 20 μL blood was respectively collected through eye orbit after 2 min (t_1) and 10 min (t_2) and added to 2.0 ml 0.1% Na_2CO_3 at once and its absorbance was determined at 620 nm (Jayathiitha and Mishra, 2004). At the same time, rats were sacrificed by femoral bloodletting; the spleens and liver were weighed. The carbon particle clearance's capacity was expressed by phagocytic index. Calculate out the phagocytic index (k) and phagocytic activity (a) as follow:

$$k = \frac{1gOD_1 - lgOD_2}{t_2 - t_1}$$

$$\alpha = \sqrt[3]{k} \times \frac{\text{Body weight}}{\text{Liverweight + Spleen weight}}$$

Production performance indexes: From days 1 ~ 30 production − index of each rat was tested:

$$\text{ADG (Averagedailygain)} = \frac{\text{(Post − weight) − (Pre − weight)}}{\text{Days}}$$

$$\text{ADI (Average daily intake)} = \frac{\text{Experimental period int ake}}{\text{Days}}$$

$$G/T = \frac{ADG}{ADI}$$

Statistical analysis: All the data were expressed as mean ± SD. Statistical differences were analyzed by Analysis of Variance (ANOVA) followed by LSD's Multiple Range test using SPSS Version 17.0 Software. The treatment effects were considered significant if the p − value was at or below 0.05.

Results and Discussion

Effects of YPS on spleen and thymus indexes in rats: The spleen and thymus indexes were slightly increased in the rats treated with YPS compared with blank control group ($P > 0.05$ or $P < 0.01$). A significant increase in weight of thymus and spleen was observed at medium-dose (100 mg/kg) group when compared with blank control group ($P > 0.05$ or $P < 0.01$). The test results for details are shown in Table 1. Thymus is the organ in which T lymphocytes develop, differentiate and mature while spleen contains T - cells and B - cells. The weight of the immune organs can reflect immune system state in a certain extent. The weight of the immune organs increase is the performance of immunity enhancement (Grossman, 1985). Thymus and spleen is the important immune organs and participate in the humoral immune and cellular immune. Polysaccharide molecules enter into cells through combining polysaccharide receptors at surface of cells and gather in the organs of spleen, thymus, etc. (Abadi *et al.*, 1998). The results showed that YPS medium - dose group could improve the weight of immune organ of rats.

Table 1 Effects of YPS on spleen and thymus indexes in rats

Groups	Quantity of polysaccharide (mg/kg)	Spleen index	Thymus index
I	200	4.12 ± 0.52^{A}	2.95 ± 0.32^{ab}
II	100	4.45 ± 0.43^{aA}	3.06 ± 0.38^{a}
III	50	4.19 ± 0.62^{a}	3.01 ± 0.43^{ab}
IV	0	3.94 ± 0.68^{bB}	2.83 ± 0.30^{b}

Effect of YPS on serum IgA, IgG, AKP and LZM of rats: Effect of YPS on serum IgA and IgG levels were determined by ELISA. As shown in Table 2, the serum IgA was significantly increased in YPS groups at Ⅰ - Ⅲ ($P > 0.05$, $P < 0.01$ and $P > 0.05$, respectively) compared with that in blank control Group Ⅳ. The similar observation also indicated in the serum IgG ($P > 0.05$, $P > 0.05$ and $P > 0.05$, respectively) compared with the Ⅳ group. The levels of IgA and IgG increased slightly in the YPS groups except serum IgA in Ⅱ group. YPS markedly augmented serum IgA of rats in dose dependent manner. IgA, IgG and IgM are the most abundant immunoglobulin produced in mammals and their role is against pathogenic and non - pathogenic microorganisms (Macpherson *et al.*, 2001). The level of serum immunoglobulins improved in a range can maintain health. This result is agreement with that yeast glucan can improve the level of serum immunoglobulins in pregnant sows (Krakowski *et al.*, 2002). Effect of YPS on serum AKP and LZM levels were determined by colorimetry. The serum AKP was significantly increased in Ⅰ and Ⅱ groups ($P > 0.05$ and $P < 0.01$) compared with that in Ⅳ group and Ⅰ group was higher than Ⅱ group ($P < 0.01$). The serum LZM was significantly increased in Ⅰ and Ⅱ groups ($P < 0.01$) compared with that in Ⅳ group but the highest value was in Ⅱ group ($P < 0.01$).

Table 2 Effect of YPS on serum IgA, IgG, AKP and LZM of rats

Croups	Ouantity of polysaccharide (mg/kg)	IgA (μg/ml)	IgG (μg/ml)	AKP (mg/100ml)	LZM (U/ml)
Ⅰ	200	8.60 ± 0.88^{b}	3.83 ± 0.25^{a}	98.71 ± 22.44^{bB}	265.62 ± 23.88^{B}
Ⅱ	100	9.36 ± 1.05^{aA}	3.90 ± 0.19^{a}	119.57 ± 11.57^{aA}	296.71 ± 21.63^{A}
Ⅲ	50	8.64 ± 0.49^{b}	3.68 ± 0.16^{b}	86.6 ± 10.21^{cB}	218.51 ± 34.64^{c}
Ⅳ	0	8.25 ± 0.59^{cB}	3.36 ± 0.20^{b}	82.36 ± 15.87^{cB}	219.86 ± 22.12^{c}

* In the same column, values with different lowercase superscripts mean significant difference ($P < 0.05$), values with different capital letter superscripts mean signifcant difference ($P < 0.01$).

YPS could increase the content of serum AKP and LZM of rats in a dose dependent manner. AKP and LZM play an important role in phagotrophy ability and sterilization ability of macrophages. AKP activity can reflect the growth performance of animals, improving AKP activity is help to improve Average Daily Gain (ADG) (Zhou *et al.*, 2010) and AKP can enhance the nonspecific immunity function. LZM is an effector of specific immune and mainly secreted by macrophages. The concentration of serum LZM is important index of nonspecific immune. This increase of the content of IgA, IgG, AKP and LZM may be reached the maximum effect when the rats were administered with the middle dose of YPS (100 mg/kg).

Effect of YPS on the phagocytic index (k) and phagocytic activity (a) of macrophages: To establish the effect of YPS on carbon particle clearance test, blood samples were taken at different time intervals. There was an enhancement in phagocytic index (k) and phagocytic activity (a) after YPS treatment for 10 days (Table 3). At dose of high and medium - YPS (200 and 100 mg/kg), a significant increase ($P < 0.01$) in phagocytic index (k) was observed but there was no significant difference at low - YPS (50 mg/kg) group. Phagocytic activity (a) also increased significantly at any dose of YPS treatment when compared with blank control group ($P < 0.01$, $P < 0.01$ and $P > 0.05$, respectively). In these results the values of k and a were characterized in a dose dependent manner.

Tahle 3 Effect of YPS on the phagocytic index k and phagocytic activity a of macrophages

Groups	Quantity of palysaccharide (mg/kg)	k	a
I	200	0.0089 ± 0.0012^{B}	5.03 ± 0.64^{aA}
II	100	0.0106 ± 0.0014^{A}	5.16 ± 0.42^{aA}
III	50	0.0061 ± 0.0005^{C}	4.21 ± 0.71^{b}
IV	0	0.0032 ± 0.0003^{C}	2.95 ± 0.48^{cB}

This result is agreement with that of Yang *et al.* (2007) and they found that pollen polysaccharides could increase the ability of lymphocyte proliferation of tumorbearing mice in a dose dependent manner. The test of carbon clearance could reflect the phagocytosis function of monocyte. This study found that oral administration of YPS was associated with significant improvement in macrophage phagocytic function (Table 3). One of the most important immune responses of the body is carried out by macrophages which is called as phagocytic function. Phagocytosis represents the final and most indispensable step of the immunological defense system (Zhao *et al.*, 2005). Activated macrophages plays an essential role inbuilding and consolidating immunological defense systems against malignancies like cancer (Blander and Medzhitov, 2004).

Effects of YPS on production performance: Average Daily Gains (ADG) inthe Ⅱ groups was significantly higher than other Groups (Ⅰ, Ⅲ and Ⅳ) ($P < 0.01$). There was no significant difference in Average Daily Intake (ADI) between each group ($P > 0.05$). But the G/I in the Ⅱ groups was significantly higher than other groups (Table 4). ADG in the polysaccharide groups were 5.63 g/day (maximum) higher than it in the blank control group. It was hard to obtain accurate data if daily measurement was taken so researchers calculated average daily gains with 30 days considered as one experimental period.

Tahle 4 Effects of YPS on production performance indexes of rats

Groups	Quantity of polysaccharide (mg/kg)	ADG (g/day)	ADI (g/day)	G/I
I	200	4.53 ± 0.64^{B}	22.09 ± 2.51	0.20 ± 0.03^{b}
II	100	5.63 ± 0.54^{A}	23.12 ± 2.78	0.24 ± 0.02^{a}
III	50	4.73 ± 0.68^{B}	22.35 ± 2.23	0.21 ± 0.02^{b}
IV	0	4.78 ± 0.57^{B}	22.18 ± 1.97	$0.22^{A} \pm 0.04^{b}$

In the same column, values with different lowercase superscripts mean significant difference ($P > 0.05$), values with different capital letter superscripts mean sigpificant difference ($P > 0.01$).

This result is disagreement with that of Hiss and Sauerwein (2003) and they found that Average Daily Gains (ADG) of β – glucan treated pigs were not different from the controls feed intake was tendentiously increased at β – glucan without alteration of feed efficiency. The growth – promoting mechanism of YPS has a bearing on the improvement of different body systems, especially digestive system. Relevant indexes are worthy of the deep researth.

Conclusion

Any dose of YPS can elevate spleen and thymus indexes, serum IgA, IgG, AKP and LZM level and phagocytic index (k) and phagocytic activity (a) of macrophages of the rats. Meanwhile it can increase Average Daily Gain (ADG) and G/I (G/I = Average daily gain/average daily intake) of rats and 100 mg/kg, is the best. YPS can improve immunologic function and production performance of rats in general which ensures its application as an ideal immune potentiator.

Acknowledgements

The study was supported by Gansu province Scientific Support Project, China (0708NKCA082, 090NKCA070) and Agricultural Biotechnology Industrialization Project of Gansu province, China (GNSW – 2010 – 07).

References

[1] Abadi J J, Friedman R A, Mageed R, Jefferis M C, Rodriguez – Barradas and L. Pirofski, Human antibodies elicited by a pneumococcal vaccine express idiot YPSic determinants indicative of V $(H)^3$ gene segment usage. J. Infect. Dis., 1998. 178: 707 – 716.

[2] Blander J M and R Medzhitov, Regulation of phagosome maturation by signals from toll – like receptors. Science, 2004. 304: 1 014 – 1 018.

[3] Chattopadhyay K T, Ghosh C A, Pujol M J, Carlucci E B, Damonte and B. Ray, Polysaccharides from *Gracilaria corticata*: Sulfation, chemical characterization and anti – HSV activities. Int. J. Biol. Macromol., 2008. 43: 346 – 351.

[4] Dai Z H, Zhang Y, Zhang and H Wang, Chemical properties and immunostimulatory activity of a water-soluble polysaccharide from the clam of *Hyriopsis cumingii* Lea. Carbohyd. Polym., 2009. 77: 365 – 369.

[5] De Barba F F M, M L L, Silveira B U, Piloni S A, Furlan and M S L, Pinho, Influence of *Pleurotus djarraor* bioactive substances on the survival time of mice inoculated with sarcoma 180. Int. J. Pharmacol.,

2011. 7：478 - 484.

[6] Farkas V. Biosynthesis of cell walls of fungi. Microbiol. Rev.，1979. 43：117 - 144.

[7] Grossman C J. Interaction between the gonadal steroids and the immune system. Science，1985. 227：257 - 261.

[8] Han Q，Z J，Ling P M，He and C Y Xiong，Immunomodulatory and antitumor activity of polysaccharide isolated from tea plant flower. Prog. Biochem. Biophys.，2010. 37：646 - 653.

[9] Hiss B S and H Sauerwein，Influence of dietary β - glucan on growth performance，lymphocyte proliferation，specific immune response and haptoglobin plasma concentrations in pigs. J. Anim. Physiol. Anim. Nutr.，2003. 87：2 - 11.

[10] Jayathirtha M G and S H，Mishra，Preliminary immunomodulatory activities of methanol extracts of *Eclipta albs* and *Centella asiatica*. Phytomedicine，2004. 11：361 - 365.

[11] Khalikova T A，S Y，Zhanaeva T A，Korolenko V I，Kaledin and G Kogan，2005. Regulation of activity of cathepsins B，L，and D in marine lymphosarcoma model at a combined treatment with cyclophosphamide and yeast polysaccharide. Cancer Lett.，2005. 223：77 - 83.

[12] Khotimchenko M Y，E P，Zueva K A，Lopatina Y S，Khotimchenko and N V Shilova，Gastroprotective effect of pectin preparations against indomethacin - induced lesions in rats. Int. J Pharmacol.，2006. 2：471 - 476.

[13] Kogan G and A Kocher，Role of yeast cell wall polysaccharides in pig nutrition and health protection. Livestock Sci.，2007. 109：161 - 165.

[14] Kogan G A，Stasko K，Bauerova M，Polovka and L Soltes *et al.*，Antioxidant properties of yeast（1 > 3）- β - D - glucan studied by electron paramagnetic resonance spectroscopy and its activity in the adjuvant arthritis. Carbohydr. Polym.，2005. 61：18 - 28.

[15] Krakowski L J，Krzyzanowski Z，Wrona K，Kostro and A K，Siwicki，The influence of nonspecific immunostimulation of pregnant sows on the immunological value of colostrum. Vet. Immunol. Immunop.，2002. 87：89 - 95.

[16] Li S P，G H，Zhang Q，Zeng Z G，Huang Y T，Wang T T X，Dong and K W K，Tsim，Hypoglycemic activity of polysaccharide，with antioxidation，isolated from cultured *Cordyceps* mycelia. Phytomedicine，2006. 13：428 - 433.

[17] Macpherson A J L，Hunziker M C，Kathy and L Alain，IgA responses in the intestinal mucosa against pathogenic and non - pathogenic microorganisms. Microbes Infect.，2001. 3：1 021 - 1 035.

[18] Murray H W. Current and future clinical applications of interferon - gamma in host antimicrobial defense. Intensive Care Med.，1996. 22：S456 - S461.

[19] Nemunaitis J.，A comparative review of colony - stimulating factors. Drugs，1997. 54：709 - 729.

[20] Ohno N M，Furukawa N N，Miura Y，Adachi M，Motor and T Yadomae，Antitumor β - glucan from the cultured fruit body of *Agaricus blazei*. Biol. Pharm. Bull.，2001. 7：820 - 828.

[21] Pang Z J Y，Chen and M Zhou，The effect of polysaccharide krestin on GPx gene expression in macrophages. Sheng Wu Hua Xue Yu Sheng Wu Wu Li Xue Bao（Shanghai），1999. 31：284 - 288.

[22] Smestad P B.，Plant polysaccharides with immunostimulatory activities. Curr. Org. Chem.，2001. 5：939 - 950.

[23] Sohail M N F，Rasul A，Karim U，Kanwal and IH. Attitalla，Plant as a source of natural antiviral agents. Asian J. Anim. Vet. Adv.，2011. 6：1 125 - 1 152.

[24] Sun L C，Wang Q，Shi and C Ma，Preparation of different molecular weight polysaccharides from *PorpJzyridium cruenturra* and their antioxidant activities. Int. J. Biol. Macromol.，2009. 45：42 - 47.

[25] Thetsrimuang C S，Khasnmuang and R Sarnthima，Antioxidant activity of crude polysaccharides from edible

fresh and dry mushroom fruiting bodies of *Lentinus* sp. strain RJ -2. Int. J. Pharmacol. , 2011. 7: 58 -65.

[26] Tzianabos A O. , Polysaccharide immunomodulators as therapeutic agents: Structural aspects and biological function. Clin. Microbiol. Rev. , 2000. 13: 523 -533.

[27] Wasser S. , 2002. Medicinal mushrooms as a source of antitumor and immunomodulating polysaccharides. Applied Microbiol. Biotechnol. , 2002. 60: 258 -274.

[28] Yang X P, D Y, Guo J M, Zhang and M C Wu, Characterization and antitumor activity of pollen polysaccharide. Int. Immunopharmacol. , 2007. 7: 427 -434.

[29] Zhang H N, J H, He L, Yuan and Z B, Lin, *In vitro* and *In vivo* protective effect of *Ganoderma lucidum* polysaccharides on alloxaninduced pancreatic islets damage. Life Sci. , 2003. 73: 2 307 -2 319.

[30] Zhao G J, Kan Z, Li and Z Chen, Characterization and immunostimulatory activity of an (1 - >6) -a-D-glucan from the root of Ipomoea batatas. Int. Immunopharmacol. , 2005. 5: 1 436 -1 445.

[31] Zhou Y Q Y, Diao Y, Tu Q, Yun, Effects of yeast β - glucan on growth performance, serumbiochemical and gastrointestinal characteristics in pre - ruminant calves. Chin. J. Anim. Sci. , 2010. 46: 47 -51.

(Published the article in Journal of Animal and Veterinary Advances affect factor: 0. 39)

Efficacy of *trans* – cinnamaldehyde against *Psoroptes cuniculi* in vitro

Shen Fengge, Xing Mingxun, Liu Lihui, Tang Xudong, Wang Wei, Wang Xiaohong, Wu Xiuping, Wang Xuelin, Wang Xinrui, Wang Guangming, Zhang Junhui, Li Lei, Zhang Jiyu, Yu Lu

(Lanzhou Institute of animal and Veterinarian Phamaceutical Science, CAAS, Lanzhou 730050, China)

Abstract: The acaricidal activity of *trans* – cinnamaldehyde was evaluated in vitro on *Psoroptes cuniculi*. In this study, different concentrations of *trans* – cinnamaldehyde were tested, and the observed mites mortality was compared with that observed in untreated and treated (Acacerulen R®) controls. The morphological changes in *P. cuniculi* treated with *trans* – cinnamaldehyde were examined with light microscopy. By the analysis of variance one – way test, up to 8 μg/ml of *trans* – cinnamaldehyde gave highly significant ($P<0.01$) per – centages of mite mortality compared with the untreated controls, but only up to 256 μg/ml, it showed the same efficacy of Acacerulen R®. At the same time, a bioassay was conducted by exposing mites to varying doses of *trans* – cinnamaldehyde in vitro cultures. The resulting data were analyzed by using a time - dose - mortality modeling tech – nique, yielding the parameters for time and dose effects of *P. cuniculi*. The β value was 2. 01, indicating that *trans* – cinnamaldehyde had a good activity to kill *P. cuniculi* adults. Based on the time – dose – mortality relationships fitted and the virulence indices estimated, *trans* – cinnamaldehyde is a promising microbial agent for mites control.

Introduction

Psoroptes cuniculi (*P. cuniculi*) is a common worldwide parasite of rabbits and causes infestation primarily in the ears, which is of importance with respect to general animal health and hygiene as well as economic concerns. The infestation can cause considerable weight loss, less favor – able feed conversion rates, vestibular dysfunction, and meningitis, which is frequently fatal when complicated by secondary bacterial pathogens (Ulutas *et al.* 2005).

Therapy and control of both human scabies and animal mange are based mainly on the use of effective drugs and chemicals. However, the use of drugs to control these parasitic arthropods presents several problems including drug resistance (Synge *et al.* 1995; Clark *et al.* 1996) and environmental damage (Halley *et al.* 1993; O'Brien 1999). These problems have lead to research efforts to discover new effective compounds. Plants and other natural materials may prove to be valuable sources of chemotherapeutic agents, thereby reducing the frequency of resistance phenomena and providing alternative drugs with greater acceptance, especially in terms of environmental safety (Alawa *et al.* 2003). Prior researches have shown that neem, thymol, 1, 8 – cineole, linalool, and extracts of *Baccharis flabellata* and *Minthostachys verticillata* had an acaricidal activity against mites (Nina *et al.* 2010; Natalia *et al.* 2009; Natalia *et al.* 2011).

Indigenous cinnamon (*Cinnamomum*) is an endemic tree in Taiwan, and *trans* – cinnamaldehyde (Fig. 1) is a chemical present as a major component of bark extract of cinnamon (Liu *et al.* 1998). Our previous studies have demonstrated that *trans* – cinnamaldehyde had excellent antifungal activity (Shreaz *et al.* 2011), antibacterial activity (Chang *et al.* 2008), antibiofilm activity (Amalaradjou and Venkitanarayanan 2011), antitermitic activity (Chang and Cheng 2002), antimite activity (Chen *et al.* 2002), antimildew (Chen and Chang 2002), mosquito larvicidal activity (Cheng *et al.* 2009), antipathogenic activity (Bowles and Miller 1993), anti – inflammatory activity (Ballabeni *et al.* 2010), anti – red imported fire ants activity (Cheng *et al.* 2008), and antihyper – uricemic activity (Wang *et al.* 2008). However, no previous investigations were done on its effect on *P. cuniculi*. Therefore, the main target of this study was to evaluate the acaricidal effect of *trans* – cinnamaldehyde on the adult stages of *P. cuniculi*. In addition, we determine the time – concentration – mortality relationships between *trans* – cinna – maldehyde and *P. cuniculi*, estimate the virulence indices, and evaluate the potential of *trans* – cinnamaldehyde for use in *P. cuniculi* control.

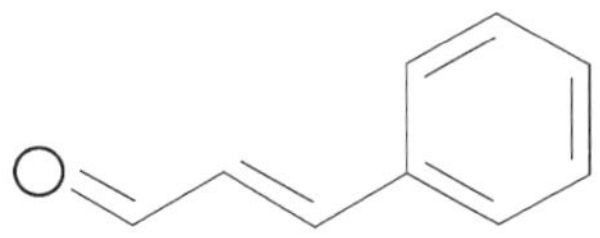

Fig. 1 Chemical formula of *trans* – cinnamaldehyde

Materials and methods

Materials

Trans – cinnamaldehyde was purchased from the National Institute of the Control of Pharmaceutical and Biological Products, Beijing, China. It was dissolved in ethanol at a concentration under sterile conditions. Twenty – five percent of pyrethrins was obtained from Sigma – Aldrich (St. Louis, MO). Pyrethrins was dissolved paraffin oil.

Mite collection

P. cuniculi mites were isolated from naturally infected rabbits. The scabs and the cerumen were collected from the infected ears. The Petri dishes were then incubated at 37℃ for 30 min. A stereomicroscope was used to isolate all motile stages, i. e., larvae, nymphs, and both sex. Motile *P. cuniculi* adults were used in all experiments.

Acaricidal activity in vitro

Mites were collected with a needle and placed in 6 – cm Petri dishes (approximately 100 adult mites/dish). *Trans* – cinna – maldehyde was diluted from the concentration of 4 096 to 8 μg/ml in water, and 2. 5 ml of each solution was directly added to Petri dishes. Five replications were made for each concentration. As negative control, five Petri dishes containing 2. 5 ml water were used, while five Petri dishes containing 2. 5 ml of a pyrethrum extract which contains 25% of pyrethrins (Acacerulen R®) were regarded as positive control; the drug used for the topical treatment of *P. cuniculi* represented the experimental control. Plates of untreated and treated controls were placed in separate humidity chambers in saturated humidity conditions at 22℃ (High Performance Incubator 2 800, Galli, Milan, Italy). After 24 h, mites were placed in clean Petri dishes containing 2. 5 ml of water, and after further 24 h, all the motionless mites were stimulated with a needle. Lack of reactions and persistent immobility indicated their death.

Toxicity evaluation

Trans – cinnamaldehyde which showed stronger acaricidal was used in the toxicity tests (median lethal concentration, LC_{50} and median lethal time, LT_{50}) against the *P. cuniculi* adults. *Trans* – cinnamaldehyde was diluted to concentrations of 4 096, 2 048, 1 024, 512, 256, 128, 64, 32, 16, and 8 μg/ml with distilled water. Polystyrene plates containing mites (20 mites per plate) were dipped in the various concentrations. Distilled water was used as the control. Each treatment was replicated three times. All the plates containing mites were incubated under the following conditions: 22℃ and saturated humidity, and the number of dead mites was calculated every 2 h. Each plate was observed for 5 min and was then replaced in the test chamber. The total incubation period was 24 h.

Statistical analysis

The data were evaluated using the analysis of variance one – way test, and LC_{50} and LT_{50} values were calculated by the complementary log – log (CLL) model (Preisler and Robertson 1989) using a special microcomputer program (Tang and Feng 2002).

Results

Acaricidal activity in vitro

Trans – cinnamaldehyde showed a good acaricidal activity against *P. cuniculi*in vitro (Table 1). With respect to the untreated controls, the percentage of mites mortality result is highly significant ($P<0.01$), up to the concentration of 8 μg/ml. With respect to the treated controls, *trans* – cinnamalde – hyde showed the same efficacy of Acacerulen® with the exception of the concentrations of 128, 64, 32, and 8 μg/ml ($P<0.01$). Morphological changes of mites with 256 μg/ml *trans* – cinnamaldehyde treated at different times were observed by a light microscope. The structure of the untreated mites was intact, with polypide darkness and opacity (Fig. 2a). Treated with *trans* – cinnamaldehyde after 10 min, mites became active. Some period of time later, mites'movement changed, with heavy exercise increased. The exercise capacity gradually weakens, and they even scarcely moved. The picture of a drug – treated mite at 2 h (Fig. 2b) showed that the mite lost its glossiness, parts of the structure turn to be red, and integral structure was indistinct as well. In Fig. 2c, we found the red section permeates; polypide became more transparent when treated with *trans* – cinnamal – dehyde at 6 h. Figure 2d showed that the cover of the mite was gradually becoming transparent and was losing its distinct outlines because of disintegration. Meanwhile, dark tissue diminished when treated with *trans* – cinnamaldehyde at 24 h.

Table 1 Percentage of mortality of *P. cuniculi* treated at 24 h with scalar concentrations (4096 – 16 μg/ml) of *trans* – cinnamaldehyde, with Acacerulen R and untreated

Untreated control	4.60*	Untreated control	4.60*
256 μg/ml	100.00**	16 μg/ml	29.60*
128 μg/ml	78.60*	8 μg/ml	13.20*
64 μg/ml	59.20*	Acacerulen R	100.00**
32 μg/ml	37.40*		

* $P<0.01$; ** $P>0.05$

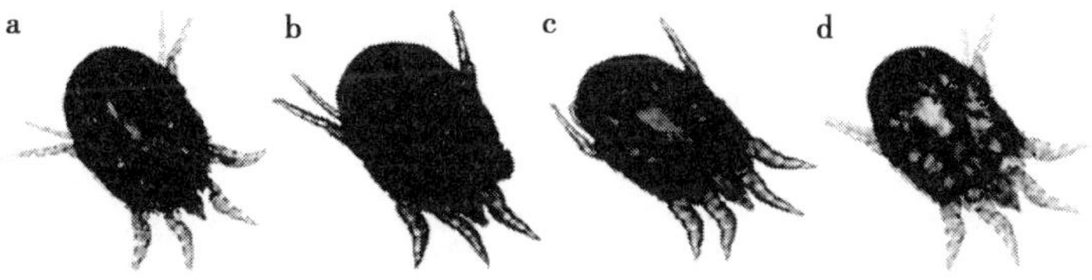

Fig. 2 The changes of mite under *trans* – cinnamaldehyde treatment were observed by Olympus
BX53 microscope (10 × magnification): a untreated mite, b 256 μg/ml – treated mite at 2 h, c 256 μg/ml – treated mite at 6 h, and d 256 μg/ml – treated mite at 24 h.

Toxicity evaluation

The time – dose – mortality data in Fig. 3 were adjusted using the background mortality records and then fitted to the conditional mortality probability model, yielding parameters for conditional dose and time effects (β, γ_j) of *P. cuniculi* assay (Table 2). The parameters for cumulative time effects, τ_j, were thus calculated using the values of γ_j. The *t* statistics for the parameters estimated were all significant ($P<0.001$), indicating the significance of the time – dose – mortality relationship of *trans* – cinnamaldehyde against *P. cuniculi*. The Hosmer – Lemeshow test for heterogeneity of the goodness of fit (Nowierski *et al.* 1996) was insignificant for *trans* – cinnamaldehyde (Hosmer – Lemeshow $\chi^2=4.88$, $P=0.77$). Thus, the data of *trans* – cinnamaldehyde against *P. cuniculi* adults fit well to the time – dose – mortality model (Pearson chi – square $=89.94$, $P=0.84$).

Table 2 Parameters estimated by fitting the time – dose – mortality model to the bioassay (trans – cinnamaldehyde against *P. cuniculi*) data

Conditional mortality model				Cumulative mortality model			
Parameter[a]	Value	SE	t^b	Parameter[a]	Value	Var (τ)	Cov (τ, β)
β	2.01	0.18	11.00	β	2.01	0.01	0.01
γ1	–11.03	1.23	8.95	τ1	–11.03	0.65	–0.05
γ2	–8.08	0.63	12.79	τ2	–8.03	0.17	–0.05
γ3	–7.55	0.62	12.16	τ3	–7.07	0.16	–0.05
γ4	–7.84	0.64	12.16	τ4	–6.69	0.15	–0.05

(Continued)

Conditional mortality model				Cumulative mortality model			
γ5	-7.45	0.61	12.17	τ5	-6.31	0.15	-0.04
γ6	-7.43	0.61	12.18	τ6	-6.02	0.14	-0.04
γ7	-7.61	0.65	11.62	τ7	-5.84	0.13	-0.04
γ8	-7.40	0.64	11.53	τ8	-5.65	0.13	-0.04
γ9	-7.18	0.63	11.49	τ9	-5.45	0.12	-0.04
γ10	-6.32	0.54	11.79	τ10	-5.10	0.11	-0.04
γ11	-6.24	0.53	11.79	τ11	-4.82	0.10	-0.04
γ12	-6.85	0.66	10.36	τ12	-4.70	0.10	-0.04
γ13	-5.89	0.54	10.94	τ13	-4.43	0.09	-0.03

* *SE* standard error.

a: represents the number of the hours or the/th hour after the exposure; b: represents the t statistics were highly significant for all parameters estimated ($P<0.0001$); Var (τ): represents the variance for τj can be expressed; Cov (τ, β): represents the covariance between τj and β can be expressed; β: is the parameter that describes the average number of toxicant molecules per receptor according to the hypothetical key receptors of the organism; τ: is the categorical variable corresponding with time τj; t: represents the value of the t - text.

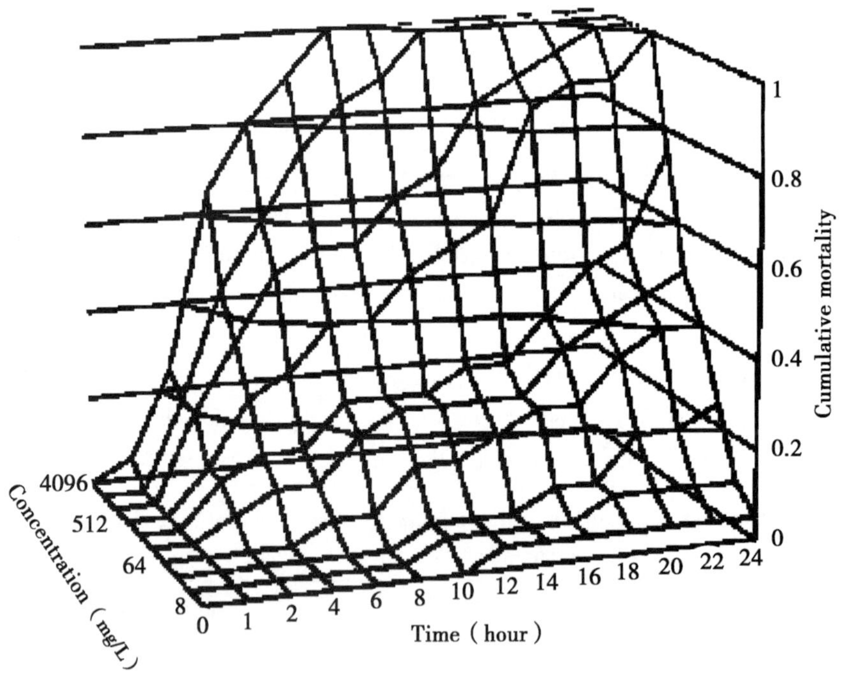

Fig. 3 Time - dose - mortality relationship for *trans* - cinnamaldehyde on *P. cuniculi* (plot for cumulative mortality probability model)

The parameters estimated for the cumulative time - dose - mortality models were used to compute virulence indices using the formulae given by Feng *et al.* (1998) and Nowierski *et al.* (1996). The values of log (LC_{50}) with standard errors were a function of the time after exposure to mites'showers (Fig. 4). The acaricidal activity of *trans* – cinnamaldehyde showed the relation of time and concentration dependent (Figs. 4 and 5). The LC_{50} values were 912, 661, 525, 427, 339, 229, 166, 145, and 107 μg/ml on 8, 10, 12, 14, 16, 18, 20, 22, and 24 h, respectively. However, during the same period, the dosage required to kill 90% of the aphids ranged from 417 to 3 631 μg/ml (Fig. 4). Dosages below 128 μg/ml caused 50% mortality at the end of the bioassay, thus generating no estimate of the LT_{50}. The estimate of the LT_{50}s decreased from 12.37 h for 128 μg/ml to 2.40 h for 4 096 μg/ml, when LT_{90}s decreased from 12.37 h for 512 μg/ml to 4.71 hours for 4 096 μg/ml. This corroborated the observation that almost all of the deaths caused by *trans* – cinnamaldehyde occurred on 2 to 14 h after the mites dipping (Fig. 5).

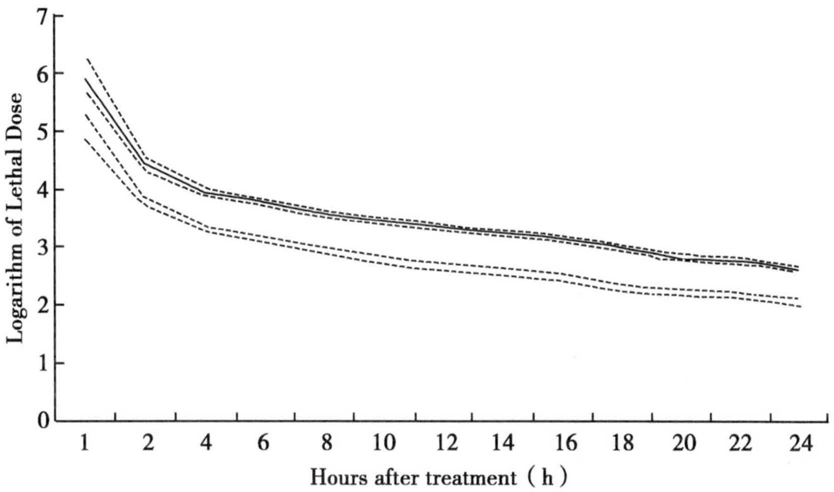

Fig. 4　The log (LC_{50}), log (LC_{90}), and their 95% confidence intervals values of *trans* – cinnamaldehyde estimated by CLL model at different time. Upper line represents Log (LC_{90}), lower line represents Log (LC_{50})

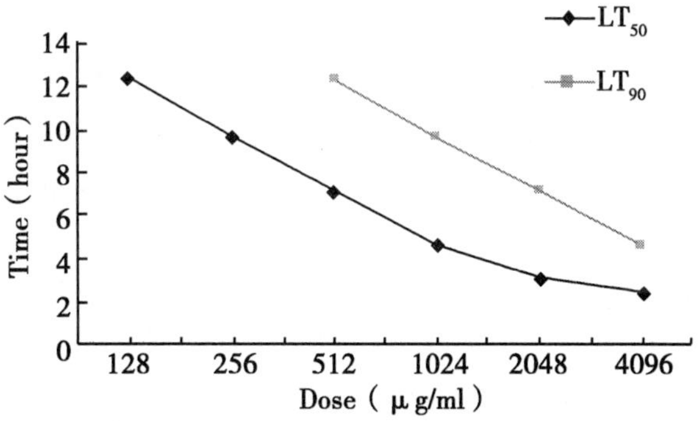

Fig. 5　The LT_{50} and LT_{90} values estimated by CLL model at different concentrations of *trans* – cinnamaldehyde

The mite *P. cuniculi* is a worldwide obligatory ectoparasite, for example, in rabbits, goats, horses, and sheep (Perrucci *et al.* 2005). In rabbits, lesions are limited to the ears (Saunders 1979). The mite causes intense pruritus with formation of crusts and scabs, which can completely fill the external ear canal and internal surface of the pinna (Bates 1999). It is a serious threat of commercial rabbit husbandry.

Previous reports have shown that the neem product had good activity against the red poultry mite, and the average mortality was 94.64% at a dose of 0.25 mg/cm^2 (Nina *et al.* 2010). Thymol, 1, 8 – cineole, and linalool had a good acaricidal activity against *Varroa destructor*; the LC_{50} values were 4.65, 8.74, and 8.55 μg/Petri dish at 24 h, respectively (Natalia *et al.* 2009). The botanical extracts from *B. flabellata* and *M. verticillata* had a good activity to kill *V. destructor*; the LC_{50} values were 1.57% and 2.16% at 24 h, respectively (Natalia *et al.* 2011). To our knowledge, this is the first time that *trans* – cinnamaldehyde is tested against *P. cuniculi* in vitro, and the results obtained were concerned with the high inhibitory rate for *trans* – cinnamaldehyde. On the other hand, *trans* – cinnamaldehyde has been reported to possess antimicrobial activity including gram – positive and gram – negative bacteria (Bowles *et al.* 1995; Friedman *et al.* 2002). Therefore, infection of secondary bacterial pathogens may be decreased in the rabbit with mites. The concentration of *trans* – cinnamaldehyde that caused the 50% and the 90% mortalities of the adult was decreased with the increase in treatment time, and the lethal time of *trans* – cinnamaldehyde shortened with the rise in the LC_{50} and the LC_{90} values. *Trans* – cinnamaldehyde demon – strated stronger toxicity against adult mites, and its activity was more affected by concentration and time.

In this experiment, high concentration of *trans* – cinnamaldehyde (greater than 256 μg/ml) showed a strong stimulation and lethal effect to mites. We found that the activity of the mites was heightened in a short time, followed with the mite convulsed, and finally polypide died. The insecticidal mechanism of *trans* – cinnamaldehyde was presumed to have a role in the neuromuscular system of mites with typical symptoms of excitement, convulsion, and death.

In previous studies, complementary log – log model was mainly used for estimating the effects of microbial pesticides, but scarcely for chemical pesticides (Nowierski *et al.* 1996; Feng and Poprawski 1999; Feng and Pu 2005; Shi *et al.* 2008). In addition, the parameter β in complementary log – log model was influenced significantly by the type of pesticides; the β value for chemical pesticides ranged from 1.2 to 4.5 (Preisler and Robertson 1989) and was significantly lower for microbial pesticides. In our study, the complementary log – log model demonstrated well in modeling time – concentration – mortality relationships of *trans* – cinnamaldehyde against *P. cuniculi* adults. The β value was 2.01, indicating that *trans* – cinnamaldehyde had a good activity to kill *P. cuniculi* adults.

The time – concentration – mortality model was established based on a Gompertz distribution and gave more appropriate data for evaluating the effectiveness of *trans* – cinnamaldehyde on *P. cuniculi* than probit analysis or logistic regression analysis. The results of the Hosmer – Lemeshow test indicated that the complementary log – log model provided a reasonable fit to the data, which was obtained in this study. In this study, the values of LC_{50} were more realistic estimates of the *P. cuniculi* mortality as a function of *trans* – cinnamaldehyde concentration and time by considering the control mortality. Moreover, the values of LC_{50} estimated the virulence more accurately than the unconsidered mortality of control.

Trans – cinnamaldehyde, the main ingredient in cassia oil, is generally classified as safe and is approved for use in foods (21 CFR 182.60) by the Food and Drug Administration. Plant essential oils have been suggested as an alternative source for *P. cuniculi* control. On the other hand, biodegradation to nontoxic products has no or little harmful effects on nontarget organisms and on the environment. Nevertheless, they have been shown to be highly effective against insecticide - resistant insect pests, and they are likely to be useful in resistance management strategies.

In conclusion, cassia oil appears to be effective in the control of *P. cuniculi*. For the practical use of products derived from cassia oil as novel *P. cuniculi* control agents to proceed, further research is necessary on the safety issues of this oil on human health. Other areas requiring attention are formulations for improving acaricidal efficacy, stability, and reducing the cost.

Acknowledgments Financial supports for this work came from the National Key Technology R&D Program (2008BADB4B05), the National Nature Science Foundation of China (No. 30871889), the Specialized Research Fund for the Doctoral Program of Higher Education (SRFDP) (No. 200801831051), the Fund for Science and the earmarked fund for China Agriculture Research System (cars - 38), and the Fundamental Research Funds for the Central Universities.

References

[1] Alawa CBI, Adamu AM, Gefu JO, Ajanusi OJ, Abdu PA, Chiezey NP, Alawa JN, Bowman DD. In vitro screening of two Nigerian medicinal plants (*Vernonia amygdalina and Annona senegalensis*) for anthelmintic activity. Vet Parasitol 113: 73 - 81. (2003)

[2] Amalaradjou MA, Venkitanarayanan K. Effect of trans - cinnamaldehyde on inhibition and inactivation of *Cronobacter sakazakii* biofilm on abiotic surfaces. J Food Prot 74 (2): 200 - 208. (2011)

[3] Ballabeni V, Tognolini M, Giorgio C, Bertoni S, Bruni R, Barocelli E. Ocotea quixos Lam. essential oil: in vitro and in vivo investi - gation on its anti - inflammatory properties. Fitoterapia 81 (4): 289 - 295. (2010)

[4] Bates PG. Inter - and intra - specific variation within the genus *Psoroptes* (Acari: Psoroptidae). Vet Parasitol 83: 201 - 217. (1999)

[5] Bowles BL, Miller AJ. Antibotulinal properties of selected aromatic and aliphatic aldehydes. J Food Prot 56: 788 - 794. (1993)

[6] Bowles BL, Sackitey SK, Williams AC. Inhibitory effects of flavor compounds on *Staphylococcus aureus* WRRC B124. J Food Saf 15: 337 - 347. (1995)

[7] Chang ST, Cheng SS. Antitermitic activity of leaf essential oils and components from *Cinnamomum osmophloeum*. Journal of Agriculture and Food Chemistry 50: 1 389 - 1 392. (2002)

[8] Chang CW, Chang WL, Chang ST, Cheng SS. Antibacterial activities of plant essential oils against *Legionella pneumophila*. Water Resources 42: 278 - 286. (2008)

[9] Chen PF, Chang ST. Application of essential oils from wood on the manufacture of environment - friendly antimicrobial paper products. Quarterly Journal of Chinese 35: 69 - 74. (2002)

[10] Chen PF, Chang ST, Wu HH. Antimite activity of essential oils and their components from *Cinnamomum osmophloeum*. Quarterly Journal of Chinese Forestry 35: 397 - 404. (2002)

[11] Cheng SS, Liu JY, Lin CY, Hsui YR, Ju MC, Wu WJ, Chang ST. Terminating red imported fire ants using Cinnamomum osmo - phloeum leaf essential oil. Bioresour Technol 99: 889 - 893. (2008)

[12] Cheng SS, Liu JY, Huang CG, Hsu YR, Chen WJ, Chang ST. Insecticidal activities of leaf essential oils from *Cinnamomum osmophloeum* against three mosquito species. Bioresour Technol 100: 457 - 464. (2009)

[13] Clark AM, Stephen FB, Cawley GD, Bellworthy SJ, Groves BA. Resistance of the sheep scab mite *Psoroptes*

ovis to propetamphos. Vet Rec 139：451.（1996）

[14] Feng MG，Poprawski TJ. Robustness of the time – dose – mortality model in bioassay data analysis of microbial control agents and chemical agents for insect control. Subtrop Plant Sci 51：36 –38.（1999）

[15] Feng MG，Pu XY. Time – concentration – mortality modeling of the synergistic interaction of Beauveria bassiana and imidacloprid *against Nilaparvata lugens*. Pest Manag Sci 61：363 –370.（2005）

[16] Feng MG，Liu CL，Xu JH，Xu Q. Modeling and biological implication of time – dose – mortality data for the entomophthoralean fungus，*Zoophthora anhuiensis*，on the green peach aphid *Myzus persicae*. J In vertebr Pathol 72：246 –251.（1998）

[17] Friedman M，Henika PR，Mandrell RE. Bactericidal activities of plant essential oils and some of their isolated constituents against *Campylobacter jejuni*，*Escherichia coli*，*Listeria monocytogenes*，and *Salmonella enterica*. J Food Prot 65：1 545 –1 560.（2002）

[18] Halley BA，Vandenheuvel WJA，Wislock PG，Herd R，Strong L，Wardhaugh K. Environmental effects of the usage of avermectines in livestock. Vet Parasitol 48：109 –125.（1993）

[19] Liu YC，Lu FY，Ou CH. Trees of Taiwan；Monographic Public No. 7，College of Agriculture，National Chung – Hshing University：Taichung，Taiwan.（1998）

[20] Natalia D，Liesel BG，Pedro B，Jorge AM，Martín JE. Acaricidal and insecticidal activity of essential oils on *Varroa destructor*（Acari：Varroidae）and *Apis mellifera*（Hymenoptera：Apidae）. Parasitol Res 106：145 –152.（2009）

[21] Natalia D，Liesel BG，Matías DM，Sara P，Jorge AM，Martín JE. Repellent and acaricidal effects of botanical extracts on *Varroa destructor*. Parasitol Res 108：79 –86.（2011）

[22] Nina L，Khaled ASA，Fathy AG，Heinz M. In vitro and field studies on the contact and fumigant toxicity of a neem – product（Mite – Stop®）against the developmental stages of the poultry red mite *Dermanyssus gallinae*. Parasitol Res 107：417 –423.（2010）

[23] Nowierski RM，Zeng Z，Jaronski S，Delgado F，Swearingen W. Analysis and modeling of time – dose – mortality of *Melanoplus sanguinipes*，*Locusta migratoria migratorioides*，and *Schistocerca gregaria*（Orthoptera：Acrididae）from *Beauveria*，*Metarhizium*，and *Paecilomyces* isolates from Madagascar. J Invertebr Pathol 67：236 –252.（1996）

[24] O'Brien DJ. Treatment of psoroptic mange with reference to epidemiology and history. Vet Parasitol 83：177 – 185.（1999）

[25] Perrucci S，Rossi G，Fichi G，O'Brien DJ. Relationship between *Psoroptes cuniculi* and the internal bacterium *Serratia marcescens*. Exp Appl Acarol 36：199 –206.（2005）

[26] Preisler HK，Robertson JL. Analysis of time – dose – mortality data. J Econ Entomol 82：1 534 –1 542.（1989）

[27] Saunders EB. Ear – mite infestation. Vet Med Small Anim Clin 74：218 –219.（1979）

[28] Shi WB，Zhang L，Feng MG. Time – concentration – mortality responses of carmine spider mite（Acari：Tetranychidae）females to three hypocrealean fungi as biocontrol agents. Biol Control 46：495 –501.（2008）

[29] Shreaz S，Sheikh RA，Bhatia R，Neelofar K，Imran S，Hashmi AA，Manzoor N，Basir SF，Khan LA. Antifungal activity of α – methyl trans cinnamaldehyde，its ligand and metal complexes：promising growth and ergosterol inhibitors. Biometals April 8.（2011）

[30] Synge BA，Bates PG，Clark AM，Stephen FB. Apparent resistance of Psoroptes ovis to flumethrin. Vet Rec 137：51.（1995）

[31] Tang QY，Feng MG. DPS Data processing system for practical statistics. Science Press Beijing，pp. 193 – 201.（2002）

[32] Ulutas B，Voyvoda H，Bayramli G，Karagenc T. Efficacy of topical administration of eprinomectin for treat-

ment of ear mite infestation in six rabbits. Vet Dermatol 16：334 – 337. (2005)
[33] Wang SY, Yang CW, Liao JW, Zhen WW, Chu FH, Chang ST. Essential oil from leaves of *Cinnamomum osmophloeum* acts as a xanthine oxidase inhibitor and reduces the serum uric acid levels in oxonate – induced mice. Phytomedicine 15：940 – 945. (2008)

(Published the article in Parasitol Res
affect factor：0. 39)

Hypericum perforatum Extract Therapy for Chickens Experimentally Infected with Infectious Bursal Disease Virus and its Influence on Immunity

Shang Ruofeng[1], He Cheng[2], Chen Jiongran[1], Pu Xiuying[3], Liu Yu[1], Hua Lanying[1], Wang Ling[1], Liang Jianping[1]

(1. Key Laboratory of New Animal Drug Project, Gansu Province; Key Laboratory of Veterinary Pharmaceutics Discovery, Ministry of Agriculture; Lanzhou Institute of Animal and Veterinary Pharmaceutics Science, Chinese Academy of Agricultural Sciences, Lanzhou 730050, China; 2. College of Veterinary Medicine, China Agricultural University, Beijing 10094, China; 3. Lanzhou University of Technology, Lanzhou 730050, China)

Abstract: *Hypericum perforatum* extract (HPE) has been proved a drug effective to many viral diseases. The purpose of this paper was to investigate the therapeutic efficacy and immuno-enhancement of HPE for chickens which were already challenged with infectious bursal disease virus (IBDV BC-6/85). Chickens infected with IBDV were treated with HPE for 5 consecutive days, the observation of immune organ indexes and pathological changes index, determination of IFN-a and detection of IBDV with RT-PCR were employed to assess *in vivo* whether or not HPE had the certain therapeutic efficacy on infectious bursal disease (IBD), and if HPE was able to improve the immunologic function. The results showed that 1 330 and 667.9 mg/kg body weight (BW) per day of HPE had significant therapeutic efficacy and improvement immunologic functions for chickens infected experimentally with IBDV.

Introduction

Infectious bursal disease (IBD), induced by infectious bursal disease virus (IBDV), has developed into one of the most severe infectious diseases of chickens. The virus causes acute morbidity and death through invasion of the bursae in chickens, especially those 4 to 6 weeks old; it kills mature B lymphocytes and those which are still differentiating (1, 2). For IBDV-infected chickens, it is easier to be infected with Newcastle disease, Marek's disease, infectious bronchitis, or other diseases (3). In an early stage, IBDV caused lower mortality (≤10%) in chickens; however, new epidemics show characteristics of genetic variation, and some of variants can result in 100% morbidity and mortality (4, 5). Because of the higher mutation rates of IBDV, it is difficult to use current vaccines to prevent disease caused by a new variant, especially a very virulent one.

Researchers'attention, therefore, has been drawn to the development of new drugs to prevent the outbreak and spread of IBD.

Hypericum perforatum, is one species of the genus Guttiferae that grows in Europe, West Asia, North Africa, and North America. It is used in Traditional Chinese Medicine for depression, tumors, detoxification, to stop bleeding, and for other functions (6). Recent studies show that there are many active components in *Hypericum perforatum*, such as hypericin, pesudohypericin, hyperforin, proan - thocyanidin, and rutoside. Hypericin, in particular, has been proven to be highly effective in virus inhibition (7). Many studies, therefore, have focused on the antiviral activity of HPE on lipid - enveloped and non - enveloped DNA and RNA viruses, such as human immunodeficiency virus (HIV), hepatitis virus, labialis virus, and varicellavirus (8, 9). Besides the research on antiviral activity, immunity improvement of HPE for piglets infected with porcine reproductive and respiratory syndrome virus (PRRSV) also has been done (10). There are, however, no reports on using HPE for inhibition of IBDV and immunity improvement for chickens infected with the virus. The aim of this study, therefore, was to determine whether or not HPE inhibits IBDV *in vivo* effectively and improves the immunologic function by recovering immune organs damaged by virus and inducing secretion of IFN - α.

Materials and methods

Drugs

Hypericum perforatum extract was provided by Lanzhou Institute of Animal Husbandry & Veterinary Pharmaceutics Science, Chinese Academy of Agricultural Sciences. Because hypericin is considered the main active ingredients in the extract, the standards of HPE are usually evaluated by it. The reversed - phase high performance liquid chromatography (RP - HPLC) was used to detect the content of hypericin in HPE. Hypericin was separated by a Symmetry C18 column (4. 6mm × 250mm, 5ml) with methanol - acetonitile - 1. 0% sodium dihydrogen phosphate (170 : 10 : 10) as mobile phases. The wavelength used for detection was 588 nm, the flow rate was 1. 0 ml/min, the column temperature was room temperature, and the injection volume was 20μL. In this condition, the content of hypericin was 0. 3% within HPE.

Virus

The IBDV (BC - 6/85) whose median infective dose (ID_{50}) for chickens was $10^{5.0}$/0. 05ml (eye - drop) was provided by China Institute of Veterinary Drugs Control, Beijing. The virus was diluted to 100 ID_{50}/0. 1ml and the suspension was added into 1 000 U/ml penicillin and 100 μg/ml streptomycin, respectively, and stored at - 70℃ for later use.

Animals

Twenty day - old Hailanhe chickens, weighing 130g to 150g, were purchased from a chicken farm in Gaobidian City, Hebei Province, China. To ensure that the chickens were BIDV - free, the agar gel precipitation (AGP) test was used to detect the virus at 10 - days old and 18 - days old, respectively. Chickens with virus or clinical signs of disease or abnormalities were not used in the study. All the chickens were tagged, allowed free access to water, and fed nonmedicated rations.

Treatments

Three hundred chickens were randomly divided into 6 groups. Chickens in the positive control group were only challenged with IBDV and no treatment was done in the negative control group. Chickens in the other groups were administered various dosages and drugs for 5 consecutive days and ensured that all drugs were taken. The treat – ment regimens are listed in Table 1 All chickens had free access to water during the treatments. Chickens in each group were completely isolated from the other groups and reared in isolated pens. After infection with IBDV, the feed intake, diarrhea, death rate, and the other clinical manifestations of all chickens were recorded daily over the 17 – day experiment. All procedures were performed strictly according to the legislation on the use and care of laboratory animals and the guidelines established by Beijing Laboratory Animal Research Centre.

Table 1 Animal grouping and treatment regimens

Groups	n^a	IBDV challenge			Drugs treatments		
		Times (d)	Dosage (ml)	Route	Times (d)	Dosage[b]	Route
HPE high – dose group	50	20	0. 1	eye – drop	22 ~ 26	1 330 mg/kg BW q24h	drench
HPE middle – dose group	50	20	0. 1	eye – drop	22 ~ 26	667. 9 mg/kg BW q24h	drench
HPE low – dose group	50	20	0. 1	eye – drop	22 ~ 26	333. 9 mg/kg BW, q24h	drench
Egg-yolk antibody control group	50	20	0. 1	eye – drop	22 ~ 26	0. 5 ml per chicken, q48h	IM
Positive control group	50	20	0. 1	eye – drop	—	—	—
Negative control group	50	—	—	—	—	—	—

[*] BW – body weight; IM – intramuscular.

a: At 25 d (5 d after IBDV challenge), 10 chickens from each group were selected and killed for use; the remainder from each group were treated continuously with the drugs until 26 d; b: 1 330, 667. 9, and 333. 9 mg of H. *perforatum* extract in 3 groups contained 4. 0, 2. 0, and 1. 0 mg of hypericin, respectively.

On days 5, 7, 9, 12, and 16 after virus challenge, 10 chickens (including chickens which died during the experiment) were randomly selected from each group. Each chicken was weighed and recorded, followed by collecting blood samples without contamination for IFN – α analysis. Then all the selected chickens were euthanized. The bursae, spleens, and thymuses of each chicken were obtained aseptically for determining immune organ indices, pathological change indices, and viral isolation.

Viral isolation and detection with RT – PCR

The bursae, which were obtained aseptically from 10 chickens on days 5, 7, 9, 12, and 16 after IBDV challenge, were frozen at – 40℃ and then thawed at room temperature 3 times to sufficiently destroy the cells. The bursae were ground and diluted into 10% to 20% suspensions with sterilized saline. Then suspensions were centrifugated at 3 000 r/min for 10 min, and the supernatants were collected followed by addition of antibiotics and storage at – 70℃ for reverse – transcriptase polymerase chain reaction (RT – PCR).

Reverse – transcriptase polymerase chain reaction was used to detect IBDV in the supernatants. A primer pair was designed based on the sequences of a fragment of STC gene in the GenBank using the software package DNAStar (Version 5. 0; DNAStar, Madison, Wisconsin, USA). The primers were synthesized by the Shanghai Sangon Biological Engineering Technology & Services Co., with the following sequences: IBDV - P1: 5′ - GGACCGGCGTCCATTCCG - 3′, IBDV - P2: 5′ - TGTGCTTCACCTCACTGTG - 3′. The extraction of viral RNA was performed using a Total RNA Kit (Sino - American Biotechnology Co.) according to the manufacturer's instructions. The RT - PCR was performed by routine methods. And the amplified PCR products were examined with 2.0% agarose gel (containing 0.5 mg/L EB) electrophoresis.

Determination of immune organ indices and pathological change indices

The water on surface of bursae, spleens, and thymuses of each chicken were randomly selected on days 5, 7, 9, 12, and 16 after virus challenge was soaked up, followed by weighing with electronic balance. Then bursa index, spleen index, and thymus index were calculated using the formula:

immune organ index (mg/g) = weight of the organ (mg) /body weight (g)

Meanwhile more observations were made on hemorrhage of bursae, and muscles of thoraxes and legs. The pathological changes were assessed according to the degree of hemorrhage. The total pathological change index (TPCI) of bursae or muscles was the sum of all chicken pathological change indices (PCI) in one group. The PCI of bursae were marked according to degrees of hemorrhage: severe hemorrhage count as 4 points; moderate hemorrhage count as 3 points; slight hemorrhage count as 1 point; and focal hemorrhage is 0.2. Muscle PCIs were set as: severe hemorrhage count as 3 points per side, moderate hemorrhage count as 2 points per side, and slight hemorrhage count as 1 point per side. If 2 sides both had severe hemorrhage, it was counted as 6 points, and so on. Then on days 5 to 7, 9, 12, and 16 after IBDV challenge, the TPCI of muscle of the selected 10 chickens in each group were obtained and the integrative pathological change indices (IPCI) were established according to the formula:

IPCI = TPCI of bursa + TPCI of muscle

Then the mean values of IPCI were calculated according to the data of IPCI obtained from days 5, 7, 9, 12, and 16.

IFN - α analysis

The samples which were collected on days 5, 7, 9, 12, and 16 after virus challenge were left at room temperature for 10 to 15 min followed by centrifugation at 1 500 rpm for 15 min. The sera were collected carefully for IFN - α determination by using enzyme - linked immunosorbent assay (ELISA) according to the manufacturer's instructions (RapidBio Lab, USA).

Statistical analysis

The data obtained were analyzed together with statistical software package (SAS Version 9.0; SAS Institute, Cary, North Carolina, USA) to account for the effects of HPE for chickens'PCI on day 16 and IFN - α on day 5, 7, 9, 12, and 16 after IBDV challenge. A parametric one - way analysis of variance (ANOVA) for repeated measures was used to examine intergroup differences. When a significant difference ($P < 0.05$) was found; Tukey's test was used to compare the means.

Results

Clinical findings

After infection with IBDV, a small number of chickens in positive control group showed slight clinical signs such as decreased feed intake, fluffed feathers, and mess, somnolence, catalepsies. Only a few chickens discharged white or yellow watery feces, but there were almost no symptoms in other groups. Throughout the whole experiment, no chickens died in any group.

Viral detection by RT – PCR

The supernatants used to detect virus were extracted for RNA and submitted to PCR amplification. The RT – PCR products were visualized by the agarose gel electrophoresis. The gel image acquired by Image Acquisition and Analysis System (GDS8000PC, UVP, America) were used to determine the presence of IBDV. Supernatants containing the virus should give a fragment with size of about 267 base pairs (bp).

The results of virus detection by RT – PCR were as follows: on days 5, 7, and 9 after infection, a fragment was amplified in the supernatants in all groups, except the negative control group, indicating the presence of the IBDV in the supernatants. On day 12 after infection, positive results were obtained from the supernatants made from HPE low – dose group and the positive control group, while negative results were obtained for supernatants made from the other groups. On day 16 after infection, no virus was detected in all groups (Table 2).

Table 2 Viral detection of bursae collected from the chickens after challenge

Groups	Day 5	Day 7	Day 9	Day 12	Day 16
HPE high – dose group	+ [a]	+	+	–	–
HPE middle – dose group	+	+	+	–	–
HPE low – dose group	+	+	+	+	–
Egg – yolk antibody Igy control group	+	+	+	–	–
Positive control group	+	+	+	+	–
Negative control group	–	–	–	–	–

[*] a: Symbols: + represents positive viral isolation; – represents negative viral isolation.

Results of immune organ indexes and PCI

On days 5, 7, 9, 12, and 16 after IBDV challenge, the bursa, spleen, and thymus of each chicken were selected randomly from each group and weighed respectively by electronic balance. The weight (mg) of those organs was divided by body weight (g) according to the calculation methods of the organ index. Then the data of the organ indices and the significant difference of the organ indices between all groups on day 16 after virus challenge were obtained (Figures 1 to 3). Figure 1 is a graph of bursa index of chickens in all groups at different dates after challenge. The data showed that the bursa indices of chickens in all groups gradually declined in general. But the bursa indices of chickens in the positive control group declined more significantly than those in the other groups, the

HPE low - dose group, and other groups except the negative control group. On day 16 after challenge, the bursa index of chickens in the HPE high - dose group, HPE middle - dose group, and egg - yolk antibody control group were not significantly different ($P < 0.01$) from the negative control group and HPE low - dose group. Figure 2 shows an increasing tendency of the spleen indices of the chickens in all groups, but the increasing degrees in 3 HPE dosage groups and egg - yolk antibody control group were remarkably higher than those in the negative control group and significantly lower than those in the positive control group. On day 16 after challenge, the spleen index of chickens in the HPE middle - dose group and egg - yolk antibody control group were significantly lower ($P < 0.01$) than indices in the positive control group and higher than those in HPE high - dose group, but no difference ($P < 0.01$) from those in the HPE low - dosage group. The spleen index of chickens in the negative control group was significantly lower ($P < 0.01$) than that in HPE high - dose group. Figure 3 shows a decreasing tendency in the thymus indices of chickens in all groups, but the data in the positive control group were always lower than those in the HPE low - dose group and remarkably lower than that in the other groups. On day 16 after challenge, there was no significant difference ($P < 0.01$) among HPE high - dose group, HPE middle - dose group, and egg - yolk antibody control group, except for the difference ($P < 0.05$) between the HPE high - dose group and the egg - yolk antibody control group. The thy - mus index of chickens in the HPE high - dose group was significantly lower ($P < 0.01$) than that in negative control group. The HPE low - dose group was significantly higher ($P < 0.01$) than that in positive control group, but was not significantly lower ($P < 0.01$) than that in egg - yolk antibody control group.

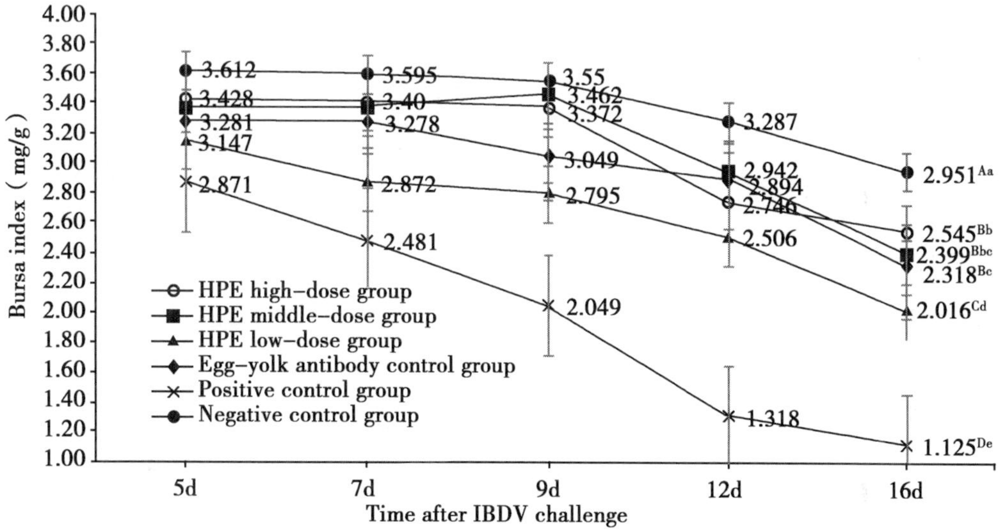

Fig. 1 The bursa index of chickens in all groups on days 5, 7, 9, 12, and 16 after IBDV challenge.

Each point and vertical bar represents the mean and standard error (n = 10). Different capital letters indicated significant difference ($P < 0.01$) and different small letters indicated difference ($P < 0.05$) at day 16 after challenge, but the same letter indicated not

The PCI of bursa and muscle were marked respectively according to the degrees of hemorrhage on days 5, 7, 9, 12, and 16 after IBDV challenge. The TPCI of different days and IPCI mean values were obtained and all the groups were arranged according to the descending order of IPCI mean

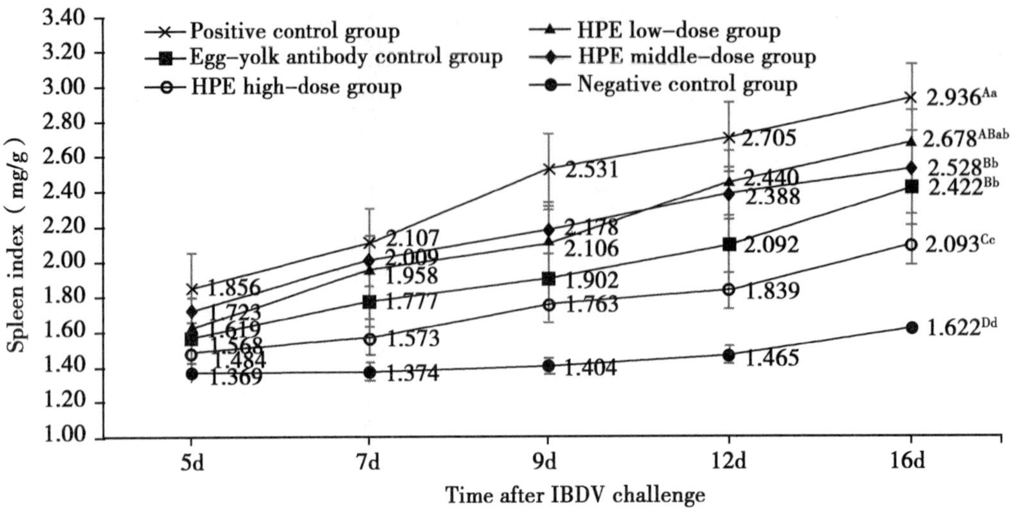

Fig. 2 The spleen index of chickens in all groups on days 5, 7, 9, 12, and 16 after IBDV challenge.

Each point and vertical bar represents the mean and standard error (n = 10). Different capital letters indicated significant difference ($P<0.01$) and different small letters indicated difference ($P<0.05$) at day 16 after challenge, but the same letter indicated not

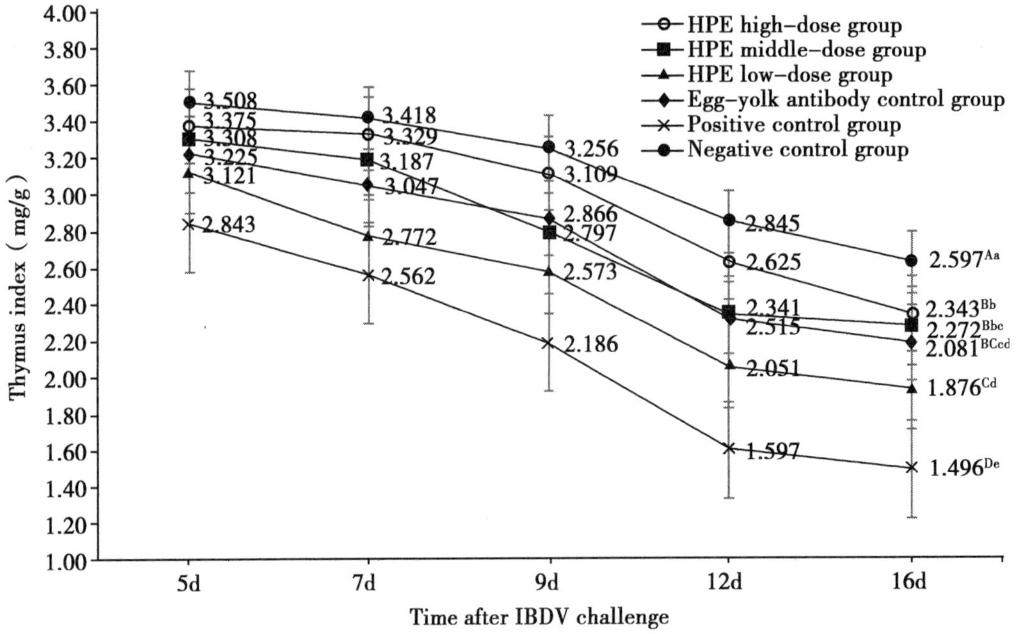

Fig. 3 The thymus index of chickens in all groups on days 5, 7, 9, 12, and 16 after IBDV challenge.

Each point and vertical bar represents the mean and standard error (n = 10). Different capital letters indicated significant difference ($P<0.01$) and different small letters indicated difference ($P<0.05$) at day 16 after challenge, but the same letter indicated not

values (Table 3). Data showed that the TPCI of bursa and muscle in each group rose slightly at the beginning then began to slide on day 7 in general except the negative control group. Of course, there are some exceptions. The TPCI and IPCI mean values in positive control group were largest. And there were significant difference （$P<0.01$） in IPCI mean values between positive control group and other groups. Between the HPE low – dose group and the egg – yolk antibody control group, IPCI mean values were no significant difference （$P<0.01$） but difference （$P<0.05$）. The IPCI mean values in the egg – yolk antibody control group were significant higher （$P<0.01$） than those in the HPE middle – dose group and the HPE high – dose group.

Table 3 Data of TPCI and IPCI

Groups	n	TPCI of bursae					TPCI of muscles					IPCI mean values[a]
		Day5	Day7	Day9	Day12	Day16	Day5	Day7	Day9	Day12	Day16	
Positive control group	10	26.2	30.0	30.0	28.0	29.0	45	48	46	47	45	74.64^{Aa}
HPE low – dose group	10	25.0	28.0	26.2	24.4	22.0	38	37	36	39	35	62.12^{Bb}
Egg – yolk antibody control group	10	21.2	25.0	24.0	23.6	20.4	39	36	38	36	33	59.24^{Bc}
HPE middle – dose group	10	15.6	15.4	17.4	17.0	16.6	35	33	36	36	33	51.00^{Cd}
HPE high – dose group	10	15.4	18.0	18.2	15.4	16.0	32	30	34	37	32	49.60^{Cd}
Negative control group	10	0.0	0.0	0.0	0.0	0.0	0	0	0	0	0	0.00^{De}

*a: IPCI mean values were marked with different capital letters were significantly different （$P<0.01$）; different small letters indicated difference （$P<0.05$）, but the same letter did not.

Results of IFN – α analysis

On days 5, 7, 9, 12, and 16 after IBDV challenge, sera of 10 chickens in each group were collected and IFN – α were determined with an ELISA. On day 7, the IFN – α in 3 HPE dosage groups and the egg – yolk antibody control group were the highest compared with that of the other days. The IFN – α in the positive control group, however, decreased over time and was significantly lower （$P<0.01$） than that in the other groups on days 5 and 7, except the negative control group which had no significant change in IFN – α through – out the experiment. On day 9, IFN – α in the positive control group was not significantly lower （$P<0.01$） than that in HPE low – dose group, and on day 12 it was significantly higher than that in other groups. The IFN – α in the HPE high – dose group were significant higher （$P<0.01$） than that in the other groups on days 5, 7, 9. On day 16 there was no significant difference among all groups （Figure 4）.

Discussion

Hypericin is considered the most active moiety in HPE and exhibits many pharmacological effects, especially in antiretroviral activity （11）. In this experiment, although the content of hypericin in HPE was no more than 0.3%, it improved the anti – IBDV effect and immuno – enhancement. It is still unknown if those effects are achieved by the hypericin only or by synergistic action of hypericin with other compounds in HPE.

Throughout the experiment, no chicken died in all groups after infection with BC – 6/85 IBDV, which indicated that the virus was unable to result in chickens' death. But BC – 6/85 IBDV caused damage or hemorrhage to bursae, muscles of thoraxes and legs, spleens, and thymuses. Therefore CPI and the detection of virus were deployed to assess the therapeutic or preventive effect

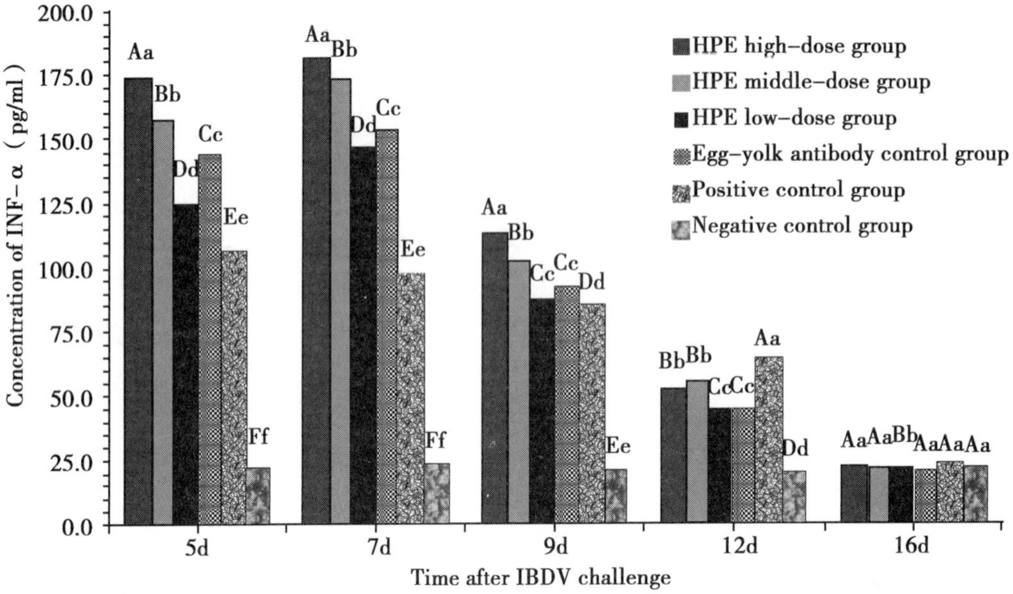

Fig. 4　Concentration of IFN－α at days 5，7，9，12，and 16 after IBDV challenge

Values are expressed as means and 5% error bars. Different capital letters indicated significant difference （$P<0.01$） and different small letters indicated difference （$P<0.05$），but the same letter indicated not

produced by drugs or vaccines against BC－6/85 IBDV.

Sensitivity and accuracy are the main benefits of RT－PCR over conventional methods （12）. There remain a number of problems associated with its use，however，especially the inherent variability of RNA （13）. Based on the conserved motifs of IBDV reported by Mundt and Muller （14），a valid pair of primers was designed and ideal PCR products were obtained. The results of viral detection showed that inhibitions of IBDV in chickens with higher dosage of HPE and egg－yolk antibody were provided to be effective.

Bursae，spleens，and thymuses are damaged to different extents when IBDV infects chickens; bursae and thymuses shrink and spleens swell. This was especially noticeable in the bursae in which B lymphocytes are destroyed and the bursae shrank with virus proliferation （15，16）. So it is necessary to assess the immune organs damage made by IBDV with those organs indices （17）. The data on bursa and thymus indices showed that the damage of bursae and thymuses of the positive group was much greater than that in the HPE dose groups and egg－yolk antibody control group on day 16 after virus challenge. The bursae and thymuses in the HPE dose groups and egg－yolk antibody control group were damaged to different extents by IBDV according to the results. The bursa and thymus indices in those groups were significantly lower （$P<0.01$） on day 16 after virus challenge than those in the negative group. Although the bursa and thymus indices in the negative group slightly decreased，it was normal for the rate of body weight gain which was higher than that of bursa or thymus over time. Spleen，served as one of peripheral immune organs，will be swollen with the activation and proliferation of T－ and B－cells induced strongly by the infection with virus. According to the spleen index in the positive and negative control groups，HPE and egg－yolk antibodies could protect the spleen to a certain extent, but not completely, from damage by IBDV.

The results of PCI showed that, except for the negative control group, the TPCI of bursa and muscle in each group rose slightly at the beginning then began to slide on day 7, but the downward trend of the positive control group was less obvious than that in the HPE treatment groups and egg - yolk antibody control group. A possible explanation for this was that HPE and egg - yolk antibody had suppressed IBDV and those tissues damaged by the virus began to recover. In addition, there was a linear relationship between the dosage of HPE and the TPCI of the bursa and muscle. The results showed no significant difference between IPCI mean values in the HPE middle - dose group compared with the HPE high - dose group, but there was a significant difference between the HPE middle - dose group and the egg - yolk antibody control group, which suggests that the HPE middle dosage [667.9 mg/kg · body weight (BW) per day] would be a candidate to treat diseases or repair the tissues damaged by IBDV.

Interferon is secreted by immune cells when stimulated by virus or other inducer and serves as an antivirus, enhancing immune response (18, 19). Determination of INF - α showed that the INF-α in the positive control group was significantly higher than that in negative control group on days 5, 7, 9, 12, and 16 after virus challenge, which indicated that the immune cells were stimulated by IBDV and had secreted a large amount of INF - α. However, the INF - α was significantly lower ($P < 0.01$) than the HPE dose groups and the egg - yolk antibody control group on days 5 and 7, which indicated that at 24 h after the last administration, INF - α was induced not only by IBDV but by HPE or egg - yolk antibody. Furthermore, the ability of the HPE middle - dose to induce INF - α was significantly higher ($P < 0.01$) than that of the egg - yolk antibody. On day 9 INF-α in HPE dose groups and egg - yolk antibody control group declined sharply, but in the positive control group it declined slightly. On day 12 INF - α in positive control group was significantly higher ($P < 0.01$) than that in other groups. The sharp decline of INF - α in HPE dose groups was likely associated with HPE concentration and the amount of IBDV in chickens.

The results of this study demonstrate that HPE has a certain therapeutic efficacy on IBD infected artificially with the virus and immuno - enhancement, especially at the dosage of 1 330 or 667.9 mg/kg BW per day.

Acknowledgments

This work was supported by National Science & Technology Pillar Program during the Eleventh Five - year Plan Period: The development and application of safety and environmental - protection veterinary drugs (2006BAD31B05). The authors thank the College of Veterinary Medicine, China Agricultural University which offered isolated pens and other experimental facilities.

References

[1] Müller H, Schnitzler D, Bernstein F, Becht H, Cornelissen D, Lütticken DH. Infectious bursal disease of poultry: Antigenic structure of the virus and control. Vet Microbiol 1992; 33: 175 - 183.

[2] Reddy SK, Silim A. Comparison of neutralizing antigens of recent isolates of infectious bursal disease virus. Arch Virol 1991; 117: 287 - 2 101.

[3] Jackwood DJ, Saif YM, Hughes JH. Nucleic acid and structural proteins of infectious bursal disease virus isolates belonging to serotypc I and II. Avian Dis 1984; 28: 990 - 1 016.

[4] Rosenberger JK, Cloud SS. Isolation and characterization of variant infectious bursal disease viruses. Proc 123rd Ann Meet Amer Vet Med Assoc Abst 1986: 81.

[5] Jackwood DH, Saif YM. Antigenic diversity of infectious bursal disease viruses. Avian Dis 1987; 31: 766 - 770.

[6] Lavagna SM, Secci D, Chimenti P, Bonsignore L, Ottaviani A, Bizzarri B. Efficacy of Hypericum and Calendula oils in the epithelial reconstruction of surgical wounds in childbirth with caesarean section. Farmaco 2001; 56: 451 - 453.

[7] Di Matteo V, Di Giovanni G. Effect of acute administration of hypericum perforatum CO_2 extract on dopamine and serotonin release in the rat central nervous system. Pharmacopsychiatry 2000; 1: 14 - 18.

[8] Tang J, Colacino JM, Larson SH, Spitzer W. Virucidal activity of hypericin against enveloped and nonenveloped DNA and RNA viruses. Antiviral Res 1990; 3: 313 - 325.

[9] Serkedjieva J, Manolova N, Zgorniak - Nowosielska I, Barbara Zawilińska, Jan Grzybek. Antiviral activity of the infusion (SHS - 174) from flowers of *Sambucus nigra* L., aerial parts of *Hypericum perforatum* L., and roots of *Saponaria officinalis* L. against influenza and herpes simplex viruses. Phytotherapy Research 1990; 4: 97 - 100.

[10] Pu Xiuying, Liang Jianping, Shang Ruofeng, *et al*. Influence of *Hypericum perforatum* extract on piglet infected with porcine respiratory and reproductive syndrome virus. Agricultural Science in China. 2009; 8: 34 - 38.

[11] Miskovsky P. Hypericin - A new antiviral and antitumor photo - sensitizer: Mechanism of action and interaction with biological macromolecules. Current Drug Targets 2002; 3: 55 - 84.

[12] Huggett J, Dhedal K, Bustin S, Zumla A. Real - time RT - PCR normalization; Strategies and considerations. Genes Immun 2005; 6: 279 - 284.

[13] Bustin SA, Nolan T. Pitfalls of quantitative real - time reverse transcription polymerase chain reaction. J Biomol Tech 2004; 15: 155 - 166.

[14] Mundt E, Muller H. Complete nucleotide sequences of 5′ - and 3′ - noncoding regions of both genome segments of different strains of infectious bursal disease virus. Virology 1995; 209: 10 - 18.

[15] Rodenberg J, Sharma JM, Balzer S, Nordgren RM, Naqi S. Flow cytometric analysis of B - cell and T - cell subpopulation in specific pathogen - free chickens infected with infectious bursal disease virus. Avian Dis 1994: 3 816 - 3 821.

[16] Tankuura N, Sharma JM. Appearance of T - cells in the bursa of Fabrieius and cecal tonsils during the acute phase of infectious bursal disease virus infection in chickens. Avian Dis. 1997; 41: 638 - 645.

[17] Rivas AL, Fabricant J. Indications of immunodepression in chickens infected with various strain of Marek's disease virus. Avian Dis 1988; 32: 1 - 8.

[18] Siegal FP, Kadowaki N, Shodell M, *et al*. The nature of the principal type I interferon - producing cells in human blood. Science 1999; 84: 1 835 - 1 837.

[19] Barnes BJ, Moore PA, Pitha PM. Virus - specific activation of novel interferon regulatory factor IRF - 5, results in the induction of distinct interferon alpha genes. Biol Chem 2001; 276: 23 382 - 23 390.

(Published the article in The Canadian Journal of Veterinary Research affect factor: 0.939)

N- (5-Chloro-1, 3-thiazol-2-yl) -2, 4-difluorobenzamide

Liu Xiwang, Li Jianyong, Zhang Han, Yang Yajun, Zhang Jiyu
(Lanzhou Institute of Animal and Veterinary Pharmaceatics Science, CAAS, Lanzhou730050, China)

Comment

Nitazoxanide, (2 - acetyloloxy - *N* - (5 - nitro - 2 - thiazolyl) benzamide), belonged to nitrothiazole analogue, was developed as a promising compound to treat both human and animal diseases (Ballard *et al.*, 2011). In this paper, we report the synthesis and structure of the title compound, which is a derivative of nitazoxanide. The conjugation between benzene ring and thiazole moiety confirmed the existance of amide anion, which is considered to directly inhibit the *PFOR* enzyme (key enzyme of central intermidiary matabolism in anaerobic organisms). The classical intermolecular hydrogen bonds N1—H1 ··· $N2^{i}$ forms centrosymmetrical dimers (Table 1). The non - classical intermolecular hydrogen bonds C4—H4···$F2^{ii}$ and C4—H4···$O3^{ii}$ stabilize molecular packing in crystal. Symmetry codes: (i) $-x+2$, $-y+1$, $-z+1$; (ii) $x+1$, y, z.

Experimental

The title compound was obtained according to routine method: to a solution of 5 - chlorothiazol - 2 - amine (1 mmol) in distilled pyridine was added a equimolar amount of 2, 4 - difluorobenzoyl chloride with stirring. When addition was complete, the reaction mixture was allowed to stand at room temperature and stirred over night. The reaction was judged complete by *TLC* analysis. The crude product then seperated on dilution was filtered out, washed with 10% $NaHCO_3$ solution, then several times with water. The dry solid was purified by chromatography to give pure compound and the crystals were obtained by recrystalization from CH_3OH.

Refinement

The positions of all H atoms were determined geometrically and refined using a riding model with C—H = 0.93Å, N—H = 0.86Å and U_{iso} (H) = 1.2U_{eq} (C, N).

Computing details

Data collection: *APEX2* (Bruker, 2008); cell refinement: *SAINT* (Bruker, 2008); data reduction: *SAINT* (Bruker, 2008); program (s) used to solve structure: *SHELXS97* (Sheldrick, 2008); program (s) used to refine structure: *SHELXL97* (Sheldrick, 2008); molecular graphics: *SHELXTL* (Sheldrick, 2008); software used to prepare material for publication: *SHELXTL* (*Sheldrick*, 2008) (Fig. 1).

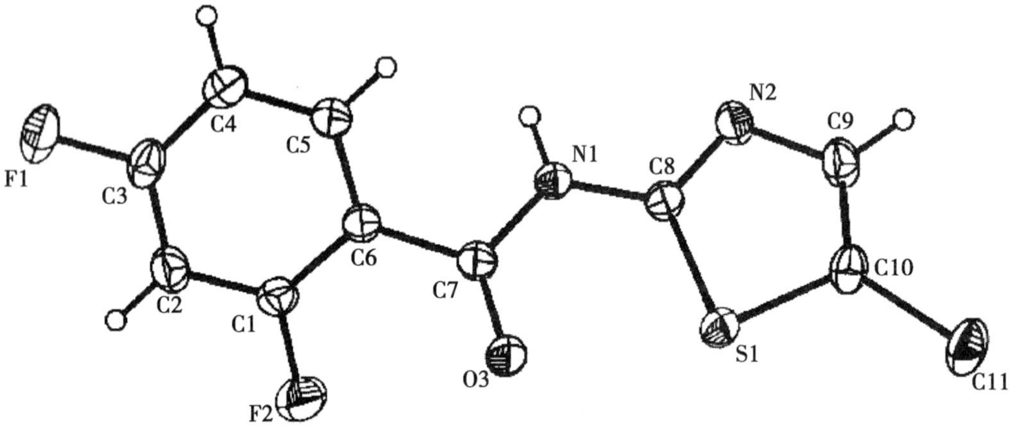

Fig. 1 The molecular structure of title compound with the atom labels

Displacement ellipsoids are drawn at the 30% probability level.

H atoms are presented as a small spheres of arbitrary radius

N- (5-Chloro-1, 3-thiazol-2-yl) -2, 4-difluorobenzamide

Crystal data

$C_{10}H_5ClF_2N_2OS$	$Z = 2$
$M_r = 274.68$	F (000) = 276
Triclinic, $P\bar{1}$	$D_x = 1.671$ Mg m^{-3}
Hall symbol: -P 1	Melting point = 428 - 429 K
$a = 6.929$ (2) Å	Mo $K\alpha$ radiation, $\lambda = 0.71073$ Å
$b = 7.330$ (2) Å	Cell parameters from 2870 reflections
$c = 12.179$ (4) Å	$\theta = 3.1 - 28.2°$
$\alpha = 101.669$ (3)°	$\mu = 0.55$ mm^{-1}
$\beta = 98.277$ (3)°	$T = 296$ K
$\gamma = 111.796$ (3)°	Block, colourless
$V = 545.9$ (3) Å^3	0.35mm × 0.33mm × 0.27 mm

Data collection

Bruker APEXII CCD	3930 measured reflections
diffractometer	1998 independent reflections

Radiation source: fine - focus sealed tube
Graphite monochromator
φ - and ω - scans
Absorption correction: multi - scan
(*SADABS*; Sheldrick, 1996)
T_{min} = 0.831, T_{max} = 0.866
1693 reflections with $I > 2\sigma(I)$
R_{int} = 0.028
θ_{max} = 25.5°, θ_{min} = 3.1°
$h = -8 \rightarrow 8$
$k = -8 \rightarrow 8$
$l = -14 \rightarrow 14$

Refinement

Refinement on F^2
Least - squares matrix: full
$R[F^2 > 2\sigma(F^2)] = 0.047$
$wR(F^2) = 0.136$
$S = 1.06$
1998 reflections
155 parameters
0 restraints
Primary atom site location: structure - invariant direct methods
Secondary atom site location: difference Fourier map
Hydrogen site location: inferred from neighbouring sites
H - atom parameters constrained
$w = 1/[\sigma^2(F_o^2) + (0.0689P)^2 + 0.4949P]$
where $P = (F_o^2 + 2F_c^2)/3$
$(\Delta/\sigma)_{max} < 0.001$
$\Delta\rho_{max} = 1.25$ e Å^{-3}
$\Delta\rho_{min} = -0.33$ e Å^{-3}
Extinction correction: *SHELXL97* (Sheldrick, 2008), $Fc^* = kFc[1 + 0.001xFc^2\lambda^3/\sin(2\theta)]^{-1/4}$
Extinction coefficient: 0.53 (3)

Special details

Geometry. All s. u. 's (except the s. u. in the dihedral angle between two l. s. planes) are estimated using the full covariance matrix. The cell s. u. 's are taken into account individually in the estimation of s. u. 's in distances, angles and torsion angles; correlations between s. u. 's in cell parameters are only used when they are defined by crystal symmetry. An approximate (isotropic) treatment of cell s. u. 's is used for estimating s. u. 's involving l. s. planes (Table 1, 2, 3, 4).

Refinement. Refinement of F^2 against ALL reflections. The weighted R - factor wR and goodness of fit S are based on F^2, conventional R - factors R are based on F, with F set to zero for negative F^2. The threshold expression of $F^2 > \sigma(F^2)$ is used only for calculating R - factors (gt) *etc*. and is not relevant to the choice of reflections for refinement. R - factors based on F^2 are statistically about twice as large as those based on F, and R - factors based on ALL data will be even larger.

Table 1 Fractional atomic coordinates and isotropic or equivalent isotropic displacement parameters (Å^2)

	x	y	z	$U_{iso}{}^*/U_{eq}$
C1	1.0517 (5)	0.7233 (5)	0.9365 (3)	0.0438 (7)
C2	1.2260 (5)	0.8083 (5)	1.0282 (3)	0.0494 (8)
H2	1.2296	0.8968	1.0958	0.059*
C3	1.3958 (5)	0.7579 (5)	1.0168 (3)	0.0467 (8)
C4	1.3945 (5)	0.6268 (5)	0.9182 (3)	0.0464 (7)
H4	1.5116	0.5954	0.9126	0.056*

(Continued)

	x	y	z	U_{iso}*/U_{eq}
C5	1.2153 (5)	0.5432 (5)	0.8280 (3)	0.0413 (7)
H5	1.2120	0.4530	0.7612	0.050*
C6	1.0387 (4)	0.5896 (4)	0.8336 (2)	0.0358 (6)
C7	0.8392 (5)	0.4900 (4)	0.7407 (2)	0.0390 (7)
C8	0.6940 (4)	0.3522 (4)	0.5350 (2)	0.0365 (6)
C9	0.5284 (5)	0.2099 (5)	0.3514 (3)	0.0505 (8)
H9	0.5166	0.1733	0.2722	0.061*
C10	0.3587 (5)	0.1551 (4)	0.3969 (3)	0.0447 (7)
C11	0.09389 (14)	0.01753 (14)	0.32522 (8)	0.0662 (4)
F1	1.5714 (3)	0.8434 (3)	1.10528 (18)	0.0673 (6)
F2	0.8890 (3)	0.7783 (4)	0.9458 (2)	0.0796 (8)
N1	0.8646 (4)	0.4602 (4)	0.62983 (19)	0.0376 (6)
H1	0.9916	0.5106	0.6191	0.045V*
N2	0.7224 (4)	0.3246 (4)	0.4302 (2)	0.0449 (6)
03	0.6609 (3)	0.4314 (4)	0.75914 (18)	0.0553 (6)
S1	0.43196 (11)	0.24360 (11)	0.54607 (6)	0.0421 (3)

Table 2 Atomic displacement parameters (Å^2)

	U^{11}	U^{22}	U^{33}	U^{12}	U^{13}	U^{23}
C1	0.0358 (15)	0.0438 (16)	0.0480 (17)	0.0122 (13)	0.0149 (13)	0.0095 (13)
C2	0.0483 (18)	0.0462 (17)	0.0386 (16)	0.0079 (14)	0.0089 (14)	0.0038 (13)
C3	0.0365 (15)	0.0494 (18)	0.0420 (16)	0.0035 (13)	0.0018 (12)	0.0198 (14)
C4	0.0377 (15)	0.0559 (19)	0.0504 (18)	0.0196 (14)	0.0119 (14)	0.0236 (15)
C5	0.0396 (15)	0.0447 (16)	0.0393 (15)	0.0156 (13)	0.0128 (12)	0.0123 (13)
C6	0.0329 (13)	0.0370 (14)	0.0339 (14)	0.0084 (11)	0.0107 (11)	0.0119 (11)
C7	0.0377 (15)	0.0413 (15)	0.0368 (15)	0.0135 (12)	0.0117 (12)	0.0121 (12)
C8	0.0339 (13)	0.0349 (14)	0.0372 (14)	0.0094 (11)	0.0101 (11)	0.0110 (11)
C9	0.0490 (18)	0.0455 (17)	0.0361 (16)	0.0020 (14)	0.0049 (13)	0.0052 (13)
C10	0.0414 (16)	0.0335 (15)	0.0440 (16)	0.0049 (12)	0.0006 (13)	0.0066 (12)
C11	0.0435 (5)	0.0572 (6)	0.0666 (6)	-0.0003 (4)	-0.0052 (4)	0.0058 (4)
F1	0.0444 (11)	0.0784 (14)	0.0528 (12)	0.0031 (10)	-0.0082 (9)	0.0204 (10)
F2	0.0495 (12)	0.0830 (16)	0.0895 (17)	0.0280 (11)	0.0141 (11)	-0.0104 (13)
N1	0.0304 (11)	0.0443 (13)	0.0336 (12)	0.0099 (10)	0.0091 (10)	0.0111 (10)
N2	0.0423 (13)	0.0449 (14)	0.0341 (13)	0.0058 (11)	0.0096 (11)	0.0062 (11)
03	0.0349 (11)	0.0801 (17)	0.0385 (12)	0.0114 (11)	0.0119 (9)	0.0125 (11)
S1	0.0327 (4)	0.0449 (5)	0.0426 (5)	0.0098 (3)	0.0088 (3)	0.0114 (3)

Table 3 Geometric parameters (Å, °)

C1—F2	1.343 (4)	C7—O3	1.220 (3)
C1—C2	1.366 (4)	C7—N1	1.371 (4)
C1—C6	1.393 (4)	C8—N2	1.306 (4)
C2—C3	1.374 (5)	C8—N1	1.379 (4)
C2—H2	0.9300	C8—S1	1.729 (3)
C3—F1	1.348 (3)	C9—C10	1.334 (5)
C3—C4	1.374 (5)	C9—N2	1.378 (4)
C4—C5	1.376 (4)	C9—H9	0.9300
C4—H4	0.9300	C10—C11	1.719 (3)
C5—C6	1.394 (4)	C10—S1	1.730 (3)
C5—H5	0.9300	N1—H1	0.8600
C6—C7	1.480 (4)		
F2—C1—C2	117.5 (3)	O3—C7—N1	120.7 (3)
F2—C1—C6	119.1 (3)	O3—C7—C6	123.3 (3)
C2—C1—C6	123.4 (3)	N1—C7—C6	116.0 (2)
C1—C2—C3	117.4 (3)	N2—C8—N1	121.3 (2)
C1—C2—H2	121.3	N2—C8—S1	115.8 (2)
C3—C2—H2	121.3	N1—C8—S1	122.9 (2)
F1—C3—C4	119.0 (3)	C10—C9—N2	115.1 (3)
F1—C3—C2	118.5 (3)	C10—C9—H9	122.5
C4—C3—C2	122.5 (3)	N2—C9—H9	122.5
C3—C4—C5	118.3 (3)	C9—C10—C11	127.8 (3)
C3—C4—H4	120.9	C9—C10—S1	111.6 (2)
C5—C4—H4	120.9	C11—C10—S1	120.56 (19)
C4—C5—C6	122.0 (3)	C7—N1—C8	122.4 (2)
C4—C5—H5	119.0	C7—N1—H1	118.8
C6—C5—H5	119.0	C8—N1—H1	118.8
C1—C6—C5	116.4 (3)	C8—N2—C9	110.0 (3)
C1—C6—C7	120.8 (3)	C8—S1—C10	87.44 (14)
C5—C6—C7	122.6 (3)		
F2—C1—C2—C3	-177.5 (3)	C1—C6—C7—N1	145.0 (3)
C6—C1—C2—C3	0.2 (5)	C5—C6—C7—N1	-40.5 (4)
C1—C2—C3—F1	178.6 (3)	N2—C9—C10—C11	178.5 (2)
C1—C2—C3—C4	-0.3 (5)	N2—C9—C10—S1	-0.6 (4)
F1—C3—C4—C5	-179.1 (3)	O3—C7—N1—C8	-4.4 (4)
C2—C3—C4—C5	-0.1 (5)	C6—C7—N1—C8	173.7 (2)
C3—C4—C5—C6	0.8 (4)	N2—C8—N1—C7	-179.6 (3)
F2—C1—C6—C5	178.1 (3)	S1—C8—N1—C7	-0.1 (4)
C2—C1—C6—C5	0.4 (4)	N1—C8—N2—C9	179.1 (3)
F2—C1—C6—C7	-7.1 (4)	S1—C8—N2—C9	-0.4 (3)
C2—C1—C6—C7	175.2 (3)	C10—C9—N2—C8	0.7 (4)
C4—C5—C6—C1	-0.9 (4)	N2—C8—S1—C10	0.1 (2)
C4—C5—C6—C7	-175.6 (3)	N1—C8—S1—C10	-179.4 (3)
C1—C6—C7—O3	-37.0 (4)	C9—C10—S1—C8	0.3 (3)
C5—C6—C7—O3	137.5 (3)	C11—C10—S1—C8	-178.9 (2)

Table 4　Hydrogen – bond geometry（Å,°）

D—H···*A*	*D*—H	H···*A*	*D*···*A*	*D*—H···*A*
N1—H1···N2[i]	0.86	2.15	2.988（3）	166
C4—H4···F2[ii]	0.93	2.38	3.127（4）	137
C4—H4···O3[ii]	0.93	2.56	3.329（4）	140

Symmetry codes：（i）$-x+2$，$-y+1$，$-z+1$；（*ii*）$x+1$，y，z.

（Published the article in supplementary materials affect factor：0.347）

常山总碱的亚急性毒性试验

郭志廷[1,2]，韦旭斌[2]，梁剑平[1]，郭文柱[1]，王学红[1]，尚若锋[1]，郝宝成[1]

（1. 中国农业科学院兰州畜牧与兽药研究所，农业部兽用药物创制重点实验室，甘肃省新兽药工程重点实验室，兰州 730050；2. 吉林大学 畜牧兽医学院，长春 130062）

摘　要：为评价常山总碱的安全性和今后系统研究其药理作用以及为临床安全用药提供试验依据，在小鼠急性毒性试验的基础上，进行亚急性毒性试验。80 只大鼠随机分成常山总碱高、中、低剂量组和空白对照组，药物组按照大鼠体质量 6.50，3.25，1.63g/kg 的剂量灌胃给药，连续给药 14d，所有大鼠于给药后分 2 次（7，14d）扑杀。结果表明，常山总碱低、中剂量组大鼠的体质量、脏器系数、血液学指标、血液生化指标、组织病理学变化等各项指标与对照组比较均无显著性差异（$P>0.05$）；高剂量组对上述部分指标有一定影响，需要进一步研究。本试验结果提示，常山总碱的毒性很低，临床用药较为安全。

关键词：常山总碱；亚急性毒性试验；抗球虫；大鼠

中药常山为虎耳草科植物常山（*Dichroa febrifuga* Lour）的干燥根，味苦辛，性寒，具有杀虫、截疟、祛痰之功效[1~3]。常山碱为常山中的主要有效组分包括常山碱和异常山碱，这 2 种成分占常山总碱的 90% 以上。常山碱及其衍生物（如常山酮）具有高效的抗球虫活性和良好的抗肿瘤活性，近年逐渐引起药物学家的广泛兴趣[4~7]。然而，常山及常山碱作为一种新型抗球虫药物，国内外研究仅停留在单味原药粉碎或几味中药简单组方后直接添加在饲料中防治球虫病的水平，对于常山碱的系统提取分离、有效成分间的构效关系和抗球虫作用及机理研究甚少[8~10]。前期我们已完成对常山总碱的提取分离和急性毒性试验，药理试验发现其对鸡柔嫩艾美尔球虫人工感染的球虫病具有良好的治疗作用。然而，常山是中华人民共和国兽药典（2005 年版）中规定的有毒药材，为了评价常山总碱的安全性，以便今后全面开展药理、药效研究，特进行常山总碱对大鼠的亚急性毒性试验。

1　材料与方法

1.1　试验药品

常山总碱由农业部兽用药物创制重点实验室制备，浅黄色粉末。制备主要过程为：常山粉碎后，加入 1% 盐酸，超生萃取 0.5h，过滤碱化，氯仿萃取，浓缩蒸干后即得常山总碱。试验时用三蒸水配成一定浓度的药物混悬液。

1.2　试验动物

Wistar 大白鼠 80 只，购自兰州大学医学院动物中心。体质量 100 ~ 130g，生产合格号：

[scxk（甘）2008－0075]，雌雄各半，雌性未产无孕，实验室条件下饲养观察1周，自由采食与饮水，健康状况良好。

1.3 主要仪器

Coulter－JT 全自动血细胞分析仪（美国 Coulter 公司），7150 型生化检测仪（日立公司），RC－5C Plus 离心机（德国 Heraeus 公司）。

1.4 分组与给药

所有大鼠随机分成4组，每组20只，雌雄各半，分别为常山总碱高、中、低剂量组和空白对照组，药物组按照大鼠体质量6.50，3.25，1.63g/kg 的剂量灌胃给药，连续给药14d。空白对照组不给药。试验期间所有大鼠自由采食，饮水。

1.5 临床症状与体质量检查

每天观察并记录各组大鼠的饮食状况、精神状态、行为活动及其他临床症状。所有大鼠给药前、给药后7，14d 均测体质量，计算不同阶段的平均体质量。

1.6 血液生理生化指标测定

参照有关文献[11]，所有大鼠于给药后7d，14d 扑杀，借助相关仪器对血液生理、生化指标进行分析，包括白细胞数（WBC）、红细胞数（RBC）、血红蛋白（HGB）、红细胞压积（HCT）、红细胞平均血红蛋白浓度（MCHC）、谷丙转氨酶（ALT）、谷草转氨酶（AST）、总蛋白（TP）、总胆红素（TBIL）、尿素氮（BUN）、肌酐（CREA）。

1.7 组织病理学观察

大鼠扑杀后，观察主要脏器的大小、形态、色泽、质感以及有无病变等，计算脏器系数（脏器系数＝（脏器质量/体质量）×100%）。同时重点采集肺脏、肝脏和肾脏，经10%中性福尔马林溶液固定，石蜡包埋切片，HE 染色，做病理组织学检查。

1.8 统计学分析

数据采用 $\bar{x} \pm s$ 表示，应用 SPSS13.0 软件对数据进行多重比较和方差统计分析。

2 结果与分析

2.1 常山总碱对大鼠临床症状及体质量的影响

试验中发现常山总碱高、中、低剂量组大鼠的进食量、饮水量、精神状态、活动状况、被毛光泽度和体温均无明显异常变化。与空白对照组比较，常山总碱高剂量组对大鼠体质量增长有明显的抑制作用（$P<0.05$），其余2组变化不显著（$P>0.05$）（表1）。

表1 常山总碱对大鼠体质量的影响 (g)

$t_{给药}$ (d)	性别	对照组	高剂量组	中剂量组	低剂量组
0	♂	135.21±7.21	151.63±8.45	145.03±8.26	152.41±8.37
	♀	148.07±8.42	142.39±6.12	149.74±7.31	150.06±9.31
7	♂	160.19±12.63	154.86±5.38*	162.18±11.93	161.31±10.15
	♀	163.72±12.79	152.19±6.13*	160.18±11.96	159.98±11.78
14	♂	180.29±20.17	158.27±7.21*	179.31±18.37	183.79±20.28

（续表）

$t_{给药}$（d）	性别	对照组	高剂量组	中剂量组	低剂量组
	♀	183.85 ±28.31	160.83 ±7.29*	175.65 ±16.84	185.08 ±19.73

注：与空白对照组比较，* $P<0.05$。下同。

2.2 常山总碱对大鼠血液学指标的影响

与空白对照组比较，常山总碱高、中、低剂量组对大鼠血液中 WBC、RBC、HGB、HCT、MCHC 的影响不显著（$P>0.05$）（表2）。

表 2 常山总碱对大鼠血液学指标的影响（$\bar{x}\pm s$，$n=8$）

$t_{给药}$（d）	指标	对照组	高剂量组	中剂量组	低剂量组
7	WBC/（$\times10^9\cdot L^{-1}$）	7.32 ±0.22	7.98 ±2.86	9.06 ±1.20	8.99 ±1.48
	RBC/（$\times10^{12}\cdot L^{-1}$）	6.47 ±0.48	6.19 ±0.50	6.13 ±0.86	6.16 ±0.43
	HGB/（$g\cdot L^{-1}$）	184.60 ±6.43	170.60 ±2.52	176.00 ±14.10	171.30 ±5.03
	HCT/（$L\cdot L^{-1}$）	0.47 ±0.05	0.43 ±0.03	0.48 ±0.08	0.42 ±0.02
	MCHC/（$g\cdot L^{-1}$）	349.80 ±8.31	351.20 ±9.42	357.10 ±8.47	349.10 ±10.25
14	WBC/（$\times10^9\cdot L^{-1}$）	7.79 ±1.05	5.66 ±2.03	5.72 ±2.21	6.45 ±2.56
	RBC/（$\times10^{12}\cdot L^{-1}$）	6.22 ±0.44	6.00 ±1.94	5.04 ±1.42	6.67 ±0.52
	HGB/（$g\cdot L^{-1}$）	164.21 ±7.23	165.36 ±3.25	168.19 ±6.37	169.25 ±6.52
	HCT/（$L\cdot L^{-1}$）	0.52 ±0.05	0.42 ±0.02	0.46 ±0.04	0.49 ±0.01
	MCHC/（$g\cdot L^{-1}$）	343.82 ±8.46	358.28 ±9.42	356.16 ±8.52	347.18 ±10.17

2.3 常山总碱对大鼠血液生化指标的影响

与空白对照组比较，常山总碱高剂量组大鼠血液中 ALT、TP、BUN 水平显著降低（$P<0.05$），AST、CREA 水平显著升高（$P<0.05$），TBIL 水平变化不显著（$P>0.05$）；常山碱中、低剂量组大鼠上述血液生化指标变化不显著（$P>0.05$）（表3）。

表 3 常山总碱对大鼠血液生化指标的影响（$\bar{x}\pm s$，$n=8$）

$t_{给药}$（d）	指标	对照组	高剂量组	中剂量组	低剂量组
7	TBIL（μmol/L）	3.63 ±0.68	4.33 ±1.01	4.03 ±0.25	3.30 ±0.46
	ALT（U/L）	65.00 ±7.55	58.00 ±8.00*	60.33 ±10.12	70.33 ±16.44
	AST（U/L）	228.30 ±20.50	410.30 ±41.72*	209.30 ±38.37	253.30 ±29.50
	TP（g/L）	63.50 ±1.73	55.93 ±6.36*	64.77 ±1.29	64.60 ±1.74
	CREA（μmol/L）	65.04 ±19.53	78.19 ±20.12*	70.23 ±23.17	69.17 ±20.52
	BUN（mmol/L）	10.31 ±1.53	8.64 ±2.03*	10.23 ±2.18	10.46 ±1.62
14	TBIL（μmol/L）	3.64 ±0.68	4.23 ±1.01	4.03 ±0.25	3.30 ±0.46
	ALT（U/L）	65.00 ±7.55	58.00 ±8.00*	60.33 ±10.12	70.33 ±16.44
	AST（U/L）	228.30 ±20.50	410.30 ±417.25*	209.30 ±38.37	253.30 ±29.50
	TP（g/L）	63.50 ±1.73	55.93 ±6.36*	64.77 ±1.29	64.60 ±1.74

（续表）

$t_{给药}$（d）	指标	对照组	高剂量组	中剂量组	低剂量组
	CREA（μmol/L）	65.04 ±19.53	78.19 ±20.12 *	70.23 ±23.17	69.17 ±20.52
	BUN（mmol/L）	10.31 ±1.53	8.64 ±2.03 *	10.23 ±2.18	10.46 ±1.62

2.4 常山总碱对大鼠脏器系数的影响

与空白对照组比较，给药7d后常山总碱高剂量组的心、肝、脾、肺、肾、胸腺的脏器系数有显著性改变（$P<0.05$），中、低剂量组无显著性改变（$P>0.05$）；给药14d后，常山总碱各剂量组的脏器系数均无显著性改变（$P>0.05$）（表4）。

表4　常山总碱对大鼠脏器系数的影响（$\bar{x} \pm s$，n=8）

$t_{给药}$（d）	脏器	性别	对照组	高剂量组	中剂量组	低剂量组
7	心	♂	0.391 ±0.041	0.435 ±0.021 *	0.374 ±0.025	0.386 ±0.027
		♀	0.385 ±0.031	0.442 ±0.017	0.391 ±0.028	0.357 ±0.029
	肺	♂	0.716 ±0.095	0.853 ±0.097 *	0.724 ±0.053	0.718 ±0.018
		♀	0.743 ±0.075	0.921 ±0.085 *	0.791 ±0.091	0.749 ±0.098
	肝	♂	2.870 ±0.460	4.210 ±0.510 *	3.680 ±0.610	3.170 ±0.450
		♀	3.480 ±0.320	4.510 ±0.370	3.610 ±0.290	3.570 ±0.380
	肾	♂	0.759 ±0.105	0.512 ±0.018 *	0.751 ±0.062	0.721 ±0.082
		♀	0.782 ±0.013	0.581 ±0.079 *	0.784 ±0.062	0.783 ±0.042
	脾	♂	0.219 ±0.072	0.197 ±0.062	0.238 ±0.058	0.216 ±0.028
		♀	0.213 ±0.031	0.187 ±0.071 *	0.278 ±0.048	0.246 ±0.017
	胸腺	♂	0.163 ±0.053	0.121 ±0.016	0.158 ±0.029	0.166 ±0.044
		♀	0.159 ±0.047	0.118 ±0.054 *	0.160 ±0.021	0.168 ±0.042
14	心	♂	0.391 ±0.041	0.435 ±0.021	0.374 ±0.025	0.386 ±0.027
		♀	0.385 ±0.031	0.442 ±0.017	0.391 ±0.028	0.357 ±0.029
	肺	♂	0.716 ±0.095	0.853 ±0.097	0.724 ±0.053	0.718 ±0.018
		♀	0.751 ±0.073	0.921 ±0.085	0.791 ±0.091	0.749 ±0.098
	肝	♂	2.870 ±0.460	4.210 ±0.510	3.680 ±0.610	3.170 ±0.450
		♀	3.480 ±0.320	4.510 ±0.370	3.610 ±0.290	3.570 ±0.380
	肾	♂	0.759 ±0.105	0.512 ±0.018	0.751 ±0.062	0.721 ±0.082
		♀	0.782 ±0.013	0.581 ±0.079	0.784 ±0.062	0.783 ±0.042
	脾	♂	0.219 ±0.072	0.197 ±0.062	0.238 ±0.058	0.216 ±0.028
		♀	0.213 ±0.031	0.187 ±0.071	0.278 ±0.048	0.246 ±0.017
	胸腺	♂	0.163 ±0.053	0.121 ±0.016	0.158 ±0.029	0.166 ±0.044
		♀	0.159 ±0.047	0.118 ±0.054	0.160 ±0.021	0.168 ±0.042

2.5 常山总碱对大鼠组织病理学变化的影响

总体损伤程度：常山总碱高剂量组 > 中剂量组 > 低剂量组 > 空白对照组。高剂量组的肺脏有一定程度的淤血和轻度的间质性肺炎，中剂量组肺脏轻度淤血，低剂量组有 1 例出现严重的支气管肺炎及小叶融合性肺炎；高剂量组的肾小管上皮细胞出现广泛变性坏死，中剂量组肾细胞出现不同程度的变性，低剂量组肾细胞均有轻度的颗粒变性；高剂量组的肝细胞均变性，甚至发生坏死、淤血，中剂量组的肝细胞出现不同程度的变性，低剂量组的肝细胞多数均有颗粒变性，少数可能发展到水泡变性（图）。

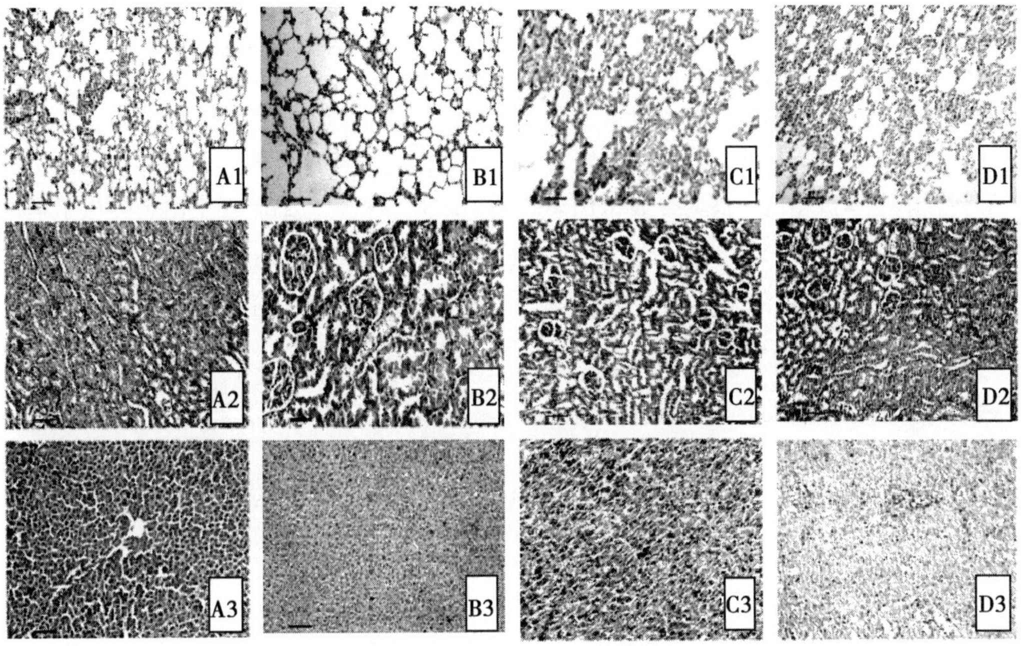

A. 空白对照组；B. 低剂量组；C. 中剂量组；D. 高剂量组。

A1 ~ D1. 肺脏；A2 ~ D2. 肾脏；A3 ~ D3. 肝脏

图　常山总碱对大鼠组织病理学变化的影响（HE 染色 ×200）

3　讨论

球虫病是由球虫寄生于畜禽的肠管和胆管上皮细胞内，引起下痢和营养障碍的一种原虫病[12~13]，以幼龄畜禽受害最严重，死亡率可达 80% 以上。随着抗球虫化学药物的广泛应用，球虫耐药性、药物残留等问题日益突出[14~15]；因此，安全、高效、环保的新型抗球虫药物研发势在必行。常山碱最早是从我国中药常山中提取出的小分子物质，属于喹唑酮类化合物，现主要用于家禽、家畜抗球虫病、抗疟疾的预防和治疗[16~18]。

亚急性毒性试验是药物安全评价的主要内容之一，是能否过渡到临床试用的主要依据。一般通过连续反复给药，观察试验动物出现的毒性反应、血液学、血液生化及病理学的改变，分析剂量 - 毒性效应的关系，主要靶器官毒性反应的性质和程度，毒性反应的可逆性以及是否有蓄积毒性等。在本试验中，与空白对照组比较常山总碱高剂量组大鼠的体质量有所减轻，肝脏、肾脏脏器系数明显增高，中剂量组和低剂量组均无显著变化；常山总碱各剂量组大鼠精神状况、大小便、饮食、体温等指标均无明显变化。实验动物的体质量变化是反映动物机体中毒效应和毒物毒性的最基本指标，脏器系数是提供受试化合物靶器官的重要指标。以上结果说明，常山总碱毒性很低，但随着给药剂量的不断加大，毒性有所增强。

药理研究表明，通过检测机体血液相关生理、生化指标在受试药物作用下有无改变，可以更加准确衡量受试药物的毒性大小。本试验中常山总碱各剂量组大鼠 RBC、WBC、HGB 等血液生理指标基本未受影响；高剂量组对大鼠血液生化指标 ALT、TP、BUN 产生有明显的抑制作用，中低剂量组均无明显异常变化，说明常山总碱剂量太高时毒性反应增强，临床用药需在安全剂量范围内。病理组织学检查可以更加直观地反映药物的毒性大小，试验发现随着常山总碱剂量不断增加，脏器损伤程度逐渐严重，甚至出现不同程度的淤血、变性和坏死。在结果判定当中，肝脏和肾脏比较准确，而肺脏相对代表性比较差。试验中个别肺脏出现严重的出血、淤血，可能是由于灌药不当造成的损伤所致。

综上分析，结合前期急性毒性试验结果，我们发现常山总碱的毒性总体上很低，临床用药安全性高。但是，并不代表常山总碱没有毒性，剂量过高或不合理用药同样会对动物健康产生不利影响。本试验的开展为今后常山总碱的系统开发利用、临床安全用药提供了理论依据，至于具体到常山总碱中某种单体成分的毒性和药理活性还不太清楚，尚需进一步研究解决。

参考文献

[1] 李燕，刘明川，金林红，等．常山化学成分及生物活性研究进展［J］．广州化工，2011，39（9）：7－9.

[2] Kaur K, Jain M, Kaur T, et al. Antimalarials from nature [J]. Bioorganic Med Chem, 2009, 17: 3 229－3 256.

[3] Jiang S, Zeng Q, Gettayacamin M, et al. Antimalarial activities and therapeutic properties of febrifugine analogs [J]. Antimicrob Agents Chemother, 2005, 49 (3): 1 169－1 176.

[4] 张雅，李春，雷国莲．常山化学成分研究［J］．中国实验方剂学杂志，2010，16（5）：40－41.

[5] Nagler A, Ohana M, Shibolet O, et al. Suppression of hepatocellular carcinoma growth in mice by the alkaloid coccidiostat halofuginone [J]. European J Cancer, 2004, 40: 1 397－1 403.

[6] Zhu S, Zhang Q, Gudise C, et al. Synthesis and biological evaluation of febrifugine analogues as potential antimalarial agents [J]. Bioorganic Med Chem, 2009, 17: 4 496－4 502.

[7] 晋爱兰，张供领，万双秀，等．中草药对肉鸡球虫病临床病例的疗效研究［J］．中国畜牧兽医，2010，37（7）：208－209.

[8] 金光明，丁希强．直杀球虫散与强力灭球王对鸡柔嫩艾美耳球虫的疗效观察［J］．中国畜牧兽医，2004，31（12）：42－44.

[9] 宋世荣，李良荣，高学军，等．常山粉对鸡球虫病的防治效果［J］．中国兽医科技，2000，30（5）：37－38.

[10] 林青，李林海，周宏超，等．中药复方制剂预防鸡巨型艾美耳球虫感染的效果观察［J］．中国兽医学报，2011，31（2）：194－197.

[11] 路浩，孙莉莎，赵宝玉，等．哈密黄芪对小鼠的亚急性毒性试验［J］．中国兽医学报，2010，30（7）：982－987.

[12] 闫文朝，王天奇，索勋，等．家兔球虫病的研究进展［J］．中国兽医科学，2010，40（11）：1 200－1 205.

[13] Baker D G. Flynns parasitology of labaoratory animals [M]. 2ed. Oxford: Blackwell Publishing Company, 2007: 454－458.

[14] Dalloul R A, Lillehoj H S. Poultry coccidiosis: recent advancements in control measures and vaccine development [J]. Expert Rev, 2006 (5): 143－163.

[15] 安健，汪明，王黎霞，等．球虫抗药性分子生物学检测技术的建立［J］. 中国兽医学报，2007，27（3）：340－342.
[16] 张越，杜会茹，王永国，等．常山碱和异常山碱的合成研究进展［J］. 河北师范大学学报：自然科学版，2008，32（4）：510－515.
[17] Ye W，You J H，Li X，et al. Design，synthesis and anticoccidial activity of a series of 3－（2－（2－methoxyphenyl）－2－oxoethyl）quinazolinone derivatives［J］. Pesticide Biochem Physiol，2010，97：194－198.
[18] Deng H，Xu S，Ye Y. A new quinazolone alkaloid from leaves of *Dichroa febrifuga*［J］. J Chin Pharm Sci，2000，9（3）：116－118.

（发表于《中国兽医学报》）

酵母多糖对环磷酰胺所致免疫损伤大鼠的拮抗作用

王　慧[1,2]，张　霞[3]，程富胜[1]，罗永江[1]，董鹏程[1]

（1. 中国农业科学院兰州畜牧与兽药研究所，农业部兽用药物创制重点实验室/甘肃省新兽药工程重点实验室，兰州 730050；2. 甘肃农业大学动物医学院，兰州 730070；3. 甘肃农业大学生命科学技术学院，兰州 730070）

摘　要：为研究酵母多糖（YP）对环磷酰胺（CTX）所致免疫损伤大鼠的拮抗作用，本实验将 80 只雄性 SD 大鼠随机分为 YP（50 mg/kg）＋CTX 组、YP（100 mg/kg）＋CTX 组、YP（200 mg/kg）＋CTX 组、CTX 对照组和正常对照组。YP＋CTX 组按剂量灌胃并称重，CTX 和正常对照组则灌胃给予等量生理盐水，连续给药 10d；在第 8d、9d，除正常对照组外，其余 4 组腹腔注射 CTX 100 mg/kg。第 11d 采血及对相关的免疫器官组织进行检测。结果显示，YP（100 mg/kg）组平均日增重和饲料报酬比正常对照组显著增加（$P<0.05$）；胸腺指数，血清中 IgA、IgG、表皮生长因子（EGF）、碱性磷酸酶（AKP）含量及空肠 SIgA 水平显著（$P<0.05$）或极显著（$P<0.01$）高于 CTX 对照组；结肠壁中前列腺素 E2（PGE2）含量与 CTX 对照组相比显著降低（$P<0.05$）。以上结果表明，YP 能提高大鼠的生长性能；YP 对 CTX 所致免疫损伤具有一定的拮抗保护作用，其中 100 mg/kg 剂量的 YP 效果最为显著。

关键词：酵母多糖；免疫增强；环磷酰胺；黏膜免疫；大鼠

多糖又称多聚糖（Polysaccharides），由 10 个以上寡糖通过糖苷键连接而成。研究发现，天然多糖具有免疫调节、抗病毒、抗肿瘤、抗凝血、抗氧化、降血糖等多种生物活性[1~4]，是医药和食品功能化学领域共同关注的研究焦点。微生物多糖是微生物发酵工业的新产品，具有安全无毒、生产周期短、理化性质独特、用途广泛、易与菌体分离及可通过深层发酵实现工业化生产等优良特性而备受人们的关注[5]。酵母是微生物多糖生产的来源之一，目前人们对酵母多糖（Yeast polysaccharide，YP）的研究多集中在抗肿瘤、抗氧化、抗诱变、营养等方面[6~8]。关于 YP 对大鼠免疫调节的作用及对肠道黏膜免疫功能影响的研究鲜见报道。本实验以免疫抑制大鼠为模型，灌服一定量的 YP 溶液，通过检测大鼠生长性能、免疫器官指数、血清中部分免疫指标以及肠黏膜免疫相关指标的变化来分析 YP 对环磷酰胺致免疫损伤大鼠免疫功能的影响，旨在明确 YP 的免疫调节作用并揭示其可能的机制，为 YP 作为饲料添加剂和免疫增强剂提供实验依据。

1　材料和方法

1.1　材料与试剂

环磷酰胺（Cytoxan，CTX）购自山西普德药业有限公司；碱性磷酸酶（AKP）、溶菌酶

(LZM) 试剂盒购自南京建成生物工程研究所; IgA、IgG、表皮生长因子(EGF)、前列腺素 E2 (PGE2) ELISA 试剂盒均为美国 Sigma 公司产品; YP 为本实验室提取、纯化。

1.2 实验分组及方法

SD 大鼠 80 只,雄性,体质量 220 g ± 10 g,由兰州大学实验动物中心提供。随机分为 YP (50 mg/kg) + CTX 组、YP (100 mg/kg) + CTX 组、YP (200 mg/kg) + CTX 组、CTX 对照组、正常对照组,每组 16 只。普通饲料适应性喂养 1 周后,YP + CTX 组按剂量每天定时灌胃,正常对照组与 CTX 模型组灌胃给予等量生理盐水,连续给药 10d,各组每天记录进食量并称重;分别在第 8d、9d,除正常对照组注射给予生理盐水外,其余 4 组腹腔注射 CTX 100 mg/kg。

1.3 样品的采集与处理

于第 11d 试验结束时,大鼠股动脉采血,分离血清置 -20℃冰箱备用。取脾脏、胸腺,剔除脂肪后称重。取 10cm 空肠末端,用 5ml PBS 反复清洗,收集清洗液,-20℃保存,用于测定空肠分泌物中 SIgA 的含量。切取远端结肠,4℃生理盐水漂洗后称取 200mg 左右,冰浴匀浆,37℃温育 15min,3 500r/min 离心 15min,将上清液 -20℃保存,用于检测 PGE2。

1.4 检测项目及方法

1.4.1 生长性能

实验鼠每天定时称重并记录采食量,获得平均日增重(ADG)、平均日采食量(ADFI)、饲料报酬(G/F)。

1.4.2 免疫器官指数

SD 大鼠迫杀后取脾脏和胸腺,剔除脂肪,用 4℃生理盐水漂洗后称重,按公式计算免疫器官指数:免疫器官指数 (mg/g) = 脏器湿重 (mg) / 体重 (g)。

1.4.3 免疫指标

比色法测定血清中 AKP 及 LZM 的含量;ELISA(双抗体夹心法)测定血清中 IgA、IgG、EGF、空肠分泌物中 SIgA 及结肠壁中 PGE2 的含量。

1.5 统计方法

试验数据应用 SPSS17.0 统计软件中 One Way ANOVA 进行分析,应用 LSD 法进行多重比较,显著水平为 $P<0.05$。

2 结果

2.1 YP 对大鼠生长性能的影响

饲喂 1 周不同剂量的 YP 及基础饲料后,各组 ADFI 无显著性差异 ($P>0.05$);YP (100mg/kg) 组 ADG 显著 ($P<0.05$) 或极显著 ($P<0.01$) 高于其他组,达到 5.63g/d,比正常对照组提高 15.36%;G/F YP (100mg/kg) 组也显著高于其他组 ($P<0.05$)(表 1)。

表 1 YP 对大鼠生长性能的影响 ($\bar{X}\pm SD$)

组别	平均日增重 (g/d)	平均日采食量 (g/d)	饲料报酬
YP (50mg/kg)	4.73 ± 0.68^{ABb}	22.35 ± 2.23	0.21 ± 0.02^{b}
YP (100mg/kg)	5.63 ± 0.54^{Aa}	23.12 ± 2.78	0.24 ± 0.02^{a}

（续表）

组别	平均日增重（g/d）	平均日采食量（g/d）	饲料报酬
YP（200mg/kg）	4.43 ±0.64[Bc]	22.09 ±2.51	0.20 ±0.03[b]
CTX（100mg/kg）	4.81 ±0.52[ABb]	22.16 ±2.30	0.22 ±0.03[b]
Control	4.88 ±0.57[ABb]	22.18 ±1.97	0.22 ±0.04[b]

注：表中数据为平均值±标准差，同一指标各组之间进行比较，含有不同小写字母表示差异显著（$P<0.05$），不同大写字母表示差异极显著（$P<0.01$），有相同字母或不标注为差异不显著（$P>0.05$）。以下各表、图相同

2.2 YP 对 CTX 致免疫低下大鼠免疫器官指数的影响

各实验组大鼠连续灌胃 YP 10d，分别在第 8d、9d 腹腔注射 CTX 后，于第 11d 观察 YP 对 CTX 致免疫低下大鼠免疫器官指数的影响。结果显示，正常对照组免疫器官指数极显著高于其他组（$P<0.01$），表明 CTX 所致免疫抑制模型可行。其中，YP（50mg/kg）+CTX 组脾脏指数（Spleen index）略高于 CTX 对照组，但差异不显著（$P>0.05$）；YP 组胸腺指数（Thymus index）高于 CTX 对照组，其中，YP（100mg/kg）+CTX 组比 CTX 对照组提高 33.33%，差异显著（$P<0.05$）（图 1）。

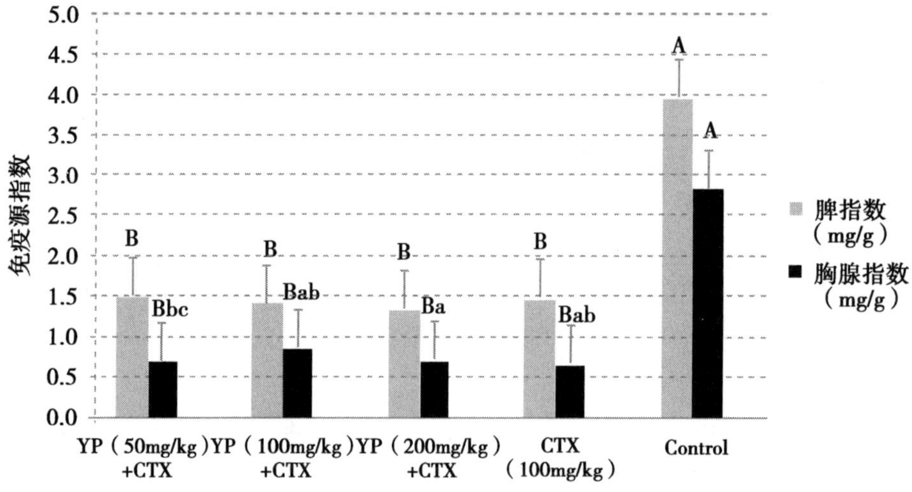

图 1 YP 对 CTX 致免疫低下大鼠免疫器官指数的影响

2.3 YP 对 CTX 致免疫低下大鼠血清中部分免疫指标的影响

大鼠腹腔注射 CTX 后，各试验组血清中除 IgA 外，IgG、AKP、LZM 含量均有不同程度下降（$P<0.05$ 或 $P<0.01$）。YP 组有提高各检测免疫指标的趋势，其中，YP（100mg/kg）+CTX 组 IgA 含量极显著高于 CTX 对照组（$P<0.01$），IgG、AKP 含量显著高于 CTX 对照组（$P<0.05$），LZM 含量 YP（100mg/kg）+CTX 组比 CTX 对照组有所提升但差异不显著（$P>0.05$）（表 2）。

2.4 YP 对 CTX 致免疫低下大鼠肠黏膜免疫部分相关指标的影响

大鼠腹腔注射 CTX 后，结果显示，YP（100mg/kg）+CTX 组空肠黏膜分泌物中 SIgA 水平及血清中 EGF 含量分别极显著（$P<0.01$）和显著（$P<0.05$）高于其他组，其他各组

之间无显著性差异（$P>0.05$）；结肠壁中 PGE2 含量 CTX 对照组最高，与正常对照组差异极显著（$P<0.01$），YP 组含量逐渐降低，其中，YP（100mg/kg）+CTX 组含量最低，与 CTX 对照组差异极显著（$P<0.01$）（图 2）。

表 2　YP 对 CTX 致免疫低下大鼠血清中 IgA、IgG、AKP、LZM 含量的影响（$\bar{X}\pm SD$）

组别	IgA（μg/ml）	IgG（μg/ml）	碱性磷酸酶（mg/100ml）	溶菌酶（U/ml）
YP（50mg/kg）+CTX	8.04 ± 0.49^{ABb}	3.19 ± 0.06^{Ba}	38.64 ± 6.11^{Bbc}	162.56 ± 34.94^{Bbc}
YP（100mg/kg）+CTX	8.86 ± 1.05^{Aa}	3.18 ± 0.11^{Ba}	51.57 ± 10.47^{Ba}	181.71 ± 21.83^{ABb}
YP（200mg/kg）+CTX	7.60 ± 0.48^{Bbc}	3.14 ± 0.25^{Bb}	46.71 ± 12.64^{Bab}	147.62 ± 23.28^{Bc}
CTX（100mg/kg）	7.70 ± 0.52^{Bbc}	3.12 ± 0.05^{Bb}	38.39 ± 3.58^{Bbc}	165.67 ± 28.78^{Bbc}
Control	8.25 ± 0.59^{ABab}	3.36 ± 0.20^{A}	82.36 ± 15.87^{A}	219.86 ± 22.02^{Aa}

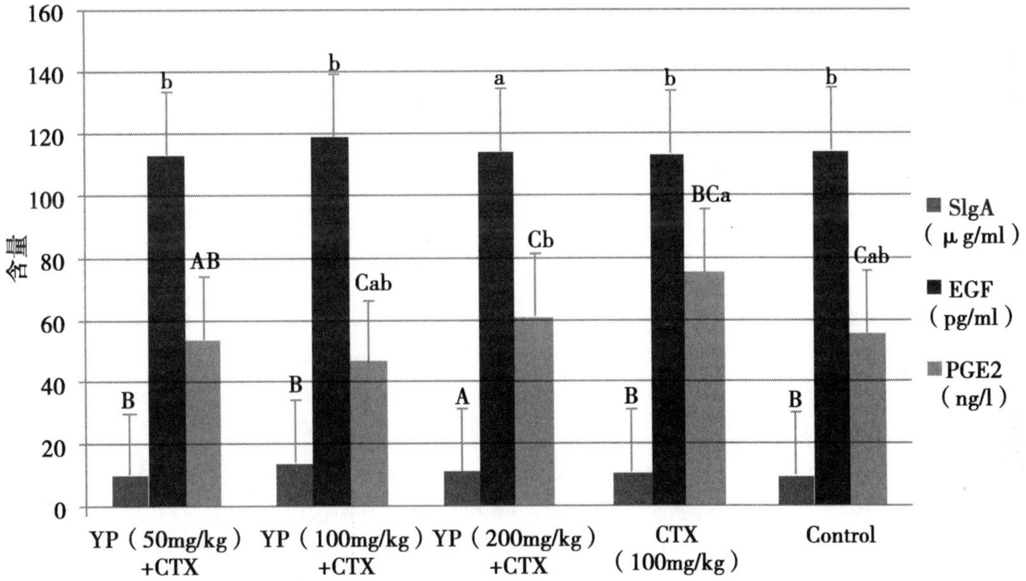

图 2　肠黏膜免疫部分相关指标的含量

3　讨论

目前，关于 YP 对动物生长性能的影响的报道较多，研究认为 YP 可以显著提高生长速度和改善饲料转化效率[9]，但也有研究表明 YP 对 ADG 和 G/F 均无显著影响，只有提高 ADFI 的趋势[10]。本研究表明，日粮中添加 100mg/kg 的 YP 可显著增加大鼠 ADG 和 G/F（$P<0.05$）。由此可见 YP 对动物生长性能的影响有剂量关系，选择合适的剂量非常重要。

CTX 是一种免疫抑制剂。免疫器官的大小在一定程度上反映机体免疫系统的状态[11]。本研究结果表明，CTX（100mg/kg）注射大鼠 2d 后可极显著降低免疫器官指数（$P<0.01$），YP 对提高脾脏指数效果不明显（$P>0.05$），但胸腺指数 YP（100mg/kg）+CTX 组比 CTX 对照组提高 33.33%（$P<0.05$），表明 YP 在一定程度上具有保护免疫不受损伤和促进免疫损伤后的恢复作用。

本实验表明，大鼠注射 CTX 后，CTX 对照组血清中 IgA、IgG 水平均降低（$P<0.01$），表明注射 CTX 后严重损坏了大鼠的免疫机制，但 YP 组可拮抗 CTX 的免疫抑制作用，结果表明，YP 对注射 CTX 的大鼠有显著的免疫恢复作用，可以提高大鼠的体液免疫水平。这与 Krakowski 报道的酵母葡聚糖能提高怀孕母猪血清中免疫球蛋白水平一致[12]。

本实验显示，YP 有提高血清中 AKP、LZM 含量的趋势，其中 YP（100mg/kg）+CTX 组 AKP 浓度显著高于 CTX 对照组（$P<0.05$），表明一定量的 YP 有助于提高大鼠吞噬细胞的吞噬能力及抗感染能力。

SIgA 是黏膜免疫应答过程中的主要体液因子。肠道中 SIgA 含量的变化直接反映机体黏膜的局部免疫状态[13]。本实验中，YP（100mg/kg）+CTX 组空肠黏液中 SIgA 的含量极显著高于正常对照组和 CTX 对照组（$P<0.01$），表明 YP（100mg/kg）可以有效增强肠黏膜的屏障作用。

EGF 具有促进上皮增殖、抑制胃酸分泌、增加 DNA 与蛋白质合成的作用[14]。PGE2 通过抑制白细胞介素 2 分泌及淋巴细胞膜表面铁蛋白受体的表达对免疫系统产生广泛的抑制作用[15]。本研究发现，与 CTX 对照组相比，YP（100mg/kg）可以显著提高血清中 EGF 水平（$P<0.05$）、极显著降低结肠壁中 PGE2 含量（$P<0.01$）。表明 YP 可以促进大鼠肠黏膜损伤的修复，并呈一定的量效关系。YP 促进肠黏膜修复可能与提高大鼠血清 EGF 的水平及降低肠道中 PGE2 含量有关。

参考文献

[1] 顾笑梅，孔健，王富生，等．一株乳酸菌所产胞外多糖对荷瘤小鼠机体免疫功能影响的研究［J］．微生物学报，2003，43（2）：251-256.

[2] Guo Shou-dong，Mao Wen-jun，Han Yin，*et al*. Structural characteristics and antioxidant activities of the extracellular polysac-charides produced by marine bacterium *Edwardsiella tarda*［J］. Bioresour Technol，2010，101（12）：4 729-4 732.

[3] Vanea CH，Drage TC，Snape CE. Bark decay by the white-rot fungus *Lentinula edodes*：Polysaccharide loss，lignin resistance and the unmasking of suberin［J］. Int Biodeterior Biodegrad，2006，57（1）：14-23.

[4] Zhu Xiao-ling，Chen Alex-F，Lin Zhi-bin. *Ganoderma lucidum* polysaccharides enhance the function of immunological effector cells in immunosuppressed mice［J］. J Ethnopharmacol，2007，111（2）：219-226.

[5] 郭敏，张宝善，金晓辉．微生物发酵生产多糖的研究进展［J］．微生物学通报，2008，35（7）：1 084-1 090.

[6] Kogan G，Stasko A，Bauerova K，*et al*. Antioxidant properties of yeast（1→3）-β-D-glucan studied by electron paramagnetic resonance spectroscopy and its activity in the adjuvant arthritis［J］. Carbohydr Polym，2005，61（1）：18-28.

[7] Cipák L，Miadoková E，Dingová H，*et al*. Comparative DNA protectivity and antimutagenicity studies using DNA-topology and Ames assays［J］. Toxicol In Vitro，2001，15（6）：677-681.

[8] Kogan G，Kocher A. Role of yeast cell wall polysaccharides in pig nutrition and health protection［J］. Livest Sci，2007，109（1）：161-165.

[9] 周怿，刁其玉，屠焰，等．酵母 β-葡聚糖对早期断奶犊牛生产性能和血液生理生化指标的影响［J］．中国畜牧杂志，2010，46（13）：47-51.

[10] Hiss S，Sauerwein H. Influence of dietary β-glucan on growth performance，lymphocyte proliferation，specific immune response and haptoglobin plasma concentrations in pigs［J］. J Anim Physiol Anim Nutr（Berl），2003，87（1-2）：2-11.

[11] Grossman CJ. Interaction between the gonadal steroids and the immune system [J]. Science, 1985, 227 (4684): 257 - 261.

[12] Krakowski L, Krzyzanowski J, Wrona Z, *et al*. The influence of nonspecific immunostimulation of pregnant sows on the immunological value of colostrums [J]. Vet Immunol Immunop, 2002, 87 (1 - 2): 89 - 95.

[13] 张殿新，钟秀会，程晶晶．芪蓝四君子汤对雏鸡小肠 IgA 分泌细胞和 SIgA（分泌型 IgA）的影响 [J]. 中国兽医学报，2010，30（12）：1 659 - 1 662.

[14] Pillai SB, Hinman CE, Luquette MH, *et al*. Heparin-binding epidermal growth factor-like growth factor protects rat intestine from ischemia/reperfusion injury [J]. J Surg Res, 1999, 87 (2): 225 - 231.

[15] Roberts P, Morgan K, Miller R, *et al*. Neuronal COX-2 expression in human myenteric plexus in active inflammatory bowel disease [J]. Gut, 2001, 48 (4): 468 - 472.

（发表于《中国预防兽医学报》）

猪戊型肝炎病毒 swCH189 株衣壳蛋白基因 CP239 片段的表达、纯化及抗原性分析

郝宝成[1,2]，梁剑平[1,3]，兰　喜[2]，刑小勇[3]，项海涛[3]，
温峰琴[3]，胡永浩[3]，柳纪省[2,3]

（1. 中国农业科学院兰州畜牧与兽药研究所，新兽药重点开放实验室，甘肃省新兽药工程重点实验室，兰州 730050；2. 中国农业科学院兰州畜牧与兽药研究所，家畜疫病病原生物学国家重点实验室，农业部畜禽病毒学重点开放实验室，农业部草食动物疫病重点开放实验室，兰州 730046；3. 甘肃农业大学 动物医学学院，兰州 730070）

摘　要：根据已克隆的戊型肝炎病毒株 swCH189 株结构蛋白基因（ORF2）的抗原性及水溶性分析，在其序列上设计套式引物，采用 RT－PCR 技术扩增猪戊型肝炎病毒（HEV）结构蛋白基因 ORF2 主要抗原基因片段，长度为 728bp，将其克隆于 PMD18－T 载体，测序结果表明插入的片段属于猪戊型肝炎病毒结构蛋白基因 ORF2 部分；将该片段插入 pet32a 表达载体，经酶切、测序鉴定证实获得了带有目的基因的重组表达质粒。将重组质粒转化 Rosetta 菌，经 0.3mmol/L IPTG 诱导得到融合表达，得到相对分子质量为 45 000的重组蛋白，以包涵体形式存在。Western blot 分析表明，该蛋白可以与猪戊型肝炎阳性血清反应，表明该蛋白具有良好的反应原性。并用 SDS 电泳切胶纯化和透析浓缩方法进行纯化，用纯化的蛋白免疫兔子，经每 2 周免疫 1 次，连续 3 次免疫后，对免疫组和对照组，分别在45d，60d，90d 采血，每只兔子采集 1 份。用夹心 ELISA 法测定免疫血清抗体，结果表明，免疫组在45d 后均产生具有 HEV 抗原结合活性的抗体。本试验对猪 swCH189 株 CP239 基因的克隆和表达研究，为进一步研究结构蛋白基因 ORF2 的生物学功能奠定了基础。

关键词：猪戊型肝炎病毒；swCH189 株；克隆；原核表达；纯化；抗原性分析

戊型肝炎（HE）是由戊型肝炎病毒（Hepatitis Evirus，HEV）引起的一种经消化道传播的急性接触性人畜共患传染病。人普遍易感。急性戊型肝炎病情较重，病死率为 2.5%，明显高于甲型（0.1%）和乙型（0.9%）肝炎，孕期妇女死亡率可高达 15% ~20%[1]，尤其是怀孕最后 3 个月的孕妇病死率更高。1986—1988 年我国新疆南部地区曾发生 HE 水源性大流行，共计有 119 280人发病，死亡 707 人，是迄今世界上最大、最严重的一次戊型肝炎流行[2]。HEV 不仅感染人，而且可以感染猪、野猪、鸡、鸟[3]、鼠[4]、猫[5~6]及鹿等动物，人也可以通过生吃鹿肉而染病[7]，表明人畜是 HEV 的共同宿主。

HEV 基因有 3 个开放阅读框（open readingframes，ORFs），ORF2 为主要的结构基因编码区，长约 2kb，其核苷酸序列也最保守。HEV ORF2 编码蛋白（PORF2）抗原表位数量多，结构复杂，除 N 端25 ~38aa 外，大多数分布在 C 端2/3 部分，包括了 PORF2 25 ~38、319 ~340、

341～354、390～470、422～437、517～530、546～580、631～648、641～660aa 的 9 个抗原活性区，研究还证实在 394～470aa 至少含有 5 个不同免疫活性的抗原表位[8～10]。Li 等[11～12]在大肠杆菌中表达了 ORF2 全长编码的 GST－ORF2 蛋白，发现其 c 端抗原性可被 N 端遮盖。Pudry 等[13]在研究大肠杆菌表达的 HEV－trpE 融合蛋白 C2（24～660aa）时，也指出 ORF2C 端 2/3 的部分其抗原表位与高级结构有关。Meng 等[14]在此基础上首先发现并确定 ORF2 编码蛋白 C 末端 452～617aa 存在中和抗原表位。本试验利用 protean 软件对 HEV swCH189 株的 ORF2 进行潜在抗原位点分析。在 ORF2C 端 2/3 处，选取381～623aa 段进行原核表达。

1　材料与方法

1.1　主要试剂与器材

1.1.1　病毒株

经 RT－PCR 检测 HEV RNA 为阳性的 swCH189 株，基因型为 HEV Ⅳ 型，由中国农业科学院兰州兽医研究所传染病室保存。

1.1.2　菌种和载体

菌种为大肠杆菌 *E. coli* JM109，克隆载体 pMD18－T 载体购自大连宝生物工程公司。表达菌 rosetta 及表达载体 pET32a（＋）由中国农业科学院兰州兽医研究所传染病实验室保存。

1.1.3　试剂

PCR 产物胶回收试剂盒、质粒提取试剂盒购自 TaKaRa 公司。RNase、低分子量蛋白质 Marker、琼脂糖购自联创公司。Fermentas 蛋白 Mark 购自鹏程公司。限制性内切酶 Nco Ⅰ 和 Not Ⅰ 购自大连宝生物。辣根过氧化物酶标记的羊抗猪 IgG 抗体、DAB 为鹏程公司产品。完全弗氏佐剂不完全弗氏佐剂均为 Sigma 公司产品。猪戊型肝炎病毒 IgG 抗体诊断试剂盒，TMB 底物，北京万泰生物药业有限公司提供。其余试剂均为分析纯或生化试剂。

1.1.4　引物设计

使用 DNAstar 软件中的 Edit－seq 与 Protean 对 ORF2 序列的翻译产物进行分析，根据推断的抗原活性位点分布，在 ORF2 区设计 1 对引物，同时，在引物 5′端添加限制性酶切位点及保护碱基，以利于酶切及与表达载体连接。引物由 TaKaRa 公司合成。引物序列如下：

5′端引物（U1）：5′－CATG <u>CCATGG</u> GTATTGCGCTAACCTTGTTTAATCTTGCTGATA－3′

3′端引物（L1）：5′－ATTT <u>GCGGCCGC</u>TCAATACTCCCGGGTTTTACCCACCTT－3′

加下划线部分为限制性内切酶识别位点，其 5′端为 4 个保护碱基。CCATGG 为 Nco Ⅰ 识别位点，GCGGCCGC 为 Not Ⅰ 识别位点。

1.2　表达基因的扩增和克隆

PCR 模版为已克隆到的质粒 PMD－ORF2，PCR 反应体系为 50μl：PMD－ORF2 全长质粒 1μl，10×Pfu Buffers 5μl，上游引物 U1（50μmol/L）1μl，下游引物 L1（50μmmol/L）1μl，dNTP（10mmol/L）1μl，Pfu 聚合酶 1μl，H_2O 40μl。PCR 仪上设置以下程序：95℃预变性 2min，94℃变性 30s，50℃退火 40s，72℃延伸 1min，共 30 个循环。72℃延伸 10min，电泳观察 PCR 结果。以 TaKaRa 公司胶回收试剂盒进行 PCR 产物回收纯化。

1.3　表达质粒的构建与鉴定

将 pET32a（＋）与 PCR 扩增片段 CP239 分别以 Not Ⅰ/Noc Ⅰ 进行消化，并进行琼脂糖凝胶电泳，以 TaKaRa 公司胶回收试剂盒进行目的片段的回收。将酶切处理好的 pET32a（+）与扩增片段 CP239 进行连接，构建可在大肠杆菌中表达的重组质粒，命名为 pET32a（+）-CP239。将连接产物转化大肠杆菌 JM109，挑取转化子，小提质粒，用 Not Ⅰ/Noc Ⅰ 酶切鉴定。

1.4 目的基因在大肠杆菌 rosetta 中的诱导表达

将重组质粒 pET32a（+）-CP239 转化大肠杆菌 rosetta。加入终浓度为 0.3mmol/L 的 IPTG，37℃，诱导 4h，同时设空白载体对照。将诱导菌液离心，超声波裂解，然后上 12% SDS-PAGE 凝胶电泳。

1.5 表达蛋白的纯化

经诱导条件的优化后，对表达菌进行大量诱导，4℃，5 000r/min 离心 30min，收集菌体。用细菌裂解液悬浮菌体，冻融 3 次后超声破碎菌体。12 000r/min 离心 10min，弃上清。使用最小体积的 4mol/L 尿素将沉淀溶解，4℃，12 000r/min 离心 20min。沉淀中透明若仍有未溶解物质，再用 8mol/L 尿素溶解 4℃，12 000r/min 离心 20min 离心，弃沉淀。加上样缓冲液，煮沸 10min。经 SDS-PAGE 电泳，采用切胶的方法，将目的条带胶块盛入透析袋，经透析电泳，去除离子杂质，用浓缩胶浓缩法的进行纯化。

1.6 Western blot

配制 12% 分离胶和 5% 浓缩胶，电泳上样体积为 100μL，样品位于浓缩胶时设定电压为 80V，样品进入分离胶后电压改为 200V 待溴酚兰色带跑至凝胶下缘时停止电泳。首先将待检样品进行 SDS-PAGE 电泳，并将 SDS-PAGE 胶、滤纸及硝酸纤维素（NC）膜在转移缓冲液中浸泡 15~30min，按 Bio. Rad 湿转使用说明将蛋白转至硝酸纤维素膜上，200mA，105min 完成湿转。用封闭液室温封闭 2h。

将硝酸纤维素膜放入干净的塑料袋中，加用封闭液稀释的人抗-HEV 阳性血清(1∶1 000),封闭塑料袋，室温摇摆平台过夜。100ml PBST 洗膜 5 次，每次 15min。加入用封闭液以1∶1 000稀释的辣根过氧化物酶标记的羊抗猪 IgG 抗体，室温温育 2h。100ml PBST 洗膜 5 次，每次 15min。加底物和反应液显色，10min 内蒸馏水冲洗终止显色。最后用滤纸吸干膜上残余液体，拍摄照片。

1.7 免疫动物

用纯化的 ORF2-CP239 蛋白免疫兔子，随机选择健康家兔 8 只，年龄 3 个月以上，体质量 2~3kg 兔子，分为免疫组和对照组 2 组，数量分别为 6 只和 2 只，纯化蛋白加等体积的完全弗氏佐剂，充分混匀后经背皮下注射，每只 300μg，对照组注射等量加佐剂的蒸馏水。经每 2 周免疫 1 次，连续 3 次免疫后，对免疫组和对照组，分别在 45d、60d、90d 采血，每只兔子采集 1 份。

1.8 双抗原夹心法 ELISA（DS-ELISA）

采用改良过碘酸钠法 HRP（Sigma 公司）标记 pET32a（+）-CP239 蛋白[15]。将经 SDS-PAGE 电泳，切胶浓缩纯化的 pET32a（+）-CP239 抗原，溶解于 0.05mol/L 碳酸包被缓冲液（pH 值 9.5）中，100μl/孔包被 96 孔聚乙烯微量滴定板，37℃吸附 2h，4℃过夜；PBST 洗液（pH 值 7.4）洗涤 1 遍，200μl/孔封闭液（含 2% 明胶、0.2% 酪蛋白和 2% 蔗糖的 PBS）37℃封闭 2h，甩尽、控干后真空封闭，4℃保存备用。检测时，于每孔中加入 50μl

待检血清，轻拍混匀后每孔加入50μl稀释好的pET32a（+）-CP239-HRP，轻拍混匀后37℃温育30min；洗涤6次，控干，加入显色剂（TMB底物，北京万泰生物药业有限公司提供），37℃温育10min后，加终止液50μl（2mol/LH_2SO_4）终止，于酶标仪（TECAN公司）上读取D_{450}/D_{620}的读值来判定所测血清不同时间样品的阳性率。

1.9 双抗原夹心法ELISA对猪血清样品的检测

利用DS-ELISA对甘肃兰州市城关区某猪屠宰场采集的109份猪血清进行了检测，并将检测结果与本传染病室建立的nRT-PCR方法检测结果进行对比分析。

2 结果与分析

2.1 CP239表达蛋白基因片段的扩增和克隆

利用PCR方法成功克隆出基因片段CP239，扩增片段长度为728bp（图1），位于ORF2 1 142～1 870bp。

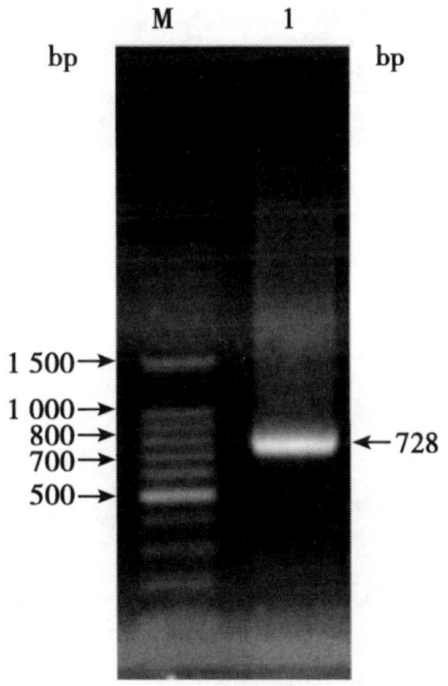

1. 扩增ORF2获得cp239片段；M. 100bp DNA Marker

图1 目的片段PCR结果

2.2 原核表达重组质粒的构建

将pET32a（+）-CP239进行Not Ⅰ/Noc Ⅰ双酶切，鉴定结果见图2，经上海生工测序证明该序列正确，表明目的片段正确插入。

2.3 目的基因的原核表达及检测

经IPTG诱导后，该片段在大肠杆菌rosetta中得到融合表达，表达蛋白的相对分子质量为45 000左右（图3）。

2.4 表达蛋白的存在形式与纯化结果

将表达菌进行超声波裂解处理，经SDS-PAGE分析发现表达蛋白主要以包涵体形式存在（图4）。

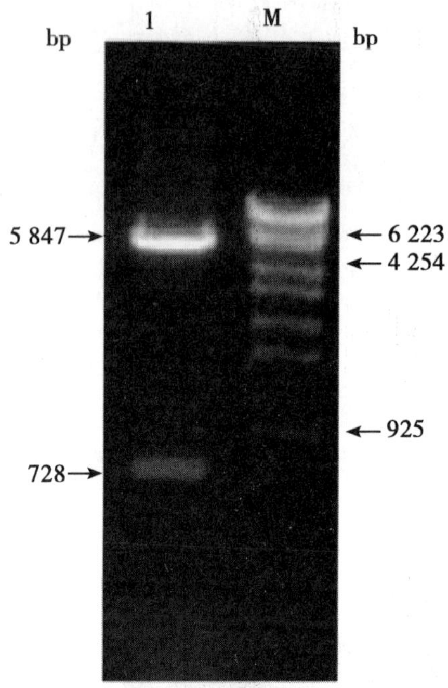

1. 酶切 pE32a（+）－CP239；M. λ－Ecot14 marker

图 2　pET32a（+）－CP239Not Ⅰ/Noc Ⅰ 酶切鉴定

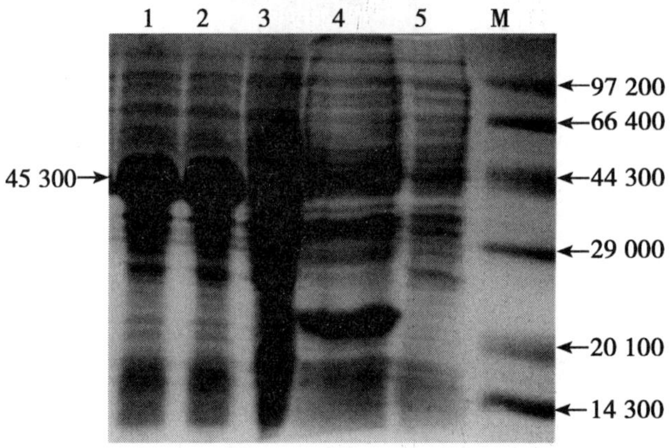

1. 0. 3mmol/L 4h 表达菌样；2. 1mmol/L 4h 表达菌样；3. 诱导表达未离心菌样；4. pET32a（+）菌样；5. 诱导表达离心上清样；M. 低分子量蛋白 Marker

图 3　目的表达蛋白 SDS 聚丙烯酰胺凝胶电泳图片

2.5　Western blot 鉴定结果

将纯化蛋白经电转移至 NC 膜，经 Western blot 表明，蛋白可与 HEV 阳性血清反应（图 5）。

2.6　双抗夹心 ELISA 检测血清阳性率

对免疫组和对照组，分别在 45d，60d，90d 采血，每只兔子采集 1 份。应用双抗夹心酶联免疫测定法（DS－ELISA）进行血清阳性率测定。免疫组 45d，60d，90d 阳性率分别为 83.3%，100%，100%（表 1）。

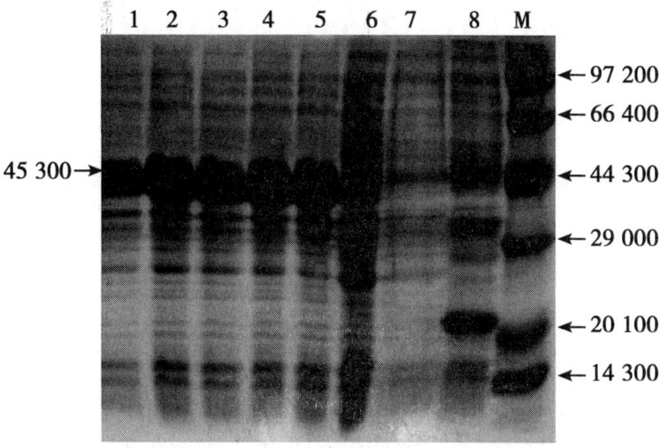

1. 0. 1mmol/L 4h 表达菌样；2. 0. 3mmol/L 4h 表达菌样；3. 0. 5mmol/L 4h 表达菌样；
4. 0. 7mmol/L 4h 表达菌样；5. 1mmol/L 4h 表达菌样；6. 未诱导表达菌样；
7. Rosetta 空菌样；8. pET32a（+）菌样；M. 低分子量蛋白 Marker

图 4　目的表达蛋白 SDS 聚丙烯酰胺凝胶电泳图片

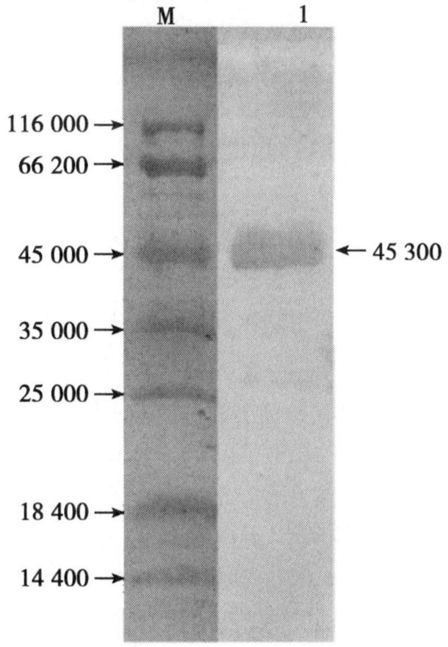

M. Fermentas 蛋白 Marker；1. 目的蛋白

图 5　目的蛋白 Western blot 检测结果

表 1　双抗夹心检测免疫动物血清阳性率

组别	数量（只）	45d 采血			60d 采血			90d 采血		
		样品数（份）	阳性数（份）	阳性率（%）	样品数（份）	阳性数（份）	阳性率（%）	样品数（份）	阳性数（份）	阳性率（%）
免疫组	6	6	5	83. 3	6	6	100	6	6	100
对照组	2	2	0	0	2	0	0	2	0	0

注：% D% 值 >0. 1500 为阳性，% D% 值 <0. 1500 为阴性

2.7 双抗原夹心法 ELISA（DS－ELISA）与 nRT－PCR 检测结果对比分析

利用 DS－ELISA 和 nRT－PCR 方法分别对从我国甘肃兰州市城关区某猪屠宰场采集的109份猪血清进行了检测，检测结果见表2。

表2 双抗原夹心法 ELISA（DS－ELISA）与 nRT－PCR 检测结果

方法	检测数（份）	阳性数（份）	阳性率（%）
DS－ELISA	109	4	3.7
nRT－PCR	109	4	3.7

从检测结果中可以看出，利用 DS－ELISA 检测猪血清抗体检出的阳性猪戊型肝炎病毒份数和本室建立的 nRT－PCR 方法检出的阳性数一样。但是在实际操作过程中，用本室建立的 nRT－PCR 方法操作比较繁多，且在试验中容易扩增出非特异条带，若不经过测序对比鉴定，会造成假阳性判定，而用 DS－ELISA 则操作简便，检测周期短，灵敏度高，且不容易产生假阳性结果。

3 讨论

本研究利用大肠杆菌 Rosseta 成功表达了 pET32a（+）－CP239 质粒，最佳表达条件为诱导时间4h，诱导温度37℃，IPTG 终浓度为0.3mmol/L。另外本研究曾尝试了 BL21，DH5@，JM109 等几种表达菌，均未获得目的蛋白的表达。

1992年 Kaur 等[16]定位了 pORF2 25～38、341～354、517～530，3个抗原活性区。Khudyakov 等[17]1993年定位了319～340、631～648、641～660，3个抗原活性区，1994年又发现在390～470和546～580aa 上还有2个抗原决定簇区段，进一步深入研究证实390～470aa 之间至少含有5个不同免疫活性的抗原表位。Li 等[18]发现由于肽链的折叠，基因表达的 ORF2 全长蛋白其 C 端的抗原表位被遮盖，而切去 N 端部分肽链的 ORF2 抗原性很好。Pudry 等[19]在研究大肠杆菌表达的 HEV－trpE 融合蛋白 C2（24～660aa）时，也指出 ORF2 蛋白 C 端2/3部分其抗原表位与高级结构有关。Meng 等[20]在此基础上首先发现并确定 ORF2 编码蛋白的 C 末端452～617aa 是 HEV 中和抗原表位的位置，并且血清学试验证明，用不同基因型含有452～617aa 片段的表达蛋白来检测血清中的抗体，能够发生交叉中和反应，提示有共同中和抗原表位的存在，但进一步的研究也发现，不同基因型抗原检测的同一基因型的 ELSIA 抗体滴度要高于不同基因型的抗体滴度，造成这种型间不同敏感性的原因是不同基因型除有共同抗原表位外还具有各自型的特异性，这是临床上用单一基因型抗原来诊断 HE，造成误诊的主要原因。HEV 不同基因型和基因亚型 HEVORF2 编码蛋白 p166（452～617aa）存在多种类型抗原表位，其中包括基因型Ⅰ、Ⅱ、Ⅲ、Ⅳ共同的，基因型Ⅲ、Ⅳ共有的和基因型Ⅲ特异的等，这些表位与天然 HEV 颗粒上的抗原表位具有相同的免疫学特征。Guo H 等[21]对禽 HEV ORF2 蛋白进行研究表明，结构域Ⅱ（可能在477～492aa 之间）至少有一个表位是禽 HEV 特有的，结构域Ⅰ（389～410aa）存在一个禽、人和猪 HEV 共有的表位，结构域Ⅳ（583～600aa）存在一个或更多表位是禽和人 HEV 共有的。

1997年，Meng 等[22]发现猪群中存在着与美国人戊型肝炎病毒基因组高度相似的病毒。2009年，郝宝成等[23]研究发现 swCH189 株与人源性 HEV（如 Xinjiang 标准株等）的 ORF2 核苷酸同源性最高为87.9%，氨基酸同源性最高为97.8%；与猪源性 HEV（如 HeBei 株等）的 ORF2 核苷酸同源性最高为 91.8%，氨基酸同源性最高为 98.8%，发现它与猪源性 HEV 的同源性更高一些。利用北京万泰生物公司戊型肝炎夹心 ELISA 诊断试剂盒，中国农业科学院哈尔滨兽医研究所生物技术重点实验室对全国 22 个省市地区进行的血清学流行病学研究，几乎所有被检地区都存在猪戊型肝炎病毒感染[24]。李文贵等[25]从各地养猪场、屠宰场共采集 270 份血清，应用双抗夹心酶联免疫测定法（DS－ELISA）进行血清学调查。检测结果 174 份样品呈 HEV－IgG 抗体阳性，总阳性率 64.4%，其中，养殖场血清抗体阳性率为 55.6%，屠宰厂为 68.9%。比较养殖场的各年龄阶段猪群显示，母猪群血清抗体阳性率最高，为 71.4%。本试验也参考了上述的试验方法，对免疫组和对照组，分别在 45d，60d，90d 采血，每只兔子采集 1 份。应用双抗夹心酶联免疫测定法（DS－ELISA）进行血清阳性率测定。最终测得免疫组 45d，60d，90d HEV－IgG 抗体阳性，阳性率分别为 83.3%，100%，100%。说明表达蛋白 pET32a（+）－CP239 免疫效果良好。

本试验采用 pET 系统，成功表达了猪源 HEVswCH189 株 ORF2C 端 381～623aa，该融合蛋白主要以包涵体形式存在，4mol/L 尿素溶解后，再用 8mol/L 尿素完全溶解，SDS－PAGE 电泳可见有 45 300的蛋白条带，Western blot 显示 45 300处出现明显反应条带。有报道称人源 HEV ORF2 表达蛋白产生聚合现象[26~29]，本试验的表达产物并没有观察到有多条反应带的出现。葛胜祥等表达了 HEV 衣壳蛋白片段 NEZ（394～604aa），可以形成良好的构象性中和表位，以弗氏佐剂免疫恒河猴具有良好的免疫原性，并诱导出高效价保护性抗体，使免疫猴获得对 HEV 的良好抵抗力[30~31]。虽然原核表达系统不能对表达产物进行糖基化、磷酸化等化学修饰作用，但是，表达产物仍可以作为抗原或用于抗体检测及制备单克隆抗体，是研究其免疫学功能或研制基因工程疫苗的侯选材料。本试验的表达产物拥有 ORF2 的大多数抗原表位包括中和抗原表位，Western blot 的结果表明，该融合蛋白可被猪的 HEV 阳性血清所识别，表明它具有 HEV 天然毒株的主要抗原表位。作为免疫原，该重组蛋白对动物的抗 HEV 免疫保护作用及其作为 HEV 诊断抗原的应用价值正在研究之中。

参考文献

[1] Purdy M A，Carson D，McCaustland K A，et al. Viral specificity of hepatitis E virus antigens identified by fluorescent antibody assay using recombinant HEV proteins [J]. J Med Virol，44：212－214.

[2] Zhuang H，Cao X Y，Liu C B，et al. Epidemiology of hepatitis E in China [J]. Gastroenterologia Japanica，1991，26：135－138.

[3] Sun Z F，Larsen C T，Dunlop A，et al. Genetic identification of from healthy chicken flocks and characterization of the capsidg from chickens with hepatitis splenomegaly syndrome in different United States [J]. J Gert Virol，2004，8：693－700.

[4] Hirano M，Ding X，Li T C，et al. Evidence for widespread among wide rats in Japan [J]. Hepatol Res，2003，27（1）：1－5.

[5] Okamoto H，Takahashi M，Nishizawa T，et al. Presence of an Japanese pet cats [J]. Infection，2004，32（1）：57－58.

[6] Choi C，Ha S K，Chae C. Development of nested RT－PCR hepatitis E virus in formalin－fixed，paraffin－embedded tissues hybridization [J]. J Virol Methods，2004，115（1）：67－71.

[7] Tei S，Kitajiama N，Takahashi K，et al. Zoonotic transmission to human beings [J]. Lancet，2003，362（9381）：371－373.

[8] Kaur M，Hymas K C，Purdy M A，et al. Human linear B－cell epitopes encoded by the hepatitis E virus in-

clude determinants in the RNA – dependent RNA Polymerase [J]. Proc Natl Acda Sci USA, 1992, 89 (9): 3 855 – 2 858.

[9] Khudyakov Yu E, Khudyakova N S, Fields H A, et al. Epitope mapping in proteins of hepatitis E virus [J]. Virol, 1993, 194 (1): 89 – 96.

[10] Khudyakov Yu E, Favorov M O, Jue D L, et al. Immunodominant antigenic regions in a structural Protein of the hepatitis E virus [J]. Virol, 1994, 198 (1): 390 – 393.

[11] Li F, Torresi J, Locarnini S A, et al. Amino – terminalepitopes are exposed when full – length open reading frame 2 of hepatitis E virus is expressed in. Eschericha coli, but carboxy – terminal epitopes are masked [J]. J Med Virol, 1997, 52 (3): 289 – 300.

[12] Li T, Yamakawa Y, Suzuki K, et al. Expression and self – assembly of empty virus – Iike particles of hepatitis E virus [J]. J Virol, 1997, 71: 7 207 – 7 213.

[13] Purdy M A, Mccaustland K A, Krawczynski K, et al. Preliminary evidence that a trpE – HEV fusion protein protects cynomolgus macaques against challenge with wild – type hepatitis E virus (HEV) [J]. J Med Virol, 1993, 41 (1): 90 – 94.

[14] Li F, Torresi J, Locarnini S A, et al. Amino – terminal epitopes are exposed when full – length open reading frame 2 of hepatitis E virus is expressed in Escherichia coli, but carboxy – terminal epitopes are masked [J]. J Med Virol, 1997, 52 (3): 289 – 300.

[15] Tijssen P, Kurstak E. Highly efficient and simple methods for the preparation of peroxidase and active peroxidase – antibody conjugates for enzyme immunoassays [J]. Anal Biochem, 1984, 136: 451 – 457.

[16] Kaur M, Hymas K C, Purdy M A, et al. Human liner B – cell epitopes encoded by the hepatitis E virus include determinants in the RNA – dependent RNApoly – merase [J]. Proc Natl Acad Sci USA, 1992, 89 (9): 89 – 96.

[17] Khudyakov Yu E, Khudyakova N S, Fields H A, et al. Epitope mapping in proteins E virus [J]. Virology, 1993, 194 (1): 89 – 96.

[18] Li F, Torresi J, Locarnini S A, et al. Amino – terminal epitopes are exposed when full – length open reading frame 2 of hepatitis E virus is expressed in Escherichia coli, but earboxy – terminal eptitopes are masked [J]. J Med Virol, 1997, 52 (3): 289 – 300.

[19] Purdy M A, McCaustland K A, Krawczynski K, et al. Preliminary evidence that a trpE – HEV fusion protein protects cynomolgues against challenge with wildtype hepatitis E virus (HEV) [J]. J Med Virol, 1993, 41 (1): 90 – 94.

[20] Meng J, Dai X, Chang J C, et al. ldentifieation and chaeterization of the neutralization epitopes of the hepatitis E virus [J]. Virol, 2001, 288 (2): 203 – 211.

[21] Guo H, Zhou E M, Sun Z F, et al. ldentification of Bcell epitopes in the capsid protein of avian hepatitis E virus (avian HEV) that are common to human and – swine HEVs or unique to avian HEV [J]. J Gen Vir – ol, 2006, 87 (Ptl): 217 – 223.

[22] Emerson S U, Nguyen H, Graff J, et al. In vitro replication of hepatit is E virus (HEV) genomes and of an H EV replicon expressing green fluorescent protein [J]. Virology, 2006, 78: 4 838 – 4 846.

[23] 郝宝成，兰喜，柳纪省，等．猪戊型肝炎病毒 swCH – GS189 株 ORF2 基因的克隆及序列分析［J］．生物技术通报，2009（5）：122 – 126.

[24] 王佑春，张华远，李河民，等．戊型肝炎病毒Ⅳ型全基因序列的分析［J］．中华微生物学和免疫学杂志，2001，21（2）：210 – 213.

[25] 李文贵，李精华，杨贵树，等．云南省猪戊型肝炎血清流行病学调查［J］．云南农业大学学报，2009，24（6）：917 – 919.

[26] Meng J, Dai X, Chang JC, et al. Identification and characterization of the neutraliza – tion epitope (s) of

the hepatitis E virus [J]. Virology, 2001, 288 (2): 203 -211.

[27] Zhang J Z, Ng M H, xia N S, et al. Conformational antigenic determiants generated by interactions between a bacterially expressed recombinant peptide of the hepatitis E virus structural protein [J]. J Med Virol, 2001, 64: 125 -132.

[28] Li T, Yamakawa Y, Suzuki K, et al. Expression and self - assembly of empty virus - like particles of hepatitis E virus [J]. J Virol, 1997, 71: 7 207 -7 213.

[29] 李少伟，张军，何志强，等. 人肠杆菌表达的戊型肝炎病毒 ORF2 片段的聚合现象研究 [J]. 生物工程学报，2002，18 (4): 463 -467.

[30] Li X F, Mohammad Z, Faizan A, et al. A C - terminal hydrophobic region is required for homo - oligomerization of the hepatitis E virus capsid (ORF2) protein [J]. J Biomed Biotechnol, 2001, 1 (3): 122 -128.

[31] 葛胜祥，张军，黄果勇，等. 大肠杆菌表达的戊型肝炎病毒 ORF2 多肽对恒河猴的免疫保护研究 [J]. 微生物学报，2003，43 (1): 35 -42.

（发表于《中国兽医学报》）

酵母锌对肉鸡生长性能及生理功能的影响

王 慧[1,2a]，张 霞[2b]，辛蕊华[1]，罗永江[1]，董鹏程[1]，程富胜[1]，胡振英[1]

（1. 中国农业科学院兰州畜牧与兽药研究所，甘肃省新兽药工程重点实验室，兰州 730050；2. 甘肃农业大学，a. 动物医学院，b. 生命科学技术学院，兰州 730070）

摘 要：选用 1 日龄白羽肉鸡 100 只，随机分成Ⅰ、Ⅱ、Ⅲ、Ⅳ、Ⅴ组，每组 20 只。Ⅰ、Ⅱ为酵母锌组，锌水平分别为 50，35mg/kg，Ⅲ为硫酸锌组，锌水平为 50mg/kg；Ⅳ为空白酵母组，酵母添加剂量为 100mg/kg；Ⅴ为基础日粮对照组。各组饲喂相同基础日粮，50d 后，对生长性能、血常规、血液生化（Crea、Urea、AST、ALT、ALB）及抗氧化（抗 O_2^-、GSH－Px、GSH）指标进行检测。结果显示，平均日增重和料重比，21d 时各组之间无显著差异（$P>0.05$），50d 时与Ⅴ组相比，Ⅰ、Ⅱ、Ⅳ组生长性能显著提高（$P<0.01$），其中Ⅰ组最高。各组血液常规指标均处于正常范围内，但Ⅰ组红细胞数极显著高于其他组（$P<0.01$）。血清中肌酐、尿素含量以及天门冬氨酸氨基转移酶和丙氨酸氨基转移酶活性各组间差异不显著（$P>0.05$），Ⅰ、Ⅱ、Ⅳ组白蛋白含量较其他组显著提高（$P<0.01$）。血清抗 O_2^- 活力、GSH 含量Ⅰ组显著高于其他组（$P<0.01$），GSH－Px 活力Ⅰ、Ⅱ组显著高于其他各组（$P<0.01$）。结果提示，日粮添加适量酵母锌（Zn 50mg/kg）可以提高肉鸡生长性能，在一定程度上增强机体免疫机能，提高肝脏功能和机体抗氧化功能。

关键词：酵母锌；肉鸡；生长性能；血液指标；抗氧化

锌是维持动物正常生理功能和生化代谢所必需的微量元素，与机体发育、骨骼生长、免疫机能、酶的活性、蛋白质、DNA、RNA 及其他物质的合成与代谢等均有密切的关系，因此被称为“生命元素”。畜禽体内锌含量不足时，不仅影响生长发育，而且会引起代谢紊乱，免疫器官萎缩，从而导致免疫功能低下，易受细菌、病毒和真菌的侵入和感染。目前，现代家禽的生长潜力与 NRC 依据 20 世纪 60 年代以前的试验结果制定的肉仔鸡锌推荐量完全不同[1]，饲料中锌的含量已不能满足机体的需要。长期以来，人们多采用价格便宜的无机锌作为畜禽的主要锌源添加剂，但无机锌有许多不利因素，如毒性大、不易被肠道吸收、未被吸收的锌通过粪便排出污染环境等。与无机锌相比，有机锌具有吸收率高、生物活性强、毒性低和环境污染小等特点，因而成为目前锌研究的一个热点。在饲料生产上，目前较为广泛使用的有机锌主要为酵母锌。酵母锌就是在培养酵母的过程中加入锌元素，通过酵母在生长过程中对锌元素的吸收和转化，使锌与酵母体内的蛋白质和多糖有机结合，它避免了无机锌易结块、吸湿性强、生物学效价低等缺点，不仅适口性好，而且具有较高的生物学效价。Ashmead 等[2]曾经报道哺乳动物对氨基酸螯合锌的生物利用率要高于氯化锌、硫酸锌和硝酸锌等无机锌，其原因是氨基酸和锌形成的不溶性螯合物保护了锌在消化道的传输，促进了锌在肠黏膜的转运。本试验以肉鸡为试验对象，以硫酸锌和酵母锌为锌源，设计不同的添加水平，探讨酵母锌对肉鸡生长性能及生理功能的影响，为酵母锌在饲料中的应用提供依据。

1 材料与方法

1.1 酵母锌

由中国农业科学院兰州畜牧与兽药研究所甘肃新兽药工程重点实验室自主研发，经原子吸收分光光度计（Z－8000 日本日立）测定其锌含量为 37 580μg/g。

1.2 试验设计与饲养管理

将 100 只体质量为（35.17±1.23）g 的 1 日龄健康白羽肉鸡（兰州希望种鸡厂），随机分为Ⅰ（酵母锌 50mg/kg）、Ⅱ（酵母锌 35mg/kg）、Ⅲ（硫酸锌 50mg/kg）、Ⅳ（酵母 100mg/kg）、Ⅴ（基础日粮对照）组，每组 20 只。试验基础日粮以玉米、豆粕为主，各组除了依照试验要求饲喂添加含有酵母锌、空白酵母、硫酸锌的日粮外，具体营养标准参照 NRC，基础日粮组成及营养水平见表 1。试验鸡均自由采食，试验全期采用笼养方式，光照、温度、湿度等条件按饲养规程控制。所有雏鸡进行常规免疫，试验期为 50d。

表 1 基础日粮组成及营养水平 （%）

原料	日粮组成		组成	营养水平	
	0～21d	22～50d		0～21d	22～50d
玉米	59.16	63.50	CP	20.88	18.54
豆粕	27.00	19.87	AME	3.00	3.00
鱼粉	4.00	3.00	Ca	1.01	0.90
麦麸	2.00	3.00	AP	0.45	0.36
菜籽饼	3.00	5.50	dLys	1.09	0.91
脂肪	2.00	1.70	dMet	0.51	0.40
L－lys	0.17	0.15	dS－AA	0.79	0.67
DL－Met	0.18	0.11	dThr	0.77	0.67
磷酸氢钙	0.10	0.70	dTrp	0.24	0.20
石粉	0.95	1.03	dIle	0.72	0.62
氯化胆碱	0.30	0.30	dArg	1.22	1.05
食盐	0.26	0.26	dHis	0.48	0.43
矿物质	0.50	0.50			
复合多维	0.05	0.05			
小苏打	0.20	0.20			
盐霉素	0.06	0.06			
总计	100	100			

1.3 样品采集与制备

试验结束时，测体质量，从每组中分别随机选取 10 只鸡，心脏无菌采血 10ml，取 1ml

测定血液常规指标，其余血液用于制备血清，冰箱保存，备用。

1.4 测定指标和方法

1.4.1 生长性能

逐日按组定量给料，并回收剩料，称重，统计耗料量，耗料量用初始料重减去末料重来计算。分别于21、50d清晨逐只称重，以组为单位来计算平均日增重（ADG）、料重比（F/G）等。

1.4.2 血液常规指标

血液常规指标用全自动血细胞检测仪分析测定。

1.4.3 血清生化指标

用迈瑞BS－420型全自动生化分析仪测定血清中肌酐（Crea）、尿素（Urea）、白蛋白（ALB）含量，天门冬氨酸氨基转移酶（AST）和丙氨酸氨基转移酶（ALT）活性。血清抗氧化指标（GSH、GSH－Px、抗 O_2^-）采用T6紫外分光光度计（北京普析通用仪器公司）测定，试剂盒购自南京建成生物工程有限公司。

1.5 数据处理

使用SPSS17.0版统计软件中ANOVA过程进行数据分析，LSD法进行多重比较，显著水平为0.05，所有结果用 $\bar{x} \pm s$ 表示。

2 结果

2.1 酵母锌对肉鸡生长性能的影响

由表2可知，1～21d时，与其他组相比Ⅲ组平均日增重较低、料重比较高（$P<0.05$）。22～50d平均日增重Ⅰ组显著高于其他组（$P<0.01$），Ⅱ组和Ⅳ组之间无显著差异（$P>0.05$），但显著高于Ⅲ、Ⅴ组（$P<0.01$）；料重比Ⅰ组最低（$P<0.01$），Ⅱ组和Ⅳ组也极显著低于Ⅲ、Ⅴ组（$P<0.01$），两者之间差异显著（$P<0.05$）。1～50d平均日增重Ⅰ组显著高于其他组（$P<0.01$），Ⅱ、Ⅳ组也显著高于Ⅲ、Ⅴ组（$P<0.01$），后者之间无显著差异（$P>0.05$）；料重比Ⅲ、Ⅴ组＞Ⅳ组＞Ⅱ组＞Ⅰ组。

表2 酵母锌对肉鸡生长性能的影响

组别	1～21d		22～50d		1～50d	
	日增重/（g/d）	料重比	日增重/（g/d）	料重比	日增重/（g/d）	料重比
Ⅰ	24.07±3.05	1.71±0.64	45.44±2.68^{C}	2.38±0.11^{C}	37.33±1.49^{C}	1.96±0.05^{D}
Ⅱ	23.97±2.57	1.72±0.05	41.20±2.70^{B}	2.51±0.10^{Bb}	36.47±0.31^{B}	2.09±0.05^{C}
Ⅲ	20.42±3.66^{a}	1.83±0.31^{a}	36.84±3.02^{A}	2.68±0.22^{A}	31.67±4.50^{A}	2.25±0.28^{A}
Ⅳ	23.39±3.12	1.73±0.25	42.85±2.54^{B}	2.60±0.29^{B}	35.21±0.96^{B}	2.16±0.23^{B}
Ⅴ	22.65±2.06	1.78±0.33	36.81±3.58^{A}	2.65±0.34^{A}	31.23±4.94^{A}	2.23±0.30^{A}

注：同列（以后表格称为同行）数值右上角大写字母不同表示差异极显著（$P<0.01$），不同小写字母表示差异显著（$P<0.05$），相同字母或无字母为差异不显著（$P>0.05$）。下表同

2.2 酵母锌对肉鸡血液常规指标的影响

由表3可知，白细胞数各试验组之间无显著差异（$P>0.05$），红细胞数Ⅰ组极显著高

于其他组（$P<0.01$），其他各组之间无显著差异（$P>0.05$）。白细胞中淋巴细胞所占比例与其他组相比Ⅰ组最低（$P<0.01$），Ⅳ组最高（$P<0.01$）；单核细胞所占比例Ⅰ组最高（$P<0.01$），Ⅳ组最低（$P<0.01$），其他组之间无显著差异（$P>0.05$）；中性粒细胞Ⅰ组显著高于其他组（$P<0.01$），Ⅱ、Ⅳ组最低（$P<0.01$），Ⅲ、Ⅴ组之间无显著差异（$P>0.05$）。

表3　酵母锌对肉鸡血液常规指标的影响

指标	Ⅰ组	Ⅱ组	Ⅲ组	Ⅳ组	Ⅴ组
白细胞数（10^9/L）	41.67±14.31	36.76±6.75	44.12±10.82	30.72±4.77^A	37.53±12.05
红细胞数（10^{12}/L）	10.61±0.22^A	0.85±0.21	0.77±0.33	0.96±0.35	0.99±0.30
淋巴细胞（%）	75.81±9.29^A	88.05±7.32^a	84.41±6.98^a	91.73±1.44^b	85.20±6.52^a
单核细胞（%）	13.23±4.06^A	8.17±4.88^B	9.00±3.06^B	4.94±0.71^C	8.51±2.25^B
中性粒细胞（%）	10.96±2.63^A	3.78±2.77^B	6.59±1.22^C	3.33±0.95^B	6.29±1.52^C

2.3　酵母锌对肉鸡血液生化指标的影响

血清中ALT、AST活性，Crea、Urea含量各试验组之间没有显著差异（$P>0.05$），Ⅰ、Ⅱ、Ⅳ组ALB含量极显著高于Ⅲ、Ⅴ组（$P<0.01$）。结果见表4。

表4　不同锌源对肉鸡血液生化指标的影响

指标	Ⅰ组	Ⅱ组	Ⅲ组	Ⅳ组	Ⅴ组
ALT（μg/L）	17.12±3.49	16.02±2.67	17.22±3.03	15.90±2.53	16.07±2.25
AST（U/L）	276.50±8.19	283.21±15.6	279.33±13.50	273.33±9.89	277.05±16.08
ALB（g/L）	17.6±1.53^A	17.3±0.80^A	16.73±1.82^a	17.3±1.23^A	15.73±2.44^b
Crea（μmol/L）	23.05±5.66	21.56±4.45	20.33±4.04	23.67±5.81	22.12±3.46
Urea（mmol/L）	2.43±0.67	2.38±0.84	2.40±0.91	2.36±0.45	2.25±0.83

2.4　酵母锌对肉鸡抗氧化指标的影响

在表5中，血清中抗$O_2^{-\cdot}$活力Ⅰ组显著高于其他组（$P<0.01$）。GSH-Px活力Ⅰ、Ⅱ组显著高于其他组（$P<0.01$），且Ⅱ组高于Ⅰ组，2组之间差异显著（$P<0.05$）。血清中GSH的含量Ⅰ、Ⅱ组显著高于其他组（$P<0.01$），Ⅲ组高于Ⅳ、Ⅴ组（$P<0.01$）。

表5　酵母锌对肉鸡抗氧化指标的影响

指标	Ⅰ组	Ⅱ组	Ⅲ组	Ⅳ组	Ⅴ组
抗$O_2^{-\cdot}$（U/L）	201.81±11.21^A	188.17±13.01	187.06±8.54	182.25±5.74	180.16±6.25
GSH-Px（U/ml）	1 167.47±77.80^{Ab}	1 270.35±78.38^A	1 014.27±54.86	969.02±61.79	987.61±97.16
GSH（mg/L^1）	21.79±1.24^A	16.96±2.66^B	7.55±1.80^C	3.59±1.24	3.06±0.67

3 讨论

3.1 酵母锌对肉鸡生长性能的影响

在鸡的日粮中添加适量酵母菌可显著提高鸡的日增重、饲料转化率、成活率和上市体质量等生产指标[3~5]。Najib 等[6]研究表明在蛋鸡饲粮中添加 0.3% 的酵母菌，其生长性能和饲料转化率都比对照组好，肉鸡日粮中添加 0.3% 酵母菌可以改善肉仔鸡生长性能。本试验结果表明酵母锌、空白酵母对 1 ~21 日龄肉仔鸡的生产性能无显著影响，但能显著提高22 ~50 日龄肉仔鸡生长性能，与对照组及硫酸锌组相比酵母锌组可显著提高肉鸡的生长性能。

3.2 酵母锌对肉鸡血液常规指标的影响

动物体内血液学指标是衡量动物机体机能状况好坏的一个重要指标。白细胞中的淋巴细胞、单核细胞等具有一定的免疫、吞噬、杀菌的作用。血液中 80% ~90% 的淋巴细胞属于 T 淋巴细胞，执行细胞免疫的功能。本试验结果表明，50 日龄时，Ⅳ组白细胞含量较低（$P<0.01$），Ⅰ组外周血白细胞数高于其他组但差异不显著（$P>0.05$）。由于白细胞中含大量的淋巴细胞、单核细胞等免疫细胞，因此，酵母锌（Zn 50mg/kg）在提高肉鸡的免疫机能方面可能有较好的效果。红细胞不仅运送氧气和二氧化碳，同时也具有重要的免疫功能。它具有识别、黏附、浓缩、杀伤抗原、清除免疫复合物的能力，参与机体免疫调控。而酵母锌Ⅰ组（Zn 50mg/kg）红细胞数极显著高于其他组（$P<0.01$），说明酵母锌对于肉鸡的细胞免疫具有增强作用，参与机体免疫调控的细胞数量增多，有利于提高机体免疫力和清除体内的病原微生物。

3.3 酵母锌对肉鸡血液生化指标的影响

临床上常用血清 AST、ALT 作为检查肝功能的指标。本试验结果显示，酵母锌组的血清 ALT、AST 与对照组相比差异不显著（$P>0.05$），说明试验中酵母锌的添加剂量在一个饲养期内对肉鸡肝脏功能无损害。白蛋白（ALB）由肝脏合成，正常肝脏合成 ALB 的速度快、效率高，其生理功能除了维持血浆渗透压，提供机体蛋白质来源和用于修补组织和提供能量外，还是营养物质的载体。肝脏功能损害时，肝脏合成 ALB 的量明显减少，并与肝脏病变的严重程度相平行。血清 ALB 在一定程度上代表了饲粮中蛋白质的营养水平及动物对蛋白质的消化吸收程度。Vernon 等[7]研究认为，血清 ALB 含量是衡量动物蛋白质需要量很敏感的标识。本试验结果显示，酵母组 ALB 含量极显著高于硫酸锌组和对照组（$P<0.01$），表明肉鸡的肝脏功能活性较强，酵母菌可提高肉鸡的肝脏功能及对蛋白质的消化吸收能力。

肌酐（Crea）是反映肾功能的主要指标之一。内生肌酐是肌肉代谢的产物。在肌肉中，肌酸主要通过不可逆的非酶脱水反应缓缓地形成肌酐，再释放到血液中，随尿排泄。因此血肌酐与体内肌肉总量关系密切，不易受饮食影响。肌酐是小分子物质，可通过肾小球滤过，在肾小管内很少吸收，每日体内产生的肌酐，几乎全部随尿排出，一般不受尿量影响。肾功能不全时，肌酐在体内蓄积成为对机体有害的毒素。本试验结果显示各组之间肌酐含量均无显著差异（$P>0.05$），表明酵母锌不影响肉鸡的肾功能。

通过测定尿素氮（Urea），可以了解肾小球的滤过功能。它主要是经肾小球滤过，并随尿液排出体外。当肾实质受损害时，肾小球滤过率降低，血液中血清尿素氮的浓度就会增加。本试验结果显示，各试验组之间无显著差异（$P>0.05$），表明酵母锌对肉鸡肾小球的滤过功能无影响。

3.4 酵母锌对肉鸡抗氧化指标的影响

锌参与机体抗氧化防御系统[8]，通过影响自由基和改变抗氧化酶与底物的状态对机体抗氧化体系产生重要的作用，它能防止细胞膜氧化、减少超氧阴离子形成。机体在锌缺乏时，会造成脂质过氧化增加，可能诱发机体产生氧化损伤，而补加锌能使该损伤得以恢复[9~11]。$O_2^{-\cdot}$是生物体多种自由基的源头，$O_2^{-\cdot}$有多种反应性，它既可以作为还原剂，又可以作为氧化剂，还可以作为碱、亲核物和配体参与反应。$O_2^{-\cdot}$的损伤效应主要是使核酸断裂，多糖解聚，不饱和脂肪酸过氧化，造成损伤等。$O_2^{-\cdot}$虽毒性小，但是存在寿命长[12]，因而损伤细胞和机体作用不小。本试验结果表明，酵母锌（Zn 50mg/kg）能够极显著提高血清中抗$O_2^{-\cdot}$的活性。可见，日粮中添加酵母锌（Zn 50mg/kg）可以在一定程度上提高机体抗$O_2^{-\cdot}$自由基活性，减轻活性氧自由基对机体的损伤。GSH－Px是体内重要的抗氧化酶之一，可防止因自由基产生的脂质过氧化物堆积所造成的对生物膜的损害，降解过氧化氢，从而维持细胞的完整及功能的正常。GSH为体内含巯基非蛋白化合物，维持细胞内氧化还原状态，保护蛋白质及酶分子免受内源性和外源性氧化作用，维持细胞功能状态[13]，抑制细胞内自由基的产生，特别是活性氧的中间产物，过量的自由基可能会造成GSH的消耗[14]。本试验结果表明，日粮中添加酵母锌（Zn 50mg/kg）极显著提高50日龄肉鸡的血清GSH－Px活性；与对照组相比补锌组能显著增加血清GSH的含量，且酵母锌优于硫酸锌，其中锌水平为50mg/kg酵母锌效果最佳；说明日粮中添加酵母锌可增强肉鸡的抗氧化功能。

参考文献

[1] Wardroup P W. Dietary nutrient allowances for chickens and turkeys [J]. Feedstuffs, 2004, 76 (38): 42－47.

[2] Ashmead H D. The roles of amino acid chelates in animal nutrition [M]. Park Ridge N J: Noyes Publiccations, 1992: 479.

[3] Bradley G L, Savage T F, Timm K I. The effects of supplementing diets with *Saccharomyces cerevisiae* varboulardii on male poult performance and ileal morphology [J]. Poult Sci, 1994 (73): 1 766－1 770.

[4] 周淑芹，孙文志，魏树龙．YC对肉仔鸡生长性能和免疫机能的影响研究［J］．饲料博览，2003（6）：1－3.

[5] 刘观忠，赵国先，安胜英，等．YC对蛋雏鸡生产性能、肠壁结构及免疫机能的影响［J］．山东家禽，2004（10）：10－13.

[6] Najib H. Effect of incorporating yeast culture *Saccharomyces cerevisiae* into the Saudi Baladi and White Leghorn layer's diet [J]. J Appl Anim Res, 1996, 10 (2): 181－186.

[7] Vernon R, Marchini J S, Cortiella J, et al. Assessment of protein nutritional status (reviews) [J]. J Nutr Suppl, 1990, 120 (11): 1 496－1 501.

[8] Powell I. The antioxidant properties of zinc [J]. J Nutr, 2000, 130: 1 447－1 454.

[9] Yousef M I, El－Hendy H A, El－Demerdashi I, et al. Dietary zinc deficiency induced－changes in the activity of enzymes and the levels of free radicals, lipids and protein electrophoretic behavior in growing rats [J]. Toxi－cology, 2002, 175: 223－234.

[10] Patricia I O, Michael S C, Zago M P, et al. Zinc deficiency induces oxidative stress and AP－1activation in 3T3 cells [J]. Free Rad Biol Med, 2000, 28 (7): 1 091－1 099.

[11] 张彩英，胡国良，郭小权，等．日粮锌源和锌水平对断奶仔猪免疫功能和抗氧化酶活性的影响［J］．中国兽医学报，2011，31（9）：1 354－1 357.

[12] Seal C J, Heaton F W. Chemical factors affecting the intestinal absorption of zinc *in vitro* and *in vivo* [J]. J Nutr, 1983, 50 (2): 317 - 324.

[13] Reid M, Jahoor F. Glutathione in disease [J]. Curr Opini Clini Nutr Metab Care, 2001, 4 (1): 65 - 71.

[14] Cao J, Henry P R, Guo R, et al. Chemical characteristics and relative bioavailability of supplemental organic zinc sources for poultry and ruminants [J]. J Anim Sci, 2000, 78: 2 039 - 2 054.

（发表于《中国兽医学报》）

常山总碱亚急性毒性试验研究

郭志廷，梁剑平，王学红，尚若锋，郭文柱，都宝成，华兰英

（中国农业科学院兰州畜牧与兽药研究所，农业部兽用药物创制重点实验室，
甘肃省新兽药工程重点实验室，兰州 730050）

摘　要：为了评价常山总碱的安全性，为今后系统研究其药理作用和临床安全用药提供试验依据，在小鼠急性毒性试验的基础上，进行亚急性毒性试验。将 80 只大鼠随机分成常山总碱高、中、低剂量组和空白对照组，药物组按照大鼠体重 6. 50g/kg，3. 25g/kg 和 1. 63g/kg 灌胃给药，连续给药 14d，所有大鼠分 2 次捕杀（第 7 天和第 14 天）。结果表明，常山总碱低、中剂量组大鼠体重、脏器系数、血液生理指标、血液生化指标、组织病理学变化等与对照组比较均无显著性差异（$P>0.05$）；高剂量组对上述部分指标有一定影响。结果提示，常山总碱毒性很低，临床用药安全性高。

关键词：常山总碱；亚急性毒性；抗球虫

（发表于《中国兽医学报》）

阿司匹林丁香酚酯的 Amse 试验

孔晓军，李剑勇，刘希望，杨亚军，张继瑜，周旭正，
魏小娟，李　冰，牛建荣

（中国农业科学院兰州畜牧与兽药研究所，农业部兽用药物创制重点实验室，
甘肃省新兽药工程重点实验室，兰州 730050）

摘　要： 对阿司匹林丁香酚酯进行鼠伤寒沙门氏菌回复突变试验，评价该化合物是否具潜在致突变性。以鼠伤寒沙门氏菌突变型菌株 TA97、TA98、TA100、TA102 为试验菌株，以及由多氯联苯（PCB）诱导的大鼠肝匀浆（S－9）制备的 S－9 混合液为代谢活化系统，经平板掺入法后计数各菌株回复突变菌落数。各菌株中，TA97 和 TA98 可以检测各种移码型诱变剂；TA100、TA1535 可以检测引起碱基对置换的诱变剂；TA102 可以检测其他测试菌株不能检出或极少检出的诱变剂：S－9 混合液则用以代谢活化某些需要代谢酶才能使沙门氏菌突变型产生回复突变的诱变剂。各菌株经特性鉴定后均符合 Ames 试验标准，S－9 生物活性鉴定合格。AEE 的 LD_{50} 为10.39g/kg，故最高剂量设为 50mg/ml，中间剂量为 25mg/ml、12.5mg/ml、6.25mg/ml，低剂量为 3.125mg/ml：同时设阳性诱变剂（叠氮钠、2－氨基芴）对照组、溶剂（DMSO）对照组、空白对照组、阿司匹林对照组、丁香酚对照组，各组均包括加 S－9 和不加S－9 两种情况，各剂量分别做二个平行平皿。试验时将培养至对数生长期的增菌培养液 100μl 加入 2ml 顶层培养基中，混匀；加各浓度 AEE100μl，活化时需要加入 S－9 混合液 500μl，再混匀，迅速倒入底层培养基上，转动平皿使顶层培养基均匀分布，固化后放入 37 ℃恒温培养箱培养 48h，观察结果。双盲法阅片，以以下公式计算突变率：突变率＝诱发回变菌落数/自发同变菌落数。在不同浓度下，无论加或不加 S9，各剂量组均未引起回复突变菌落数明显增加，亦无线性关系。Amse 实验是国际上普遍认可的化合物致突变检测方法，其灵敏度和可靠性毋庸置疑，本实验将 AEE 与标准菌液混合后在最低营养条件下培养，计算突变率，并且分别做了＋S9 和－S9 两种情况，在实验当中严格无菌操作，排除了各种外界环境导致的假阴性结果。结果显示，各受试物剂量组回复突变菌落数在有无 S9 情况下其突变率均未达到自发回变菌落数的二倍，其表明 AEE 无致突变性。

关键词： 阿司匹林丁香酚酯；Ames 试验

（发表于《中国毒理学会兽医毒理学与饲料毒理学学术讨论会暨兽医毒理专业委员会第四次全国代表大会会议论文录》）

阿司匹林丁香酚酯对大鼠血液学和血液生化指标影响

李剑勇，于远光，杨亚军，张继瑜，周旭正，魏小娟，李 冰，刘希望

（中国农业科学院兰州畜牧与兽药研究所，农业部兽用药物创制重点实验室，
甘肃省新兽药工程重点实验室，兰州 730050）

摘 要：观察阿司匹林丁香酚酯对大鼠血液学成分及血液生化指标成分的影响，为其临床开发应用提供依据。选择成年健康 wistar 大鼠随机分为四组：空白对照组、阿司匹林丁香酚酯组、阿司匹林与丁香酚药物对照组，连续灌服相应药物两周，在最后一次灌药后每组大鼠各取 1/2 股动脉取血，留下的 1/2 动物停止给药后继续观察 1 周再取血。采用全自动血液分析仪与全自动血液生化分析仪测定大鼠的红细胞、血小板、甘油三酯、总胆固醇、碱性磷酸酶等血液学、血清生化各项指标的变化。阿司匹林丁香酚酯对大鼠血液成分及血液生化有一定的影响，且影响较阿司匹林小；具有显著降低大鼠血清甘油三酯、总胆固醇含量的作用，且强于阿司匹林与丁香酚。阿司匹林丁香酚酯对血液及血液生化指标影响较低，较对照药具有更好的药用价值。

关键词：阿司匹林丁香酚酯；血液学指标；血液生化指标

（发表于《中国毒理学会兽医毒理学与饲料毒理学学术讨论会暨兽医毒理专业委员会第四次全国代表大会会议论文录》）

超临界 CO_2 萃取凤眼草挥发油化学成分

刘　宇，梁剑平，华兰英，尚若峰，王学红，吕嘉文，李　冰

（中国农业科学院兰州畜牧与兽药研究所，甘肃省新兽药工程重点实验室，兰州 730050）

摘　要：为了研究凤眼草挥发油的化学成分，本文采用超临界 CO_2 萃取技术从凤眼草中提取挥发油，并用 GC－MS 法采用最佳分析条件对化学成分进行鉴定，用峰面积归一法测定各化合物在挥发油中的相对百分含量；通过研究，鉴定出 42 种化合物，其峰面积相对含量约占挥发油总量的 92.91 %。凤眼草挥发油的主要组分为邻苯二甲酸异丁辛酯（7.05 %），棕榈酸（8.74 %），（Z，Z）－9，12－十八碳二烯酸(20.47 %)，（Z）－9，17－十八碳二烯醛（15.14 %），22，23－二氢豆甾醇（14.09 %），3，5，6，7，8，8a－六氢－4－8a－二甲基－6－［1－甲乙烯基］－2（1H）萘酮(6.86 %)，α－香树脂醇（2.04 %），羽扇烯酮（3.53 %）等。

关键词：凤眼草；超临界二氧化碳萃取；挥发油；气相色谱－质谱

（发表于《食品研究与开发》）

大鼠及小鼠口服氢溴酸槟榔碱片剂的急性毒性试验

周绪正，张继瑜，李金善，李剑勇，魏小娟，牛建荣，李　冰

（中国农业科学院兰州畜牧与兽药研究所，农业部兽用药物创制重点实验室，甘肃省新兽药工程重点实验室，兰州 730050）

摘　要：为了研究氢溴酸槟榔碱片剂对小鼠及大鼠的口服急性毒性，试验以小鼠及大鼠为实验动物，经口一次性灌胃给药，统计死亡率，用改良寇氏法计算半数致死量（LD_{50}）并观察毒性作用。结果表明：氢溴酸槟榔碱片剂对小鼠的口服 LD_{50} 为 6 545 mg/kg，其 95% 可信区间为 5 534 ～7 740 mg/kg；对大鼠的最大耐受量（LD_0）大于 10.0 g/kg，则其 LD_{50} 必大于 10.0 g/kg。说明氢溴酸槟榔碱片剂属于低毒性的或实际无毒的药物，以 20 ～50 mg/kg 的剂量应用于临床非常安全。

关键词：氢溴酸槟榔碱；小鼠；大鼠；口服；急性毒性

（发表于《黑龙江畜牧兽医》）

高效液相色谱－串联质谱法测定血浆中氢溴酸槟榔碱的含量

李　冰，周绪正，杨亚军，李剑勇，牛建荣，魏小娟，李金善，张继瑜

（中国农业科学院兰州畜牧与兽药研究所，农业部兽用药物创制重点实验室，
甘肃省新兽药工程重点实验室，兰州 730050）

摘　要：建立测定氢溴酸槟榔碱血药浓度的 HPLC－MS/MS 定量分析方法。采用 Zorbax SB C18（2.1mm×30mm，3.5μm）分析柱，以甲醇－甲酸铵水溶液（pH 值 4.5）（15：85）为流动相，流速为 0.4 ml·min^{-1}；采用 ESI＋MRM 进行离子监测。用于定量分析的母/子离子对为 m/z 156.2→53.0（槟榔碱）。氢溴酸槟榔碱片剂检测方法的线性范围为 5～120ng·ml^{-1}，最低检测限为 4ng·ml^{-1}。平均加样回收率为 99.36%，RSD 为 1.06%，日内、日间精密度 RSD 均小于 3%。本方法灵敏、准确，可用于氢溴酸槟榔碱血药浓度检测，为其在动物体内药代动力学的研究奠定基础。

关键词：槟榔碱；液相色谱－质谱联用；血药浓度

（发表于《第十一届全国青年药学工作者最新科研成果交流会论文集》）

苦参碱抗病毒作用研究进展

程培培[1,2]，李剑勇[1]，杨亚军[1]，刘希望[1]

（1. 中国农业科学院兰州畜牧与兽药研究所，农业部兽用药物创制重点实验室，
甘肃省新兽药工程重点实验室，兰州 730050；2. 甘肃农业大学动物医学院）

苦参碱是从苦参、山豆根、苦豆子等主要槐属植物中提取的主要生物碱之一，1958 年首次被分离出来[1]，分子式为 $C_{15}H_{24}N_2O$。近年来对苦参碱大量的药理和临床研究发现，其具有抗病毒、抗纤维化、抗炎、抗肿瘤等作用；对中枢神经系统有镇静、镇痛、解热、降温作用；还可强心、降压、抗心律失常；对皮肤疾病如过敏性皮炎、湿疹以及妇科病等疗效显著，特别是对各种肿瘤、慢性肝炎等不良反应较少，研究开发的各类制剂已被广泛用于临床[2]。虽然用苦参碱治疗病毒性肝炎的研究报道很多，但其抗病毒机制并不十分明确；关于苦参碱对其他病毒的作用机制的研究也鲜见报道。笔者等对苦参碱的抗病毒作用、机制以及应用作一综述。

（发表于《中兽医医药杂志》）

牛双芽巴贝斯虫 rap－1 蛋白的真核表达

韩　琳[1,2]，张继瑜[2]，袁莉刚[1]，李　冰[2]

（1. 甘肃农业大学动物医学院，兰州 730070；2. 中国农业科学院兰州畜牧与兽药研究所/农业部兽用药物创制重点实验室/甘肃省新兽药工程重点实验室，兰州 730050）

摘　要：以 846bp 的牛双芽巴贝斯虫兰州株潜在药物靶标 pGEM－T－*rap*－1 质粒为模板，选择真核表达载体 GFP，构建真核表达质粒。采用脂质体转染法，将质粒转染绵羊成纤维细胞，通过 G418 筛选，筛选阳性克隆。结果显示牛双芽巴贝斯虫 *rap*－1 蛋白在绵羊成纤维细胞内成功表达。

关键词：牛双芽巴贝斯虫；*rap－1* 蛋白；转染

（发表于《江苏农业科学》）

牛源金黄色葡萄球菌的耐药性及耐甲氧西林金黄色葡萄球菌的检测

苏　洋[1,2]，蒲万霞[1]，陈智华[2]，邓海平[1]

（1. 中国农业科学院兰州畜牧与兽药研究所/农业部兽用药物创制重点实验室/甘肃省新兽药工程重点实验室，兰州 730050；2. 甘肃农业大学动物医学院，兰州 730070）

摘　要：了解内蒙古地区奶牛乳房炎金黄色葡萄球菌耐药性和耐甲氧西林金黄色葡萄球菌（MRSA）感染的情况，为奶牛乳房炎的防治提供理论依据。采用 K－B 纸片扩散法，检测分离自内蒙古地区 38 株金黄色葡萄球菌对 17 种药物的敏感性，同时用琼脂稀释法检测苯唑西林、万古霉素对金黄色葡萄球菌的最小抑菌浓度（MIC）；再用头孢西丁、苯唑西林纸片扩散法、苯唑西林盐琼脂筛选法和 PCR 方法扩增 mecA 耐药基因对分离菌株进行全面 MRSA 检测。分离菌株对每种抗生素都有不同程度抗性，对氨苄西林、头孢拉丁、青霉素、复方新诺明、新生霉素和链霉素的耐药率都高于 45%，而对氧氟沙星、丁胺卡那霉素、万古霉素、环丙沙星、庆大霉素和头孢唑林的敏感性高于 90%，2 株细菌的万古霉素 MIC≥16μg · ml^{-1}；其中 8 株细菌的苯唑西林 MIC≥8μg · ml^{-1}，而其他菌株的苯唑西林 MIC≤2μg · ml^{-1}；分离菌株多重耐药情况严重，耐受 3 种及 3 种以上药物的菌株占 84.21%，其中 4 株细菌能同时耐受 9 种不同抗菌药物；16（42.11%）株细菌被检测携带 mecA 耐药基因，而仅有其中 7 株的苯唑西林 MIC≥4μg · ml^{-1}；头孢西丁、苯唑西林纸片扩散法和苯唑西林盐琼脂筛选法分别检出 7 株、10 株和 7 株表型为 MRSA 的菌株。分离菌株的耐药性和多重耐药现象较为严重，被调查地区奶牛场中已经存在 MRSA 和 OS－MRSA 感染情况，且感染率高。

关键词：金黄色葡萄球菌；耐甲氧西林金黄色葡萄球菌；耐药性；mecA 基因；牛乳房炎

（发表于《中国农业科学》）

青蒿琥酯纳米乳中青蒿琥酯的分光光度法和HPLC法分析

李均亮，李　冰，周绪正，张继瑜

（中国农业科学院兰州畜牧与兽药研究所/农业部兽用药物创制重点实验室/甘肃省新兽药工程重点实验室，兰州 730050）

摘　要：比较了HPLC法和紫外分光光度法测定青蒿琥酯的含量，HPLC法和紫外分光光度法的线性范围分别为5～50 μg/ml（R^2 =0.9992）和5～100μg/ml（R^2 =0.9967），两种方法的分析结果无明显差异，表明这两种方法可较为准确地测定青蒿琥酯纳米乳中青蒿琥酯的含量，其中，紫外分光光度法的操作简单，适合分析杂质含量低的样品，而HPLC法分辨率高，适用于成分复杂青蒿琥酯制剂的检测。

关键词：高效液相色谱；分光光度法；青蒿琥酯；定量分析

（发表于《湖北农业科学》）

塞拉菌素透皮制剂的皮肤刺激性试验

周绪正，张继瑜，汪　芳，李金善，李剑勇，李　冰，牛建荣，魏小娟，杨亚军，刘希望

（中国农业科学院兰州畜牧与兽药研究所/农业部兽用药物创制重点实验室/甘肃省新兽药工程重点实验室，兰州 730050）

塞拉菌素透皮制剂是兼具抗体内寄生虫和抗体外寄生虫双重作用的新一代大环内脂类抗寄生虫药，对蜱、螨、蚤、虱、恶丝虫、部分线虫都有很好的驱杀效果，且吸收较好，半衰期较长，对靶动物无毒副作用，治疗效果确实，为进一步探讨临床用药的安全性，笔者对其进行了皮肤刺激性试验。

（发表于《黑龙江畜牧兽医》）

牛双芽巴贝斯虫 *rap* –1 基因的克隆与序列分析

韩　琳[1,2]，张继瑜[2]，袁莉刚[1]，李　冰[2]

（1. 甘肃农业大学动物医学院，兰州 730070；2. 中国农业科学院兰州畜牧与兽药研究所/农业部兽用药物创制重点实验室/甘肃省新兽药工程重点实验室，兰州 730050）

摘　要：对牛双芽巴贝斯虫（*Babesia bigemina*）潜在药物靶标 Rap –1 的基因进行克隆与序列分析。以兰州株牛双芽巴贝斯虫基因组为模板，经 PCR 扩增获得了 *rap* –1 基因，将该基因克隆到 pGEM – T Easy Vector 上，对重组质粒进行 PCR 扩增，对目的基因进行酶切鉴定及序列测定分析。结果表明，*rap* –1 基因长度为 846bp，同源性分析结果显示，克隆序列与 GenBank 收录的巴西株牛双芽巴贝斯虫的核苷酸序列同源性为 99.74%，说明兰州株牛双芽巴贝斯虫的 *rap* –1 与 GenBank 公布的参考基因具有高度同源性，*rap* –1 基因具有很高的保守性。

关键词：牛双芽巴贝斯虫（*Babesia bigemina*）；*rap* –1 基因；克隆；序列分析

（发表于《湖北农业科学》）

不同中药方剂防治鸡传染性支气管炎的疗效研究

王 玲[1]，陈灵然[1]，郭天芬[1]、李宏胜[1]，杨 峰[1]，胡广胜[2]，王旭荣[1]

（1. 中国农业科学院兰州畜牧与兽药研究所，农业部兽用药物创制重点实验室，甘肃省新兽药工程重点实验室，兰州 730050；2. 甘肃农业大学动物医学院，兰州 730070）

摘 要：介绍防治鸡传染性支气管炎（Infectious Bronchitis，IB）的中药方剂与临床效果，总结防治鸡 IB 中药的种类，探讨中药防治鸡 IB 研究存在的问题与应用前景。综述了以清热解毒为主，化痰止咳、清利咽喉为辅组成的中药复方制剂在防治鸡 IB 方面的应用现状及效果观察，归纳了防治呼吸型、肾形鸡 IB 的中药方剂种类，中西药结合防治鸡 IB 以及中西药临床疗效比较研究。鸡传染性支气管炎以呼吸道症状、产蛋量下降或肾脏病变为主要特征，现多采用疫苗免疫和西药治疗，由于 IBV 毒株具有高度易变异和多血清型的特点，现有疫苗缺乏交叉免疫或完全无交叉免疫，给鸡传染性支气管炎的诊断和防治带来了很大困难，发病后尚无特效药物治疗。中兽医学虽然没有鸡传染性支气管炎这一病名，但其所表现的临床特征与中兽医学肺、脾、肾三脏的病理病机密切相关，按照中兽医理论，鸡 IB 乃疫疠毒邪犯肺，痰浊壅塞，导致肺失宣肃、气滞血凝、津液失于布散所致，其病之根本在肺。中医治疗鸡传支坚持祛邪与扶正两大原则，祛邪是祛除有害病因——疫毒之邪，即抑制或消除致病因子，排除病理产物；扶正是通过增强机体的抗病力，提高免疫功能，扶止普遍应用于各种传染病的始终，基本原则是养阴保津。现代药理研究表明，有些中药不但具有抑杀病毒的作用，且能调节机体的免疫功能，具有非特异性的抗病毒功能。中药在防治鸡 IB 方面有着很大潜力和优势，采用中兽医之理、法、方、药防治，特别是同一方剂可防治鸡不同的呼吸道疾病，更证明了中兽医学“异病同治”辨证论治原则的应用价值。清热解毒类、凉血活血类、补益类中药能直接降解病原体的毒素，还能增强机体的免疫功能，因此，防治 IB 的方剂中把清热解毒类药作为主药用来抑制病毒，辅以清热化痰、平喘止咳药，达到祛邪的目的，且协同补益类中药共同增强机体免疫力，达到扶正的目的。中药避免了抗病毒药、抗生素类药物的不合理应用和禽蛋产品中的药物残留，用于临床收效甚好，所用方剂与西药相比具有独特的优势，经过科学配制可对禽病防治起到很大作用。在此基础上探讨了中药防治鸡 IB 的可能性，研发抗鸡 IB 中药所面临的问题及途径，以期进一步完善并扩大中药在防治畜禽传染性疾病上的应用。

关键词：中药；防治；鸡传染性支气管炎；疗效

（发表于《中国毒理学会兽医毒理学与饲料毒理学学术讨论会暨兽医毒理专业委员会第四次全国代表大会会议论文录》）

蛋鸡球虫病中西医结合治疗措施

郭志廷[1]，梁剑平[1]，罗晓琴[2]

（1. 中国农业科学院兰州畜牧与兽药研究所，农业部兽用药物创制重点实验室，甘肃省新兽药工程重点实验室，兰州 730050；2. 兰州市动物卫生监督所，兰州 730050）

（全文省略）

（发表于《中国家禽》）

动物性食品中兽药残留分析检测技术研究进展

李　冰，李剑勇，周绪正，杨亚军，牛建荣，魏小娟，李金善，张继瑜

（中国农业科学院兰州畜牧与兽药研究所，农业部兽用药物创制重点实验室，甘肃省新兽药工程重点实验室，兰州 730050）

摘　要：兽药作为动物疾病防治的重要保障，对改善动物生产性能和产品品质、保持生态平衡等多方面具有重要功能，是畜牧业健康发展、食品安全和公共卫生的重要保障。但不合理使用和滥用兽药及饲料药物添加剂的情况普遍存在，既造成了动物食品中有害物质残留，又对人类健康造成损害，同时还威胁到环境和畜牧业持续健康发展。为保证人类健康，迫切需要开发简洁、快速、高灵敏度、高通量且低成本的兽药残留检测技术。本文对动物性食品中兽药残留分析的样品前处理方法和检测技术进行了研究探讨。

关键词：动物性食品；兽药残留；样品前处理；检测技术

（发表于《畜牧与兽医》）

复方茜草灌注液的急性毒性试验

王学红，梁剑平，郭文柱，刘　宇，郝宝成

（中国农业科学院兰州畜牧与兽药研究所/农业部兽用药物创制重点实验室/
甘肃省新兽药工程重点实验室，兰州 730050）

摘　要：为了评价复方茜草灌注液的安全性，试验以小白鼠为试验动物进行了急性毒性试验。结果表明：小白鼠口服复方茜草灌注液的半数致死量（LD_{50}）为 6.59g/kg，其 95% 可信范围为 6.06～7.01g/kg，根据中药毒性的分级标准可以确定为实际无毒。

关键词：复方茜草灌注液；小白鼠；急性毒性

（发表于《黑龙江畜牧兽医》）

复方茜草灌注液对小鼠蓄积毒性试验

王学红，梁剑平，郭文柱，刘　宇，郝宝成

（中国农业科学院兰州畜牧与兽药研究所，农业部兽用药物创制重点实验室，
甘肃省新兽药工程重点实验室，兰州 730050）

摘　要：为评价复方茜草灌注液的安全性，采用 20d 剂量递增连续染毒法对小鼠进行蓄积毒性试验，观察给药期间小鼠的体重变化和死亡情况，以蓄积系数评价其蓄积毒性。结果显示，各试验组小鼠未发现有明显毒性反应，无小鼠死亡，但对小鼠的体重增长有一定影响。表明复方茜草灌注液在小鼠体内无明显的蓄积毒性作用。

关键词：复方茜草灌注液；蓄积毒性；小鼠

（发表于《中兽医医药杂志》）

金丝桃素对免疫抑制小鼠免疫功能和抗氧化能力的影响

胡小艳[1,2]，尚若峰[1,2]，刘　宇[1]，王学红[1]，华兰英[1]，石广亮[3]，梁剑平[1]

（1. 中国农业科学院兰州畜牧与兽药研究所，农业部兽用药物创制重点实验室，甘肃省/中国农业科学院新兽药工程重点实验室，兰州 730050；2. 中国农业科学院研究生院，北京 100081；3. 甘肃农业大学动物医学院，兰州 730070）

摘　要：为观察金丝桃素对免疫抑制小鼠的免疫功能和抗氧化能力的影响，给小鼠皮下注射地塞米松（Dex）1.25mg·kg^{-1}，建立免疫抑制模型，分别以600和300mg·kg^{-1}两个剂量的金丝桃素混悬液（含金丝桃素分别为3.6mg·kg^{-1}和7.2mg·kg^{-1}）对小鼠灌胃，测定胸腺指数和脾脏指数、丙二醛（MDA）、H_2O_2、NO以及对自由基产生和清除相关的6种酶活性。结果表明，高剂量金丝桃素能显著提高免疫抑制小鼠的脾脏指数（$P<0.05$）；高低剂量金丝桃素均能极显著降低免疫抑制小鼠的MDA、NO和黄嘌呤氧化酶（XOD）浓度（$P<0.01$），极显著提高免疫抑制小鼠的总超氧化物歧化酶（T－SOD）、总抗氧化力（T－AOC）、谷胱苷肽过氧化物酶（GSH－PX）和过氧化物酶（POD）浓度（$P<0.01$），对H_2O_2的水平没有影响。说明，金丝桃素通过提高脾脏指数，降低过氧化物产物MDA和NO水平，改变机体内对自由基产生和清除相关酶活性来影响体内自由基的产生和清除，使机体免疫器官免受损伤，以拮抗Dex所致的免疫抑制，增强机体的免疫功能。

关键词：金丝桃素；免疫抑制小鼠；免疫器官指数；自由基；酶活性

（发表于《西北农业学报》）

金丝桃素粉剂对人工感染传染性囊病病毒鸡的疗效试验

尚若锋，刘　宇，陈灵然，郭文柱，郭志廷，梁剑平

（中国农业科学院兰州畜牧与兽药研究所，农业部兽用药物创制重点实验室，甘肃省新兽药工程重点实验室，兰州 730050）

摘　要：金丝桃素是从贯叶连翘中提取的一种极具抗病毒活性成分。为了研究金丝桃素粉剂对腔上囊的治疗效果，对20日龄雏鸡人工感染传染性囊病病毒（IBDV BC－6/85）后，以不同的剂量连续4d口服金丝桃素粉剂。通过观察腔上囊和肌肉的出血情况，以及病原分离，评价该药物对鸡传染性囊病的治疗效果。结果表明：以667.9mg/kg体重的金丝桃素粉剂连续口服给药4d，能有效的治疗传染性囊病，效果优于对照药物高免卵黄抗体。

关键词：金丝桃素粉剂；传染性囊病病毒；雏鸡；治疗效果

（发表于《中国兽医杂志》）

苦椿皮提取物对小鼠腹泻防治作用研究

程富胜[1]，刘　宇[1]，张　霞[2]，王　慧[1]，董鹏程[2]，曹会萍[2]

（1. 中国农业科学院兰州畜牧与兽药研究所，农业部兽用药物创制重点实验室，甘肃省新兽药工程重点实验室，兰州 730050；2. 甘肃农业大学生命科学技术学院）

摘　要：为了探讨苦椿皮提取物对小鼠腹泻防治作用，取体重相近的实验小鼠64只，随机分为低剂量组、高剂量组、模型组和空白对照组，每组16只，采用番泻叶灌胃建立动物腹泻模型，测定各组小鼠碱性磷酸酶、乳酸脱氢酶、淀粉酶和脂肪酶活性。结果低剂量组与高剂量组各种酶的活性极显著高于模型组（$P<0.01$），低剂量组与高剂量组酶活性差异不显著（$P>0.05$）。与正常小鼠相比，苦椿皮提取物能够显著增强腹泻小鼠各种酶活性，表明苦椿皮提取物对小鼠腹泻具有显著的防治作用。

关键词：苦椿皮；番泻叶；腹泻模型；消化酶

（发表于《中兽医医药杂志》）

青蒿琥酯治疗泰勒焦虫病的研究进展

张　杰[1,2]，张继瑜[1]，李　冰[1]，周绪正[1]

（1. 中国农业科学院兰州畜牧与兽药研究所/中国农业科学院新兽药工程重点开放实验室，兰州 730050；2. 甘肃农业大学动物医学院，兰州 730070）

摘　要：青蒿琥酯（Artesunate）是抗疟药青蒿素（Artemisinin）的衍生物，在兽医临床上主要用于治疗牛、羊的泰勒焦虫病（Theileriosis）以及双芽焦虫病（Double buds piroplasmosis）。作者对近 20 年来青蒿琥酯在不同动物的药物代谢动力学特征、泰勒焦虫病的临床治疗以及泰勒焦虫病对中国畜牧业的危害进行了总结。

关键词：青蒿琥酯；泰勒焦虫病；临床症状；临床治疗

（发表于《湖北农业科学》）

塞拉菌素溶液对家兔的皮肤刺激性试验

周绪正，张继瑜，汪　芳，李金善，李剑勇，李　冰，牛建荣，魏小娟，杨亚军，刘希望

（中国农业科学院兰州畜牧与兽药研究所，农业部兽用药物创制重点实验室，甘肃省新兽药工程重点实验室，兰州，730050）

摘　要：研究塞拉菌素透皮制剂对日本大耳白兔皮肤的刺激胜试验。以《化学药物刺激胜、过敏胜和溶血胜研究技术指导原则》为指导对日本大耳白兔进行一次性给药，评价药物对家兔皮肤的刺激性。该制剂对皮肤的刺激程度为轻度。塞拉菌素制剂可以作为外用药物。

关键词：塞拉菌素制剂；日本大耳白兔；皮肤刺激胜试验

（发表于《第三届中国兽医临床大会文集》）

赛拉菌素溶液对犬的安全性试验研究

周绪正，张继瑜，李金善，李　冰，李剑勇，魏小娟，
牛建荣，杨亚军，刘希望
（中国农业科学院兰州畜牧与兽药研究所，农业部兽用药物创制重点实验室，
甘肃省新兽药工程重点实验室，兰州 730050）

摘　要：赛拉菌素溶液为浙江海正药业股份有限公司生产的抗动物体内外寄生虫的大环内脂类药物新制剂，临床通过透皮给药主要用于犬等宠物体内线虫和体外节肢昆虫的预防与治疗；本实验为检验赛拉菌素溶液对靶动物犬的安全性，按照中国农业部《兽药临床试验技术规范》之要求，2009 年 9 月 1 日至 2010 年 12 月 13 日中国农业科学院兰州畜牧与兽药研究所完成赛拉菌素溶液对靶动物德国牧羊犬的安全性研究；40 只德国牧羊犬（公母各 20 只）随机分为 5 组，每组 8 只犬（公母各 4 只）。第一组为赛拉菌素溶液（0.75ml：45mg）正常给药组（6mg/kg BW），第二、第三、第四组为试验组，依次为：低剂量组（18mg/kg BW）、中剂量组（30mg/kg BW）、高剂量组（60mg/kg BW），第五组为空白对照组（0mg/kg BW），按量在试验犬的颈背部肩胛前给予赛拉菌素溶液；分别于给药前和给药后第五天，第十二天，第三十天前肢股静脉采集全血和抗凝血各 4ml/犬和 2ml/犬，进行血液肝肾功能指标生化检测和血液学检测；并对试验 30 天后处死的犬，剖检取有关组织器官（心、肝、肺、肾）进行病理学检查。结果显示：所有试验犬在给药前后各项血液学指标均在正常值范围内，没有明显变化，各组内给药前及给药后不同时间段比较，各组间同一指标相同时间段比较，差异均不显著，$P>0.05$；给药前后肝功能及肾功能的血液生化的主要检测的指标数据比较，各试验组与对照组的各项指标均在正常范围值内，并且给药前后各项指标变化差异不显著，$P>0.50$；各组试验犬心、肝、肺、肾等实质器官病理检测结果无明显的异常；赛拉菌素制剂在 6～60mg/kg BW 的剂量范围内使用时，对机体的肝、肾未造成明显的毒副作用，提示在试验剂量范围内本品对犬无明显临床可观察毒性作用。

关键词：赛拉菌素溶液；德国牧羊犬；安全性试验

（发表于《中国毒理学会兽医毒理学与饲料毒理学学术讨论会暨兽医毒理专业委员会第四次全国代表大会会议论文录》）

银翘蓝芩注射液的稳定性研究

王兴业[1,2]，杨亚军[2]，李剑勇[2]，张继瑜[2]，刘希望[2]，刘治岐[3]，
牛建荣[2]，周绪正[2]，魏小娟[2]，李　冰[2]

（1. 甘肃农业大学动物医学院，兰州 730070；2. 中国农业科学院兰州畜牧与兽药研究所/农业部兽用药物创制重点实验室/甘肃省新兽药工程重点实验室，兰州 730050；3. 兰州理工大学生命科学与工程学院，兰州 730050）

摘　要：以性状、pH 值、相对密度和有效成分含量等为考察项目，通过加速试验对银翘蓝芩注射液的稳定性进行考察。结果表明，随着放置时间推移，银翘蓝芩注射液中绿原酸、黄芩苷和连翘苷含量基本没有变化，其他各项检查指标均无明显变化，未出现不溶性微粒，符合药品质量标准各项规定，其质量可控、稳定性较好、可满足临床用药需求。

关键词：银翘蓝芩注射液；加速试验；稳定性

（发表于《江苏农业科学》）

中兽医药现代化技术平台——血清药理学

程富胜[1]，张　霞[2]，赵朝忠[1]，王华东[1]

（1. 中国农业科学院兰州畜牧与兽药研究所，农业部兽用药物创制重点实验室，甘肃省新兽药工程重点实验室，兰州 730050；2. 甘肃农业大学生命科学技术学院）

（全文省略）

（发表于《中兽医医药杂志》）

自拟中药对人工感染 IBV 雏鸡免疫器官及血清 IgG 的影响

陈灵然[1]，严作廷[1]，王　萌[2]，王东升[1]，张世栋[1]，
叶得河[2]，王　玲[1]，于远光[2]，魏兴军[3]

（1. 中国农业科学院兰州畜牧与兽药研究所，农业部兽用药物创制重点实验室，
甘肃省新兽药工程重点实验室，兰州 730050；2. 甘肃农业大学
动物医学院；3. 兰州华陇家禽育种有限公司）

摘　要：研究自拟中药方剂对鸡传染性支气管炎的防治效果。首先选取 SPF 鸡胚测定 IBV－M41 的 ELD_{50}，以确定攻毒剂量。然后将 400 只 19 日龄无 IB 抗体的健康雏鸡随机分为 8 组，每组 50 只，分别为预防高（2.0ml/次）、中（1.0ml/次）、低(0.5ml/次)组，治疗高（4.0ml/次）、中（2.0ml/次）、低（1.0ml/次）组，阳性对照组和阴性对照组，19 日龄时预防组开始用药，连用 5d。21 日龄时除阴性对照组外其余各组均以 $100ELD_{50}$ 剂量的 IBV－M41 人工感染雏鸡。待症状出现 24h 后，治疗组开始给药，连用 5d，期间进行临床观察。考察该自拟中药的临床预防及治疗效果，并测定不同处理组不同日龄患病雏鸡免疫器官指数及血清 IgG 的质量浓度。在用药过程中，自拟复方中药的预防和治疗不同剂量组免疫器官指数与阴性对照组相比差异显著，组间差异不显著；用药疗程结束后，各组间差异不显著。但该中药方剂对不同日龄不同处理组雏鸡血清 IgG 的质量浓度影响明显，且与阴性、阳性对照组相比差异极显著。该中药方剂对雏鸡法氏囊及脾脏发育有促进作用，且诱生 IgG，使其血清含量显著提高，对机体产生免疫保护，从而达到对鸡传染性支气管炎的防治效果。

关键词：鸡传染性支气管炎；自拟中药方剂；法氏囊指数；脾脏指数；IgG

（发表于《中兽医医药杂志》）

临床病料采集、保存、送检及安全防护方法

周绪正[1,2]，蔺红玲[1,2]，李　冰[1,2]，牛建荣[1,2]，魏小娟[1,2]，李金善[2]，
李剑勇[2]，杨亚军[2]，刘希望[2]，张继瑜[1,2]

（1. 国家肉牛牦牛产业技术体系疾病控制功能研究室，兰州 730050；2. 中国农业科学院
兰州畜牧与兽药研究所，农业部兽用药物创制重点实验室/甘肃省
新兽药工程重点实验室，兰州 730050）

摘　要：本文对病料采集前的准备、采集注意事项、常见动物病料的采集方法、病料的固定和保存、病料的送检以及工作人员的安全防护 6 个方面进行了系统阐述。

关键词：病料；采集；保存；送检；人员防护

（发表于《养殖与饲料》）

牛皮蝇蛆病的综合防控

周绪正[1,2]，李　冰[1,2]，牛建荣[1,2]，魏小娟[1,2]，李金善[2]，
李剑勇[2]，杨亚军[2]，刘希望[2]，张继瑜[1,2]

（1. 国家肉牛牦牛产业技术体系疾病控制研究室，药物与临床用药岗位；
2. 中国农业科学院兰州畜牧与兽药研究所，农业部兽用药物创制重点实验室，甘肃省新兽药工程重点实验室，兰州 730050）

牛皮蝇蛆病（HyPedennosis）是由双翅目、皮蝇科、皮蝇属幼虫寄生于牛体内而引起的、对养牛业危害严重的一种国际性的人畜共患寄生虫病。现已报道的病原分属于 10 个属中的 39 个种，常见的有牛皮蝇（*H. bovis*）、纹皮蝇（*H. lineatum*）和中华皮蝇（*H. sinense*），特异性的宿主为牛属动物的黄牛、牦牛、犏牛和水牛，但有时也可感染马等其他家畜和人，而且能够发育成熟。本病的主要特征是动物消瘦、贫血、发育受阻、体重减轻、产肉产奶产绒产毛量下降、皮肤穿孔，感染强度高的可致动物死亡，鉴于本病对畜牧业的严重危害，世界上许多国家在 20 世纪 20 ~ 30 年代便启动了皮蝇蛆病控制与扑灭行动，并将消灭皮蝇蛆病立法加以强制执行。皮蝇蛆病流行范围广，在北纬 18° ~ 60°范围内的 55 个国家都有本病的流行。我国牛皮蝇蛆病的地理分布主要在东北、华北、西北等地区，其中，内蒙古自治区（全书称内蒙古）、甘肃、青海、新疆维吾尔自治区（全书称新疆）和西藏 5 个省（区）的农牧区尤为严重，在流行区内，感染率达 80% 以上，严重感染地区高达 98% ~ 100%，特别是 1 ~ 3 岁的犊牛易感，4 岁以上牛发病率降低。本文对牛皮蝇蛆病的综合防控进行综述，为该病的有效预防与治疗提供技术支撑。

（发表于《中国食草动物科学》）

肉牛运输应激综合症药物防治

张继瑜[1,2]，周绪正[1,2]，李　冰[1,2]，牛建荣[1,2]，魏小娟[1,2]，
李金善[2]，李剑勇[2]，杨亚军[2]，刘希望[2]

（1. 国家肉牛牦牛产业技术体系疾病控制研究室；2. 中国农业科学院
兰州畜牧与兽药研究所，农业部兽用药物创制重点实验室，甘肃省
新兽药工程重点实验室，兰州 730050）

随着肉牛养殖向标准化、规模化、集约化的发展和市场的需求，尤其是犊牛和架子牛的集中育肥，导致肉牛异地运输非常频繁，但由于长途运输、饥渴、拥挤、混群、环境突变等容易诱发运输应激综合症，加之用药不合理，导致较高的死亡率，给许多养殖户造成巨大的损失。据研究，引起运输应激综合症的病原主要是支原体，大肠杆菌、沙门氏菌、链球菌、葡萄球菌、真菌及血液原虫等继发感染进一步加重了病情。如何利用现有流行病学资料和药物，保障需要运输牛运输前、运输途中及抵达后的安全，想尽一切办法提高被运牛的免疫力和抵抗力，减少运输应激综合症的发生则是本文的目的。采用启运前和抵达目的地后肌内注射敏感抗菌素，全程限量饮用添加口服补液盐或中药“菌毒清”或黄芪多糖溶液的办法，能有效降低肉牛运输应激综合征的发病率。

（发表于《中国食草动物科学》）

中药防治鸡传染性支气管炎存在的问题及对策

王　玲[1]，陈灵然[1]，郭天芬[1]，李宏胜[1]，杨　峰[1]，
胡广胜[2]，周绪正[1]，牛建荣[1]

（1. 中国农业科学院兰州畜牧与兽药研究所，农业部兽用药物创制重点实验室，甘肃省新兽药工程重点实验室，甘肃兰州 730050；2. 甘肃农业大学动物医学院，兰州 730070）

摘　要：鸡传染性支气管炎（IB）疫苗缺乏交叉免疫或完全无交叉免疫，发病后无特效药物治疗。有些中药具有抑杀病毒的作用，且能调节机体的免疫功能，具有非特异性的抗病毒功能，用于鸡 IB 的临床防治收效甚好，所用方剂与西药相比具有独特的优势。本文针对研发防治鸡传支中药存在的主要问题，探讨了研究的思路和途径，应避免仅对草拟方剂做简单的临床验证，需要对中药方剂的药效学、药理学、药代动力学、毒理学、中药化学及作用机理做深入研究，以及科学严谨的试验设计支持，大量试验数据证明，以期进一步完善并扩大中药在防治畜禽传染性疾病上的应用。

关键词：鸡传染性支气管炎；中药方剂；防治

（发表于《中国动物检疫》）

中兽医与临床兽医学科

Complete Genome Sequence of a Mink Calicivirus in China

Yang Bochao[1,2], Wang Fengxue[2], Zhang Shuqin[2], Xu Guicai[2], Wen Yongjun[2], Li Jianxi[1], Yang Zhiqiang[1], WuI Hua[3]

(1. Lanzhou Institute of Animal and Veterinary Pharmaceutics Science, Chinese Academy of Agricultural Sciences, Lanzhou730050, China; 2. State Key Laboratory for Molecular Biology of Special Economic Animals, Institute of Special Economic Animal and Plant Sciences, Chinese Academy of Agricultural Sciences, Chuangchun, China; 3. Sinovet (Beijing) Biotechnology Co., Ltd., Beijing, China)

Mink calicivirus (MCV), a member of genus *Vesivirus* in the *Caliciviridae* family, is a single – stranded positive – sense RNA virus first isolated in the United States from normal clinically fine mink in 1964 (1, 3). A new strain of MCV, MCV – DL, was isolated from a disease outbreak in mink in 2007 in Shenyang Province, northeastern China. Although MCV has previously been detected (2), the complete genome sequence has not yet been reported.

To provide more information about MCV and to aid researchers studying calicivirus taxonomy, we determined the entire genome sequence of the Chinese MCV isolate. Complete genome sequencing was performed using strategies reported for other positive – sense single – stranded RNA viruses (4 ~ 6). Genomic RNA was extracted from MCV – infected cell cultures and converted to cDNA by a combined random – priming and oligo (dT) – priming strategy. First – round PCR primers were designed from multiple genome alignments of available vesiviruses (4 ~ 6), and further sets of primers were designed on the basis of the partial genomic sequences obtained. Overlapping fragments covering the entire genome were generated and sequenced. PCR products were cloned into vector pMD 18 – T (TaKaRa) and sequenced by Invitrogen. The 5′ and 3′ ends of the viral genome were amplified by rapid amplification of cDNA ends. All fragments were sequenced in both directions. Overlapping consensus sequences were assembled and manually edited to produce the final genome sequence using the software program BioEdit 7.0.9. Phylogenetic trees were constructed using MEGA 4.0 (7).

The complete genome of MCV – DL was found to be 8 409 nucleotides (nt), excluding the polyadenylate tract, with a GC content of 45.6%. Three open reading frames (ORFs) were predicted from the nucleotide sequence with DNAStar (version 5.0; DNAStar Inc., Madison, WI) and by comparing the results with the genomeorganization and ORFs of other caliciviruses (2, 5). ORF1, stretching from nt 14 to nt 5851, encodes a large putative polyprotein of 1 946 amino acids (aa). ORF2 ranges from nt 5857 to nt 7899, and ORF3 ranges from nt 8130 to nt 8306, encoding proteins of 681 and 59 aa, respectively. The 5′ and 3′ untranslated regions are 13 and 103 nt long, respectively.

Phylogenetic analysis of the available genome sequences of representative vesiviruses showed that MCV - DL clustered with calicivirus 2117 (accession No. AY343325, unknown origin) and canine calicivirus (CaCV; accession No. NC_ 004542, isolated from a canine) rather than with the other vesiviruses. Across the whole genome, the highest sequence identity (60.0%) was with calicivirus 2117. Alignment of the sequences of the predicted gene products of MCV and other vesiviruses showed that the highest amino acid sequence identities were with calicivirus 2117 and CaCV. The proteins encoded by ORF1, ORF2, and ORF3 of MCV were found to have 63.5, 59.5, and 40.8% identity with calicivirus 2117 and 42.6, 49.2, and 50.8% amino acid sequence identity with CaCV.

This is the first report of a complete MCV genome sequence which is critical for further investigation of the molecular characteristics and should allow elucidation of its phylogenetic relationship with other vesiviruses.

Nucleotide sequence accession number. The complete genome sequence of MCV strain MCV - DL/2007/CN has been deposited in GenBank under accession No. JX847605.

Acknowledgment

This work was supported by the National High Technology Research and Development Program of China (863 Program) (2011AA10A21).

References

[1] Evermann JF, Smith AW, Skilling DE, McKeirnan AJ. 1983. Ultrastructure of newly recognized caliciviruses of the dog and mink. Arch. Virol. 76: 257 - 261.

[2] Guo M, Evermann JF, Saif LJ. 2001. Detection and molecular characterization of cultivable caliciviruses from clinically normal mink and enteric caliciviruses associated with diarrhea in mink. Arch. Virol. 146: 479 - 493.

[3] Long GG, Evermann JF, Gorham JR. 1980. Naturally occurring picornavirus infection of domestic mink. Can. J. Comp. Med. 44: 412 - 417.

[4] Martín - Alonso JM, et al. 2005. Isolation and characterization of a new vesivirus from rabbits. Virology 337: 373 - 383.

[5] Oehmig A, et al. 2003. Identification of a calicivirus isolate of unknown origin. J. Gen. Virol. 84: 2 837 - 2 845.

[6] Roerink F, Hashimoto M, Tohya Y, Mochizuki M. 1999. Organization of the canine calicivirus genome from the RNA polymerase gene to the poly (A) tail. J. Gen. Virol. 80 (Pt 4): 929 - 935.

[7] Tamura K, Dudley J, Nei M, Kumar S. 2007. MEGA4: molecular evolutionary genetics analysis (MEGA) software version 4.0. Mol. Biol. Evol. 24: 1 596 - 1 599.

(Published the article in GENOME ANNOUNCEMENT affect factor: 5.402)

Ethno - veterinary Survey of Medicinal Plants in Ruoergai Region, Sichuan Province, China

Shang Xiaofei[1], Tao Cuixiang[2], Miao Xiaolou[1], Wang Dongsheng[1], Tangmuke[3], Dawa[3], Wang Yu[1], Yang Yaoguang[1], Pan Hu[1]

(1. Engineering and Technology Center of Traditional Chinese Veterinary Medicine of Gansu, Key Laboratory of Veterinary Pharmaceutical Development, Ministry of Agriculture, Lanzhou Institute Animal and Veterinary Pharmaceutical Science, Chinese Academy of Agricultural Science, Lanzhou 730050, China; 2. Department of Pharmacy, The Affiliated Hospital of Gansu College of T. C. M, Lanzhou 730000, China; 3. Animal Husbandry and Veterinary Bureau of Ruoergai County, Sichuan Province, Ruoergai 624500 China)

Abstract: *Aim of study*: In this study we aimed to survey and investigate the medicinal plants which are used to treat the veterinary diseases in Ruoergai region, Sichuan province, China. Meanwhile, the important medicinal plants were collected and identified for the further study.

Materials and methods: Twenty folk veterinary practitioners from 8 township animal husbandry and veterinary stations in Ruoergai region were investigated and interviewed. The important local medicinal materials, including plants, animals and mineral drugs, were collected by scientific methods and identified by the pharmacognosist of Lanzhou University, China.

Results: According to the investigation, only 20 folk veterinary practitioners still used 129 species of traditional medicine to treat the livestocks diseases. In these medicine, 93 species were native and Ranunculaceae (12, 12.90%), Compositae (11, 11.83%), Papaveraceae (7, 7.53%) were the predominant families. At the same time, herbs (36.56%) are the most widely used part of plant, and respiratory diseases (21.02%) and gastrointestinal diseases (19.89%) were the main animal's diseases in this region.

Conclusion: Ethno - veterinary medicine made an extraordinary contribution to the sound development of animal husbandry in Ruoergai. But the inherit, protect and development should be paid more attentions in the future, and the species which have not been studied should be developed priority to find biological activities and new bioactive compounds further.

Key words: Ethno - veterinary; Medicinal plants; Ruoergai region; Survey

1 Introduction

As the capital animal husbandry production base of Tibetan in China, Ruoergai is known for its large numbers of livestock. It is located in the northeastern edge of Qinghai - Tibet plateau at an average altitude of 3 500 m. The total population of this region is 73 400, 91.5% of them is Tibetan people and 5.5% is Han nationality people. Two thousand years ago, the local ancestors have started to tame and breed animals, such as yak and horse. In this process, people not only obtained abundant foods from the animals, but also considerable incomes. 86.6% of the residents are herdsmen and 90% of the incomes are from animal husbandry (Fig. 1).

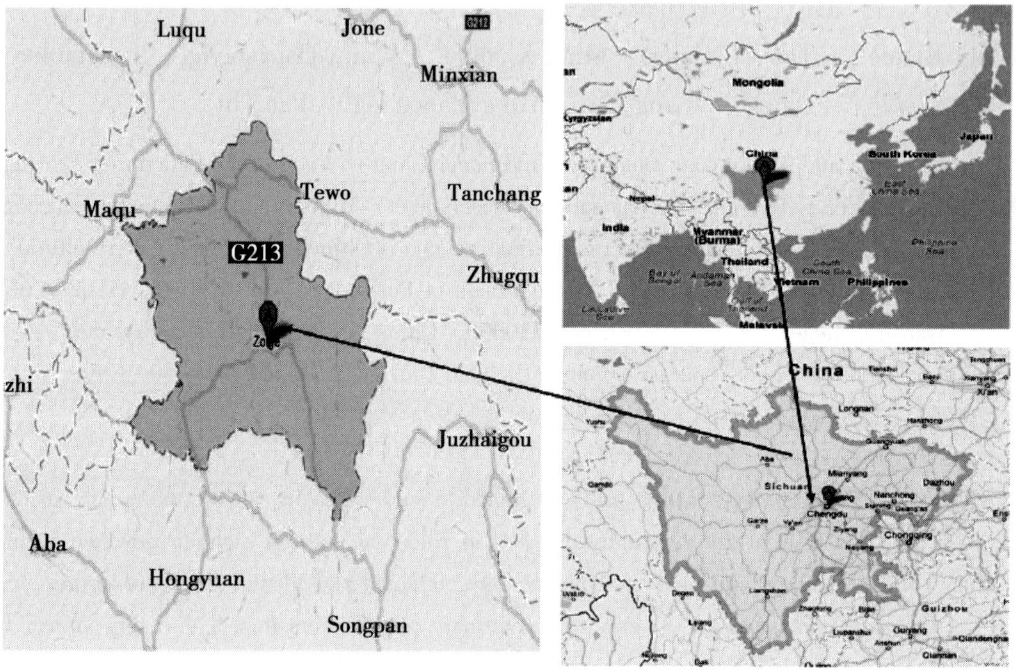

Fig. 1 Location of study areas

In order to prevent and treat the diseases of animals, local people have gotten practical experiences abundantly. Some experiences were transmitted orally from generation to generation, and others were recorded in books in the history. Especially in 1960s ~ 1980s, along with nationwide field reconnaissance of medicinal materials, ethno - veterinary get the comprehensive development. A lot of folk veterinary books were published, such as 'Tibet used Chinese herbal medicine', 'Plateau herbal treatments manual', 'Tibetan veterinary experience documents' (Tibetan veterinary experience documents editorial committee, 1979), etc. At the same time, many medicinal materials, which have the marked therapeutic effects on the respiratory diseases, digestive system diseases, gynecology and obstetrics diseases, etc., were found and used in veterinary clinic. Despite the fact that ethno - veterinary medicine was very important for the animal healthcares, but along with many chemical drugs, especially antibiotics, being widely used in the animal husbandry, local people ignored the heritage and development of ethnoveterinary medicines gradually. The numbers of veterinary practitioners were rapid decreased, and up to now there was no scientific survey on the ethno - veterinary in Ruoergai region.

In the current study, we aim to collect and document indigenous ethno - veterinary knowledge by surveying and investigating 20 folk veterinary practitioners from 8 township animal husbandry and veterinary stations, and collecting more than 129 species of medicinal materials, which were used to treat the livestocks diseases in Ruoergai region.

2 Methods

2.1 *Research area*

As one of the important National plateau pasturing areas, Ruoergai region has a large number of yak, Tibet - sheep and horse. The number of stock is more than 1 000 000, and 90% the incomes of peoples were from animal husbandry. It is located in the northwestern part of Sichuan province and the northeastern edge of Qinghai - Tibet plateau between 320 561 to 340 191 North Latitude, and 1 020 081 to 1 030 391 East Longitude with an average altitude of 3 500m, and embraces 17 towns and 96 villages with 10 620km^2 of total area. The annual mean temperature and annual precipitation were 1.1℃ and 648.5 mm, respectively. Meanwhile, there are 12 nationalities, including Tibetan, Han, Muslim, Qiang, Yi, Mongolia, Tujia, Lisu, Man, Naxi, Buyi, Bai, etc., live in this multiracial inhabit region, in those, the population of Tibetan nationality was 69 104, and the local language were Tibetan and Chinese (Fig. 1) (http://map.google.com).

Ruoergai region is abundant in grassland resource with 12 126 300 acreages of the total area, and the type of grassland is major sorted to alpine meadow, mountain meadow, alpine halfswamp, alpine swamp, mountain shrub, woodland meadow, etc. In those, alpine meadow and alpine half - swamp were the main components, and the percent of usable grassland area was 72.09%. Embracing 2 498 600 acreages of Wetland National Nature Reserve, Ruoergai is also called as the most beautiful alpine wetland grassland of China and Wetland of International Importance. Meanwhile, Ruoergai region belongs to the Yellow River system, and is the headstream of several sub - rivers. The developable hydropower resources are more than 33 700 kW, and named as the kidney of western plateau in China.

At the same time, Ruoergai region was abound in wildlife resources, including more than 250 species of vertebrates and 1 100 species of wild plants. In those, about 200 species have been used as the medical materials in the history, and about 100 species have been used to treat the livestock diseases in veterinary clinic, such as Notopterygium incisum Ting ex H. T Chang, *Rheum palmatum L. var. tanguticum* Maxim. Ex Regel, *Berberis julianae* Schneid, *Ephedra minuta* Florin *var. dioeca* C. Y. Cheng, *Rhodiola algida* (Ledeb) Fisch. et Mey. *var. tangutica* (Maxim.) S. H. Fu, *Dracocephalum tanguticum* Maxim, *Saussurea* medusa Maxim and *Arenaria kansrensis* Maxim, etc. These medical materials have contributed to the development of animal husbandry in Tibetanplateau area (Ruoergai government, 2011).

2.2 *History and medical theory of* Ruoergai

As above said, Ruoergai has 91.5% Tibetan people and 5.5% Han nationality people, and the local languages were Tibetan and Chinese. This condition was consistent with the development of local history. Since 710 year, along with two Tang Princesses Wen - Cheng and Jin - Cheng married Tibet chief, respectively, the cooperation and communication between Chinese and Tibet have became more and more frequently. Tibetan people immigrated largely to the part areas of Sichuan, Qinghai and Gansu provinces. Under this specific historic condition, Tibetan settled down in the Ruoergai region gradually. In this process, the traditional Tibetan veterinary medicine (TTVM) was brought in and became the dominant medical theory in Ruoergai.

As one of the important folk veterinary medicines in China, TTVM has the special veterinary theory, which was integrated Chinese, India, Nepal and other regions veterinary theory, and was created and formed in the struggle with animals diseases. The theory contains 'Five elements' and 'Three essences'. 'Five elements' was explained as the five different elements of the nature world, including Earth (Tu), Water (Shui), Fire (Huo), Wind (Qi) and Space (Kongjian) (Chinese: gold, wood, water, fire and earth); 'Three essences' was defined as physio - essences of body, including 'Long', 'Chiba', 'Baigan' (Tibetan veterinary experience documents editorial committee, 1979).

Before 8 century, Tibetan medicinal knowledge was obtained from practical experience, and transmitted orally from generation to generation. Following with these two marital alliances, lots of Han medicines (Chinese medicines) were transmitted to Tibet. The medicinal knowledge began to be recorded in literatures, and a series of classical medical books were published. As the oldest Tibetan literature, 'YueWangYaoZhen' was written in 8 century, which collected 780 herbs and animal medicines. In 12 and 18 century, 'SiBuYiDian' and 'JinZhuBenCao' were created, which recorded 1 002 and 2 294 species, respectively (Northwest Plateau Institute of Biology, Chinese Academy of Sciences, 1991) (Chinese materia editorial committee, 2002). Even though some folk medicines recorded in these books have been not only used to treat human's diseases but also animal's for thousand years. But most of books are main incline to introducing folk medicines at the human diseases aspect, and the books for TTVM are little.

In Ruoergai, local people mainly relied on medicinal materials to treat diseases and keep healthy of livestock for thousand years. In this process, they accumulated practical experiences and obtained abundant ethno - veterinary medicines. Especially in 1960s ~ 1980s, every township animal husbandry and veterinary stations has established the center of traditional Tibetan veterinary medicine. The numbers of folk veterinary practitioners were more than 200, and the medical materials are up to 90% amount of veterinary medicines in clinic. Many good prescriptions and methods were created. But now, only about 8 townships animal husbandry and veterinary stations reserved the center of TTVM. The ethnobotanical knowledge, traditional medical theory, processing technology, folk veterinary clinic methods and so on, were abandoned step by step. Some reasons might be interpreted this phenomenon. First, many chemical drugs were widely used to control the livestock diseases recently, such as antibiotics, antiparasitic drugs, etc. Second, along with the urbanization, youngsters are more willing to work in cities than to learn the ethno - veterinary knowledge in village. Meanwhile, some excellent old folk herbalists or veterinaries died, and some folk veterinary theory and methods have not been inherited well. In our investigation, we tried to visit more folk practitioners and obtain more ethno - veterinary knowledge. Unfortunately, only 20 folk veterinary practitioners were found and willing to use traditional medicine to treat the livestock's diseases.

2.3 *Field interview methods*

Twenty folk veterinary practitioners (men) from 8 township animal husbandry and veterinary stations (Hongxing, Axi, Tangke, Banyou, Xiaman, Qiuji, Jiangzha, Reer), which major distributed in the grassland of Ruoergai and serviced to treat and prevent the animals diseases, were interviewed. All people were over 50 years old and dedicated to treat animal's diseases using the local medicinal materials for a long time. Information was obtained by semi – structured interviews, personal conservations and guided fieldtrips with the help of traditional Tibetan veterinary medicine association of Ruoergai. Meanwhile, Tangmuke and Dawa, served in the Animal husbandry and veterinary bureau of Ruoergai County and well known by local people, as the translators help us to interview local people with the Tibetan language and poor Chinese language. Local names, parts used, processing methods, methods of preparation, types of diseases were carefully documented during each interview, and this survey was carried out from January 2011 to December 2011 (Fig. 2).

Fig. 2 The procession of undertaking the ethno – veterinary survey of medicinal materials in Ruoergai region

2.4 *Voucher specimen collection*

To exhibit and protect the medicinal materials in Ruoergai preferably, we collected the voucher specimen between April and October 2011 (Fig. 3). Voucher specimens were collected and prepared under the directions of herbalists and Dawa. Plants were identified by pharmacognosy professor Zhigang Ma, Department of pharmacy, Lanzhou University, and deposited in Traditional Chinese Medicine Voucher Herbarium, Lanzhou Institute of Husbandry and Pharmaceutical Sciences, Chinese Academy of Agricultural Science (CAAS). All data were collected in a database (Access v. 7.0).

3 Results and discussion

3.1 *Native ethnobiological survey*

3.1.1 *Families and medicinal plants*

From 20 key informants of 93 plant species, which belong to 30 families, were cited to be used in traditional veterinary medicine by folk veterinarian practitioners and peoples. *Ranunculaceae* was the family with the largest number of medicinal species (12, 12.90%), followed by *Compositae* (11, 11.83%), *Papaveraceae* (7, 7.53%) and *Labiatae* (6, 6.54%). These four families represent 38.8% of the total of medicinal plants. The remaining 57 species belongs to other 27

Meconopsis punicea Maxim.（ZSY028） Saussurea medusa Maxim.（ZSY131）

Fig. 3 The procession of collecting voucher specimen and purchasing material medicines in Ruoergai

families with less than 5 species.

Of the 93 species, *Bupleurum chinense* DC., *Rheum palmatum* L., *Rhodiola algida* (Ledeb) Fisch. et Mey. *var. tangutica* (Maxim.) S. H. Fu., *Saussurea medusa* Maxim. and *Scopolia japonica* Maxim were cited by all informats. Followed by *Oxytropis falcate* Bunge (18), Ephedra minuta Florin var. dioeca C. Y. Cheng (18), *Hyoscyamus niger* L. (17), *Rhododendron anthopogonoides* Maxim. (16), *Hippophae rhamnoides* L. *subsp. gyantsensis* Rousi. (16) and *Hippophae thibetana* Schlecht. (16). The results indicated that these species were widely used in veterinary clinic, and as the local idiomatical materials, *Rhodiola algida* (Ledeb) Fisch. et Mey. *var. tangutica* (Maxim.) S. H. Fu. and *Scopolia japonica* Maxim. etc. were inclined to be introduced by folk herbalist.

At the same time, the name of some medicinal plants was still confused. The same plant has the different name in the different region, and some different plants have the same Tibetan name. Such as *Pedicularis torta* Maxim., *Pedicularis oederi* Vahl *var. sinensis* (Maxim.) Hurus., Pedicularis longiflora Rudolph var. tubiformis (Klotz.) Tsoong. and *Pedicularis kansuensis* Maxim., has only one Tibetan name Mei Duo Lang Lang, were used in TTVM. And Corydalis dasyptera Maxim., *Corydalis melanochlora* Maxim. and *Corydalis straminea* Maxim. has one Tibetan name 'Dang Ri Si Wa'. *Gentiana straminea* Maxim. and *Gentiana dahurica* Fisch. has one name 'Ji Jie Ga Bao', etc. (Table 1).

Meanwhile, the results of our survey also demonstrated that the most habitat of collected plant was meadow (59, 63.44%), followed by scrub (15.05%), rock cracks (11.83%), alongside river (5.38%) and roadsides (4.30%). The result was consistent to the ecology environment (Table 1).

3.1.2 *Plant parts and mode of preparation*

Herbs (34, 36.56%) is the most widely part of plant used to treat the animal diseases in Ruoergai, followed by flowers (18, 19.35%), roots (18, 19.35%), leaves (13, 13.98%), fruits (9, 9.68%), aerial parts (8, 8.60%), rhizomes (6, 6.45%), seeds (4, 4.30%), root tubers (3, 3.2%), branches (2, 2.15%) and seedlings (1, 1.08%). In these, 24 species (25.81%) has the two and more than two parts used in veterinary clinic, and the different parts may have different effects. Such as the roots of *Aconitum grmnandrum* Maxim were used to treat the pains, but the leaves were applied to treat the parasite diseases, etc. Mean-

Table 1 Native plant used as traditional medicine in Ruoergai

Scientific name(Voucher no.)	Family	Common name[a]	Part used	Disease[b]	Chinese or Tibetan[c]	Pharmacopoeia and reference[d]	Info. no.	Habit	Altitude	Rank of endangered species
Aconitum flavum Hand. - Mazz. (ZSY102)	*Ranunculaceae*	Tie Bang Chui(C), Bang A Na Bao(T)	Seedlings, root tubers	Fever, internal; Mange, external	B	W. R.	5	Meadow	3 300m	—
Aconitum grmnandrum Maxim. (ZSY020)	*Ranunculaceae*	Wu Tou(C)	Roots, leaves	Pain(root), Parasite disease(leaf) internal	T	W. R.	1	Meadow	3 200m	—
Aconitum kongboense. (ZSY135)	*Ranunculaceae*	Xue Shan Yi Zhi Hao(C,T)	Herbs	Inflammation, Pain, internal	T	N. S. R.	2	Scrub	3 850m	—
Aconitum tanguticum (Maxim.) Stapf. (ZSY104)	*Ranunculaceae*	Wu Tou(C), Bang Ga(T).	Herbs	Pneumonia, internal	T	W. R.	5	Rock cracks	3 250m	—
Adonis coerulea Maxim. (ZSY103)	*Ranunculaceae*	Lan Hua Ce Jin Zhan(C), Jia Zi Dou Luo(T)	Herbs	Mange, herpes. external	T	Zhang et al. (1991), Dai et al. (2010)	10	Meadow	3 300m	—
Ajuga lupulina Maxim. (ZSY004)	*Labiatae*	Jin Gu Cao(C), Sheng Dou (T).	Herbs	Fever and anthrax, internal	T	N. S. R.	1	Meadow	3 327m	—
Allium prattii C. H. Wright. (ZSY040)	*Liliaceae*	Ye Cong(C), Ri Cong(T)	Herbs	Dyspepsia, internal	B	N. S. R.	7	Meadow	3 900m	—
Anemone rivularis Buch. - Ham. Ex DC. (ZSY016)	*Ranunculaceae*	Cao Yu Mei(C), Su Ga(T).	Fruits	Dyspepsia, internal	B	W. R.	3	Meadow	3 250m	—
Arenaria kansrensis Maxim. (ZSY003)	*Caryophyllaceae*	Gan Su Xue Lin Zhi(C), AZhong Ga Bao(T)	Herbs	Xeropulmonary cough, internal	T	N. S. R.	2	Rock cracks	3 900m	—
Aster diplostephioides (DC.) C. B. Clarke. (ZSY076)	*Compositae*	Chong Guan Zi Wan(C), Mei Duo Lu Mei(T).	Flowers	Bronchitis and cough, internal	T	N. S. R.	11	Meadow	3 200m	—
Astragalus licentanus Hand. - Mazz. (ZSY033)	*Leguminosae*	Gan Su Huang Qi(C)	Roots	Qi asthenia, chronic diarrhea, internal	B	W. R.	4	Meadow	3 639m	—
Berberis julianae Schneid (ZSY052)	*Berberidaceae*	Ci Hong Zhu(C,T), Jie Ba (T).	Roots, mediopellis of rhizomes	Dysentery, internal	B	W. R. P.	15	Meadow	3 550m	—
Bupleurum chinense DC. (ZSY137)	*Umbelliferae*	Chai Hu(C,T)	Roots	Common cold and upper respiratory tract infection, internal	B	W. R. P.	20	Roadside	3 270m	I

(Continued)

Scientific name(Voucher no.)	Family	Common name[a]	Part used	Disease[b]	Chinese or Tibetan[c]	Pharmacopoeia and reference[d]	Info. no.	Habit	Altitude	Rank of endangered species
Catabrosa aquatic (L.) Beauv. (ZSY120)	*Hippuridaceae*	Yan Gou Cao(C)	Herbs	Pneumonia, hepatitis, internal	T	N. S. R.	3	Alongside rivers	3 280m	—
Carum carvi L. (ZSY079)	*Umbelliferae*	Zang Hui Xiang(C), GuoNiao (T)	Fruits	Inflammation, internal	B	W. R.	9	Meadow	4 000m	Ⅲ
Clematis tangutica (Maxim.) Korsh. (ZSY055)	*Ranunculaceae*	Tie Xian Lian(C), Ye Meng (T)	Stems, leaves	Dysentery, dyspepsia, internal	B	Zhang, et al. (2009), Wu et al. (2008), Li et al. (2008), Zhang et al. (2006)	12	Meadow	3 200m	—
Corydalis dasyptera Maxim. (ZSY088)	*Papaveraceae*	Die Lie Zi Jin(C), Dang Ri Si Wa(T)	Herbs	Pain, stomach ailment, internal	T	N. S. R.	3	Roadside	3 200m	—
Corydalis melanochlora Maxim. (ZSY083)	*Papaveraceae*	An Lv Zi Jin(C), Dang Ri Si Wa(T)	Herbs	Pain, stomach ailment, internal	T	N. S. R.	1	Roadside	3 300m	—
Corydalis straminea Maxim. (ZSY072)	*Papaveraceae*	Cao Huang Hua Zi Jin(C), Dang Ri Si Wa(T)	Herbs	Pain, stomach ailment, internal	T	N. S. R.	1	Meadow	3 961m	—
Daphne tangutlca Maxim. (ZSY008)	*Thymelaeaceae*	Shan Gan Rui Xiang(C), Seng Xin Na Ma(T)	Flowers, fruits, roots, leaves	Parasitic diseases, internal	B	W. R.	15	Scrub	3 420m	—
Delphinium caeruleum Jacq. ex Camb. (ZSY105)	*Ranunculaceae*	Lan Cui Que Hua(C), Qia Gang(T)	Aerial parts	Dysentery, internal	T	Pan et al. (1992)	5	Meadow	3 370m	—
Delphinium candelabrum var. monanthum (Hand. - Mazz.) W. T. Wang. (ZSY108)	*Ranunculaceae*	Dan Hua Que Hua(C), Xia Gang Wa(T)	Herbs	Dysentery, internal	T	N. S. R.	4	Meadow	3 370m	—
Dracocephalum tanguticum Maxim. (ZSY085)	*Labiatae*	Tang Gu Te Qin Lan(C), Zhi Yang Gu(T)	Herbs	Stomach ailment, internal	B	W. R.	13	Meadow	3 200m	—
Elsholtzia densa Benth. (ZSY067)	*Labiatae*	Mi Hua Xiang Ru(C), Qi Rou Se Bu(T)	Aerial parts	Diuresis, inflammation, Parasitic diseases, internal; Itch of skin, external	T	W. R.	11	Meadow	4 000m	—
Ephedra minuta Florin *var. dioeca* C. Y. Cheng. (ZSY058)	*Ephedraceae*	Ma Huang(C), Ce Dun Mu (T)	Herbaceous stems and roots	Cough, internal; hemorrhage, external	B	W. R.	18	Rock cracks	3 680m	Ⅱ

(Continued)

Scientific name (Voucher no.)	Family	Common name[a]	Part used	Disease[b]	Chinese or Tibetan[c]	Pharmacopoeia and reference[d]	Info. no.	Habit	Altitude	Rank of endangered species
Equisetum arvense L. (ZSY082)	*Equisetaceae*	Wen Jing(C,T)	Herbs	Cough, hemostasis, internal	B	W. R.	3	Roadside	3 200m	—
Eruca sativa Mill. (ZSY093)	*Cruciferae*	Zhi Ma Cai(C), Gai Cai(T)	Herbs	Lymphadenitis, internal	T	N. S. R.	2	Meadow	3 300m	—
Gentiana dahurica Fisch. (ZSY127)	*Gentianaceae*	Qin Jiao(C). Ji Jie Ga Bao (T)	Flowers and roots	Inflammation, internal	B	W. R. P.	15	Meadow	3 200m	—
Gentiana straminea Maxim. (ZSY46)	*Gentianaceae*	Ma Hua Jiao(C). Ji Jie Ga Bao(T)	Flowers	Diarrhea, inflammation, internal	B	W. R.	8	Meadow	3 900m	Ⅰ
Gentianopsis grandis (H. Smith) Ma. (ZSY010)	*Gentianaceae*	Bian Lei(C), Ji He Di(T)	Herbs	Pain, cough, and young diarrhea, internal	T	N. S. R.	5	Scrub	3 360m	—
Gymnadenia conopsea (L.) R. Br. (ZSY049)	*Orchidaceae*	Shou Zhang Shen(C), Wang Bao La Ba(T)	Tubers	Body empty, lung disease, internal	B	W. R.	9	Meadow	3 650m	Ⅲ
Heracleum millefolium Diels. (ZSY057)	*Umbelliferae*	Du Huo(C), Zhu Ga(T)	Roots, fruits	Pain, inflammation and parasitic diseases, internal	B	Rao et al. (1995)	13	Meadow	3 200m	—
Hippophae thibetana Schlecht. (ZSY056)	*Elaeagnaceae*	Xi Zang Sha Ji(C), Da Er Bu (T)	Fruits	Cough, tuberculosis, dyspepsia and stomach ulcer, internal	B	Qi and Li (2011), Chen et al. (2004)	16	Scrub	3 200m	—
Hippophae rhamnoides L. subsp. gyantsensis Rousi. (ZSY126)	*Elaeagnaceae*	Jiang Zi Sha Ji(C), Da Er Bu (T)	Fruits	Cough, tuberculosis, dyspepsia and stomach ulcer, internal	B	Xiao(2007)	16	Scrub	3 450m	—
Hippuris vulgaris L. (ZSY024)	*Hippuridaceae*	Shan Ye Zao(C), Dan Bu Ga La(T)	Herbs	Tuberculosis, cough. internal	T	Shao et al. (2008), Pu et al. (2011)	3	Alongside rivers	3 200m	—
Hypecoum leptocarpum Hook. f. et Thoms. (ZSY068)	*Papaveraceae*	Jiao Hui Xiang(C), Ba Er Ba Da(T)	Herbs	Inflammatory, pain, internal	B	W. R.	8	Meadow	3 650m	—
Hyoscyamus niger L. (ZSY109)	*Solanaceae*	Tian Xian Zi(C), Tang Chong (T)	Seeds	Pain, parasitic disease, internal	B	W. R. P.	17	Scrub	3 200m	Ⅱ
Incarvillea compacta Maxim. (ZSY029)	*Bignoniaceae*	Mi Sheng Bo Luo Hua(C), Ou Qu(T)	Flowers, seeds and roots	Dyspepsia, internal	T	N. S. R.	8	Meadow	3 600m	—
Ixeris chinensis (Thunb.) Nakai. (ZSY098)	*Compositae*	Ku Cai(C), Za Chi(T)	Herbs	Jaundice, cholecystitis, internal	B	W. R.	5	Alongside rivers	3 200m	—

(Continued)

Scientific name(Voucher no.)	Family	Common name[a]	Part used	Disease[b]	Chinese or Tibetan[c]	Pharmacopoeia and reference[d]	Info. no.	Habit	Altitude	Rank of endangered species
Lamiophlomis rotate (Benth.) Kudo. (ZSY027)	*Labiatae*	Du Yi Wei (C), Da Ba Ba (T)	Aerial parts	Inflammatory, pain, internal	T	W. R.	8	Meadow	3 531m	Ⅰ
Lamium amplexicaule L. (ZSY095)	*Labiatae*	Bao Gai Cao(C)	Herbs	Mumps, icteric hepatitis, internal	T	W. R.	3	Meadow	3 300m	—
Leontopodium leontopodioides (Willd.) Beauv. (ZSY107)	*Compositae*	Huo Rong Cao (C), Zha Tuo Ba(T)	Aerial parts	Influenza, internal	B	W. R.	4	Meadow	3 650m	—
Lris goniocarpa. (ZSY015)	*Iridaceae*	Rui Guo Yuan Wei (C), Zhe Ma(T)	Seeds	Lumbricus and pinworm, internal	T	N. S. R.	2	Meadow	3 277m	—
Malva crispa Linn. (ZSY018)	*Malvaceae*	Dong Kui (C), Ma Neng Jian Mu Ba(T)	Tender leaves	Urine astringent, anuresis, internal	B	W. R.	5	Meadow	3 200m	—
Meconopsis horridula Hook. f. et Thoms. (ZSY036)	*Papaveraceae*	Duo Ci Lv Rong Hao (C), A Qia Cai Wen(T)	Flowers, herbs	Bone fracture, internal	T	W. R.	15	Rock cracks	3 850m	—
Meconopsis punicea Maxim. (ZSY028)	*Papaveraceae*	Hong Hua Lv Rong Hao(C), Wu Bai Ma Bu(T)	Herbs	Pulmonary disease, inflammatory, internal	T	W. R.	10	Meadow	3 531m	—
Morina kokonorica Hao. (ZSY066)	*Dipsacaceae*	Ci Shen(C), Jiang Ci Ga Bao (T)	Young herbs	Gastric disease, internal	B	Lv(2008)	6	Meadow	4 000m	—
Nardostachys chinensis Batal. (ZSY037)	*Valerianaceae*	Gan Song(C), Bang Bei(T)	Herbs	Inflammatory, internal	B	W. R. P.	11	Meadow	3 960m	Ⅱ
Notopterygium incisum Ting ex H. T Chang. (ZSY019)	*Umbelliferae*	Qiang Huo(C), Zhu Na(T)	Roots and rhizomes	Pain, hemorrhage, parasitic disease, internal	B	W. R. P.	15	Meadow	3 200m	Ⅱ
Orostachys fimbriatus (Turcz.) Berg. (ZSY097)	*Crassulaceae*	Wa Song(C), Ke Xiu Ba(T)	Aerial parts	Diuresis, calculus, internal	B	W. R. P.	1	Rock cracks	3 200m	—
Oxytropis ochrocephala Bunge. (ZSY032)	*Leguminosae*	Huang Hua Ji Dou(C), SaiGa (T)	Herbs	Edema, lung – heat, internal	T	W. R.	4	Meadow	3 639m	—
Oxytropis falcata Bunge. (ZSY101)	*Leguminosae*	Lian Xing Ji Dou(C), Sai Ga (T)	Herbs	Edema, lung – heat, internal	T	W. R.	18	Meadow	3 450m	—

(Continued)

Scientific name(Voucher no.)	Family	Common name[a]	Part used	Disease[b]	Chinese or Tibetan[c]	Pharmacopoeia and reference[d]	Info. no.	Habit	Altitude	Rank of endangered species
Papaver rhoeas L. (ZSY017)	*Papaveraceae*	Li Chun Hua (C), Jia Men (T)	Flowers and herbs	Dysentery, pain, internal.	B	W. R.	1	Meadow	3 200m	—
Paraquilegia microphylla. (ZSY062)	*Ranunculaceae*	Ni Lou Dou Cai (C), Yi Mu De Jin(T)	Branchs and leaves	Dystocia, retained after birth, internal	T	N. S. R.	14	Rock cracks	4 100m	—
Pedicularis kansuensis Maxim. (ZSY077)	*Scrophulariaceae*	Gan Su Ma Xian Hao (C), Mei Duo Lang Lang(T)	Flowers	Inflammation, edema, urinary obstruction, internal.	T	Bao et al. (2008)	12	Meadow	3 200m	—
Pedicularis longiflora Rudolph var. tubiformis (Klotz.) Tsoong. (ZSY009)	*Scrophulariaceae*	Ban Chun Ma Xian Hao(C), Mei Duo Lang Lang(T)	Flowers	Urinary obstruction, edema, internal	T	Zhang and He(1981)	8	Meadow	3 350m	—
Pedicularis oederi Vahl var. sinensis(Maxim.)Hurus. (ZSY063)	*Scrophulariaceae*	Hua Ma Xian Hao (C), Mei Duo Lang Lang(T)	Flowers	Urinary obstruction, edema, internal	T	N. S. R.	8	Rock cracks	4 170m	—
Pedicularis torta Maxim. (ZSY092)	*Scrophulariaceae*	Niu Xuan Ma Xian Hao(C), Mei Duo Lang Lang(T)	Flowers	Inflammation, Urinary obstruction, internal	T	N. S. R.	8	Meadow	3 300m	—
Pegaeophyton scapiflorum(Hook. f. et Thoms.) Marq. et Shaw. (ZSY114)	*Cruciferae*	Gao Shan La Gen Cai (C), Suo Luo Ga Bao(T)	Herbs	Cough, lung disease, and fever, internal	T	N. S. R.	6	Meadow	3 850m	Ⅱ
Phlomis younghusbandii Mukerjee. (ZSY151)	*Labiatae*	Pang Xie Jia(C), Lou Mu Er (T)	Root tubers	Common cold, tracheitis, internal	T	W. R.	5	Meadow	3 350m	—
Plantago depressa Willd. (ZSY110)	*Plantaginaceae*	Che Qian(C), Ta Rang(T)	Herbs	Urinary obstruction, dysentery, internal	B	W. R. P.	15	Alongside rivers	3 200m	—
Plantago major L. (ZSY111)	*Plantaginaceae*	Da Ye Che Qian(C), Na Ran Mu(T)	Aerial parts	Dysentery, internal	B	W. R.	3	Alongside rivers	3 200m	—
Polygonum capitatum Buch. - Ham. ex D. Don Prodr. (ZSY039)	*Polygonaceae*	Tou Hua Liao(C)	Herbs	Dyspepsia, lung disease, dysentery, internal	B	W. R.	4	Meadow	3 962m	—
Potentilla anserine L. (ZSY078)	*Rosaceae*	Jue Ma(C), Zhuo Er Ma(T)	Root tubers	Body empty, cough, dysentery, internal	B	W. R.	9	Meadow	3 200m	—
Potentilla fruticosa L. (ZSY007)	*Rosaceae*	Jin Lu Mei (C), Ban Na Er (T)	Flowers and leaves	Lung disease and dyspepsia, internal	B	W. R.	4	Scrub	3 420m	—

(Continued)

Scientific name(Voucher no.)	Family	Common name[a]	Part used	Disease[b]	Chinese or Tibetan[c]	Pharmacopoeia and reference[d]	Info. no.	Habit	Altitude	Rank of endangered species
Potentilla glabra Lodd. (ZSY006)	*Rosaceae*	Yin Lu Mei(C), Ban Ga Er (T)	Flowers and leaves	Lung disease, internal	B	N. S. R.	4	Scrub	3 420m	—
Pterocephalus hookeri (C. B. Clarke) Huck. (ZSY113)	*Dipsacaceae*	Yi Shou Cao (C), Bang Zi Duo Wu(T)	Herbs	Dysentery, internal	T	W. R.	6	Meadow	3 650m	Ⅰ
Ranunculus tanguticus Maxim. Ovcz. (ZSY073)	*Ranunculaceae*	Gao Yuan Mao Gen(C), Jie Cha(T)	Leaves and flowers	Stomach cold and edema, internal	T	N. S. R.	7	Meadow	4 200m	—
Rheum glabricanle G. San. (ZSY053)	*Polygonaceae*	Guang Jing Da Huang(C), Jun Mu Zha(T)	Roots and rhizomes	Abdominal distension, constipation, stomach pain, internal	B	Wei et al. (2004), Wei et al. (2005), Wei et al. (2006), Wang et al. (2006).	15	Meadow	3 550m	—
Rheum palmatum L. (ZSY125)	*Polygonaceae*	Zhang Ye Da Huang(C), Jun Mu Zha(T)	Roots and rhizomes	Abdominal distension, constipation, stomach pain, internal	B	W. R. P.	20	Meadow	3 200m	Ⅱ
Rhodiola algida (Ledeb) Fisch. et Mey. var. tangutica(Maxim.) S. H. Fu. (ZSY060)	*Crassulaceae*	Hong Jing Tian(C), Suo Luo Ma Bu(T)	Roots	Lung disease, cough, pain and inflammation, internal	B	W. R.	20	Rock cracks	3 750m	—
Rhodiola kirilowii (Regel) Maxim. (ZSY091)	*Crassulaceae*	Xia Ye Hong Jing Tian(C), Ga Du Er(T)	Roots	Pneumonia, fever, diarrhea and limb edema, internal	T	W. R.	1	Meadow	3 961m	—
Rhododendron anthopogonoides Maxim. (ZSY042)	*Ericaceae*	Lie Xiang Du Juan(C), Da Le Ga Bu(T)	Flowers and leaves	Cough and inflammation, internal	B	W. R.	16	Scrub	3 900m	—
Rhododendron capitatum Maxim. (ZSY041)	*Ericaceae*	Tou Hua Du Juan(C), Ta Le Na He(T)	Leaves and flowers	Lung disease, cough and inflammation, internal	B	W. R.	12	Scrub	3 900m	Ⅲ
Rhododendron latoucheaeo Franch. (ZSY044)	*Ericaceae*	Da Ye Du Juan(C)	Leaves	Lung disease, cough and inflammation, internal	B	N. S. R.	6	Scrub	3 900m	—
Rumex acetosa Linn. (ZSY011)	*Polygonaceae*	Suan Mo(C), Jun Zha(T)	Roots and rhizomes	Fever, internal	B	W. R.	5	Scrub	3 350m	—
Rupus irritans Focke. (ZSY031)	*Rosaceae*	Zi Se Xuan Gou Zi(C), Gan Za Ga Ri(T)	Branchs	Common cold, fever and cough, internal	T	Wang and Jia(1999)	8	Meadow	3 620m	—
Sabina chinensis (Linn.) Ant. (ZSY096)	*Cupressaceae*	Yuan Bai(C), Xiu Ba(T)	Leaves and fruits	Lung disease, internal	B	W. R.	12	Rock cracks	3 200m	—

(Continued)

Scientific name(Voucher no.)	Family	Common name[a]	Part used	Disease[b]	Chinese or Tibetan[c]	Pharmacopoeia and reference[d]	Info. no.	Habit	Altitude	Rank of endangered species
Sabina squamata (Buch. - Hamilt.) Ant. (ZSY038)	*Cupressaceae*	Gao Shan Bai (C), Xiu Ba (T)	Fruits and leaves	lung disease, internal	T	N. S. R.	3	Rock cracks	3 960m	—
Saussnrea arenaria Maxim. (ZSY030)	*Compositae*	Sha Sheng Feng Mao Ju(C), Za Chi(T)	Flowers	Jaundice, cholecystitis, internal	T	Chen et al. (1992a, b), Huang et al. (2007), Wang et al. (2011).	2	Meadow	3 200m	—
Saussurea epilobioides Maxim. var. cana Hand. - Mazz. (ZSY050)	*Compositae*	Hui Mao Liu Ye Feng Mao Ju (C), Ye Ge Xing Na Bao(T)	Herbs	Pain, intoxication, internal	T	N. S. R.	1	Meadow	3 550m	—
Saussurea frondosa Hand. - Mzt. (ZSY070)	*Compositae*	Xia Yi Feng Mao Ju(C), Za-Chi(T)	Flowers	Jaundice, cholecystitis, internal	T	N. S. R.	9	Meadow	3 900m	—
Saussurea graminea Dunn. (ZSY035)	*Compositae*	He Ye Feng Mao Ju(C), Za-Chi(T)	Aerial parts	Jaundice, cholecystitis, internal	T	N. S. R.	3	Meadow	3 650m	—
Saussurea medusa Maxim. (ZSY131)	*Compositae*	Shui Mu Xue Lian Hua(C), Xi Chen Qia Gui Su Ba(T)	Herbs	Gynaecopathia and inflammation, internal	T	W. R.	20	Rock cracks	4 351m	I
Scopolia japonica Maxim. (ZSY026)	*Solanaceae*	Dong Lang Dang (C), Tang Chong(T)	Rhizomes	Inflammation, internal; Parasitic disease, external	B	W. R.	20	Meadow	3 200m	—
Senecio diversipinnus Ling. (ZSY054)	*Compositae*	Yi Yu Qian Li Guang(C), Ye Ge Xing(T)	Aerial parts	Dysentery, eczema, internal	T	N. S. R.	7	Meadow	3 550m	—
Sinopodophyllum hexandrum (Royle) Ying. (ZSY112)	*Berberidaceae*	Tao Er Qi (C), Ao Mao Sai (T)	Fruits, stems	Gynaecopathia and inflammation, internal	T	W. R.	6	Scrub	3 900m	II
Soroseris hookeriana (C. B. Clarke) Stebb. (ZSY034)	*Compositae*	Tang Jie Juan Mao Ju (C), Suo Gong Se Bao(T)	Herbs	Inflammation, pain, internal	T	N. S. R.	5	Meadow	3 639m	—
Stellera chamaejasme L. (ZSY001)	*Thymelaeaceae*	Rui Xiang Lang Du (C), Ri Jia Ba(T)	Roots	Inflammation, ulcer, internal; Mange, external.	B	W. R.	13	Meadow	3 224m	—
Taraxacum mongolicum Hand. - Mazz. (ZSY074)	*Compositae*	Pu Gong Ying (C), Ku Er Mang(T)	Herbs	Ulcer, fever, internal	B	W. R. P.	4	Meadow	3 200m	—
Thlaspi arvense L. (ZSY021)	*Cruciferae*	Xi Ming(C), Zhai Ka(T)	Seeds	Pneumonia, internal	B	W. R.	3	Meadow	3 200m	—

(Continued)

Scientific name(Voucher no.)	Family	Common name[a]	Part used	Disease[b]	Chinese or Tibetan[c]	Pharmacopoeia and reference[d]	Info. no.	Habit	Altitude	Rank of endangered species
Trollius ranunculoidea Hemsl. (ZSY003)	*Ranunculaceae*	Mao Gen Zhuang Jin Lian(C)	Roots	Swelling of throat, internal	B	N. S. R.	5	Meadow	3 260m	—
Valeriana fauriei Brig. (ZSY012)	*Valerianaceae*	Kuo Ye Xie Cao(C), Zhu Ma (T)	Roots	Inflammation, internal	B	W. R.	4	Scrub	3 322m	—

a. Chinese(C), Tibetan(T). b. Internal application medicine(internal), external application medicine(external). c. Both traditional Chinese veterinary medicine and traditional Tibetan veterinary medicine(B), traditional Tibetan veterinary medicine(T). d. N. S. R. no scientific report; W. R. widely researched; Parmacopoeias(P).

while, herb, roots, rhizome and root tuber (all 65.56%) were extensive applied to treat the animal's diseases, which may cause the death of the plant and harm the local ecology environment. In spite of some species have being cultivated by the support of local government, such as *Gentiana dahurica* Fisch. and *Rheum palmatum* L., but in the concept of local herbalists that the effects of cultural materials were poorer than wild plant. The further study should be carried to compare the activities between cultural plant and wild plant (Table 1).

At the same time, in our investigation we found most of species being reported by local people did not be used as a single medicine but in preparations. So in order to survey the mode of preparation, we consulted it as a single question to local veterinarian practitioner. The results demonstrated that powders (70%) were the most mode of preparation, followed by decoctions (15%), pills (10%), unguentum (3%), tincture (1%), and others (1%) (Fig. 4). For one thing, this kind form of medication is more convenient to administer to animals; For another thing, it is relative to simple processing equipment, and save the cost price of preparation.

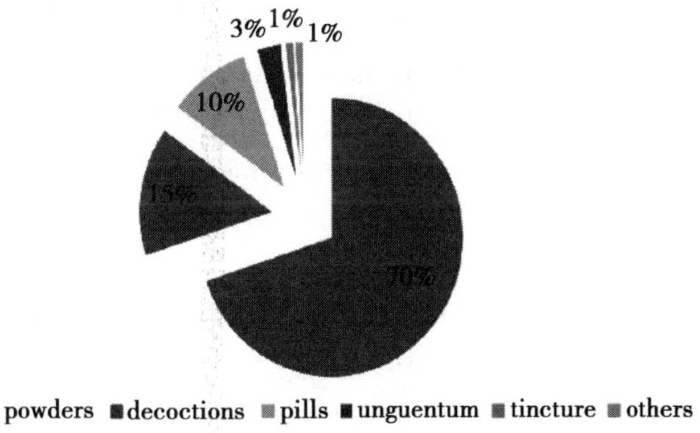

Fig. 4 The mode of preparation of ethno – veterinary in Ruoergai

3.1.3 *Ethnopharmacological uses*

According to the method developed by Cook (1995), various ailments affected a system of the body were divided into 12 'usage categories' (Collins et al., 2006; Zheng and Xing, 2009). These categories are respiratory system disorders (RES), digestive system disorders (DIG), injuries (INJ), infections/infestations (INF), muscular – skeletal system disorders (MUS), parasitic diseases (PAR), genitourinary system disorders (GEN), nervous system disorders (NER), circulatory system disorders (CIR), pregnancy/birth disorders (PRE), skin diseases (SKI) and poisonings (POI). From Fig. 5, respiratory diseases (21.09%) and digestive system disorders (19.89%) were mentioned most by local herbalists. Followed by injuries (17.61%), infections/infestations (9.66%), muscular – skeletal system disorders (9.66%), etc. (Fig. 5).

Meanwhile, 87 plants of 93 species can be used as internal medicines, and 1 plant can be only used as external application drug. Other 5 plants can be used as internal and external application medicines. 48 species (51.61%) could be used as both traditional Chinese veterinary medicine (TCVM) and traditional Tibetan veterinary medicine (TTVM), and others (45, 48.39%) were only used as Tibetan medicine. In all native species, 52 plant species (55.91%) were widely

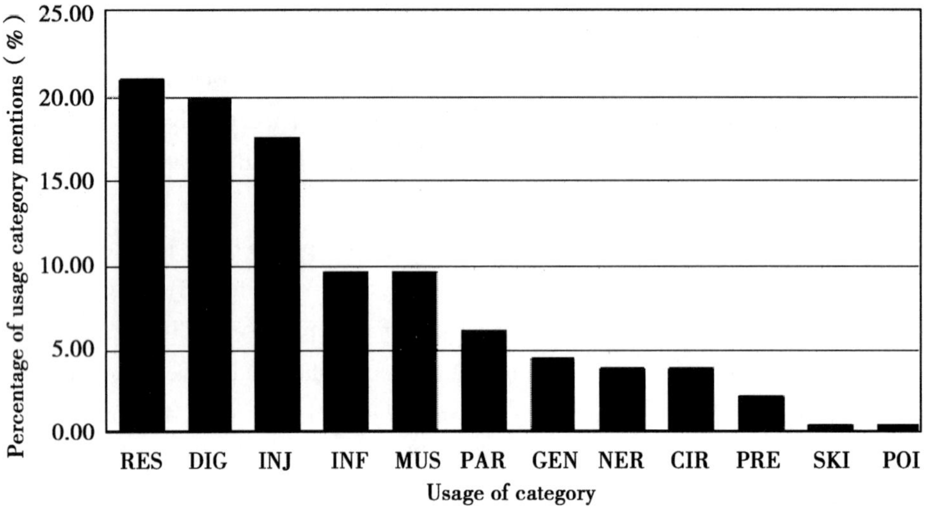

Fig. 5 Frequency of usage category mentions for medicinal plants in Ruoergai

studied, 13 species (13.98%) with 1 - 5 relative scientific literature, and only 10 was cited in Chinese veterinary pharmacopoeias (10.73%). On the other side, many species (28, 30.11%) were applied in clinic by folk veterinaries through experience and handed down over generation, which have not been studied by the modern technology and lacked of quality control.

3.1.4 *The resources and biodiversity of plant species*

In the history of Ruoergai, the traditional veterinary medicine was excavated and used abundantly to treat animals diseases. Recently, along with many chemical drugs having been widely used in the animal husbandry, the usage amounts of natural veterinary medicine were rapidly decreased. Meanwhile, increasing agriculture, over-grazing and disorder tourism have become the main influencing factors for the resources and biodiversity decrease of plant species in Ruoergai. According to the contents of 65 endangered Tibetan medicine established by Conference of protecting and developing endangered Tibetan medical materials in 2000 year, 5 (5.38%), 7 (7.53%) and 4 (4.30%) species of the 93 plant species recorded in this paper were list in Ⅰ, Ⅱ and Ⅲ grade of endangered species, respectively. Now, the decrease of resources and biodiversity in Ruoergai has attracted the local governments attention, and some species have being cultivated by the support of local government, such as *Gentiana dahurica* Fisch. and *Rheum palmatum* L. Some policy of limitation the overgrazing and disorder tourism have been published and carried out to recover the good ecological environment. Obviously, how to balance the natural resources protection and the development, including economy, veterinary medicine and tourism in Ruoergai region should be paid more attention.

3.2 *Brief introduction of imported medicinal materials*

In our survey, we also investigated 29 imported medicinal materials in Ruoergai, which belongs to 19 botanical different families. Zingiberaceae was the family with the largest number of medicinal species (4, 13.79%), followed by *Umbelliferae* and *Leguminosae* (3, 10.34%). *And Areca catechu* L., *Carthamus tinctorius* L., *Glycyrrhiza uralensis* Fisch., *Terminalia chebula* Retz. were cited by all informats as the most used plant species. In these species, fruits (11, 37.93%)

are the most widely part of plant, followed by rhizomes (4, 13.79%), flowers, roots, seeds and heartwoods (2, 6.90%). These medical plants were used to gastrointestinal diseases mainly (17, 58.62%). Comparing to native plant species in Ruoergai, 22 species (75.86%) were list in Chinese veterinary pharmacopoeias, 27 species (93.10%) were studied widely (Table 2) (Veterinary Pharmacopoeia Commission, 2010).

Table 2 Imported medicinal material used in Ruoergai

Scientific name	Family	Part used	Local name[a]	Diseases	Parmacopoeias and reference[b]	Info. no.
Alpinia officinale Hance.	*Zingiberaceae*	Rhizomes	Gao Liang Jiang (C), Ga Jia (T).	Common cold and cough, internal	W.R., P	12
Amomum kravanh Pierre ex Gagnep.	*Zingiberaceae*	Fruits	Bai Dou Kou (C), Jia Na Su Men (T).	Stomach cold, dyspepsia, internal	W.R., P	7
Amomum tsao - ko Crevost et Lemarie	*Zingiberaceae*	Fruits	Cao Guo (C), Ga Gao La (T).	Stomach cold, dyspepsia	W.R., P	15
Areca catechu L.	*Palmaceae*	Fruits	Bing Lang (C), Guo Yu (T).	Parasitic diseases	W.R., P	20
Aquilaria sinensis (Lour.) Gilg	*Thymelaceae*	Heartwood with resins	Chen Xiang (C), A Er La (T).	Qi asthenia	W.R., P	5
Carthamus tinctorius L.	*Compositae*	Flowers	Cao Hong Hua (C), KuKong (T).	Pneumonia	W.R., P	20
Cinnamomum cassia Presl.	*Lauraceae*	Barks	Rou Gui (C), Xin Cha (T).	Dyspepsia, diarrhea	W.R., P	8
Coptis chinensis Franch.	*Ranunculaceae*	Rhizomes	Huang Lian (C), Niang Zi Zhe (T).	Dysentery, Enteritis	W.R., P	12
Coriandrum sativum L.	*Umbelliferae*	Fruits and Herbs	Yuan Sui (C), Wu Su (T).	Stomach diseases	W.R.	7
Dolomiaea souliei (Franch.) Shih [*Vladimiria souliei* (Franch.) Ling]	*Compositae*	Roots	Chuan Mu Xiang (C), Bu Ga Mu La (T).	Stomach diseases, pain	W.R., P	5
Embelia laeta (L.) Mez.	*Myrsinaceae*	Fruits	Suan Teng Guo (C), Qi Dang Ga (T).	Parasitic diseases.	Lin (2006)	3
Ferula sinkiangensis K. M. Shen	*Umbelliferae*	Resins	A Wer (C), Xing Gung (T).	Parasitic diseases, dyspepsia	W.R.	9
Foeniculum vulgare	*Umbelliferae*	Fruits	Xiao Hui Xiang (C).	Dyspepsia	W.R.	11
Glycyrrhiza uralensis Fisch.	*Leguminosae*	Rhizomes	Gan Cao (C), Xin E Er (T).	Lung disease, tracheitis	W.R., P	20
Kaempferia galangal L.	*Zingiberaceae*	Rhizomes	Shan Nai (C), Ga Mu (T).	Dyspepsia, diarrhea	W.R., P	15
Lithospermum erythrorhizon Sieb. et Zucc	*Boraginaceae*	Roots	Zi Cao (C)	Purgation	W.R., P	6
Myristica fragrans Houtt.	*Myristicaceae*	Kernels	Rou Dou Kou (C)	Stomach cold, dyspepsia	W.R., P	12
Piper longum L.	*Piperaceae*	Racemations	Bi Ba (C), Bi Bi Lin (T).	Stomach cold, dyspepsia	W.R., P	4

(Continued)

Scientific name	Family	Part used	Local name[a]	Diseases	Parmacopoeias and reference[b]	Info. no.
Piper nigrum L.	*Piperaceae*	Fruits	Hu Jiao (C), Po Wa Ri (T).	Stomach cold, dyspepsia	W. R., P	7
Pterocarpus indicus Willd.	*Leguminosae*	Heartwoods	Zi Tan Xiang (C), Zai Dan Ma Bu (T).	Blood stasis	Khan and Omoloso (2003), Li (1998), Hou (1999)	3
Punica granatum L.	*Punicaceae*	Seeds	Shi Liu Pi (C), Sai Zhu (T).	Dyspepsia	W. R., P	15
Syzygium aromaticum (L.) Merr. et Perry [*Eugenia caryophyllata* Thunb.]	*Myrtaceae*	Flower	buds Ding Xiang (C), Li Xi (T).	Dyspepsia	W. R., P	16
Terminalia chebula Retz.	*Combretaceae*	Fruits	He Zi (C), A Ru Re (T).	Dyspepsia	W. R., P	20
Pyrus pashia Buch. – Ham. ex D. Don	*Rosaceae*	Fruits	Shan Zha (C), Shan Li Hong (T).	Dyspepsia	W. R., P	9
Sambucus williamsii Hance	*Caprifoliaceae*	Root barks	Jie Gu Mu (C)	Inflammation, pain, bone fracture	W. R., P	8
Santalum album L.	*Santalaceae*	Stems	Bai Tan Xiang (C)	Lung disease, inflammation	W. R., P	3
Vigna angularis (Willd.) Ohwi et Ohashi	*Leguminosae*	Seeds	Chi Dou (C)	Diuresis, diarrhea	W. R., P	2
Vitis vinifera L.	*Vitaceae*	Fruits	Pu Tao (C), Gong Zhu Mu (T).	Diuresis, lung disease	W. R.	9
Zanthoxylum bungeanum Maxim.	*Rutaceae*	Fruits	Hu Jiao (C), Ye Ma (T).	Parasitic diseases, dyspepsia	W. R.	10

* a. Chinese (C), Tibetan (T); b. W. R. widely researched. Parmacopoeias (P).

4 Conclusion

The present investigation showed that about 129 species were applied in veterinary clinic in Ruoergai. 93 Species of them were native, 29 others were imported, and 7 were mineral medicines. These species make a great contribution to the animal health and the economic development of Ruoergai. Most of species were used to treat diseases both human and livestock, meanwhile some species were only applied in veterinary clinic but not human. So these plants should be studied and developed further.

Due to special geographical position and historic condition, the communications between traditional Chinese veterinary medicine (TCVM) and traditional Tibetan veterinary medicine (TTVM) is more frequent than Xizang autonomous region. Under the guidance of TTVM, the advanced theory and technology of TCVM were adopted, formed a unique theory system.

Some low – altitude and Chinese medicinal materials were imported from middle and eastern part of China, and used as single medicine or a composition of preparation to treat various animal diseases by folk veterinarians. In our investigation, we found that 29 medicinal materials were imported from other regions, such as, *Aquilaria sinensis* (Lour.) Gilg (Chenxiang) and *Areca catechu Linn. (Binglang) from Guangxi and Yunnan provinces, were used to treat Qi asthenia and parasitic diseases, and so on. This provides an example how to apply the lowaltitude materials and methods to treat the plateau animal diseases.*

Unique processing method and technology of TTVM were also very particular. Some simple distillation equipment, stone mill, copper pan, iron pan and homemade drying cabinets were applied to pre – treatment the medical materials (Fig. 6). Comparing to TCVM, by the special and complex processing methods, many miner matters and heavy metals were pre – treated and used to treat the animal's diseases. In our survey, 7 mineral drugs were applied in veterinary clinic (5.43% in all medicinal materials), such as after 9 steps processing (Collected – Bruising – Choosing – Cleaning – Decocting and sock in waters with *mirabilite* and milk for 24h—Cleaning again – Levigating – Stir – frying – Open – air drying), *Calcite* (Hanshuishi) was used to treat the digestive system diseases of animals. And *Sulphur* (Liuhuang) was used to treat the skin diseases, *Mirabilite* (Mangxiao), *Gypsum* (Shigao), *Halite* (Qinyan) were usually used to treat the dyspepsia and dysentery of animals, etc. (Table 3).

Fig. 6 The processing equipments of ethno – veterinary medicine in Ruoergai

Table 3 Mineral medicines used in Ruoergai

Scientific name	Nature	Local name[a]	Disease	Inf. no.
Realgar	Sulphide	Xiong Huang (C), Dong Rui (T)	Parasite disease	13
Sulphur	Native sulfur	Liu Huang (C), Mu Si (T)	Parasite disease	11
Chalcanthite	Bluestone	Dan Fan (C), Pi Ban (T)	Aphthae	6
Trona	Sulfates	Jian Feng (C), Pu Duo (T)	Dyspepsia	15
Halite	Halogenide	Qin Yan (C), Jia Cha (T)	Dyspepsia	20
Calcite	Carbonates	Han Shui Shi (C), Jun Xi (T)	Dyspepsia, dysentery	20
Gypsum	Calcium sulfate	Shi Gao (C)	Fever, lung fever	10

* a. Chinese (C), Tibetan (T).

Recently, along with the chemical drugs were used widely and the rapid progress of urbanization and industrialization in China, the numbers of veterinary practitioners and application of traditional herbs is gradually decreasing. Now, all local herbalist or veterinaries were more than 50 years old, and most of them have not successors in Ruoergai. As a result, a loss of medicinal materials knowledge is inevitable. So the inherit, protect and development of its should be paid more attentions in the future, and the species which have not been studied should be priority developed to find biological activities and new bioactive compounds further.

Acknowledgments

The authors acknowledge their gratitude to the animal husbandry and veterinary bureau of Ruoergai county and traditional Tibetan veterinary medicine association of Ruoergai for the great assistance. This work was financed by The Special Fund of Chinese Central Government for Basic Scientific Research Operations in Commonweal Research Institutes (No. 1610322011009).

References

[1] Bao GS, Wang HS, Xu XX, Jia YF. Effect of aqueous extract of aerial part from *Pedicularis kansuensis* Maxim. on seed genmination and seeding growth of different forages. Chinese Qinghai Journal of Animal and Veterinary Sciences (2008) 39: 1 – 3.

[2] Chen, N. Y., Zai, J. J., Pan, H. P., He, Y. L., Song, Z. Z., Jia, Z. J., 1992a. Studies on chemical constituents of essential oils of three species of *Saussurea*. Acta Botanica Yunnanica 14: 203 – 210.

[3] Chen, T. G., Li, F. C., Cai, X., Li, R., Ji, F., 2004. The study on the biochemical characters of *Hippophae thibetana*. Journal of Gansu Forestry Science and Technology 29: 19.

[4] Chinese materia editorial committee, 2002. State Chinese Medicine Administration Bureau. Chinese materia, Tibetan volume. Shanghai Scientific and Technical Publishers, Shanghai.

[5] Collins, S., Martins, X., Mitchell, A., Teshome, A., Arnason, J. T., 2006. Quantitative ethnobotany of two east Timorese cultures. Economic Botany 60: 347 – 361.

[6] Cook, F. E. M., 1995. Economic Botany Data Collection Standard. Royal Botanic Gardens, Kew, United Kingdom 146.

[7] Dai, Y., Zhang, B. B., Xu, Y., Liao, Z. X., 2010. Chemical constituents of *Adonis coerulea* Maxim. Natural Product Research and Development 22: 594 – 596.

[8] Huang, M. C., Liao, Z. X., Chen, D. F., 2007. Chemical constituents of *Saussurea arenaria* and their cytotoxicity. Chinese Traditional and Herbal Drugs 38: 1 463.

[9] Hou, S. H., 1999. Identification of *Santanlum album*, *Pterocarpus indicus*, *Dalbergia odorifera*, *Caesalpinia sappan* with TLC and UV. Qinghai Medicine Journal 29: 54.

[10] Khan, M. R., Omoloso, A. D., 2003. Antibacterial activity of Pterocarpus *indicus*. Fitoterapia 74: 603 – 605.

[11] Li, C. X., Jiao, Y., Zhang, R., Hou, Z. J., An, H. G., 2008. Studies on the antioxidative activity of the water extract of *C. tangutica* (Maxim.) Korsh flower. Natural Product Research and Development 20: 134 – 137.

[12] Lin, Z. L., 2006. Study on nutrition and pigment of *E. laeta*. Acta agriculturae Jiangxi 18: 86 – 88.

[13] Li, S. Y., 1998. Textual research on *Pterocarpus indicus*. Journal of Chinese Medicinal Materials 21: 259.

[14] Lv, C. P., 2008. Study on the Chemical Constituents of *Morina* kokonorica. Lanzhou University, Master Thesis.

[15] Northwest Plateau Institute of Biology, Chinese Academy of Sciences, 1991. Tibetan Medicine. Qinghai People's Publishing House, Xining.

[16] Pan, Y. J., Wang, R., Chen, S. N., Chen, Y. Z., 1992b. A new diterpenoid alkaloid from *Delphinium caeruleum*. Chemical Journal of Chinese University 13: 1 418 – 1 419.

[17] Pharmacopoeia Commission, 2010. Ministry of public health. Pharmacopoeia of the People's Republic of Chi-

na, Part 1. Medical Science Press, Beijing, China.
[18] Pu, Z., Wang, J. L., Yuan, R. Y., 2011. Quality standard of a Tibetan *medicineHippuris vulgaris* L. Guide of China Medicine 9: 48 – 50.
[19] Qi, S. G., Li, L., 2011. Measuring flavonoid glycosides content in branch of *Hippophae rharmnoides* and *Hippophae thibetana* with RP – HPLC. Journal of Qinghai University (Natural science) 29: 58.
[20] Ruoergai government, 2011. <http://www.ruoergai.gov.cn>.
[21] Rao, G. X., Pu, F. D., Sun, H. D., 1995. Chemical constituents of *Heracleum millefolium*, *H. millefolium var. longilobum* and it's significance on taxonomy. Natural Product Research and Development 7: 16 – 19.
[22] Shao, H. X., Yang., J. Y., Han, J. X., Li, H. X., Qiu, J., Zhang, X. M., 2008. Phamacognosy identification of *Hippuris vulgaris* L. Li Shi Zhen Medicine and Material Medica Research 19: 1 943 – 1 945.
[23] Tibetan veterinary experience documents editorial committee, 1979. Tibetan Veterinary Experience Documents. China Agriculture Press, Beijing.
[24] Wang, B. G., Jia, Z. J., 1999. Studies on the chemical constituents of two species of raspberry (*Rubus* L.). Chinese Traditional and Herbal Drugs 30: 83 – 85.
[25] Wang, Y. J., Wei, Y. H., Wang, X. H., Zhang, J. Q., Wu, X. A., 2006. Studies on the component of essential oils and the antibacterial activity in vitro of *Rheum glabricaule*. Journal of Chinese Medicinal Materials 29: 2 072 – 2 075.
[26] Wang, Y. F., Xiao, L. N., Yang, Z. B., Li, Z. T., 2011. Study on components of essential oil from 3 species of Saussurea DC of the eastern of the Qinghai – Tibetan plateau and their systematic significance. Journal of Northwest Normal University (Natural science) 47: 80.
[27] Wei, Y. H., Zhang, C. Z., Li, C., Tao, B. Q., 2004. Chemcial constituents of *Rheum glabricaule* (I). Chinese Traditional and Herbal Drugs 35: 732 – 735.
[28] Wei, Y. H., Wu, X. A., Zhang, C. Z., Li, C., 2005. Studies on chemical constituents of *Rheum glabricaule*. Journal of Chinese Medicinal Materials 28: 658 – 660.
[29] Wei, Y. H., Wu, X. A., Zhang, C. Z., Li, C., Song, L., 2006. Studies on Chemical constituents of *Rheum glabricaule* Sam. Chinese Pharmaceutical Journal 41: 253 – 255.
[30] Wu, D. Q., Li, C. X., An, Z. J., Zhang, C. Y., H. G., 2008. Determination of microelements in *C. tangutica* (Maxim.) Korsh flower by FAAS. Natural product Research and Development 28: 228 – 230.
[31] Xiao, W., 2007. Study of *Hippophae* on Qinghai – Tibet Plateau, Pharmacognosy and Genetic Relationship. Sichuan University, Master Thesis.
[32] Zhang, B. C., He, S. X., 1981. Preliminary study on the allelopathy of the extracts of *Pedicularis longiflora* Rudolph *var. tublformis* (Klotz.) Tsoog. Acta Ecologica Sinica 1: 227 – 230.
[33] Zhang, H. D., Zhang, S. J., Chen, Y. Z., 1991. Studies on chemical constituents of *Adonis coerulea* Maxim—a Tibetan medicinal herb. Journal of Lanzhou University (Natural Science) 27: 88 – 92.
[34] Zhang, R., Zhang, Z. J., Hou, Z. J., Li, C. X., Han, D. H., Xie, Z. P., 2006. The discussion of technology of flavonoiods extraction from flowers of *Clematis tangutica* (Maxim.) Korsh. Chinese Wild Plant Resources 25: 58 – 60.
[35] Zhang, X. W., Mei, L. J., Shao, Z., Tao, Y. D., 2009. Determination of Oleanic acid in *C. tangutica* (Maxim.) Korsh flower by HPLC. Chinese Journal of Analysis Laboratory 28: 238 – 240.
[36] Zheng, X. L., Xing, F. W., 2009. Ethnobotanical study on medicinal plants around Mt. Yinggeling, Hainan Island, China. Journal of Ethnopharmacology 124: 197.

(Published the article in Journal of Ethnopharmacology affect factor: 3.014)

Effects of Fermentation Astragalus Polysaccharides on Experimental Hepatic Fibrosis

Qin Zhe[1,2], Li Jianxi[2], Yang Zhiqiang[2], Zhang Kai[2], Zhang Jinyan[2], Meng Jiaren[2], Wang Long[2], Wang Lei[2]

(1. College of Veterinary Medicine, Gansu Agricultural University, Lanzhou730070, China; 2. Lanzhou Institute of Animal and Veterinary Pharmaceutics Science, Chinese Academy of Agricultural Sciences, Lanzhou730050, China)

Abstract: This study aims to explore the antagonistic function of Fermentation Astragalus Polysaccharides (FAPS) on experimental hepatic fibrosis induced by CCl_4 and its possible mechanism. Hepatic fibrosis was induced by subcutaneous injection with carbon tetrachloride twice weekly for 8 weeks in Sprague – Dawley rats. Different does Fennentation Astragalus Polysaccharides (FAPS) (50, 100 and 150 mg/ (kg/day. and colchicines (50 mg/ (kg/day)) were administered intragastrically daily to carbon tetrachloride – treated rats, After the experiment, rats from all the groups were weighed and the following parameters were measured: Alanine Aminotransferase (ALT), Aspartate aminotransferase (AST), Alkahne Phosphatase (ALP), Hyaluronic Acid (HA), Laminin (LN), IV type of Collagen (CIV) and anunonia tenniml Procollagen β peptide (PCIII) in their sera and of Iotal Antioxidant Capacity (I – AOC), Glutathione Peroxdase (GSH – Px), Malondialdehyde (MDA) and Hydroxylproline (Hyp) in homogenate in the animal livers. Additionally, partial liver tissues were examined histopathologically as well as by Hemotoxylin and Eosin (H and E) and Masson chromoscopy. Results showed that compared with the control group, rats from the model group had a significant decrease in body weight, a significant rise in ALT, AST and ALP levels in the serum and also a remarkable hft of HA, LN, CIV and PCIII the four indicators of hepatic fibrosis. Moreover, T – AGC, GST and GSH – Px enzyme activities were reduced significantly in model rats while concentrations of MDA, Hyp and reduced GSH were raised substantially ($P < 0.01$). Histopathological examination, Hemotoxylin and Eosin (H and E) and masson chromoscopy revealed an evident change in hepatic fibrosis in the livers of the model group. Compared with the model group, animals treated with FAPS all had a greater weight, a lower activity of ALT, AST and ALP in the sera, lower levels of HA, LN, CIV and ammonia terminal PClII as well as less Hyp. In addition, FAPS significantly increased the reduced MDA, GSH – Px and GST values, elevated Total Antioxidant Capacity (T – AOC) and brought the abnormally high reduced GSH back to a normal level. Such effects on hepatic fibrosis in animals from the $FAPS_H$ group all twned out stronger than those of Col group animals ($P > 0.05$), Thus, the data indicates that FAPS can produce an apparent antagonism to hepatic fibrosis and the mechanism by which it does this may be associated with generation of oxygen free radicals.

Key words: Fermentation Astragalus Polysaccharides (FAPS); Hepatic fibrosis; Rat; Generation; Radicals; Mechanism

Introduction

Hepatic fibrosis is a type of pathological repaIr response occwring in conjllllction with liver injwy and characterized by an abnormal proliferation of collagenous fiber within the liver tissue, the denaturation and necrosis of hepatocytes and damage to hepatic structure. Hepatic fibrosis results in an aberration of liver fllllction and hampers prognosis of liver disease. It is an imperative stage by which chronic liver diseases develop into cirrhosis (Shu *et al.*, 2007; Aboutwerat *et al.*, 2003; Shimizu, 2003). Hepatic fibroses may be caused by parasitic infection (tapeworm (Ding *et al.*, 2008), nematode (Xu *et al.*, 2011) or toxin (aflatoxin (Wang and Groopman, 1999) and these sources have imposed a considerable cost on animal husbandry in China. Prevention of hepatic fibrosis has been drawing interest from researchers in China and abroad but there is still for a lack of treatments for clinical applications.

However, traditional Chinese medicine can not only act against hepatic fibrosis but also speed up an animal's recovery from chronic liver diseases by encouraging the regeneration of hepatocytes (Ao *et al.*, 2004). Astragal is extracted from the dry roots of the legwninous plants *Astragalus membraanceus* (Fisch.) Bge. *var. mongholicus* (Bge.) Hsiao and *Astragalus membranaceus* (Fisch.) Bge. It is a major medicine present in many traditional Chinese medicine prescriptions (such as Yiganning Keli, Liganlong Keli and Yinqiganfu Keli (CPC, 2010) for liver diseases. The effective components include triterpenoid saponis, flavonoids compounds and astragalus polysaccharides (CPC, 2010; Zou *et al.*, 2002) and many researchers have showed that astragalus polysaccharides were effective against hepatic fibrosis (Zou *et al.*, 2002; Zhang *et al.*, 2003; Li *et al.*, 2007). However, its application has been severely restricted by poor extraction efficiency, through water extraction and alcohol precipitation and the resulting waste of materials. With studies of Chinese medicine preparation on - going, progress has been made in medicine preparation by means of fermentation in place of traditional methods (Wang *et al.*, 2008). Through degradation by microorganism, effective components and active substances in the drug can be extracted. As a consequence, medicine cost can be reduced due to stronger drug potency and lower necessary dosage (Wang *et al.*, 2008). Researchers biologically fermented astragals using *Streptococcus alactolyticus* and elevated the yield of *Astragalus polysaccharides* from 1.8% ~3.8% and its overall content from 33.53% ~70.88%. Since, little research has been carried out on the mechanism by which FAPS antagonizes hepatic fibrosis, here we sought to explore the issue via constructing an experimental model of liver injrny by use of CCl_4 and examining the effects of FAPS on the weight of the animals, their biochemical indicators of sennn and liver and histopathological changes to the organ. This study provides useful experimental data for the development of FAPS as a medicine capable of treating liver injuries in the future.

Materials and Methods

Astragals were identified from *Astragalus membranaceus* (Fisch.) Bge received from the Veterinary Medicine and Feed Monitor in Gansu province. Colchicine was purchased from Wako Pure Chemical Industries, Ltd. Japan. The radio immune assay test kit of Hyaluronic Acid (HA), Laminin (LN), IV type of Collagen (CIV) and ammonia terminal Procollagen β peptide (PCIII) was purchased from Beijing North Institute of Biological Technology. The test kit of Alanine Aminotransferase (ALT), Aspartate aminotransferase (AST) and Alkaline Phosphatase (ALP) was bought from Shenzhen Mindray Bio - Medical Electronics Co., Ltd. The test kit of Malondialdehyde (MDA), Glutathione Peroxdase (GSHPx), Glutathione S - Transferase (GST), Glutathione (GSH), Total Antioxidant Capacity (T - AOC) and Hydroxylproline (Hyp) was from Nanjing Jiancheng Bioengineering Institute. The test kit for protein concentration using the Lowry method was from Beijing Solarbio Science and Technology Co., Ltd. CCl_4 was provided by Tianjin Zhiyuan Chemical Agent Co., Ltd. and was diluted 1 : 1 with peanut oil. The FGM9 strain of *Streptococcus alactolyticus* was prepared by the Animal Husbandry and Veterinary Medicine Research Institute, Chinese Academy of Agricultural Sciences in Lanzhou, Gansu province. Gifu Anaerobic Medium (GAM) culture medium was made by Qingdao Hope Bio - Technology Co., Ltd. Absolute ethyl alcohol and glacial acetic acid were Analytical Reagent (AR) and were purchased from Tianjin Tianxin Fine Chemicals Development Center.

Apparatuses: The microbial fermentation tank was made by Shanghai Gaoji Bio - Engineering Co., Ltd. The Allegra X - 15R centrifuge was made by Beckman in the USA. The BS - 420 automatic clinical chemistry analyzer was made by Shenzhen Mindray Bio - Medical Electronics Co., Ltd. The DPC - γC - 12 immunoradiometric counter was produced by Tianjin Depu Bio - medical Electronics Co., Ltd. The Spetralv1ax M2° microplate reader was made by Molecular Devices, USA. The Leica automatic biopsy system was manufactured by Leica, Germany and the Olympus X71 microscopic imaging system was made by Olympus, USA. ZHWY - 2102C Incubator shaker was made by Shanghai Zhicheng Amlysis Instnnnent Co., Ltd.

Extraction of FAPS: *Streptococcus alactolyticus* in the proportion of 2% was inoculated into GAM media (pH = 7) that contained 8% astragals solution. After a 72 h anaerobic fermentation in a 37℃ thermotank, the fermented solution was then freeze dried. The dried powder (4kg) of fermented astragals was extracted respectively with 40 and 20 L 80% ethanol and then underwent a 2 h reflux extraction at 78℃. After filtration, the residue was extracted, respectively with 40, 20 and 20 L water and boiled for 1.5 hat 80℃ followed by another filtration. Thereafter, the three filtrates were combined and condensed to 1 : 1 (the ratio was weight of fermented astragals freeze - dried powder to the voume of filtrate). Deproteinization was done using the Sevag method (Staub, 1965). Briefly, a 4 : 1 mixture of chloroform and normal butanol was added to the filtrate, the mixture was 80 r/min of shaking for 30 min and the mixture was centrifuged for 5 min at 3 000 r/min and the liquid supernatant was removed and reserved. This procedure was repeated three times. Following deproteinization, the combined supernatants were subjected to a 24 h flow - water dialysis and a 24 h distilled water dialysis. Then add ethyl alcohol to 4x the vohnne of the resulting dialyzed filtrate, the mixture was stirred and deposited for 12 h and then centrifuged by 3 000 r/min for 5 min. The precipitate was the washed with absolute ethyl alcohol, acetone and diethyl ether with the same volwne with concentrated filtrates and freeze dried. The yield of FPS was 160 g totally.

Grouping of experimental animals: The Experimental Animal Center located in the Department of Medicine, Lanzhou University provided us with intact male Sprague Dawley (SD) rats, certification No. SCXK (Gan) 2005 – 0007. A 7 days observation confinned that the feeding environment could come up to GB 14922.2 – 2001 requirements.

Sixty sanitary male rats were randomly divided into six groups, ten in each group. The groups included a a normal control group and a model group, a group administered a low dose of FAPS ($FAPS_e$) 50mg/ (kg/day), one administered with a mid dose ($FAPS_M$) 100 mg/ (kg/day), one administered with a high dose ($FAPS_H$) (150mg/ (kg/day)) and one administered with Colchicine (Col) 0.2 mg/ (kg/day). (Zou *et al.*, 2002). At the same time, the rats were injected hypodermically with 2 ml/kg 50% CCl_4 sterilized peanut oil solution, twice a week for 8 weeks. The exception was for the normal control group which was injected solely with a corresponding amount of peanut oil. From the first injection of CCl_4 on, animals from each of the drug groups were also intragastrically administered FAPS and according to group assignments, once a day for 8 weeks. The model group and the normal control group were intragastrically administered normal saline solution of corresponding amounts. Animals were weighed once a week.

Measurement of text indicators: After 8 weeks, the experiment concluded, deprivation of food but not water for 16 h and then measured the body weight. Blood was collected from the femoral artery of each rat and centrifuged (IS min at 2 000 r/min). Then sera preserved at –80℃ for the determination and measurement of ALT, AST, ALP, HA, LN, Cχ, PCβ, GSH – Px and GST. All the animals were sacrificed by cervical dislocation; part of the liver tissue was fetched rapidly and placed in liquid nitrogen to be used for the detection of MDA, GSH, T – AOC, Hyp and protein. The left lobe of liver was isolated and fixed in 10% neutral formalin solution for the preparation of routine paraffin slides. H and E and masson staining were performed for histological and pathological examinations of treated rats (Zhuoming, 1998).

Statistic data process: SPSS Software (Version 17.0) (SPSS, 2009) was utilized for the statistic analysis. Measurement data were expressed in the form of mean ± standard deviation ($\bar{x} \pm s$). Comparisons among groups were conducted by means of ANOVA. Difference is statistically significant when $P < 0.05$.

Results

Effects of FAPS on rats' body weight: By the end of the experiment, the control rats had a body weight siginficantly greater than the model group ($P < 0.01$) as did the $FAPS_L$ $FAPS_M$ and $FAPS_H$ groups ($P < 0.05$). The model rats, weighting 306 g on average before treatment had increased weights after being treated with FAPS (50, 100, 150mg/kg) or Col (0.2 mg/kg). All treated rats had lower weights than the control. However, the difference between the $FAPS_H$ group and the control group and the difference between the $FAPS_H$ group and the model group were not significant ($P > 0.05$) (Table 1).

Table 1 Variation of rat body weight before and after the experiment (n = 10)

Groups	Dosage (mg/ (kg · day))	Before experiment (g)	After experiment (g)
Control	–	217 ± 21	385 ± 31#
Model	–	215 ± 19	306 ± 19Δ
$FAPS_L$	50. 0	215 ± 24	329 ± 24*
$FAPS_M$	100. 0	217 ± 16	336 ± 20*
$FAPS_H$	150. 0	216 ± 16	342 ± 36
Col	0. 2	214 ± 12	331 ± 16*

* Δ$P < 0.05$ vs the control group, * $P < 0.01$ vs the control group, #$P < 0.01$ vs. the model group. $FAPS_L$ was low dose of FAPS by 50 mg/ (kg · day), $FAPS_M$ was mid dose of FAPS by 100 mg/ (kg · day), $FAPS_H$ was high dose of FAPS by 150 mg/ (kg · day), Col was colchicine by 0. 2 mg/ (kg · day) .

Effects of FAPS on serum ALT, AST and ALP of rats with hepatic fibrosis: In contrast with the control group and the normal group, the model group rats had significantly elevated serum ALT, AST and ALP which were remarkably reduced by FAPS (50 100 and 150mg/kg) or Col (0. 2mg/kg) treatments, especially by the high - dosage FAPS treatment ($P < 0.05$). This indicates that FAPS could protect rats against chronic liver damage (Table 2).

Table 2 Effects of FAPS on senun Alanine Aminotransferase (ALT), Aspartate aminotransferase (AST) and Alkaline Phosphatase (ALP) of rats induced with hepatic fibrosis by CCl_4 ($\bar{X} \pm S$, n = 10)

Groups	Dosage (mg/ (kg · day))	ALT (nmol/S · L)	ALP (μmol/S · L)	AST (nmol/S · L)
Control	–	60. 75 ± 9. 3800#	103. 67 ± 22. 79#	205. 50 ± 29. 180#
Model	–	613. 88 ± 145. 88Δ	269. 30 ± 82. 92*	698. 40 ± 119. 54Δ
$FAPS_L$	50. 0	532. 83 ± 120. 02Δ	257. 17 ± 57. 97*	636. 67 ± 182. 80Δ
$FAPS_M$	100. 0	470. 71 ± 126. 79Δ	225. 83 ± 38. 70*	597. 50 ± 127. 68*
$FAPS_H$	150. 0	371. 80 ± 73. 71Δ#	180. 40 ± 62. 12**	494. 40 ± 114. 34*#
Col	0. 2	502. 75 ± 167. 95Δ	341. 17 ± 88. 18*	478. 83 ± 113. 91*#

*. $P < 0.05$ vs. the control, z. $P < 0.01$ vs. the control, #. $P < 0.01$ vs. the model, a. $P < 0.05$ vs. $FAPS_H$. $FAPS_L$ was low dose of FAPS by 50 mg/ (kg · day), $FAPS_M$ was mid dose of FAPS by 100 mg/ (kg · day), $FAPS_H$ was high dose of FAPS by 150 mg/ (kg · day), Col was colchicine by 0. 2 mg/ (kg · day)

Effects of FAPS on the contents of PCⅢ, CX, LNand HA in rat serum and the content of Hyp in hepatic tissue: Table 3 shows that in comparison with the normal control, PCβ, Cχ, LN, HA and Hyp in model rats were all siginficantly higher ($P < 0.01$). After FAPS (50, 100 and 150mg/kg) or Col (0. 2mg/kg) treatments, model rats had significantly reduced levels of PCβ, Cχ, LN in the serum and of HA in hepatic tissue. Differences in CIV activity between the $FAPS_H$ group and the control group were not significant ($P > 0.05$). FAPS (50, 100 and 150mg/kg) and Col (0. 2mg/kg) treatments could produce a significant reduction in Hyp in the liver homogenate in model rats, particularly in contrast with the model group before treatment ($P < 0.05$).

The results of the five indicators of liver fibrosis stated above suggest that both FAPS and APS could inhibit hepatic fibrosis to a certain extent and the inhibition strengthened along with dosage augmentation along the concentration gradient.

Table 3 Effects of FAPS on the concentrations of ammonia terminal Procollagen βpeptide (PCβ), IV type of Collagen (CIV), Laminin (LN), Hyaluronic Acid (HA) in serum and the concentration of Hyp in hepatic homogenate in rats induced with hepatic fibrosis ($\bar{X} \pm S$, n = 10)

Groups	Dosage (mg/ (kg · day))	n	PCβ (ng/L)	Cχ (ng/ml)	LN (ng/ml)	HA (μg/ml)	Hyp (μg/mg prct)
Control	—	10	$16.65 \pm 3.20^{\#}$	$12.56 \pm 0.96^{\#}$	$28.03 \pm 3.87^{\#a}$	$192.80 \pm 28.33^{\#a}$	$0.07 \pm 0.01^{\#}$
Model	—	10	33.04 ± 6.41^{z}	26.49 ± 4.39^{z}	54.11 ± 13.33^{z}	389.70 ± 61.40^{z}	0.41 ± 0.02^{za}
$FAPS_L$	50.0	10	30.79 ± 3.88^{z}	$23.83 \pm 2.35^{*}$	50.09 ± 4.39^{z}	350.82 ± 78.27^{z}	0.31 ± 0.01^{za}
$FAPS_M$	100.0	10	$28.30 \pm 3.38^{*}$	$20.47 \pm 2.25^{*}$	$46.47 \pm 2.25^{*}$	$335.23 \pm 54.00^{*}$	0.22 ± 0.01^{z}
$FAPS_H$	150.0	10	$26.06 \pm 3.28^{*}$	19.10 ± 3.86	$42.10 \pm 3.86^{*}$	$308.81 \pm 44.60^{*}$	$0.12 \pm 0.02^{z\#}$
Col	0.2	10	$27.96 \pm 5.52^{*}$	$23.94 \pm 3.33^{*}$	$45.83 \pm 2.35^{*}$	$252.33 \pm 33.46^{\#}$	$0.17 \pm 0.01^{z\#}$

*. $P < 0.05$ vs. the control, z. $P < 0.01$ vs. the control, #. $P < 0.01$ vs. the model, a. $P < 0.05$ vs. $FAPS_H$. $FAPS_L$ was low dose of FAPS by 50 mg/ (kg · day), $FAPS_M$ was mid dose of FAPS by 100 mg/ (kg · day), $FAPS_H$ was high dose of FAPS by 150 mg/ (kg · day), Col was colchicine by 0.2 mg/ (kg · day)

Effects of FAPS on T－AOC, GSH, MDA and protein content in hepatic tissues of rats: Results in Table 4 shows that compared with the normal control, model rats had significantly decreased concentrations of T－AOC and protein ($P < 0.01$) but significantly increased concentrations of MDA ($P < 0.05$) and GSH ($P < 0.01$). FAPS (150mg/kg) treatment applied on model rats substantially reduced MDA content but increased T－AOC and protein contents ($P < 0.01$). FAPS (50, 100 and 150mg/kg) and Col (0.2mg/kg) treatments could both increase GSH but not significantly ($P > 0.05$).

Table 4 Effects of Tctal Antioxidant Capacity (T－AOC), Glutathione (GSH), Malondialdehyde (MDA) and protein contents in hepatic tissues of rats induced with hepatic fibrosis ($\bar{X} \pm S$, n = 10)

Groups	Dosage (mg/ (kg · day))	T－AOC (U/mg prot)	GSH (μmol/mg prot)	MDA (nmol/mg prot)	Protein (μg/μL)
Control	—	$0.23 \pm 0.02^{\#}$	$1.55 \pm 0.29^{\#*}$	0.39 ± 0.05	$11.84 \pm 0.96^{\#}$
Model	—	0.03 ± 0.01^{z}	3.50 ± 1.04^{z}	$0.90 \pm 0.21^{*}$	9.92 ± 1.10^{za}
$FAPS_L$	50.0	0.05 ± 0.01^{z}	3.35 ± 0.55^{z}	$0.86 \pm 0.25^{*}$	$10.21 \pm 1.00^{*a}$
$FAPS_M$	100.0	$0.14 \pm 0.03^{*\#}$	3.25 ± 0.37^{z}	$0.83 \pm 0.32^{*}$	11.01 ± 1.24
$FAPS_H$	150.0	$0.15 \pm 0.002^{*\#}$	3.04 ± 0.57^{z}	$0.80 \pm 0.37^{*}$	$11.47 \pm 1.47^{\#}$
Col	0.2	0.07 ± 0.01^{z}	3.05 ± 0.40^{z}	$0.88 \pm 0.23^{*}$	$10.09 \pm 1.13^{*a}$

*. $P < 0.05$ vs. the control, z. $P < 0.01$ vs. the control, #. $P < 0.01$ vs. the model, a. $P < 0.05$ vs. $FAPS_H$. $FAPS_L$ was low dose of FAPS by 50 mg/ (kg · day), $FAPS_M$ was mid dose of FAPS by 100 mg/ (kg · day), $FAPS_H$ was high dose of FAPS by 150 mg/ (kg · day), Col was colchicine by 0.2 mg/ (kg · day)

Effect of FAPS on the activities of GSH－Px and GST in serum: As shown in Table 5, model rats had a decline of GSH－Px and GST activities ($P < 0.01$) but an increase of MDA content

($P<0.01$), both as compared with the normalcontrol. $FAPS_H$ (150mg/kg) treatment reduced MDA in model rats (not significantly) but did significantly increase GSH－Px and GST activities ($P<0.05$).

Table 5 Effects of Fernentation Astragalus Polysaccharides (FAPS) on activities of Glutathione Peroxdase (GSH－Px) and Glutathione S－Transferase (GST) in senun in rats induced with hepatic fibrosis ($\bar{X}\pm S$, n=10)

Groups	Dosage (mg/ (kg · day))	GST (U/ml senun)	GSH－Px (U)	MDA (nmol/ml)
Control	—	$67.65\pm16.31^{\#}$	$2\,899.18\pm280.87^{\#}$	$1.21\pm0.15^{\#}$
Model	—	49.55 ± 8.080^{za}	$2\,212.09\pm298.37^{za}$	3.85 ± 0.60^{z}
$FAPS_L$	50.0	51.99 ± 10.39^{z}	$2\,380.84\pm361.26^{z}$	3.80 ± 0.41^{z}
$FAPS_M$	100.0	$56.87\pm15.10^{*}$	$2\,530.99\pm209.66^{*}$	3.45 ± 0.37^{z}
$FAPS_H$	150.0	64.28 ± 19.45	$2\,648.07\pm199.66^{\#}$	3.12 ± 0.55^{z}
Col	0.2	60.3 ± 19.910	$2\,487.16\pm339.60^{*}$	3.29 ± 0.74^{z}

*. $P<0.05$ vs. the control, z. $P<0.01$ vs. the control, #. $P<0.01$ vs. the model, a. $P<0.05$ vs. $FAPS_H$. $FAPS_L$ was low dose of FAPS by 50 mg/ (kg · day), $FAPS_M$ was mid dose of FAPS by 100 mg/ (kg · day), $FAPS_H$ was high dose of FAPS by 150 mg/ (kg · day), Col was colchicine by 0.2 mg/ (kg · day)

Effects of FAPS on liver histology and pathology in rats induced with hepatic fibrosis: According to the H and E staining results shown in Fig. 1 for the normal group, hepatic cells were evenly radiating outward from the central vein without collagen fiber hyperplasia and the hepatic lobule was structurally complete. For the model group, hepatic cell cords became disordered, fibrous connective tissue accreted and the number of inflammatory cells and dead cells increased in the portal area. In addition, the accreted collagen fibers formed streak fibrous septa, thickened fiber partitions appeared in hepatic portal bundle area and pseudolobules could be found. FAPS (50, 100, 150mg/kg) and APS (250mg/kg) treatments applied to model rats improved hepatic fibrosis in different degrees with inflammatory cells found only within fibrous septa. Masson staining results (Fig. 2) showed that in the case of normal rats, the hepatic lobule was structurally complete, hepatic cells were in even arrangement and no significant collagen fiber hyperplasia was in place, except for the little observed around blood vessels in the portal area. In the case of model rats however, collagen fibers increased remarkably, extending from the portal area and inflammatory or dead area, fonning fibrous septa of different thickness over the portal area and blood vessels by either cutting through or surrounding hepatic lobules.

Also, pseudolobules were clearly observed. FAPS (50, 100 and 150mg/kg) and APS (250mg/kg). treatments lessened hepatic fibrosis to different extents and made things better with fat vacuoles seen only within fibrous septa and some blue fibers running around blood vessels.

Discussion

Hepatic fibrosis is a pathophysiological process with high incidence and high mortality. It can be induced by hepatitis viruses, alcohol addiction, schistosomiasis infection, autoimmune diseases, drugs and other factors (Shu *et al.*, 2007; Aboutwerat *et al.*, 2003; Shimizu, 2003). Hepatic fi-

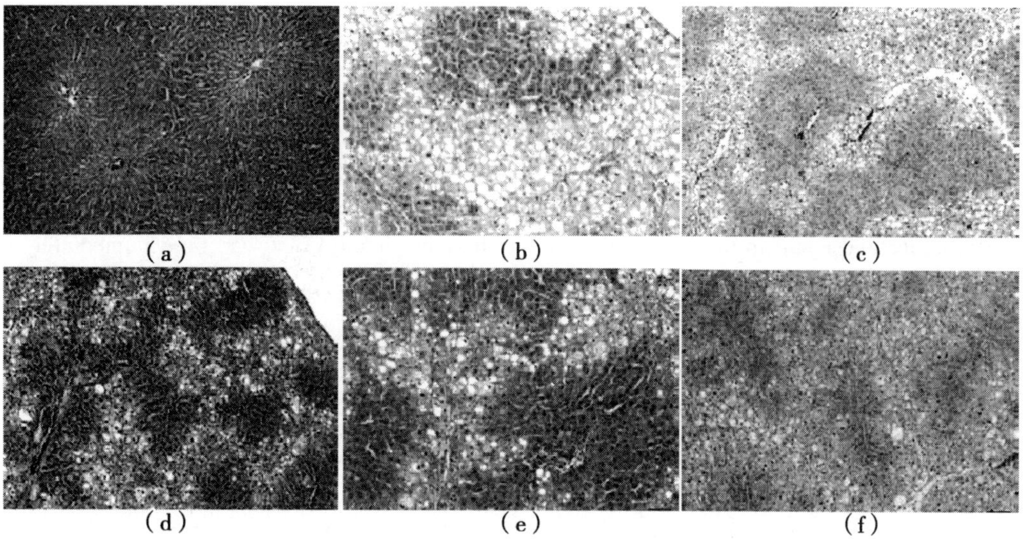

Fig. 1　HE staining of liver biopsy (HE ×100); (a) control; (b) model; (c) $FAPS_L$ (50mg/kg); (d) $FAPS_M$ (100mg/kg); (e) $FAPS_H$ (150mg/kg); (f) Col (0.2mg/kg). $FAPS_L$ was low dose of FAPS by 50mg/ (kg · day), $FAPS_M$ was mid dose of FAPS by 100mg/ (kg · day), $FAPS_H$ was high dose of FAPS by 150mg/ (kg · day), Col was colchicine by 0.2 mg/ (kg · day)

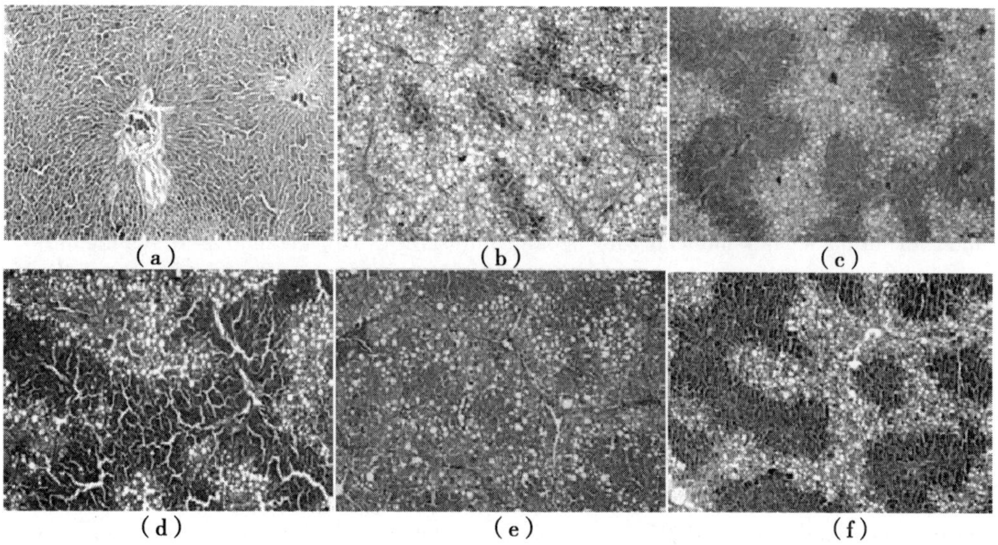

Fig. 2　Masson staining of liver biopsy (Masson ×100); (a) control; (b) model; (c) $FAPS_L$ was low dose of FAPS by 50mg/ (kg · day); (d) $FAPS_M$ was mid dose of FAPS by 100mg/ (kg · day); (e) $FAPS_H$ was high dose of FAPS by 150mg/ (kg · day); (f) Col was colchicine by 0.2 mg/ (kg · day)

brosis results from an imbalance between the synthesis and the degradation of the Extracellular Matrix (ECM) which resulting in an excessive deposition of it in the intercellular substance (Loguercio and Federico, 2003; Hung *et al.*, 2005). MDA (Malondialdehyde) can prompt an organism to produce a large amount of interstitial collagen as well as other ECM components which first deposit in hepatic clearances and ultimately give rise to overall hepatic fibrosis (Martin - Aragon *et al.*, 2001). CCI_4 is a potent hepatotropic poison that causes damage by peroxidation (Boll *et al.*, 2001). The specific mechanism goes this way: on one hand after the metabolism facilitated by the monooxygenase system on which the liver microsomal cytochrome P450 is dependent, the C - CI bonds in CCl_4 undergo hemolytic cleavage and consequently result in chloroform free radicals (. CCI_3) and chlorine free radicals (. Cl). These free radicals attack phospholipid molecules on liver cell membranes and result in lipid peroxidation. On the other hand,. CCI_3 can bind immediately with molecular oxygen and form . CCI_3O_2 which is more capable of causing lipid peroxidation and even cell mortality (Spiteller, 2003). These free radicals also attack unsaturated lipid content on cell membrane through covalent binding with macromolecules within hepatic cells and in this way, induce lipid peroxidation and damages to hepatic cells which consequently develop into liver injury and damage the structure and function of cell membranes (Maezono *et al.*, 1996; Yoshikawa *et al.*, 2002). In addition, products of the lipid peroxidation caused by . CCI_3 can induce an inflammatory response (Albano, 2002) activate Kupffer cells and Hepatic Stellate Cells (HSC) and as a result, prompt hepatic fibrosis (Xu *et al.*, 2003a). Furthermore, CCI, can inhibit the pump activity of $Ca_2{}^+$ on the cell membrane. As a result, a large amount of Ca_{2+} flows inward and deposits in cells (Ye, 1997) so much so that structure and function of mitochondria are damaged which in twn causes the activation of phospholipase and thus the degradation of membrane phospholipids, the damage of lysosome membrane and then the release of proteolytic enzymes. Consequently, xanthine dehydrogenase is twned into xanthine oxidase giving rise to a large nwnber of oxygen radicals and thus aggravating liver injury. When hepatic cells are injured, ALT, AST and ALP - components of the cytoplasm liquid are released so that the activities of ALT, AST and ALP in sennn are elevated, indicating degrees of the injury (Kew, 2000). An abnormal rise of ALT signifies cell apoptosis and higher levels of ALP indicate cholestasis (Pablo *et al.*, 2005). In this study, activities of ALT, AST and ALP in model rats induced with hepatic fibrosis by CCL_4 all increased significantly ($P <$ 0.01). But both FAPS (50, 100 and 150mg/kg) and APS (250mg/kg) treatments to these rats reduced ALT, ASP and ALP levels, especially the $FAPS_H$ treatment ($P < 0.01$). This all suggests that through its protection of hepatic cell membranes, FAPS can assist in recovery or prevent liver from injury imposed by hepatic fibrosis.

It has been demonstrated by a large amount of clinical data which HA, LN, PCⅢ and CⅣ levels are in positive correlation with the inflammatory activity and fibrosis of hepatic tissue (Shi *et al.*, 2008; Liang and Zheng, 2002). As a peptide located at the amino terminal and removed by relevant enzymes before the formation of tropocollagen by procollagen type Ⅲ, PCⅢ is a serum indicator of hepatic fibrosis. Its rise signifies the occwrence of the fibrosis. Also, it has been reported that the mRNA of procollagen type Ⅲ decreases throughout the process of hepatic fibrosis induced by CCl_4 in rats (Ding *et al.*, 2005; Liu *et al.*, 2006). HA is macromolecule glucosamine polysaccharose synthesized by interstitial cells. It is distributed in various connective tissues. Liver endothelial cells is the primary location where HA in serum is removed. In the case of chronic liver diseases, because of the injury the organ is subject to the liver, its endothelial cells' ability to take in and decompose HA is damaged and thus the level of HA will increase. Therefore, HA is an index that can report on liver endothelial cells' function and hepatic fibrosis. The application value held by HA in the diagnosis of hepatic fibrosis has been confinned by multiple experimental studies (Xu *et al.*, 2003b). LN plays a role in hepatic fibrosis by connecting macromolecular components in the matrix, participating in the formation of basilernma and the capillarization of hepatic sinusoid. Nonnally, the level of LN in hwnan serum is relatively low but during hepatic fibrosis it will rise considerably (Zheng *et al.*, 2006). CIV is a major component of the basilar membrane. It can effectively reflect the variation of the basilar membrane. Because it emerges at the first sign of hepatic fibrosis, it can be suitably used for the early diagnosis of the disease. Moreover, as the disease progresses, ClV levels will gradually increase (Altallah *et al.*, 2007). Hyp is a pathognomonic amino acid in collagen. It is indicative of the content of collagen protein and the degree of hepatic fibrosis (Shi *et al.*, 2008). The Hydroxyproline (Hyp) in the hepatic tissues has content as great as 13.4% in collagen protein while elastin contains little of it. Moreover, it does not exist in other proteins at all. At the time of hepatic fibrosis, collagen fiber will increase and as such so will Hyp. Thus, the content of Hyp measured in liver tissue can be converted into the content of collagen protein in the liver which then reflects the degree of fibrosis (Laurin *et al.*, 1996; Kim *et al.*, 2003). In this study, FAPS (50, 100 and 150mg/kg) and APS (250mg/kg) treatments decreased HA, LN, CIV, PC Ⅲ levels and Hyp content in the treated rats. The Hyp reductions observed with FAPS or APS treatments were all significant ($P<0.01$). On the other hand, decreases of CIV, LN and HA in the treated groups were not significant when compared with the model group. Yet as dosages increased, the variation between the treated group and the model group became more and more evident. The results together imply that FAPS and APS treatments can produce a certain antagonism to CCl_4 induced hepatic fibrosis.

T-AOC refers to the general status of an organism's antioxidant defense system which includes an enzymatic system and a non-enzymatic system. The former system includes SOD, GSH-Px, CAT, GST and so on while the latter system consists of vitamins, amino acids and metalloproteins such as vitamin E, vitamin C, carotene, cysteine, tryptophan, glucose, transferrin and lactoferrin protein (Kaplowitz, 2002). Antioxidant defense is performed primarily in three approaches: the first is to remove free radicals and active oxygen so as to avoid lipid peroxidation; the second is to decompose hydrogen peroxide materials by cutting off the peroxidation chain and the third is to get rid of catalytic metal ions. The experimental results show that T-AOC in model rats decreased significantly in contrast to the normal group ($P<0.01$), indicating a weakening of antioxidant defense. Nevertheless, $FAPS_H$ and $FAPS_M$ treatments increased T-AOC levels ($P<0.01$).

Histopathological examinations (HE staining, Masson staining) performed for the experiment confinned the protective effects FAPS and APS on rats with CCl_4 induced hepatic fibrosis. The process of hepatic fibrosis involves changes of ECM components and their contents and activation of matrix protein cells. Lipid peroxidation and its metabolic products are necessary for the inducement of expression of collagen genes in fat-storing cells which after being activated, synthesize mainly Collagen I and Ⅲ. These are interstitial collagens whose contents can determine the degree of cirrhosis. Oxidative Stress (OS) is present in many kinds of pathogenic processes of liver damage. As long as pathogenic causes have not been removed, they can continuously act on the liver and affect hepatic fibrosis and even liver cirrhosis (Ogeturk *et al.*, 2008; Torok, 2008; Disario *et al.*, 2007). Under normal physiological conditions, Reactive Oxygen Species (ROS) is too low to cause any pathological changes (Gebhardt, 2002; Parola and Robino, 2001). However, under pathologic conditions, ROS is produced faster than it can be removed. Thus, this leads to the deposition of active oxygen and then oxidative stress. In addition, ROS can act as an important intracellular messenger to activate signal transduction pathways and thus indirectly result in damages to cells and tissues (Poli, 2000). It was previously proposed that ROS mediated hepatic fibrosis progressed in the following manner: more ROS is produced than removed, peroxidation damages are caused to inflammatory cells such as hepatic cells, Kuppfer cells and neutrophils and finally ROS as well as other secretory products, further activate Hepatic Stellate Cell (HSC). Recently however, it has been argued that in addition to the mechanism, ROS, superoxide anion, MDA, aldehyde and arachidonic acid, an oxidized low density lipoprotein can all stimulate and proliferate HSC. Active oxygen molecules which in the sennn of model animals induced with hepatic fibrosis will increase and so will hydroxy - deoxy guanosine, MDA and relevant products. They are mainly distributed over piecemeal necrosis and around the portal vein and consequently cause significant changes in serum oxidative stress indicators, i. e., MDA, SOD and GSH - Px (Refik *et al.*, 2004). In this study, researchers examined the concentration of MDA and GDH - Px in model rats and found these rats had a concentration of MDA much higher than the control rats ($P<0.05$) suggesting oxidative stress occurred in the model rats. On the other hand, we found the GSH - Px concentration in the antioxidant defense system decreased significantly ($P<0.01$) in the model rats. This indicated a weakening of the antioxidant defense system. FAPSH treatment reduced the MDA and increased the GSH - Px activity. Thus, it is clear that lipid peroxidation and its metabolic products play an important role in mediating hepatic fibrosis. Included in cwrent treatment protocols for hepatic fibrosis is antioxidant treatment, so antioxidants have a likely role in the mechanism by which FAPSH antagonizes hepatic fibrosis.

Conclusion

By histopathological examination, researchers also found that preventative treatment of model rats with FAPS could effectively inhibit and lessen hepatic fibrosis and these positive effects increased along with the applied dosages. All the results shown here suggest FAPS can protect hepatic cells, inhibit HSC activation and prevent CCl_4 induced hepatic fibrosis in rats.

Acknowledgements

This research was supported by the Study on Mechanism of Astragalus Polysaccharides Transformation Induced by Lactobacillus FGM9 *in vitro* (N0. 31072162), by the Introduction and Application of Mastitis Diagnosis and Control in Dairy Cow (2010 - C7), by the Project of National Dairy Industry and Technology System (CARS - 37).

References

[1] Aboutwerat A, Pemberton P W, Smith A, Burrows P C, McMahon R F T, Jain S K,

Warnes T W, Oxidant stress is a significant feature of primary biliary cirrhosis. Biochim. Biophys. Acta Mol. Basis Dis., 2003, 1637: 142-150.

[2] Albano E., Free radical mechanisms in immune reactions associated with alcoholic liver disease. Free Radic. Bio. Med., 2002, 32: 110-114.

[3] Ao Q, Mak K M, Ren C, Lieber C S, Leptin stimulate tissue inhibitor of metalloproteinase-1 in human hepatic stellate cells: Respective roles of the JAK/STAT and JAK-mediated H2O2-dependant MAPK pathways. J. Biol. Chem., 2004, 279: 4 292-4 304.

[4] Attallah A M, Mosa T E, Omran M M, Abo-Zeid M M, El-Dosoky I, Shaker Y M, Irnrnunodetection of collagen types I, II, III and IV for differentiation of liver fibrosis stages in patients with chronic Hey. J. Immunoassay Immunochem., 2007, 28: 155-168.

[5] Boll M, Weber L W, Becker E, Stampfl A, Mechanism of carbon tetrachloride-induced hepatotoxicity: Hepatocellular damage by reactive carbon tetrachloride metabolites. Z Naturforsch C, 2001, 56: 649-659.

[6] CPC, Chinese Pharmacopoeia. Vol. 1, Medical Scienceand Technology Press, Beijing, China, 2010, pp: 283, 403, 761, 870.

[7] Ding X D, Wang H Q, Wu Q, Wang X L, Huang Y, Effect of astrogalosides on liver fibrosis in mice with schistosomiasis japonica. Shijie Huaren Xiaohua Zazhi, 2008, 16: 125-131.

[8] Ding X J, Li S B, Li S Z, Liu H S, Liu B, A quantitative study of the relationship between levels of liver fibrosis markers in sera and fibrosis stages of liver tissues of patients with chronic hepatic diseases. Chinese J. Herpetol, 2005, 13: 911-914.

[9] Disario A, Candelaresi C, Omenetti A, Benedetti A, Vitamin E in chronic liver diseases and liver fibrosis. Vitamins hormones, 2007, 76: 551-573.

[10] Gebhardt R, Oxidative stress, plant-derived antioxidants and liver fibrosis. Planta Medica., 2002, 68: 289-296.

[11] Hung K S, Lee T H, Chou W Y, Wu C L, Cho C L, *et al.*, Interleukin-10 gene therapy reverses thioaceta mide-induced liver fibrosis in mice. Biochem. Biophys. Res. Commun., 2005, 336: 324-331.

[12] Kaplowitz N., Biochemica land cellular mechanisms of toxic liver injwy. Semin. Liver Dis., 2002, 22: 137-144.

[13] Kew M C, Serum aminotransferase concentration as evidence of hepatocellular damage. Lancet, 2000, 355: 591-592.

[14] Kim J L, Tsujino T, Fujioka Y, Saito K, Yokoyama M, Bezafibrate Improves hypertension and insulin sensitivity in humans. Hypertens Res., 2003, 26: 307-313.

[15] Laurin J, Lindor K D, Crippin J S, Gossard A, Gores G J *et al.*, Ursodeoxycholic acid or clofibrate in the treatment of non-alcoholic steatohepatitis: A pilot study. Hepatol., 1996, 23: 1 464-1 467.

[16] Li S P, Xu X Y, Sun Z, Chen Z, Astragalus polysaccharides and astragalosides regulate cytokine secretion in LX-2 cell line. J. Zhejiang Univ. Med. Sci., 2007, 36: 543-548.

[17] Liang X H, Zheng H, Value of simultaneous determination of sennn hyaluronic acid, colla-

gen type IV and the laminin level in diagnosing liver fibrosis. Bull. Hunan Med. Univ. , 2002，27：67 -68.

[18] Liu J，Wang J Y，Lu Y，Serum fibrosis markers in diagnosing liver fibrosis. Chin. J. Internal Med. , 2006，45：475 -477.

[19] Loguercio C，Federico A，Oxidative stress in viral and alcoholic hepatitis. Free Radical Biol. Med. , 2003，34：1 -10.

[20] Maezono K，Kajiwara K，Mawatari K，Ine protects liver from injwy caused by F - galactosamine and CCl_4. Hepatology，1996，1：185 -191.

[21] Martin - Aragon S，Heras B D，Reus M I S，Pharmacological modification of endogenous antioxidant ursolic acid on carbon tetrachlorideinduced liver damage enzymes rats and pnmary cultures of rat hepatocytes. Toxicol. Pathol. , 2001，53：199 -206.

[22] Ogeturk M，Kus I，Pekrnez H，Yekeler H，Sahin S，Sihnaz M，Inhibition of carbon tetrachloridemediated apoptosis and oxidative stress by melatonin in xperimental liver fibrosis. Toxicol. Ind. Health，2008，24：201 -208.

[23] Pablo M，Mario M G，Del C M，Resolution of liver fibrosis in chronic CC14 administration in the rat after discontinuation of treatment Effect of silymarin，silibinin，colchicine and trimethylcolchicinic acid. Basic Clin. Pbannacol. Toxicol. , 2005，96：375 -380.

[24] Parola M，Robino G，Oxidative stress - related molecules and liver fibrosis. J. Hepatol. , 2001，35：297 -306.

[25] Poli G，Pathogenesis of liver fi brosis：Role of oxidative stress. Mol. Aspects Med. , 2000，21：49 -98.

[26] Refik M M，Comert B，Oncu K，Vural S A，Akay C *et al.*，Tlie effect of taurine treatment on oxidative stress in experimental liver fibrosis. Hepatol. Res. , 2004，28：207 -215.

[27] SPSS，2009. For Windows，Rel. 17. 0，2009. Chicago，SPSS Inc. , Chicago，IL.

[28] Shi H B，Fu J F，Wang C L，Clinical value of hepatic fibrosis parameters and serum ferritin in obese children with nonalcoholic fatty liver disease. J. Zhejiang Univ. , 2008，37：245 -249.

[29] Shimizu I，Impact of estrogens on the progression of liver disease. Liver Int. , 2003，23：63 -69.

[30] Shu J C，WU H，Pi X J，He Y J，LV X *et al.*，Curcumin inhibits lipid peroxidation and the expression of TGF - β1 and PDGF in the liver of rats with hepatic fibrosis. Chin. J. Pathophysiol. , 2007，23：2 405 -2 409.

[31] Spiteller G，Are lipid peroxidation processes induced by changes in the cell wall structure and how are these processes connected with diseases Med. Hypotheses，2003，60：69 -83.

[32] Staub A M，Removal of protein - sevag method：Methods in carboliydrate. Cliemistry，1965，5：5 -6.

[33] Torok N J，Recent advances in the pathogenesis and diagnosis of liver fibrosis. J. Gastroenterol. , 2008，43：315 -321.

[34] Wang J S，Groopman J D，DNA damage by mycotoxins. Mulat. Res. , 1999，424：167 -181.

[35] Wang X Z, Yang Z Q, Li J X, Study on the interaction between the coarse polysaccharide in fermentation complex of Astragalus and the probiotics isolated from chicken intestine. J. Gansu Agric. Univ., 2008, 45: 25-28.

[36] Xu H G, Fang J P, Huang S L, Diagnostic values of serum levels of HA, PCⅢ, CIV and LN to the liver fibrosis in children with beta-thalassemia major. Zliongliua Er Ke Za Zhi., 2003a, 41.603-606.

[37] Xu J, Fu Y, Chen A, Activation of peroxisome proliferator-activated receptor-γ contributes to the inhibitory effects of Curcumin on rat liepatic stellate cell growth. Am. J. Pliysiol. Gastrointest. Liver Pliysiol., 2003b, 285: 20-30.

[38] Xu X Q, Ning Y, Li P, Slien Z H, Lou L, Pathological analysis of hepatic fibrosis in pigs infected with *Taenia solium*. J. Pathogen Biol., 2011, 6: 344-345.

[39] Ye F Y, National Clinical Laboratory Procedures. 2nd Edn., Southeast University Press, USA., pp: 1997, 206-216.

[40] Yoshikawa M, Ninomiya K, Slinnoda H, Nishida N, Matsuda H, Hepatoprotective and antioxidative properties of Salacia reticulate: Preventive effects of phenolic constituents on CCl_4 induced liver injwy in mice. BioI. Pharm. Bull., 2002, 25: 72-76.

[41] Zhang X X, Yang Y, Chen M Z, Effect of total polysaccharide of Astragalus on proliferation and collagen production of HSC-T6 cells. Cliinese J. Clin. Pharmacol. Tlierapeutics, 2003, 06: 645-647.

[42] Zheng J F, Liang L J, Wu C X, Chen J S, Zhang Z S, Transplantation of fetal liver epithelial progenitor cells ameliorates experimental liver fibrosis in mice. World J. Gastroenterol., 2006, 12: 7 292-7 298.

[43] Zhuoming U, Practice of Histological Techniques. 2nd Edn., People's Health Publisliing House, Beijing, 1998, pp: 8-70.

[44] Zou L Y, Wu T, Cui L, Liver cirrhosis induced bone loss in mice and the prevention effect of astragalus polysaccharides. Chinese J. Integrated Traditional Western Med. Liver Dis., 2002, 02: 95-98.

(Published the article in J. Anim. Vet. Adv. affect factor: 0.39)

纳米铜对大鼠肝脏毒性相关蛋白过氧化氢酶的分离鉴定及生物信息学分析

董书伟[1]，高昭辉[1,3]，申小云[2]，薛慧文[3]，荔　霞[1]

（1. 中国农业科学院兰州畜牧与兽药研究所，农业部兽用药物创制重点实验，甘肃省中兽药工程技术中心，兰州 730050；2. 贵州省毕节学院，贵州毕节 551700；3. 甘肃农业大学动物医学院，兰州 730070）

摘　要：分离和鉴定纳米铜对大鼠肝脏毒性相关蛋白过氧化氢酶（catalase，CAT），探讨 CAT 在毒性发挥中的作用，为揭示纳米铜对肝脏毒性机制提供依据。应用 2 - DE 技术和 PDQuest8. 0 软件在大鼠肝脏蛋白组中筛选纳米铜对肝脏毒性差异蛋白，经质谱鉴定后进行生物信息学分析。筛选到下调的差异蛋白点 6602 和 7702 与肝毒性相关，鉴定均为 CAT 蛋白；其性质稳定，有一定亲水性，无信号肽，定位于细胞质，可能属于非分泌性蛋白，含有过氧化氢酶活性位点 64FDRERIPERVVHAKGAG80 和过氧化氢酶亚铁血红素配合基位点 354RLFAYPDTH362 等功能位点；无规则卷曲、α 螺旋和延伸链是其主要的二级结构元件，并预测了其三级结构图；同源性分析表明，大鼠的 CAT 与其他 8 个物种有较高同源性，并构建了 CAT 蛋白的系统进化树。纳米铜通过下调大鼠肝脏中 CAT 蛋白表达，引起肝细胞氧化应激损伤，可能是其发挥毒性作用的途径之一。

关键词：纳米铜；肝毒性；蛋白质组学；过氧化氢酶；生物信息学；大鼠

0　引言

随着纳米科技的飞速发展，各种优质纳米材料相继诞生，并以其独特的生物学特性被广泛应用于工业、畜牧业、化妆品以及医药等领域[1]，人们接触纳米材料的机会也日益增多，但是，在其呈现诱人的纳米生物效应的同时，其安全性问题也不容忽视，目前，虽尚无充足的证据说明纳米材料对人体有害，但初步的毒理学研究表明，它们对人类健康和生态系统有潜在的负面影响[2]。纳米铜是较为常见的人造金属纳米材料，已经被广泛作为畜禽饲料添加剂，抗骨质疏松及抗衰老的纳米药物和宫内节育器等应用于生物医学领域[3~4]，还可作为润滑油添加剂、高效催化剂、电子类产品的表面涂层应用于工业领域，给畜牧业和工业的发展带来技术上的革命[5~7]。近年来，一些学者已经初步探讨了纳米铜的毒性机制，认为其可能通过在低 pH 值的胃液中反应生成的铜离子和粒子铜两种形态的铜共同发挥作用，也有学者通过代谢组学技术研究发现纳米铜可使线粒体功能受损，酮体生成、脂肪酸 β 氧化和糖酵解过程增强，进而对动物肝、肾、脾、神经系统以及鱼类腮部等造成损伤[8~12]。目前，针对纳米铜的毒理研究仍停留在一般毒性评价水平，其毒性相关基因、蛋白和作用机制尚未明确，在蛋白水平上研究其毒性的报道还极为鲜见。本研究采用二维凝胶电泳（two - dimensional electrophoresis，2 - DE）和基质辅助激光解吸电离串联飞行时间质谱（matrix - assisted laser desorption ionization tandemtime of flight mass spectrometry，MALDI - TOF - TOFMS）等蛋白质组学技术，筛选纳米铜致大鼠肝毒性的差异表达蛋白，并分离鉴定了肝毒性相关蛋白过氧化氢酶（CAT），对其进行生物信息学分析，为研究该蛋白在肝毒性发挥机制中的作用提供依据，也为在蛋白水平揭示纳米铜毒性机制奠定基础。

1 材料与方法

1.1 试验时间、地点

本试验于2011年5~9月实施，动物病理模型建立和双向电泳试验在兰州畜牧与兽药研究所完成，蛋白质的质谱鉴定委托上海中科新生命公司完成。

1.2 主要试剂与仪器

纳米级铜粉（粒径25~100nm），购自深圳尊业纳米有限公司；羟丙甲基纤维素（hydroxy propyl methyl cellulose，HPMC，分析纯）购自上海 Colorcon Coating 技术有限公司；17cm 非线性固相化 pH 梯度胶条（CAT. No. 163 - 2009）为 Bio - Rad 公司产品；2 - D clean - up kit（CAT. No. 80 - 6484 - 51）为 GE 公司产品；Bradford 蛋白浓度测定试剂盒（碧云天生物技术研究所）。

PROTEAN IEF Cell 系统；二维电泳凝胶系统（Bio - Rad 公司）；PDQuest8.0 图像分析软件（Bio - Rad 公司）；U2800 紫外可见分光光度计（Hitachi 公司）；UMAX 扫描仪（台湾）；4800 串联飞行时间质谱仪（4800Plus MALDI TOF/TOF TM Analyzer，ABI）。

1.3 实验动物及染毒方式

SPF 级雄性 Wistar 大鼠 20 只，体重 220g 左右，购自甘肃省中医学院。自然条件下饲养，自由摄食饮水，适应性饲养一周，随机分为溶剂对照组（1% HPMC）与纳米铜组（200mg/kg）[14]，共2组，每组10只。试验前在充满氮气的手套箱中将纳米铜用1% HPMC 分散为200mg/kg，置于50ml 离心管中备用，灌胃前再超声10min 使纳米铜颗粒分散均匀。每日称取动物体重，确定灌胃染毒量，每天1次，连续5d，结束后次日麻醉剖杀，取肝脏和血液样品，冷冻保存备用。

1.4 病理组织学观察

分别取福尔马林溶液固定后的对照组和纳米铜中毒组大鼠的适量肝组织，做石蜡切片，H.E 染色，观察肝脏的病理组织学变化，如果中毒组大鼠的肝细胞发生病变视为染毒成功。

1.5 蛋白样品制备

取各组肝脏样品300mg，经生理盐水漂洗后置匀浆杯中，冰浴条件下匀浆，然后加入1.5ml 的组织裂解液于 Ependorf 管中，冰浴条件下超声裂解20min，每隔5min 振荡摇匀一次，12 000 × g 离心，吸取上清。利用2 - D clean - up 试剂盒对蛋白样品进行除杂纯化，采用 Bradford 法[13]定量蛋白，分装样品，置 -80℃保存备用。

1.6 双向凝胶电泳

第一向等电聚焦（Isoelectric focusing，IEF）电泳[15~16]：吸取制备好的400μg 蛋白样品与水化上样缓冲液混合均匀并室温放置1h，加入聚焦盘中，并将 pH3—10 IPG 胶条胶面朝下对应放于聚焦槽内，然后设定 PROTEAN IEF Cell 电泳仪程序，开始等电聚焦电泳。

第二向 SDS－PAGE 电泳[17~18]：取出 IPG 胶条，经平衡后移至 12% SDS－PAGE 凝胶顶端，加封胶液，移入电泳槽内。起始电压为 50V，待样品完全移出胶条，加大电压至 200V，待溴酚蓝指示线达到凝胶底部边缘 1cm 时停止电泳。取出凝胶，切角做记号。

1.7 图像扫描与分析

凝胶采用胶体考马斯亮蓝方法染色[19]。通过 UMAX 2100 扫描仪采集 2D 图像，然后用 PDQuest8.0 分析软件进行背景消减、斑点检测、匹配，获取斑点位置坐标等处理，以“斑点染色强度和面积三倍量的变化”为标准自动进行统计分析，筛选差异蛋白点，对部分差异蛋白点进行 3D 图像扫描。

1.8 胶内酶解及脱盐[20]

用剪过的孔径大小不同（1~5mm）的枪头从凝胶图上挖取差异蛋白质斑点，置 Eppendorf 管中，加入 200~400μl 100mmol/L NH_4HCO_3/30% ACN 脱色，冻干后，加入 5μl 2.5~10ng/μl 测序级 Trypsin（Promega）溶液，37℃反应过夜，20h 左右；吸出酶解液，移入新 EP 管中，在原管中再加入 100μL 60% ACN/0.1% TFA，超声 15min，合并前次溶液，冻干；用 Ziptip（Millipore）进行脱盐。

1.9 质谱分析

使用 4 800串联飞行时间质谱仪进行分析，激光源为 355nm 波长的 Nd：YAG 激光器，加速电压为 2kV，采用正离子模式和自动获取数据的模式采集数据，PMF 质量扫描范围为 800~4 000Da，选择信噪比大于 50 的母离子进行二级质谱（MS/MS）分析。将 MS 和 MS/MS 结果联合搜库，检索数据库为 IPI_ rat，Protein Score > 1 000并且 Total Ion Score >60 被接受。

1.10 差异蛋白 CAT 的生物信息学分析

应用 ProtParam 软件进行氨基酸序列组成、蛋白质相对分子量、疏水性、等电点等理化特征分析；TMHMNM 分析蛋白质的跨膜区域；Signal 4.0 Server 预测其信号肽；利用 ProtScale 分析其疏水性；PSORT Ⅱ server 在线工具进行该蛋白亚细胞定位预测；SMART 分析该蛋白的结构域；利用 PredictProtein 预测该蛋白的二级结构和功能位点，并用 Swiss－model 在线工具对其三级结构进行预测；用 ClustalX 2. 0 对多个物种的 CAT 进行同源性分析，并用 MEGA4.0 软件根据比较结果构建所有同源序列的进化树。上述信号肽、细胞定位和功能位点预测的参数均设置为默认值。

2 结果与分析

2.1 纳米铜中毒大鼠病理模型的建立

染毒 3d 时，纳米铜中毒组大鼠开始出现精神萎靡、食欲减退、腹泻、呕吐、呼吸减弱、振颤和弓背等异常症状，且随着染毒次数增多逐渐加重，而对照组大鼠未见异常。在病理组织学观察中发现，中毒组大鼠肝细胞出现颗粒变性，中央静脉严重淤血（图 1），而对照组大鼠肝脏结构正常，由此说明，纳米铜中毒大鼠病理模型建立成功。

2.2 差异蛋白的筛选与 CAT 的分离鉴定

将对照组和中毒组大鼠肝脏样品的总蛋白进行二维电泳，所得到的图谱上蛋白斑点分布清晰，经过 3 次重复实验，发现凝胶之间具有较好重复性，能够满足 2－DE 图谱分析的要

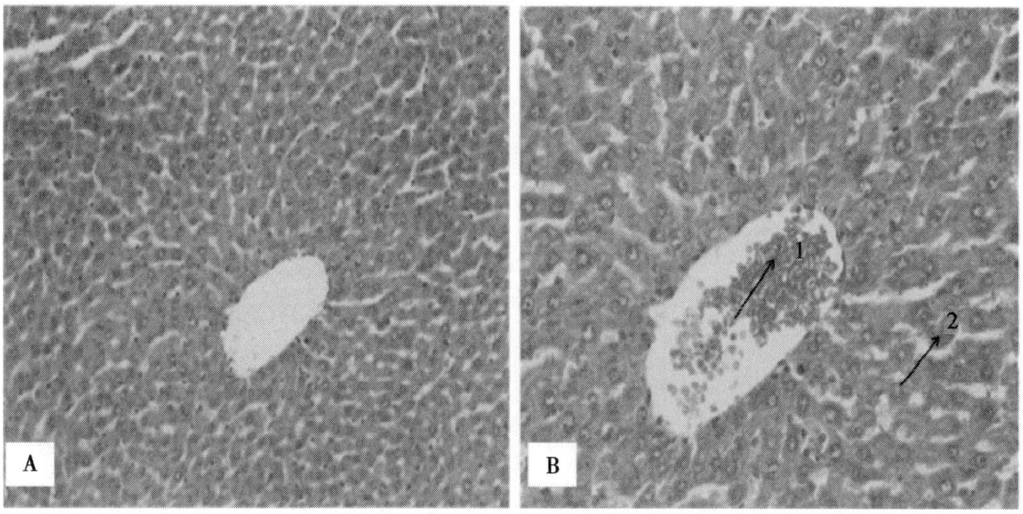

图 B 中箭头 1 所示中央静脉淤血，箭头 2 所示肝细胞颗粒变性，400 ×

图 1　对照组 A（200 ×）和中毒组 B 大鼠肝脏病理学切片（400 ×，H. E 染色）

求。图 2 为对照组和中毒组大鼠肝脏蛋白组的 2－DE 图谱，经 PDQuest8. 0 分析共发现 58 个蛋白点在表达量上有显著差异（$P<0.05$），经分离和质谱鉴定差异蛋白后，发现差异斑点 Spot6602 和 Spot7702 均为 CAT 蛋白（图 3），得分分别为 1050 和 1010。为全方位展示差异蛋白 Spot6602 和 Spot7702 的表达情况，对其进行了三维扫描，见图 4 和图 5。结果表明，大鼠染毒后，肝脏蛋白组表达模式发生了较大变化，有 58 个蛋白点表达有显著差异，包括 CAT 在内的这些差异蛋白的功能及表达差异可能与纳米铜的肝毒性有密切关系，有利于在蛋白水平进一步阐明纳米铜肝脏毒性的作用机制。

2. 3　差异蛋白 CAT 的生物信息学分析

2. 3. 1　理化性质分析

利用 ProtParam 在线工具分析 CAT 的基本理化性质，结果显示 CAT 的分子式为 C2668H4046N754O779S19，相对分子量为 59 757. 2，理论等电点 pI 为 7. 07，酸性氨基酸残基总数（Asp + Glu）为 61，碱性氨基酸残基总数（Arg + Lys）为 60，丙氨酸、脯氨酸、天冬氨酸、甘氨酸、天冬酰胺和缬氨酸含量较高，分别为 8. 2%、7. 2%、6. 7%、6. 5%、6. 3%、6. 1%。其水溶液在 280nm 处的消光系数为 64 665，不稳定系数为 33. 78，属于较稳定性蛋白（不稳定系数 <40 为稳定性蛋白），平均亲水系数为 －0. 639，表明其有一定的亲水性。

2. 3. 2　亲水性和跨膜区分析

用 ProtScale 预测 CAT 氨基酸序列的疏水性/亲水性。依据氨基酸分值越低亲水性越强、分值越高疏水性越强的规律，可以看出多肽链第 50 位 Ser 具有最低的分值 －2. 833 和最强的亲水性；第 31 位点 Gly 具有最高的分值 2. 044 和最强的疏水性，整个多肽链表现为亲水性，与 ProtParam 预测结果基本一致。

采用 TMHMM 2. 0 Server 预测其跨膜区域，结果显示该蛋白所有氨基酸位于膜外，结合亲水性分析可预测 CAT 蛋白可能不属于膜蛋白。

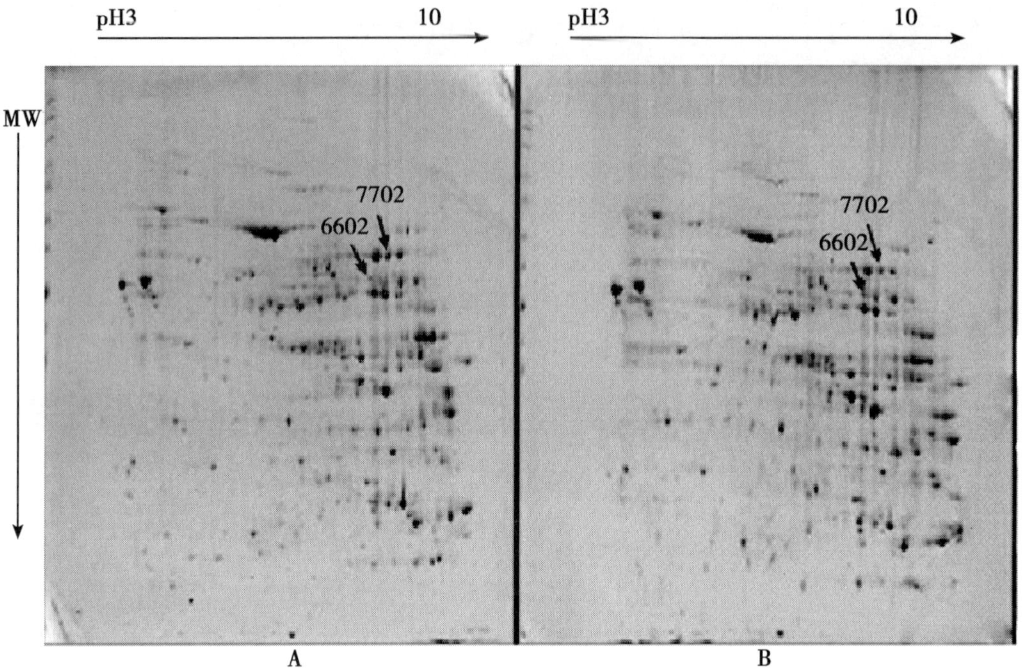

图 2　对照组大鼠肝脏（A）与中毒组大鼠肝脏（B）的 2－DE 图谱

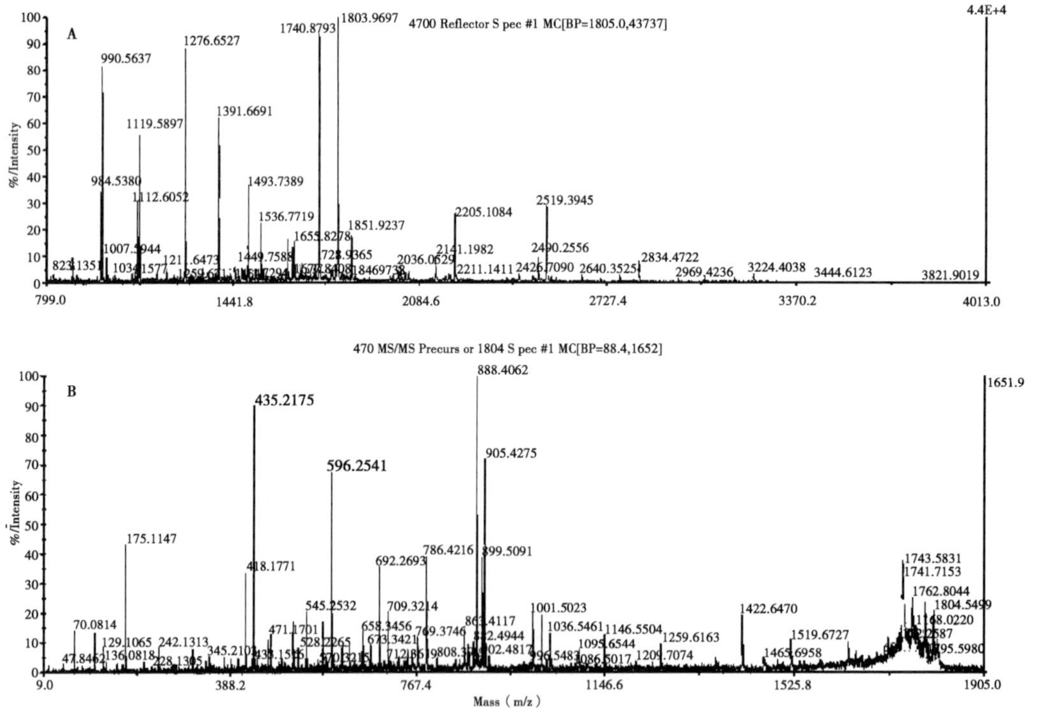

A 为 MADIL－TOF－MS 图谱；B 为母离子是质核比 1804 峰的

MADIL－TOF－MS/MS 图谱，其特征峰为 888.4，丰度为 1 652

图 3　CAT 蛋白的质谱鉴定图

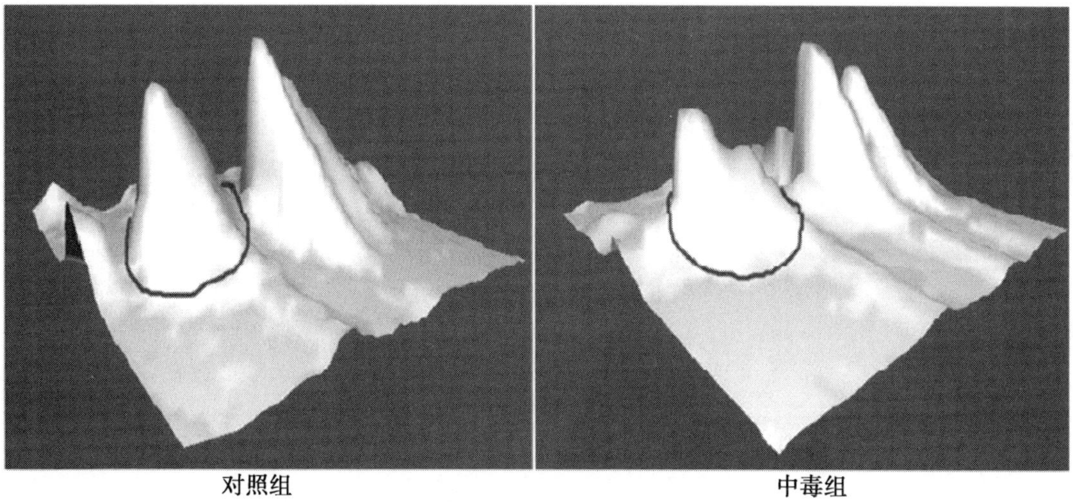

图 4　差异蛋白 Spot7702 的 3D 视图

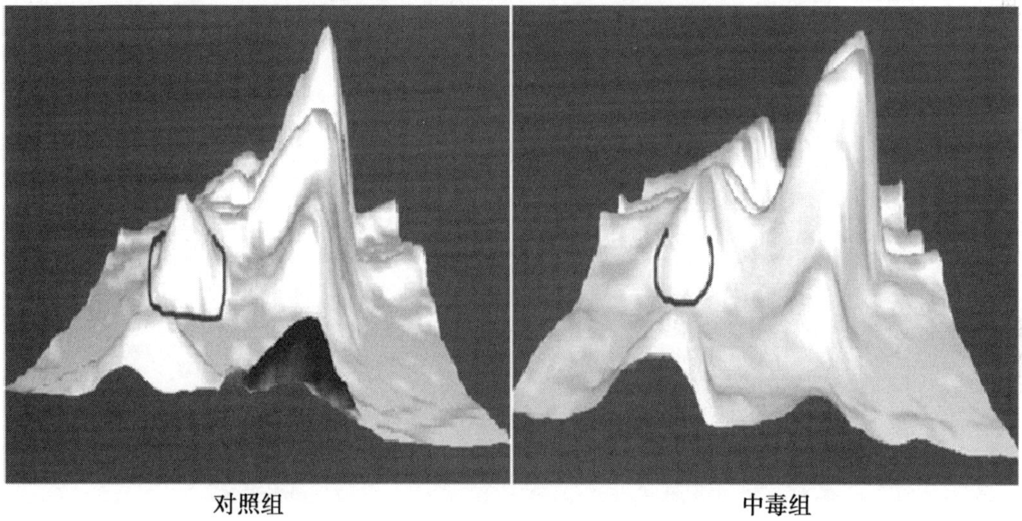

图 5　差异蛋白 Spot6602 的 3D 视图

2.3.3　信号肽和亚细胞定位预测

信号肽的预测和分析对深入认识和理解蛋白质的细胞定位及结构域具有重要的意义。运用 Signa IP 信号肽预测工具预测差异蛋白 CAT 存在信号肽的概率为 0，锚定蛋白概率为 0，结果显示 CAT 蛋白没有信号肽，可能是非分泌性蛋白。使用 TargetP 和 PSORT II 预测 CAT 的亚细胞定位，由结果（表）可知，差异蛋白 CAT 定位于细胞质（其他为 0.921），不存在信号肽（SP 值为 0.065），故推测其属于非分泌性蛋白。

2.3.4　蛋白的功能位点分析

通过 Predict Protein 对 CAT 氨基酸序列中可能存在的功能域进行预测。发现有 1 个过氧化氢酶活性位点（64 ~ 80）、2 个 N – 糖基化位点（439 ~ 442，481 ~ 484）、7 个蛋白激酶 C 磷酸化位点（125 ~ 127，167 ~ 169，201 ~ 203，212 ~ 219，285 ~ 287，361 ~ 363，502 ~ 504）、5 个酪蛋白激酶 II 磷酸化位点（125 ~ 128，285 ~ 288，410 ~ 413，417 ~ 420，434 ~

437)、1 个酪氨酸激酶磷酸化位点（252～261)、6 个 N－肉豆酰位点（32～37，121～126，204～209，272～277，399～404，516～521)、1 个酰胺化位点（103～106)、1 个过氧化氢酶亚铁血红素配合基位点（354～362)，其中过氧化氢酶活性部位位点是此酶的关键活性位点，特征序列为 FDRERIPERVVHAKGAG。

2.3.5 蛋白的二级、三级结构预测

利用 SOPMA 服务器预测蛋白的二级结构，发现该蛋白由 30.55% α 螺旋、15.94% 延伸链、6.26% β 转角、47.25% 无规卷曲组成。可推断无规则卷曲、α 螺旋和延伸链是其主要的二级结构元件，无规则卷曲散布于整个蛋白中。用 Swiss－model 将 CAT 与蛋白质结构数据库中的蛋白质三维结构进行匹配，输回模拟的三维结构图（图 6)。结果显示为 4 个区域：N－末端区域，形成一个延伸的非球形氨基酸末端手臂结构；反平行 8 股 β 桶状区域；一个缠绕成环状的非随机链接区域；E、F、G 和 H 球形螺旋蛋白相似的 C－末端螺旋区域，该结果与预测的二级结构结果一致。

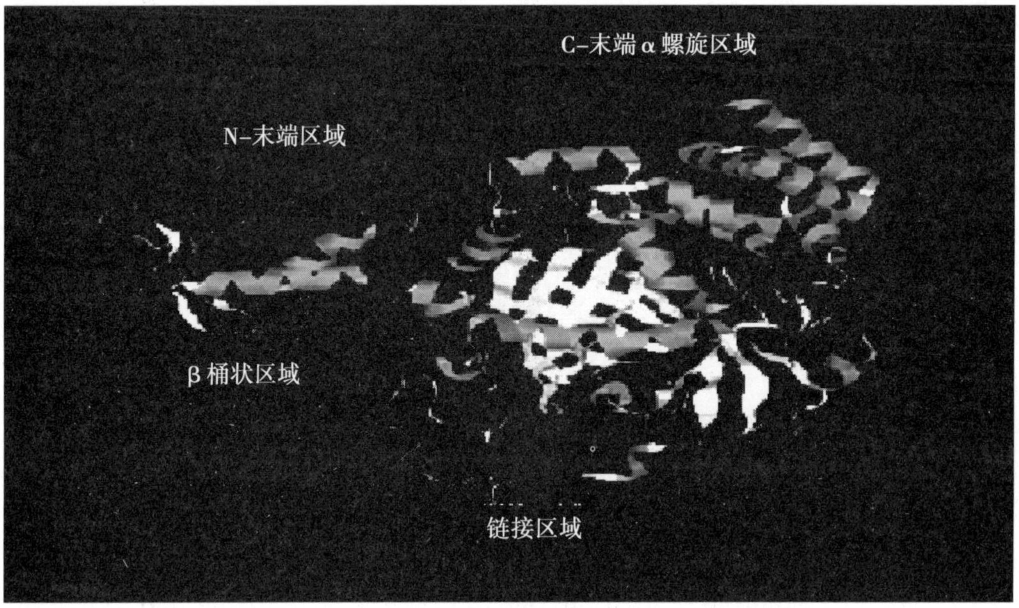

图 6 差异表达蛋白 CAT 的 3D 结构预测图

2.3.6 蛋白质的同源性分析

选取大鼠、家鼠、仓鼠、家犬、猪、牛、猩猩、猕猴和兔 9 个不同物种的 CAT 氨基酸序列，应用 ClustalX2.0 对其进行多序列比对，同时用序列着色软件 Boxshade 对生成的比对序列结果着色（图 7)。结合 NCBI 的 Blastp 程序进行同源性分析，大鼠与家鼠、仓鼠及犬 CAT 的同源性分别为 95.25%、89.72% 和 92.03%。

将该蛋白与来自 8 个不同物种的同源蛋白进行进化分析，并利用 MEGA4.0 软件构建 CAT 蛋白的 NJ 树（neighbor－joining tree，NJ tree)[21]（图 8)。从图上可以看出，大鼠和家鼠聚为一类，其次与仓鼠聚为一类，再次与犬共同为一分支中，由此说明大鼠、家鼠、仓鼠及犬的 CAT 有较近的遗传距离。

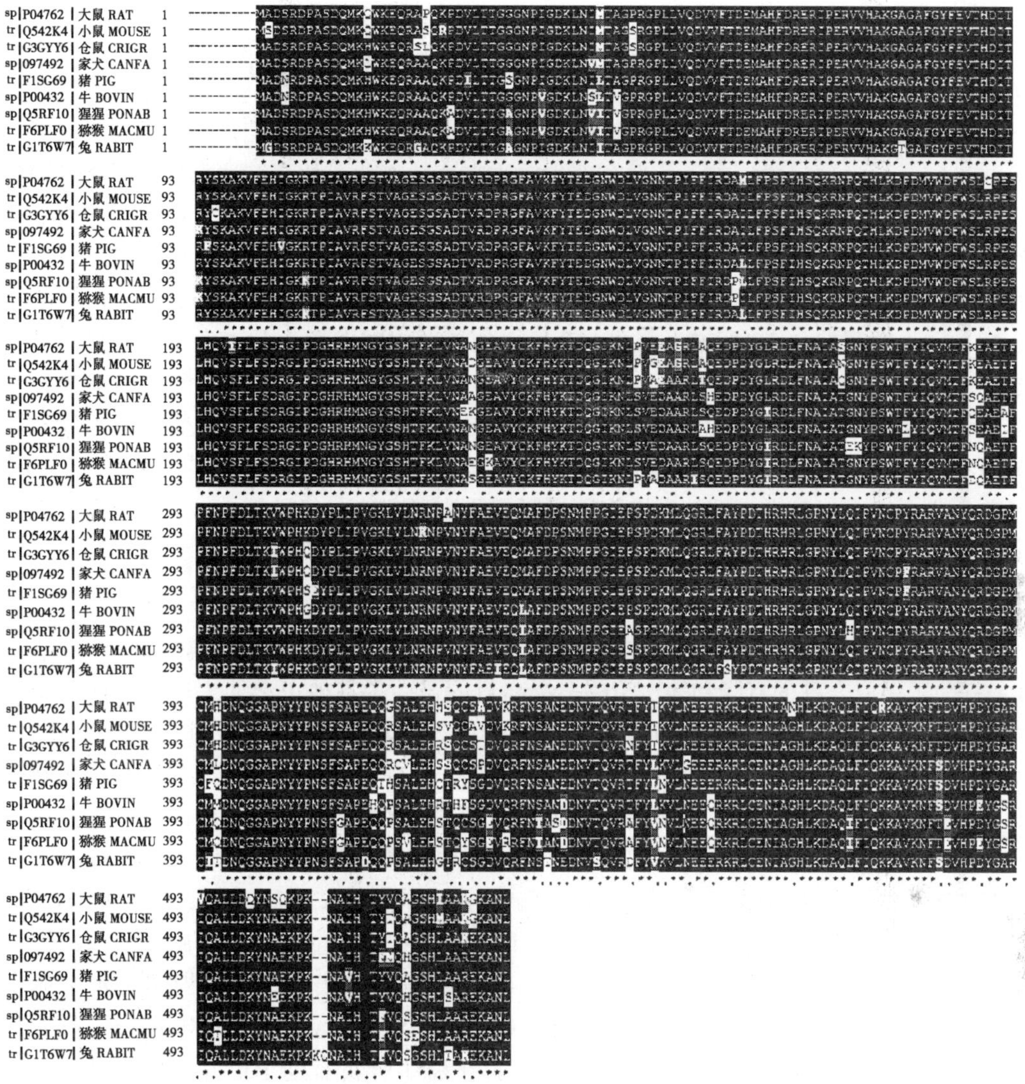

图 7　大鼠和其他 8 个物种的 CAT 蛋白的氨基酸序列同源性比对分析

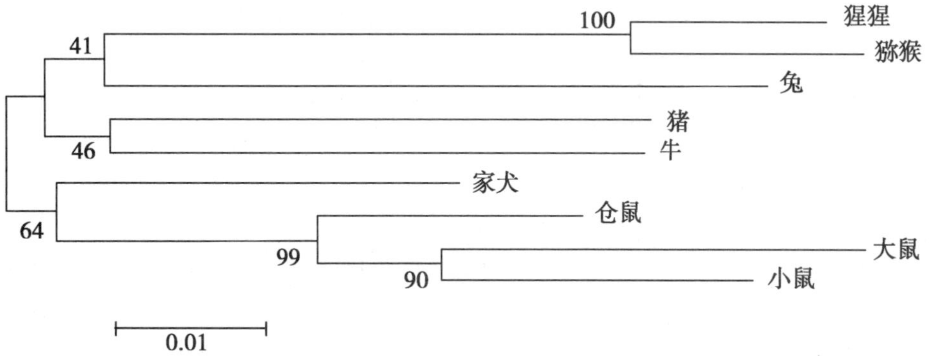

图 8　大鼠和其他 8 个物种的 CAT 蛋白序列的系统发育树

3 讨论

3.1 纳米铜的毒理蛋白质组学研究

利用蛋白质组学技术进行药物毒理学研究，即毒理蛋白质组学，是系统毒理学采用的主要方法，它是通过分析机体中毒前后全部蛋白质表达的变化，从而筛选毒性标志物并研究其毒理作用机制的学科[22~23]。本项研究通过蛋白质组学方法对纳米铜对大鼠肝脏毒性差异蛋白组学进行分析，在2－DE图谱上筛选到58个蛋白位点在表达量上有明显差异，经质谱鉴定和数据库搜索后，共鉴定出15种差异蛋白，其中，有2个差异蛋白点均被鉴定为CAT蛋白。这种同一种蛋白在2D图谱上存在多个位置的情况在双向电泳分离中十分常见，其原因也多种多样[24]，如高度保守的蛋白家族成员的存在、各种可变剪辑产生的蛋白、多种蛋白前体的存在、蛋白处理过程产生的有规则断裂、各种翻译后不同形式的修饰等，对这些原因的分析将有利于对该蛋白结构与功能的了解。本试验通过2－DE分析后获得的包括CAT的所有差异蛋白点可能与纳米铜的毒性作用机制相关，研究这些差异蛋白可为后续研究奠定良好的基础。

3.2 生物信息学分析

利用生物信息学分析蛋白质的理化性质（如等电点、分子质量、亲水性、氨基酸组成），有助于采取合理的克隆和表达策略，提高目的蛋白的高效可溶性表达，以期获得有活性的重组蛋白。此外，选择一些分析较为准确的生物信息学软件进行功能位点、二级结构、三维结构等的分析和预测，可为蛋白质的功能研究起到理论指导作用。本研究采用生物信息学对差异蛋白CAT进行分析，结果显示CAT是稳定性蛋白，有一定亲水性，可能不属于膜蛋白；无信号肽，亚细胞定位于细胞质，可能属于非分泌性蛋白；包含过氧化氢酶活性部位位点^{64}FDRERIPERVVH AKGAG^{80}，过氧化氢酶亚铁血红素配合基位点^{354}RLFAYPDTH^{362}以及2个N－糖基化位点^{439}NVTQ^{442}和^{481}NFTD^{484}等几个主要的功能位点；无规则卷曲、α螺旋和延伸链是该蛋白主要的二级结构元件，Swiss－model输回的模拟三维结构图有4个典型区域，预测结果与二级结构结果一致；同源性分析表明大鼠的CAT与其他物种有较强同源性，比对证实该蛋白有较高保守性。利用生物信息学对CAT分析有助于更好地针对目的基因和蛋白进行研究，进一步验证其生物学特性，为筛选纳米铜毒性的相关生物标志物提供依据。

3.3 过氧化氢酶

过氧化氢酶是主要的一种抗氧化物酶，几乎存在于所有的有氧呼吸生物体中，可将机体生成的H_2O_2分解为H_2O和O_2，从而使机体免受过氧化氢（H_2O_2）的损害[25]。机体存在两类抗氧化系统，一类是酶抗氧化系统，包括超氧化物歧化酶（SOD）、CAT、谷胱甘肽过氧化物酶（GPX）等，另一类是非酶抗氧化系统。当机体遭受各种有害刺激时，体内会产生较多量的高活性分子如活性氧自由基（reactive oxygen species，ROS），包括超氧阴离子（$O_2^{\cdot -}$）、羟自由基（OH·）和过氧化氢（H_2O_2）等[26]，活性氧可以刺激信号转导途径、调控细胞生长和凋亡，然而，当大量活性氧蓄积时，机体氧化程度超出氧化物的清除速度，氧化系统和抗氧化系统失衡，可导致脂类物质过氧化、蛋白质氧化、DNA损伤、膜中断以及线粒体功能障碍，对机体造成严重损害[27~30]。本研究发现，在中毒组大鼠肝脏中，纳米铜能够引起CAT表达降低，提示纳米铜可能通过氧化应激途径对大鼠肝脏造成一定损伤，具体研究需要进一步病理性相关实验验证。因此，推测CAT蛋白的表达下调可能是纳米铜引起肝细胞损伤和发挥毒性作用的途径之一，同时为揭示纳米铜毒性作用机制提供了依据。

4 结论

纳米铜染毒大鼠的肝脏蛋白组表达模式发生了较大变化；纳米铜通过下调 CAT 蛋白表达，引起肝细胞氧化应激损伤，可能是其发挥毒性作用的途径之一。

References

[1] 孔涛，郝雪琴，赵振升，等．纳米微量元素在畜牧业中的应用．饲料研究，2011，12（2）：12－13.

[2] Nel A，Xia T，Madler L，Li N. Toxic potential of materials at the nanolevel. Science，2006，311（3）：622－627.

[3] 王艳华．纳米铜和硫酸铜对断奶仔猪生长、腹泻和消化的影响及作用机理探讨［D］. 杭州：浙江大学，2002.

[4] 甄波，朱长虹，谢长生，等．纳米铜/聚合物复合材料宫内节育器对猕猴宫腔液 t－PA、PAF、PGE2 水平的影响．生殖与避孕，2006，26（8）：467－471.

[5] Liu G，Li X，Qin D，Xing Y，Guo，R Fan. Investigation of the mending effect and mechanism of copper nanoparticles on a tribologically stressed surface. Tribology Letters，2004，17（4）：961－966.

[6] Cioffi N，Ditaranto N，Torsi L，Picca R A. Sabbatini L，Valentini A，Novello L，Tantillo G，Bleve－Zacheo T，Zambonin P G. Analytical characterization of bioactive Fluoropolymer ultra－thin coatings modified by copper nanoparticles. Analytical and Bioanalytical Chemistry，2005，381（3）：607－616.

[7] 荔霞，刘永明，齐志明，等．纳米铜毒性研究进展．动物医学进展，2010，31（8）：74－78.

[8] Chen Z，Meng H，Xing G M，Chen C Y，Zhao Y L，Jia G，Wang T C，Yuan H，Ye C，Zhao F，Chai Z F，Zhu C F，Fang X H，Ma B C，Wan L J. Acute toxicological effects of copper nanoparticles in vivo. Toxicology Letters，2006，163（2）：109－120.

[9] Meng H，Chen Z，Xing G M，Yuan H，Chen C Y，Zhao F，Zhang C C，Zhao Y L. Ultrahigh reactivity provokes nanotoxicity：Explanation of oral toxicity of nano－copper particles. Toxicology Letters，2007，175（1－3）：102－110.

[10] Griffitt R J，Weil R，Hyndman K A，Denslow N D，Powers K，Taylor D，Barber D S. Exposure to copper nanoparticles causes gill injury and acute lethality in Zebrafish（Danio rerio）. Environmental Science and Technology，2007，41（23）：8 178－8 186.

[11] Lei R H，Wu C Q，Yang B H，Ma H Z，Shi C，Wang Q J，Wang Q X，Yuan Y，Liao M Y. Integrated metabolomic analysis of the nano－sized copper particle－induced hepatotoxicity and nephrotoxicity in rats：A rapid in vivo screening method for nanotoxicity. Toxicology and Applied Pharmacology，2008，232（2）：292－301.

[12] Prabhu B M，Ali S F，Murdock R C，Hussain S M，Srivatsan M. Copper nanoparticles exert size and concentration dependent toxicity on somatosensory neurons of rat. Nanotoxicology，2010，4（2）：150－160.

[13] Zhou S B，Bailey M J，Dunn M J，Preedy V R，Emery P W. A quantitative investigation into the losses of proteins at different stages of a two－dimensional gel electrophoresis procedure. Proteomics，2005，5（11）：2 739－2 747.

[14] 杨保华．利用基因组学和蛋白质组学技术研究纳米铜的肝、肾毒性及作用机制［D］. 北京：军事医学科学院，2010.

[15] Friedman D B，Hoving S，Westerneier R. Isoelectric focusing and two－dimensional gel electrophoresis. Methods Enzymology，2009，463（7）：515－540.

[16] Thierry Rabilloud，Mireille Chevallet，Sylvie Luche，Cécile Lelong. Two－dimensional gel electrophoresis in

proteomics：Past，present and future，Journal of Proteomics，2010，73（11）：2 064 –2 077.
[17] Hermann SchäggerTricine – SDS – PAGE. Nature Protocols，2006，1（1）：16 –22.
[18] 韩伟东，王栋，郝海生，杜卫华，赵学明，朱化彬．牛精子蛋白质组的双向电泳和质谱鉴定初步研究．安徽农业大学学报，2009，36（4）：538 –542.
[19] Candiano G，Bruschi M，Musante L，Santucci L，Ghiggeri G M，Carnemolla B，Orecchia P，Zardi L，Righetti P G. Blue silver：A very sensitive colloidal Coomassie G –250 staining for proteome analysis. *Electrophoresis*，2004，25（9）：1 327 –1 333.
[20] Ma Y L，Peng J Y，Huang L，Liu W J，Zhang P，Qin H L. Searching for serum tumor markers for colorectal cancer using a 2 – D DIGE approach. Electrophoresis，2009，30（15）：2 591 –2 599.
[21] Zhang W，Sun Z R. Random local neighbor joining：A new method for reconstructing phylogenetic trees. Molecular Phylogenetics and Evolution，2008，47（1）：117 –128.
[22] Yu L R. Pharmacoproteomics and toxicoproteomics：The field of dreams. Journal of Proteomics，2011，74（12）：2 549 –2 553.
[23] Veraksa A. When peptides fly：Advances in Drosophila proteomics. Journal of Proteomics，2010，73（11）：2 158 –2 170.
[24] 徐庆刚，陆健，郑建洲，等．河豚4SNc –2Tudor蛋白的鉴定及生物信息分析．安徽农业科学，2010，38（3）：1 163 –1 166.
[25] Salvi M，Battaglia V，Brunati A M，Rocca N L，Tibaldi E，Pietrangeli P，Marcocci L，MondovìB，Carlo A. Toninello R A. Catalase takes part in rat liver mitochondria oxidative stress defense. Biological Chemistry，2007，282（33）：24 407 –24 415.
[26] Nordberg J，Arner E S J. Reactive oxygen species，antioxidants and the mammalian thioredoxin system. Free Radical and Biology Medicine，2001，31（11）：1 287 –1 312.
[27] Li C H，Ni D J，Song L S，Zhao J M，Zhang H，Li NMolecular cloning and characterization of a Catalase gene from Zhikong scallop Chlamys farreri. Fish and Shellfish Immunology，2008，24，26 –34.
[28] AndradesM，Ritter C，OliveiraM R，Streck E L，Moreira J C F，Dal – Pizzol F. Antioxidant treatment reverses organ failure in rat model of sepsis：role of antioxidant enzymes imbalance，neutrophil infiltration，and oxidative stress. Journal of Surgical Research，2011，67（2）：307 –313.
[29] Yamamoto K，Banno Y，Fujii H，Miake F，Kashige N，Aso Y. Catalase from the silkworm，Bombyx mori：gene sequence，distribution，and over expression. Insect Biochemistry and Molecular Biology，2005，35：277 –283.
[30] 廖明阳，刘华钢．纳米铜对肾细胞的氧化损伤作用．中国药理学通报，2011，27（2）：239 –242.

（发表于《中国农业科学》）

奶牛乳腺炎源大肠杆菌中耶尔森菌强毒力岛相关基因的检测及序列分析

徐继英[1]，杨志强[2]，陈化琦[2]，刘俊林[1]，邢　娟[3]，李建喜[2]，李宏胜[2]

（1. 西北民族大学生命科学与工程学院，兰州 730000；2. 中国农业科学院兰州畜牧与兽药研究所，兰州 730050；3. 中国农业科学院兰州兽医研究所，兰州 730000）

摘　要：研究奶牛乳腺炎源大肠杆菌中耶尔森菌 HPI 携带情况及其与 O 血清型的关系，并对部分菌株的相关基因序列进行分析。从中国北京、内蒙古、甘肃、四川、重庆、云南、贵州等 7 个省市部分地区 1 260份临床型和隐性奶牛乳腺炎奶样中分离得到 190 株大肠杆菌，对分离菌株进行耶尔森菌强毒力岛核心区 irp2 基因、fyuA 基因及 HPI 毒力岛在大肠杆菌染色体中插入位置的鉴定，分析 HPI 毒力岛的携带情况及其与分离菌株 O 血清型之间的关系。190 株大肠杆菌分离株中，irp2 基因阳性率为 26.31%（50/190），fyuA 基因阳性率为 18.94%（36/190）。50 株 HPI + 分离株中检出 asn_ tRNA_ intB 基因 32 株，阳性率为 64%（32/50）。本试验克隆的 irp2 基因（273bp）、fyuA 基因（1 071bp）、asn_ tRNA_ intB 基因（1512bp）均与已发表序列高度同源，同源性分别在 97.1%、98.2%、97.2%以上，且其 HPI 毒力岛大多位于大肠杆菌染色体的 asn_ tRNA 位点上。耶尔森菌 HPI 在奶牛乳腺炎源大肠杆菌中广泛流行分布，但也存在差异，而不同血清型菌株携带 HPI 的倾向性可能只与特定的血清型有一定的关系。

关键词：奶牛乳腺炎；大肠杆菌；HPI 毒力岛；O 血清型

0　引言

耶尔森菌强毒力岛最早发现于耶尔森菌属，因与该属菌的小鼠致死表型密切相关，故被命名为强毒力岛（high pathogenicity island，HPI）[1~2]，鼠疫耶尔森氏菌染色体上的 HPI 毒力岛是决定耶尔森菌毒力或致病水平的一个较大染色体片段[1]。其核心功能区被称为 irp2 - fyuA 基因簇，是参与合成铁载体耶尔森杆菌素（介导铁摄取功能系统）的主要基因成分[3~5]。且有研究表明 HPI 毒力岛在耶尔森氏菌和致病性大肠杆菌之间存在水平传播[6~7]。近年来，由于国内外以菌苗预防为主的综合防治措施在生产中不断推广应用，使得葡萄球菌和链球菌引起的奶牛乳腺炎的发生率明显下降，而肠道菌尤其是大肠杆菌在奶牛乳腺炎主要病原菌中的优势地位变得逐渐明显起来[8~11]。大肠杆菌引起的乳腺炎常见于奶牛泌乳早期和牛奶体细胞数低下的高产奶牛[12]，但有关中国地区奶牛乳腺炎源大肠杆菌耶尔森菌强毒力岛的检测尚未见报道。本研究可为中国奶牛乳腺炎的预防和控制提供一定的理论指导。irp2 基因、irp1 基因是 HPI 毒力岛核心功能区 irp2 - fyuA 基因簇的主要结构基因，其中，irp2 可以作为 HPI 毒力岛的检测标志[13~14]，fyuA 是编码鼠疫菌素受体（FyuA）的鼠疫菌素受体基因[15~17]。Schubert 等于 1998 年首次报道：在 5 种致泻性大肠杆菌中，除肠出血性大肠杆菌外，均不同程度地携带耶尔森菌 HPI 基因，且其 irp2、fyuA 与耶尔森氏菌几乎相同，这提示 HPI 毒力岛可在耶尔森氏菌和致病性大肠杆菌之间水平传播[6~7,18]。目前，已在人、牛、兔腹泻致病性大肠杆菌中发现了 HPI 毒力岛的存在并与致病性密切相关[5,7,19~20]。叶长芸等发现 43 株肠产志贺样毒素且具侵袭的大肠杆菌中有 15 株的 irp2 基因、fyuA 基因呈现阳性，阳性率为 34.9%[21]。有关奶牛乳腺炎源大肠杆菌中 HPI 毒力岛的研究报道较少，目前，中国地区的相关研究报道还未见发表。本研究应用 PCR 方法检测奶牛乳腺炎源大肠杆菌中耶尔森菌强毒力岛（HPI）的分布，并对部分菌株的相关基因序列进行了比较分析，以期探讨奶牛乳腺炎源大肠杆菌中 HPI 毒力岛的携带情况及其与 O 血清型之间的关系。

1 材料与方法

1.1 试验时间、地点

本试验于 2011 年在中国农业科学院兰州畜牧与兽医研究所进行。

1.2 材料

1.2.1 菌株

分别自中国北京、内蒙古、甘肃、四川、重庆、云南、贵州等 7 个省市部分地区（2007—2009 年）临床型和隐性奶牛乳腺炎奶样中分离纯化获得大肠杆菌 190 株，所有细菌均采用麦康凯平板划线分离，然后在 LB 培养基上 37℃ 纯培养 24h，冻干保存。$irp2^+$、$fyuA^+$ 基因大肠杆菌 S433014 株、asn_ tRNA_ $intB^+$ 基因大肠杆菌 S452621 株均为扬州大学兽医微生物实验室惠赠。

1.2.2 引物

根据 GenBank 中已发表序列，分别设计了 3 对特异性引物，用于 HPI 毒力岛 irp2、fyuA、asn_ tRNA_ intB 基因的扩增，预期扩增出的目的片段大小分别为 273、1 071 和 1 512 bp，由大连宝生物（TaKaRa）公司合成，浓度均为 50mmol/L。

1.2.3 试剂

细菌全基因组 DNA 提取试剂盒购自天根生物工程有限公司；Fragment Purification Kit、Agarose Gel DNA Purification Kit、Taq 酶、dNTP、DNA Marker DL2000、RNase A、pMD18 - T 载体系统等均购自大连 TaKaRa 公司；Tris_ 平衡酚、优质琼脂糖等均购自上海华美生物工程公司。

1.3 大肠杆菌 O 血清型鉴定

大肠杆菌标准抗 O 血清购自中国兽医药品卫生监察所，参照兽医微生物学实验指导——平板凝集法和试管凝集法进行血清型鉴定。

1.4 DNA 模板的制备

所有 190 株奶牛乳腺炎源大肠杆菌均在 LB 平板上 37℃ 纯培养 24h，各挑取单个菌落，接种于 LB 液体培养基中过夜振摇培养，参照细菌全基因组 DNA 提取试剂盒说明，提取大肠杆菌的全基因组 DNA，并以之为模板，-20℃ 保存。

1.5 HPI 毒力岛 irp2、fyuA、asn_ tRNA_ intB 基因的 PCR 检测

反应体系 50μl：10 × PCRBuffer 5μl、$MgCl_2$（25mmol · L^{-1}）3μl、4 × dNTPs（10mmol ·

L^{-1}）4μl、上、下游引物（均为 50μmol·L^{-1}）各 1μl、Taq 酶（5U·$μl^{-1}$）0.5μl、DNA 模板 1μl；加 H_2O 至 50μl。引物序列及 PCR 反应条件见表 1。

表 1 试验中所用的引物

引物	序列（5′-3′）	目的基因	退火温度（℃）	PCR 片段长度（bp）
Irp2 - L	CTGTTACCGGACAACCGC	irp2	55	273
Irp2 - R	GGGCAGCGTTTCTTCTTC			
FyuA - L	CCGTCTTACAGGGACTCACAACAAT	fyuA	58	1071
FyuA - R	GGTACAGCCCAAACACCATATCAAC			
intB - L	GTGTGAAAACTCTTCTCGGTGC	asn_ tRNA_ intB	60	1512
intB - R	GTCGCTCTTTCATTCCTCTGTG			

取扩增产物 4μl 在 1% 琼脂糖凝胶中电泳 30min。以 Marker DL 2000 为参照，观察扩增片段大小。

1.6 奶牛乳腺炎源大肠杆菌 HPI 的分子流行病学

取上述制备的 190 株奶牛乳腺炎源大肠杆菌的 DNA 模板，分别采用上述方法进行 PCR 扩增，以确定 HPI 在奶牛乳腺炎源大肠杆菌中的分子流行病学。

1.7 irp2、fyuA、asn_ tRNA_ intB 基因的克隆与序列分析

随机取检测为 $irp2^+$、$fyuA^+$、$intB^+$ 大肠杆菌的 DNA 模板各 5 株，分别用于 irp2、fyuA、asn_ tRNA_ intB 基因的 PCR 扩增。将以上各菌株扩增产物经试剂盒回收纯化后连接 pMD 18 - T 载体，转化大肠杆菌 JM109，挑选阳性克隆提取质粒，送 TaKaRa 公司测序。测序结果用 DNAStar7 软件进行处理并与 GenBank 中的序列进行比较分析。

2 结果与分析

2.1 HPI 毒力岛核心区 irp2、fyuA 基因的 PCR 检测

利用建立的 PCR 方法对 190 株奶牛乳腺炎源大肠杆菌分离株的 HPI 毒力岛核心区 irp2 基因和 fyuA 基因分别进行了检测，所得条带与预期目的片段大小相符（图 1、图 2）。共计 50 株大肠杆菌检出 irp2 基因，阳性率为 26.31%（50/190）；36 株大肠杆菌检出 fyuA 基因，同时也检出 irp2 基因，阳性率为 18.94%（36/190）。这提示在 HPI 毒力岛的转移过程中，奶牛乳腺炎源大肠杆菌 fyuA 基因可能存在一定丢失。

2.2 HPI 毒力岛在阳性菌株中的定位鉴定

本试验结果中，50 株 HPI^+ 分离株检出 asn_ tRNA_ intB 基因 32 株，所得条带与预期目的片段大小相符（图 3），阳性率为 64%（32/50），这表明 asn tRNA 位点是奶牛乳腺炎源大肠杆菌 HPI 毒力岛的整合位置。

2.3 irp2、fyuA 和 asn_ tRNA_ intB 基因序列分析

阳性克隆序列测定结果用 DNAStar 软件分析显示，分离菌株克隆的 irp2 基因序列均为 273bp，与预期大小一致，与已发表的 irp2 基因序列同源性 >97.1%，且分离株 3 与已发表序

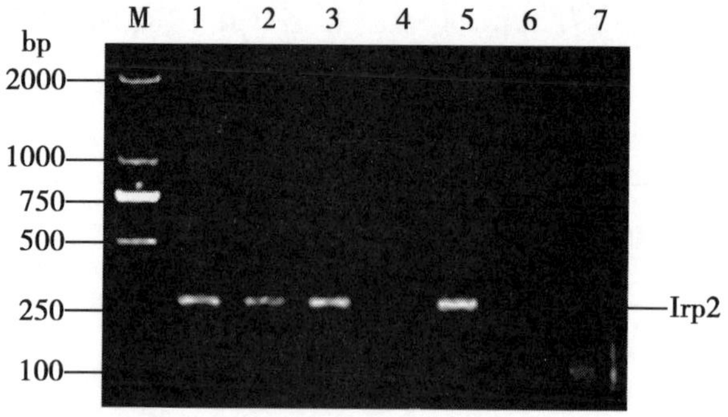

M：DNA Marker DL2000；1：S433014；2，3，5：irp2+；4，6，7：irp2

图 1　部分分离菌株 irp2⁺基因 PCR 扩增

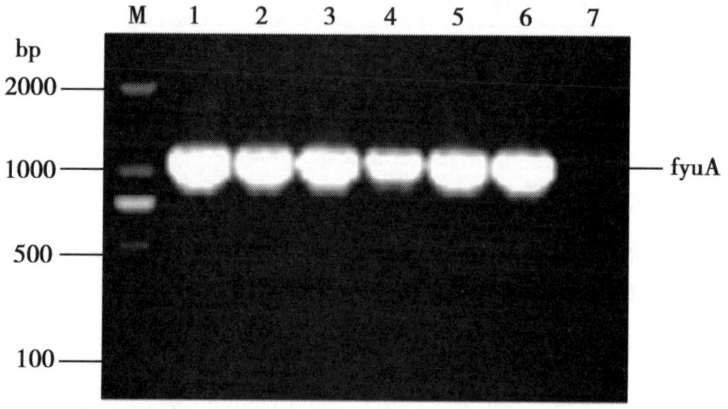

M：DNAMarkerDL2000；1：S433014；2，3，4，5，6：fyuA+；7：fyuA

图 2　部分分离菌株 fyuA⁺基因 PCR 扩增

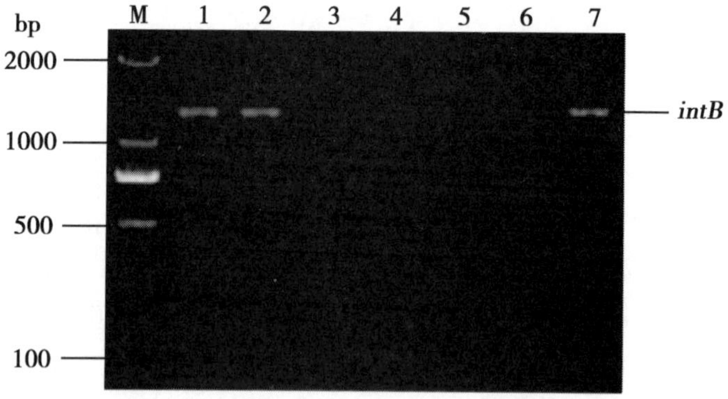

M：DNAMarker DL2000；1：S452621；

2，7：asn_ tRNA_ intB+；3，4，5，6：asn_ tRNA_ intB

图 3　部分分离菌株 asn_ tRNA_ intB⁺基因 PCR 扩增

列 Z35451 同源性达 100%；克隆的 fyuA 基因序列均为 1 071bp，与预期大小相符，与已发表的 fyuA 序列同源性>98.2%，且分离株 7 与分离株 8 同源性达 100%；克隆的asn_ tRNA_ intB基因序列均为 1 512bp，与预期大小一致，与已发表的 asn_ tRNA_ intB 基因序列同源性>97.2%。irp2、fyuA、asn_ tRNA_ intB 基因系统发育树见图 4～图 6。

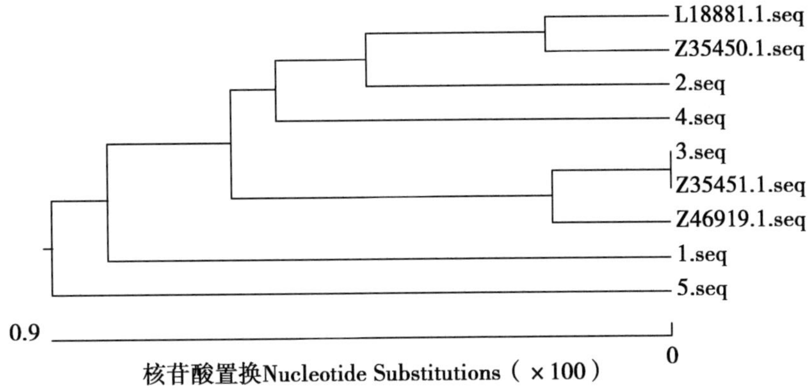

图4　irp2 基因片段核苷酸序列系统发育树

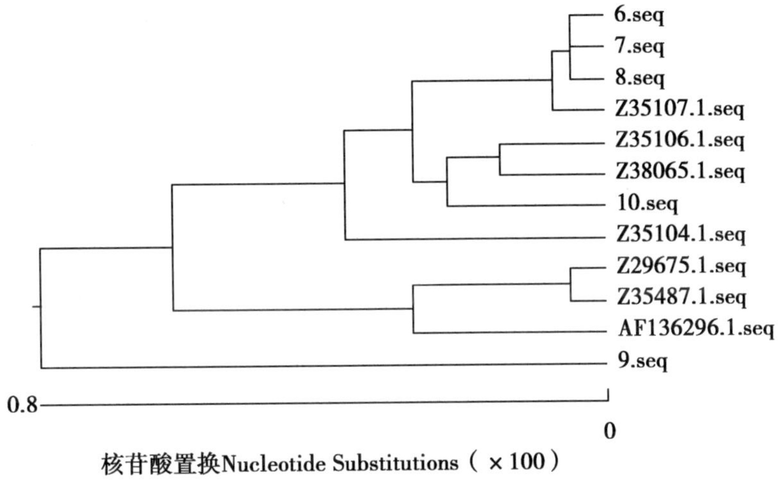

图5　fyuA 基因片段核苷酸序列系统发育树

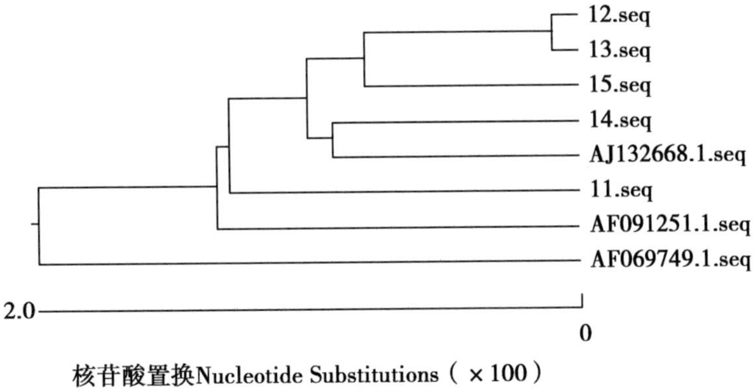

图6　asn_ tRNA_ intB 基因片段核苷酸序列系统发育树

2.4　HPI 毒力岛与 O 血清之间的关系

O 血清型鉴定结果显示，190 个大肠杆菌分离株共计鉴定出 42 种血清型，覆盖了 108 株大肠杆菌，另有 4 株自凝，78 株未鉴定出型。其中，优势血清型为 O_{93}、O_9、O_{146}、O_7、O_{74}，各有 16、13、10、7、6 株。在所有被检出的 42 种血清型中，38 株 HPI^+ 分布于 17 种不同血清型菌株中，剩余10株HPI+血清型未定，2株自凝。不同血清型HPI检出率有很大的差别。将本研究中的各优势血清型O_{93}、O_9、O_{146}、O_7、O_{74}菌株的HPI检出率进行比较，发现O_{93}血清型HPI检出率较高，有43.75%的菌株检测呈阳性，O_9、O_{74}、O_7、O_{146}血清型阳性率分别为38.46%、33.3%、28.57%和20%。而在非优势血清型菌株中，O_{149}血清型菌株有很高的HPI检出率，100%呈阳性（表2）。

表2　分离菌株HPI与O血清型相关性的比较

血清型	阳性率	血清型	阳性率
O_{93}	7/16[1)]	O_{36}	0/3
O_9	5/13	O_{61}	0/3
O_{146}	2/10	O_{66}	1/3
O_7	2/7	O_{74}	2/6
O_{10}	0/4	O_{83}	0/3
O_{21}	0/4	O_{107}	1/3
O_{22}	0/4	O_{149}	3/3
O_{32}	0/3	others	27/105

* 表中分数为HPI阳性菌株/总菌株数。

3　讨论

耶尔森菌强毒力岛是人源、牛源、兔源致泻性大肠杆菌的重要毒力因子[5,7,19~20]，大肠杆菌引起的乳腺炎常见于奶牛泌乳早期和牛奶体细胞数低下的高产奶牛[12]，但有关中国地区奶牛乳腺炎源大肠杆菌HPI毒力岛的检测尚未见报道。本研究中，190个大肠杆菌分离株中，有26.31%奶牛乳腺炎源大肠杆菌分离株携带有irp2基因，此结果低于人源大肠杆菌分离株32.25%~34.9%[21~22]、禽源大肠杆菌分离株44.9%的阳性率[23]，却明显高于猪源大肠杆菌分离株12.7%~16.66%的阳性率[24~25]。提示HPI毒力岛在奶牛乳腺炎源大肠杆菌中广泛流行分布，但各血清型之间仍存在一定差异。本研究中irp2基因的检出率较高，且序列分析证实奶牛乳腺炎源大肠杆菌的irp2、fyuA、asn_ tRNA_ intB基因序列均与GenBank中已发表序列高度同源，这为耶尔森菌HPI毒力岛在不同菌种之间的水平转移提供了一定的证据。

Irp2、fyuA基因是HPI毒力岛核心区的主要结构基因，它们通常同时存在，fyuA蛋白是鼠疫菌素受体（与摄铁有关）[13~15]。本实验中，irp2和fyuA基因并不一定都同时存在于同一株大肠杆菌中，耶氏杆菌素FyuA受体仅在72%的HPI阳性株中表达，这种现象可能与耶尔森菌HPI在菌种之间水平转移的过程中发生突变或基因重组使得fyuA片段的丢失有关，也可能是由于耶氏杆菌素摄铁系统并不是拥有HPI毒力岛最主要的优势，它的缺失可能会导致HPI阳性菌株的摄铁能力减弱，并由此可间接降低细菌的毒性[13]。有关fyuA基因表达不同的原因目前仍然还不明确。

在耶尔森菌HPI基因簇的右端有一个与P4噬菌体整合酶基因有同源性的编码序列intB基因，它位于asn tRNA位点。现有报道认为，HPI毒力岛具有不稳定性，并含有潜在的插入序列、整合酶等可移动成分，主要转移到大肠杆菌染色体的asn_ tRNA位点[26]。本研究结果表明，在 HPI⁺奶牛乳腺炎源大肠杆菌中，约有 64% HPI 携有 intB 基因且插入到 asn_ tRNA 位点，这与 HPI 毒力岛在人源、猪源大肠杆菌中大多数连接在该位点报道一致[22,25]，但大肠杆菌在通过水平转移获得 HPI 毒力岛的过程中可能对其进行了修饰或重组，使得部分 intB 基因片段丢失，asn_ tRNA_ intB－基因也可能与 HPI 毒力岛被整合到染色体的其他位置或是毒力岛的边缘被修饰有关[19,27]。国外亦曾报道从鸟中分离的 2 株致败血症的大肠杆菌中，HPI 的整合酶基因发生了相同的缺失[28]。这些菌株也可能通过有别于其他大肠杆菌的水平基因转移方式获得 HPI。其中的 HPI 可能代表一类新型、独特、整合酶部分缺失的 HPI。由于基因的缺失导致整合酶无功能，将使 HPI 稳定地插入这些菌株的基因组。产生这一现象的确切原因及这些菌株之间的关系还有待进一步研究。

奶牛乳腺炎源大肠杆菌血清型比较复杂，通过对分离菌株 O 血清型与 HPI 毒力岛的相关性分析，可以发现 O_{93}、O_{9}、O_{146}、O_{7}、O_{74}这 5 种血清型为奶牛源大肠杆菌的优势血清型，但其携带 HPI 毒力岛的阳性率却存在较大差异，而非优势血清型 O_{149}分离株却有着 100% 的 HPI 检出率，这说明不同血清型菌株其 HPI 的分布是不同的，这种分布特性可能只与特定的血清型有一定的关系，而与所属优势血清型并无明显相关性，大肠杆菌的 HPI 毒力岛也可能仅是引起奶牛乳腺炎的致病因素之一。

耶尔森菌 HPI 毒力岛可通过质粒水平传播到另一株细菌的特性，使得这些毒力因子有可能在各种菌株、动物和人类之间互相传播[7,22]，且 HPI 毒力岛已广泛地分布于人的致病性肠杆菌中，对肠道外感染发挥重要作用[29]，这将对人类的健康和公共卫生安全带来很大的问题和威胁。此外，大肠杆菌在奶牛乳腺炎主要病原菌中的优势地位逐渐凸显[8~11]，但有关中国地区奶牛乳腺炎源大肠杆菌耶尔森菌强毒力岛的检测尚未见报道。本研究可为中国奶牛乳腺炎的预防和控制提供一定的理论参考，对临床上研制有效的奶牛乳腺炎大肠杆菌疫苗具有重要的指导意义，对保障人类食品安全和公共卫生安全也具有积极的意义。

4 结果

耶尔森菌 HPI 在奶牛乳腺炎源大肠杆菌中广泛流行分布，但各血清型之间存在一定差异，部分 fyuA 基因、intB 基因片段可能存在一定丢失；不同血清型菌株携带 HPI 毒力岛的阳性率存在较大差异，而部分非优势血清型分离株却有着较高的 HPI 检出率，这种分布特性可能只与特定的血清型有一定的关系，而与所属优势血清型并无明显相关性。

References

[1] Carniel E, Guilvout I, Prentice M. Characterization of a large chromosomal 'high_ pathogenicity island' in biotype 1*B Yersinia enterocolitica*. *Journal of Bacteriology*, 1996, 178 (23): 6 743－6 751.

[2] Buchrieser C, Prentice M, Carniel E. The 102－kilobase unstable region of *Yersinia pestis* comprises a high－pathogenicity island linked to a pigmentation segment which undergoes internal rearrangement. *Journal of Bacteriology*, 1998a, 180: 2 321－2 329.

[3] Rakin A, Noelting C, Schubert S, Heesmann J. Common and specific characteristics of the high－pathogenicity island of *Yersinia enterocolitica*. *Infection and Immunity*, 1999, 67 (10): 5 265－5 274.

[4] Mokracka J, Koczura R, Kaznowski A. Yersiniabactin and other siderophores produced by clinical isolates of Enterobacter spp. and *Citrobacter spp. FEMS Immunology & Medical Microbiology*, 2004, 40 (1): 51－55.

[5] Carniel E. The Yersinia high－pathogenicity island: an iron－uptake island. *Microbes and Infection*, 2001, 3 (7): 561－569.

[6] Rakin A, Urbitsch P, Heesemann J. Evidence for two evolutionary lineages of highly pathogenic *Yersinia species*. *Journal of Bacteriology*, 1995, 177（9）：2 292 - 2 298.

[7] Bach S, de Almeida A, Carniel E. The *Yersinia* high - pathogenicity island is present in different members of the family Enterobacteriaceae. *FEMS Microbiology Letters*, 2000, 183（2）：289 - 294.

[8] Rangel P M, Marin J M. Antimicrobial resistance in Brazilian isolates of Shiga toxin - encoding Escherichia coli from cows with mastitis. *Ars Veterinaria, Jaboticabal, SP*, 2009, 25（1）：18 - 23.

[9] Sumathi B R, Gomes A R, Krishnappa G. Antibiogram profile based dendrogram analysis of *Escherichia coli* serotypes isolated from bovine mastitis. *Veterinary World*, 2008, 1（2）：37 - 39.

[10] 王娜，高学军．哈尔滨地区奶牛隐性乳房炎病原菌的分离鉴定．东北农业大学学报，2011，42（2）：29 - 32.

[11] 金兰梅，伍清林，马高民，等．规模化奶牛场奶牛乳房炎病原菌的分离与药敏试验．黑龙江畜牧兽医，2010，42（21）：101 - 103.

[12] Burvenich C, Van Merris V, Mehrzad J, Diez - Fraile A, Duchateau L. Severity of *E. coli* mastitis is mainly determined by cow factors. *Veterinary Reseach*, 2003, 34（5）：521 - 564.

[13] Lucier T S, Fetherston J D, Brubaker R R, Perry R D. Iron uptake and iron - repressible polypeptides in *Yersinia pestis*. *Infection and Immunity*, 1996, 64（8）：3 023 - 3 031.

[14] Koczura R, Kaznowski A. The Yersinia high - pathogenicity island and ironuptake systems in clinical isolates of *Escherichia coli*. *Journal of Medical Microbiology*, 2003, 52：637 - 642.

[15] Brem D, Pelludat C, Rakin A, Jacobi C A, Heesemann J. Functional analysis of yersiniabactin transport genes of *Yersinia enterocolitica*. *Microbiology*, 2001, 147（5）：1 115 - 1 127.

[16] Schubert S, Rakin A, Heesmann J. The Yersinia high - pathogenicity island（HPI）：evolutionary and functional aspects. *International Journal of Medical Microbiology*, 2004, 294（2）：83 - 94.

[17] Fetherston J. D, Lillard J. W, Perry R. D. Analysis of the Pesticin Receptor from *Yersinia pestis*：Role in Iron - Deficient Growth and Possible Regulation by Its Siderophore. *Journal of Bacteriology*, 1995：1 824 - 1 833.

[18] Rakin A, Schubert S, Guilvout I. Local hopping of IS3 elements into the A + T rich part of the high - pathogenicity island in *Yersinia enterocolitica1B, O*：8. *FEMS Microbiology Letters*, 2000, 182（2）：225 - 229.

[19] Clermont O, Bonacorsi S, Bingen E. The *Yersinia* high - pathogenicity island is highly predominant in virulence - associated phylogeneticgroups of *Escherichia coli*. *FEMS Microbiology Letters*, 2001, 196（2）：153 - 157.

[20] Penteado A S, Ugrinovich L A, Blanco J. Serobiotypes and virulence genes of *Escherichia coli* strains isolated from diarrheic and healthy rabbits in Brazil. *Veterinary Microbiology*, 2002, 89（1）：41 - 51.

[21] 叶长芸，徐建国．部分 ESIEC 菌株存在耶尔森氏菌 HPI 毒力岛．疾病监测，2000，15（2）：48 - 50.

[22] 王勇，王红，向前，等．产毒性和致病性大肠埃希菌中小肠结肠炎耶尔森菌强毒力岛的检测．第一军医大学学报，2002，22（7）：580 - 583.

[23] 金文杰，郑志明，秦爱建，等．禽致病性大肠杆菌中耶尔森菌强毒力岛的分子流行病学调查．中国兽医科学，2006，36（10）：787 - 790.

[24] 陈祥，赵娟，高崧，等. 我国部分地区猪源大肠杆菌 LEE 和 HPI 毒力岛相关基因的检测．中国人兽共患病学报，2006，22（1）：33 - 47.

[25] 成大荣，孙怀昌，徐建生，等．断奶仔猪源大肠杆菌 LEE 及 HPI 毒力岛的检测．微生物学报，2006，46（3）：368 - 372.

[26] Schubert S, Rakin A, Fischer D, Sorsa J, Heesmann J. Characterization of the integration site of Yersinia high - pathogenicity island in *Escherichia coli*. *FEMS Microbiology Letters*, 1999, 179（2）：409 - 414.

[27] Rakin A, Noelting C, Schropp P, Heesemann J. Integrative module of the high - pathogenicity island of

Yersinia. *Molecular Microbiology*, 2001, 39 (2): 407 -416.

[28] Karch H, Schubert S, Zhang D, Zhang W, Schmidt H, Olschlager T, Hacker J. A Genomic island, termed high - pathogenicity island, is present in certain non - O157 shiga toxin - producing *Escherichia coli* Clonal Lineages. *Infection and Immunity*, 1999, 67 (11): 5 994 -6 001.

[29] Schubert S, Picard B, Gouriou S, Heesemann J, Denamur E. *Yersinia* High - pathogenicity island contributes to virulence in *Escherichia coli* causing extraintestinal infections. *Infection and Immunity*, 2002, 70 (9): 5 335 -5 337.

(发表于《中国农业科学》)

奶牛蹄叶炎与血浆中矿物元素含量的相关性分析

董书伟[1,2]，荔　霞[1]，严作廷[1]，高昭辉[1,3]，王胜义[1]，
齐志明[1]，刘世祥[1]，刘永明[1]

（1. 中国农业科学院兰州畜牧与兽药研究所，农业部兽用药物创制重点实验室，
甘肃省中兽药工程技术中心，兰州 730050；2. 甘肃省牦牛繁育工程重点实验室，
兰州 730050；3. 农业大学动物医学院，兰州 730070）

摘　要： 采集 15 头健康奶牛和 36 头蹄叶炎患病奶牛血浆样品，检测其 Fe、Zn、Cu 微量元素和 Mg、Ca、P 常量元素的含量及钙磷比，结果表明：2 组间 Fe、Cu、Mg、Ca 和 P 矿物元素含量及钙磷比无显著性差异（$P>0.05$），患病牛 Zn 含量却显著低于健康牛（$P<0.05$）。因此，奶牛血浆中 Zn 含量与蹄叶炎发病显著相关，Zn 含量可能会作为蹄叶炎发病的监测指标，也可用于奶牛蹄叶炎的早期诊断和防治。

关键词： 蹄叶炎；矿物元素；相关性分析；奶牛

牛蹄病是奶牛常发的三大疾病之一，它对奶牛养种蹄病，据林为民等[2]调查，新疆石河子地区奶牛蹄殖业的影响仅次于乳房炎。而在奶牛蹄病中，41% 叶炎平均发病率为 9.12% ~19.78%，淘汰率为的病例是蹄叶炎[1]。蹄叶炎是蹄壁真皮的乳头层和 8.1% ~ 15.3%。由此可见，蹄叶炎给奶业造成了巨血管层发生的弥漫性、浆液性和无菌性炎症，可引起大的经济损失，因此，开展对蹄叶炎病因、发病机理奶牛肢蹄疼痛、精神不安、食欲减退、生产性能明显及防治措施等方面的研究，对于提高奶牛生产性能下降、饲料报酬率降低等，严重者甚至被淘汰。此研究具有重要的现实意义和经济价值。但是，关于奶牛外，蹄叶炎还会导致蹄变形，蹄底溃疡及白线病等多蹄叶炎病因及防治的综述很多，试验研究的报道却极为鲜见。

矿物元素是机体合成酶、激素、维生素等多种生物活性物质的重要原料，研究报道，在蹄病患病牛血清中相关矿物元素含量异常，是造成蹄病发生的重要原因[3]，因此，矿物元素对于维持奶牛肢蹄健康有重要作用。奶牛蹄病有众多类型，且病因繁杂，但大多数研究没有对奶牛蹄病这一大类疾病进行区分，而笼统的研究蹄病与矿物元素之间的关系，截至目前，国内尚未见到针对研究奶牛蹄叶炎与其血浆中矿物元素的相关性报道。本试验精心选取蹄叶炎患病牛，检测其血浆中几种常量、微量元素含量，并与健康奶牛对照，分析血浆中矿物元素含量与蹄叶炎发病的相关性，为奶牛蹄叶炎的临床预防和治疗提供一定的科学依据。

1　材料与方法

1.1　实验动物

在甘肃省某奶牛场选取具有蹄叶炎临床症状，经临床兽医诊断为蹄叶炎的奶牛 36 头为

患病组，然后随机抽取肢蹄健康且无其他疾病临床症状的泌乳牛 15 头作为健康对照组，常规饲养管理，自由采食，日粮按照 NRC（2001）配制全混合日粮（TMR）。

1.2 主要仪器

MARS Xpress 型号微波消解仪，CEM 公司；ZEEnit－700 原子吸收光谱仪，德国 Jena 公司；BS－420 全自动生化分析仪，深圳迈瑞公司。

1.3 试验方法

1.3.1 采样与样品处理

自奶牛颈静脉采集血液，肝素钠抗凝，待运送到实验室后，以 3 000r/min 离心 15min，吸取上层血浆，分装保存在－80℃低温冰箱中备用。

1.3.2 Fe、Zn 和 Cu 微量元素含量的测定

精确吸取 1ml 血浆样品于干燥的聚四氟乙烯防爆消化管中，然后加入 3ml 浓硝酸（70% HNO3）和 7ml 盐酸（36.5% HCl），进行微波消解，排酸，定容至 10ml，并配制各元素标准溶液，制作标准曲线，然后检测样品中各元素含量。各元素标准溶液含量 100mg/L，批号 08002，购自核工业北京化工冶金研究院。

1.3.3 Mg、Ca 和 P 常量元素的含量及钙磷比测定

Mg、Ca 和 P 的含量测定操作按照试剂盒说明书，质控和试剂盒均购自深圳迈瑞公司。Mg 测定采用二甲苯胺蓝法，Ca 测定采用偶氮胂Ⅲ法，P 测定采用磷钼酸法。

1.4 数据统计分析

运用 SPSS13.0 对各组数据作单因素方差分析比较，结果均以 $x \pm s$ 表示，并采用逐步多因素 Logistic 回归法对奶牛血浆中矿物元素水平与蹄叶炎发病作相关性分析，模型纳入标准为 $P<0.05$，排除标准为 P>0.1，以 $P<0.05$ 有统计学意义。

2 结果与分析

2.1 奶牛血浆中 Fe、Zn 和 Cu 微量元素含量比较

蹄叶炎患病组奶牛和健康组奶牛血浆中 Fe、Zn 和 Cu 元素含量统计结果见表 1。由表 1 可知，患病组奶牛血浆中 Fe 水平略高于健康组，但差异不显著（$P>0.05$）；患病组奶牛血浆中 Zn 含量显著低于健康组（$P<0.05$），Cu 含量亦低于健康组，但无显著差异（$P>0.05$）。

表 1 奶牛血浆中 Fe、Zn 和 Cu 元素含量（$x \pm s$） (mg/L)

组别	Fe	Zn	Cu
健康组	76.3 ± 32.1	2.8 ± 0.31	0.66 ± 0.18
患病组	83.5 ± 28.0	2.4 ± 0.30*	0.60 ± 0.22

注：* 表示与健康组相比差异显著（$P<0.05$）。下同。

2.2 奶牛血浆中 Mg、Ca 和 P 常量元素含量比较

健康组和蹄叶炎患病组奶牛血浆中 Mg、Ca 和 P 元素含量统计结果见表 2。由表 2 可知，患病组奶牛血浆中 Mg、Ca、P 元素含量都略高于健康组，但组间差异均不显著（$P>0.05$）；两组间钙磷比值也无显著性差异（$P>0.05$）。

表2　奶牛血浆中 Mg、Ca 和 P 元素含量（$x \pm s$）　（mmol/L）

组别	Mg	Ca	P	钙磷比值
健康组	2.51 ±0.60	1.74 ±0.45	1.83 ±0.55	0.978 ±0.17
患病组	2.54 ±0.52	1.82 ±0.44	1.88 ±0.54	0.990 ±0.18

2.3　奶牛蹄叶炎与相关矿物元素的多元 Logistic 回归分析

采用 Logistic 回归法，将血浆中 Fe、Zn、Cu、Mg、Ca 和 P 元素及钙磷比与是否患有蹄叶炎进行多元回归分析，其结果见表3。由表3可知，从标准化回归系数（B）来看，Zn 对奶牛蹄叶炎发病显著相关，有统计学意义（$P<0.05$），其他各种矿物元素水平及钙磷比对奶牛蹄叶炎发病无统计学意义（$P>0.05$），与上述统计结果一致。

表3　奶牛蹄叶炎与血浆中元素水平的多元逻辑回归分析结果

组别	B	SE	Wald	Sig	Exp	95% CI	
						Lower	Upper
铜	-1.03	1.723	0.357	0.550	0.357	0.012	10.46
锌	-2.75	1.261	4.770	0.014	0.064	0.005	0.754
铁	0.113	0.130	0.758	0.384	1.120	0.868	1.445
镁	-0.707	1.200	0.347	0.556	0.493	0.047	5.182
钙	1.290	3.353	0.148	0.700	3.632	0.005	5.893
磷	-0.105	2.967	0.001	0.972	0.900	0.003	1.853
钙磷比	-0.360	5.988	0.004	0.952	0.697	0.006	9.188

3　讨论

奶牛蹄病与乳房炎和生殖系统疾病共同称为奶牛的三大疾病，给奶牛饲养业造成了巨大的经济损失，严重制约了奶业的健康快速发展。已有研究报道：奶牛发生蹄病时体内矿物元素代谢异常，如吴树清等[3]认为，低钙、低磷、低锌、高铜是奶牛蹄病发生的主要因素。但是，蹄病是奶牛常发的一大类疾病，包括蹄叶炎、蹄变形、腐蹄病和蹄间质增生等多种形式，并且，单就蹄叶炎的病因就有摄入过多碳水化合物、不适运动、遗传、季节、环境和继发于瘤胃酸中毒、子宫炎、乳房炎疾病等多种因素[4]，因此，对奶牛蹄病的具体形式不加以分类，而笼统地研究病因与蹄病发生的关系，势必造成结果不准确，报道不一的局面。本试验只选取了有蹄叶炎临床症状的患病牛作为研究对象，并与健康奶牛作对照，通过检测健康奶牛和蹄叶炎患病牛血浆中微量和常量矿物元素含量，探索其与蹄叶炎发病的相关性，结果发现两组间 Fe、Cu、Mg、Ca 和 P 矿物元素含量无显著性差异（$P>0.05$），血浆中钙磷比值差异也不显著，但是，在患病牛血浆中 Zn 含量却显著低于健康奶牛（$P<0.05$），并采用多元逻辑回归法对各个指标与奶牛蹄叶炎发病的相关性进行分析，发现血浆中 Zn 水平与蹄叶炎发病呈显著性相关（$P<0.05$），再次验证了本次的试验结果。由此可推测，奶牛血浆中 Zn 含量可以作为蹄叶炎监测的指标之一。

3.1 矿物元素与奶牛蹄叶炎的关系

矿物元素是动物的生长发育和物质代谢的重要原料，其含量相对稳定的，且在一定的生理范围内浮动。当动物发生疾病时，血浆中某种元素的含量会超出其生理范围，因此，我们可以通过检测血浆中矿物元素含量来诊断或检测某种疾病的发生。

3.2 微量元素 Fe、Cu 和 Zn 与奶牛蹄叶炎的相关性分析

在体内铁的主要功能是作为血红蛋白和肌红蛋白中血红素的组成成分，在氧运输和电子传递链中起着核心作用[5]，健康牛血清 Fe 含量的参考值为（97.2 ±29.1）mg/L。但是，日粮中铁的过量添加会抑制铜锌的吸收、转运和利用，造成铜锌的含量降低，从而影响到奶牛蹄角质的形成和保护，容易引起各种蹄病的发生。铜是过氧化物歧化酶和细胞色素氧化酶等多种酶的组成和激活剂，作为催化剂参与体内氧化还原反应。铜在机体免疫力方面也有重要作用，铜缺乏可表现为胸腺萎缩、免疫细胞活性降低、抗体合成受阻和效价降低等[6]。健康奶牛血浆 Cu 含量应大于（0.64 ~0.7）mg/L[7]，血浆铜含量低于 0.5mg/L 是铜缺乏的标志，刘德义等[7]研究证明，奶牛日粮中缺乏铜有可能引起蹄裂、蹄底溃疡和蹄底脓肿。锌有抗脂质过氧化的功能，对生物膜有重要的保护作用，还参与角质素、角蛋白和胶原蛋白的形成，可加强牛蹄的坚硬性、完整性，并能增强机体的免疫力。当锌缺乏时，动物会发生皮肤角化不全、生长发育迟缓、繁殖机能下降，蹄壳变形、开裂，骨骼发育异常和脱毛等病理现象。试验表明，在奶牛日粮相同的条件下，每天仅添加 5g 氨基酸锌，就能有效提高奶牛产奶量，减少乳房炎和肢蹄病的发生[8]。

本试验结果表明，奶牛发生蹄叶炎时，血浆中 Fe 含量略高于健康组，但差异不显著（$P > 0.05$），这与李开江等[9]报道一致；Cu、Zn 含量较健康奶牛低，其中 Zn 含量显著低于健康组（$P < 0.05$），与林德贵等[10]研究基本一致，但与刘德义等[7]报道不一致，可能是因为检测的样品不同，他们采集的多是蹄病奶牛的被毛和蹄角质，而本试验采集的是奶牛血浆，被毛和蹄角质的生长有一定的时间周期，而血浆更能反映奶牛体内即时的矿物元素含量变化。

3.3 Mg、Ca 和 P 常量元素与奶牛蹄叶炎的相关性分析

在体内镁的作用主要是与钙、钾、钠协同维持肌肉和神经兴奋性，维持心肌的正常功能和结构。奶牛缺镁主要症状是痉挛，但 Mg 代谢紊乱与蹄病发生的关系目前尚未阐明。Ca、P 是奶牛大量需求的矿物元素，为维持泌乳和血液中 Ca、P 含量的需要，机体会动用骨骼中储存的 Ca、P，如果此时奶牛日粮中钙、磷缺乏或比例不当，容易造成钙磷代谢紊乱，奶牛出现骨质疏松、蹄部角质软化或蹄变形等，引发肢蹄病[11 ~14]。但在本试验中，健康组和患病组奶牛血浆中 Mg、Ca 和 P 常量元素及钙磷比没有显著性差异，未发现异常，这与上述研究结果不一致，可能是由于本试验选择的研究对象只是蹄叶炎奶牛，而有学者把所有蹄病类型的奶牛都放在一起研究。

综上所述，本研究通过分析奶牛血浆中矿物元素含量与蹄叶炎发病的相关性，发现蹄叶炎患病牛血浆中 Zn 含量却显著低于健康奶牛，因此，奶牛血浆中 Zn 含量可能会作为蹄叶炎发病的监测指标，可用于奶牛蹄叶炎的早期诊断和防治。但由于样本数量的限制，没有在更多的病例中证实这一推测，所以，在下一步工作中，我们要扩大样本数、进一步验证该结果的准确性。以后我们可以参考本试验结果，不要盲目地对所有疾病都采用大剂量抗生素治疗，并根据奶牛所在地区不同，科学合理地补充日粮中矿物元素，调整至最优比例，对于指

导临床用药和防治奶牛蹄叶炎有积极意义。

参考文献

[1] 齐长明. 奶牛疾病学［M］. 北京：中国农业科学技术出版社，2006：593－595.

[2] 林为民，陶岳，史文军，等. 新疆石河子地区奶牛蹄叶炎流行病学调查与综合防治［J］. 中国奶牛，2011（10）：48－51.

[3] 吴树清，马刚，王新生，等. 呼市地区奶牛蹄病与相关矿物元素比较研究［J］. 内蒙古农业大学学报，2003，24（4）：26－30.

[4] Underwood E J，Suttle N F. The Mineral Nutrition of Livestock［M］. Cambridge：CABI Publishing，2004.

[5] 张瑜，许建海，朱晓萍，等. 日粮不同铜、钼水平对辽宁绒山羊机体抗氧化性能的影响［J］. 中国畜牧杂志，2010，46（17）：30－33.

[6] Mulligan F J，Grady L，Rice D A，*et al*. A herd health approach to dairy cow nutrition and production diseases of the transition cow［J］. Animal Reprod Sci，2006，96：331－353.

[7] 刘德义，江汪洋，顾有方，等. 奶牛蹄病与微量元素铜、锰、锌关系的研究［J］. 中国奶牛，2003（4）：21－23.

[8] 刘辉放，陈昌建，肖戈. 蛋氨酸锌对牛奶产量、乳房炎和蹄病的影响［J］. 中国奶牛，2004（1）：26－27.

[9] 李开江，唐兆新，梁群超，等. 蹄病奶牛血液中6种微量元素和自由基代谢的变化［J］. 畜牧与兽医，2009，41（2）：56－58.

[10] 林德贵，温代如. 北京黑白花牛蹄底角质中25种元素含量的研究［J］. 畜牧兽医学报，1990，21（1）：43－47.

[11] 温洁，王俊东，杨宏斌，等. Ca、P营养失衡对乳牛蹄病的影响［J］. 山西农业大学学报，2005，25（2）：123－124.

[12] 邓发清. 乳牛肢蹄病与部分矿物元素代谢的相关性研究［J］. 中国畜牧杂志，2008，44（9）：45－47.

[13] 王振勇，张明江，王林. 奶牛蹄病与相关矿物元素关系的研究［J］. 中国畜牧兽医，2005，32（8）：21－22.

[14] Bergsten C. Causes，risk factors，and prevention of laminitis and related claw lesions［J］. Acta Vet Scand Suppl，2003，98：157－166.

（发表于《中国兽医学报》）

芩连液与白虎汤对气分证家兔补体经典途径活化的影响比较

张世栋，王东升，王旭荣，李世宏，李锦宇，陈炅然，李宏胜，严作廷

（中国农业科学院兰州畜牧与兽药研究所，农业部兽用药物创制重点实验室，
中国农业科学院临床兽医学研究中心，甘肃省中兽药工程技术研究中心，兰州 730050）

摘　要：为比较经方白虎汤与自拟方芩连液的分子疗效区别，以静脉注射内毒素建立了具有高热证候的气分证家兔模型，并用白虎汤和自拟方芩连液分别进行治疗，通过ELISA检测比较治疗前后血清总补体（CH50）的含量变化，并采用实时荧光定量PCR技术检测了肝组织中补体第三成分（C3）基因和C－反应蛋白（CRP）基因的表达变化。结果表明，在气分证家兔模型中，补体系统以经典途径活化。白虎汤的治疗能同时保证C3基因和CRP基因的高表达起到治愈作用，而芩连液仅以提高CRP的表达量发挥一定的保护作用。研究表明白虎汤对动物气分证的疗效优于芩连液。

关键词：气分证；补体；CH50；C3；CRP

温病是由温邪引起的以发热为主症的一类急性外感热病。气分证是由温热之邪入里，正盛邪实，正邪剧争出现的阳热亢盛的里热证，是温病典型的证候之一[1]。现代医学中的传染性与感染性疾病也多属温病范畴，临床多以发热为主症[2]。白虎汤具有清热除烦、止渴生津之功效，被现代临床广泛运用于急性传染病或非传染性急性热病，成为治疗高热急症的首选经方[3]。内毒素（LPS）是一种重要的致热原，微量LPS就能引起动物体温上升[4]，同时导致多器官及系统性损害，肝脏是LPS损害的主要靶器官之一[5]。本研究采取静脉注射LPS建立了具有高热证候的气分证家兔模型，并用白虎汤和自拟方芩连液分别进行治疗，比较治疗前后血清总补体（CH50）的含量变化及肝组织中补体第三成分（C3）基因和C－反应蛋白（CRP）基因的表达变化，从免疫学的角度比较两种组方药物的疗效差别。

1　材料与方法

1.1　主要试剂与仪器

白虎汤由石膏、知母、粳米、甘草组成；芩连液由大青叶、黄芩、地榆、连翘、甘草、紫草和八角茴香组成。常规煎煮后，取上清过滤并浓缩至1g·ml^{-1}。

脂多糖（LPS，O55：B5）购自北京拜尔迪生物技术有限公司。Trizol为美国Invitrogen公司的产品。DEPC水，gDNA Eraser试剂盒和SYBR Premix Ex*Taq* Ⅱ试剂盒均为日本TaKaRa公司产品。ELISA检测试剂盒为美国R&D公司产品，其他试剂均为国产分析纯。

ThermoScientific MK3酶标仪、ABI公司的普通PCR仪、伯乐Bio－Rad CFX 96实时荧光定量PCR仪，以及NanoDrop 2 000超微量分光光度计。

1.2 实验动物分组及处理

新西兰兔24只，普通级，雌雄不限，单笼饲养，平均体质量2.0kg，体质量变异不超过平均体质量的20%。实验动物及饲料由中国农业科学院兰州兽医研究所实验动物中心提供。试验前适应性饲养3d，以适应试验环境。将24只家兔随机分为4组，每组6只。设对照组（CN），其注射与治疗均以无热源生理盐水处理；模型组（LPS）、白虎汤治疗组（LB）和芩连液治疗组（LQ）均以静脉注射15μg/kg LPS处理，同时分别灌服生理盐水（7ml/kg）、白虎汤（7ml/kg）和芩连液（7ml/kg），每6h 1次，连续治疗2次。CN和LPS组在灌注治疗2.5h后经心脏采血并分离血清冻存于－80℃，然后注射空气栓处死家兔，取适量肝组织固定于RNA fixer溶液中；LB和LQ组在2次治疗后也做相同处理。

1.3 实验方法

1.3.1 血清总补体CH50含量检测

按照试剂盒操作说明，以固相夹心ELISA法检测血清中总补体CH50的含量，并对数据进行统计分析。

1.3.2 组织总RNA的提取及完整性检测

取50～100mg经RNA fixer固定过的肝组织置于组织匀浆器，加1ml Trizol试剂，手动研磨均匀后室温放置5min，使其充分裂解后12 000r·min^{-1}离心5min，取上清，按200μl氯仿·ml^{-1} Trizol加入氯仿，振荡混匀后室温放置15min。再以4 ℃ 12 000r·min^{-1}离心15min，取上层水相，至另一离心管中，每毫升Trizol加入0.5ml异丙醇混匀，室温放置5～10min后，4℃12 000r/min离心10min，弃上清，RNA沉于管底。每毫升Trizol加入1ml 75%乙醇，洗涤沉淀，然后4℃8 000r·min^{-1}离心5min，尽量弃上清，使沉淀室温晾干或真空干燥5～10min。用50μl DEPC水溶解RNA样品，55～60 ℃放置5～10min。取适量样品用Nano－Drop2000测OD值定量RNA浓度，样品A_{260nm}/A_{280nm}值在1.6～1.8认为提取效果良好。同时，取RNA样品进行琼脂糖凝胶电泳检测，10V/cm，20min，凝胶成像系统下观察其完整性，－80℃保存样品。

1.3.3 引物设计与合成

登录NCBI的GenBank，检索到家兔β－actin（gb | X60733）、补体C3α（gb | M32434）和CRP（gb | M13497）的基因序列，遵循PCR引物设计原则进行引物设计（表1），引物均由北京六合华大基因科技股份有限公司合成。

表1 扩增β－actin、C3α和CPR分子mRNA片段的引物

基因名称	片段大小	引物序列（5′－3′）Sequences（5′－3′）of primers	
		正向序列	反向序列
β－actin	153bp	CAATGGCTCCGGCATGTGC	CGCTTGCTCTGGGCCTCG
C3α	154bp	AACCCTGGACCCAGAGAACC	GTCGATGGCATCCTCCGTCA
CRP	159bp	CATGGAGAAGCTGCTGTGGTG	GTGAAGGCTTTGAGTGGCTTC

1.3.4 cDNA合成

将检测过完整的RNA样品进行反转录，取1μg总RNA为模板，以寡核苷酸Oligo（dT）为引物，使用TaKaRa gDNA Eraser试剂盒（DRR047s）并参照其操作说明进行反转录反应。

1.3.5 实时荧光定量 RT－PCR 的扩增

取 2μl 反转录的 cDNA，管家基因和目的基因的上下游引物各 0.5 μl，SYBR Premix Ex*Taq* Ⅱ（DRR081A）12.5μl，补去离子水至 25μl 体系，在 Bio－Rad CFX Manager 荧光定量 PCR 仪上进行反应，每个样品的扩增设置 3 个复孔，反应程序：95℃预变性 30s；95℃解链 5s；60℃退火 25s；72℃延伸 20s；65℃荧光检测 5s，循环 39 次。

1.4 结果判断及数据统计

所有数据采用 SPSS13.0 进行统计分析，各组数据以"$\bar{x} \pm s$"表示，组间差异采用 Duncans LSD 法做比较。

2 结果与分析

2.1 血清总补体（CH50）含量检测结果

血清 CH50 含量的 ELISA 检测结果显示，以 15μg/kg 体质量的剂量注射 LPS 的动物，其血清总补体含量显著高于对照组（$P<0.01$）；经白虎汤治疗后的家兔血清 CH50 含量与对照组无显著差异（$P>0.05$）；而经芩连液治疗后的家兔血清 CH50 显著低于对照（$P<0.01$），但与白虎汤治疗组相比无显著差异（$P>0.05$）。结果表明，LPS 能使动物血清 CH50 含量显著增加，而白虎汤和芩连液的治疗作用则可抑制 LPS 对 CH50 含量的影响。

表 2 各组家兔血清总补体（CH50）含量检测结果比较（$n=6$） （KU/L）

组别	CH50 含量 Content of CH50
对照组 CN	23.32 ± 2.17^{Bb}
模型组 LPS	30.42 ± 5.68^{Aa}
白虎汤治疗组 LB	22.79 ± 3.19^{Bb}
芩连液治疗组 LQ	18.35 ± 2.77^{Bc}

* 数据肩标不同小写字母表示差异显著（$P<0.05$），肩标相同小写字母者表示差异不显著（$P>0.05$）；肩标不同大写字母表示差异极显著（$P<0.01$），下同

2.2 *β－actin*、*C3α* 和 *CRP* 基因 RT－PCR 扩增产物的电泳结果

以 cDNA 为模板，分别对 *β－actin*、*CRP* 和 *C3α* 基因进行 RT－PCR 扩增，筛选反应条件（包括引物浓度、退火温度的选择、循环数的确定等），1.5% 的琼脂糖凝胶电泳检测扩增结果（图 1）。从图 1 中可以看出，4 组样品中的扩增产物片段大小均一，无杂带及拖带现象，说明设计选择的引物特异性较强，反应条件较好；扩增产物片段长度约 150bp，与引物设计结果相符。

2.3 SYBR Green Ⅰ Real－time PCR 结果

根据 Bio－Rad CFX Manager 荧光定量 PCR 仪自动分析获取的 Ct 值，以 $2^{-\Delta\Delta Ct}$ 法计算补体 *C3α* 与 *CRP* 基因相对于内参基因 *β－actin* 的表达量并作统计分析。

2.3.1 *C3α* 基因的相对表达量

以 15μg/kg 体质量的剂量静脉注射 LPS 的动物，其肝组织补体 *C3α* 基因的表达水平显著高于对照（$P<0.05$）；经白虎汤治疗后的家兔其 *C3α* 基因表达水平显著高于对照（$P<0.05$），但与模型组差异不显著（$P>0.05$）。经芩连液治疗后，家兔 *C3α* 基因的表达水平极

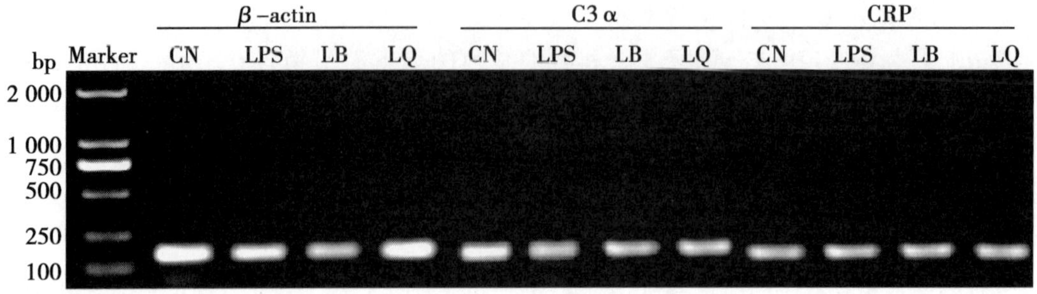

图1 β-actin、CRP 和 C3α 基因 RT-PCR 产物检测

显著低于其他3组（$P<0.01$）。结果表明，肝组织 *C3α* 基因的表达水平在气分证模型动物中会显著提高，而白虎汤的治疗对此变化无显著影响；芩连液的治疗却能显著抑制气分证动物 *C3α* 基因的表达水平，阻断 LPS 对 C3α 基因表达的影响（图2）。

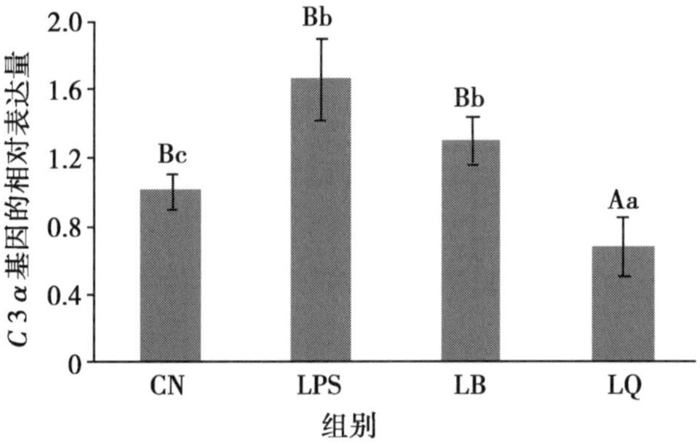

图2 不同处理组动物肝组织中补体 C3α 基因相对表达量的变化

2.3.2 *CRP* 基因的相对表达量

气分证模型动物 *CRP* 基因的相对表达量显著高于对照组（$P<0.01$），而白虎汤和芩连液治疗后的动物其基因的表达量都极显著地高于对照组和气分证组（$P<0.01$），但两者之间差异不显著（$P>0.05$）。结果表明，气分证模型动物肝组织中 *CRP* 基因的相对表达量显著升高，白虎汤和芩连液的治疗可进一步显著地提高 *CRP* 基因的表达，增强动物的先天免疫保护力（图3）。

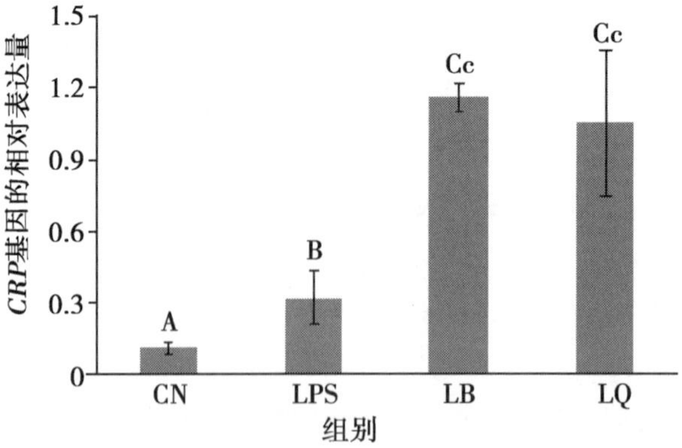

图3 不同处理组动物肝组织中 CRP 基因相对表达量的变化

3 讨论

补体系统是由 30 ~ 40 种广泛存在于哺乳动物体液和细胞膜表面具有酶活性的蛋白质组成的反应系统[6]，在预防微生物入侵机体的体液免疫防御中起到重要作用[7~8]。补体激活有经典途径、替代途径以及甘露聚糖结合凝集素（MBL）途径 3 条途径。3 条途径最终都以活化 C3 而交汇于一点[9]。C3 在补体中含量最高、能力最强，是补体系统的核心成分[10]，它的活化和裂解是启动和放大补体级联反应的关键步骤，是连接先天免疫与适应性免疫的中介，并且参与各种生物免疫调节反应[6]。C3 是由一个稳定的 β 链和一个含硫酯键的 α 链（C3α）经二硫键共价连接而成的相对分子质量 187ku 的蛋白分子。活化后的 C3 裂解为活性结构 C3b 和过敏毒素 C3a，C3b 经其活性硫酯键与靶分子结合，进而放大补体活化的级联反应[6,11]。临床上以测定血清总补体（CH50）与 C3、C4 等含量和活性水平作为判断机体免疫功能有用的指标[12]。CH50 的定量测量可用于补体经典途径的激活分析[13]。本试验中，以 *C3α* 的表达量来指示 C3 的表达水平。模型组动物血清 CH50 含量升高及肝组织 *C3α* 基因表达增强，说明气分证模型动物先天防御的补体系统可通过经典途径活化。经白虎汤和芩连液治疗后，CH50 含量和 C3 表达量均下降，说明两种药物都能抑制气分证模型中补体经典途径的活化，且芩连液的作用效果显著强于白虎汤。

CRP 是由肝脏产生的一种急性期蛋白[14]。机体遭受感染或损伤后，在急性期反应的 24 ~ 48h 内，血清中 CRP 含量可升高近 1 000倍，之后又会迅速恢复至正常水平，是先天免疫防御的重要组分[15]。临床上通过 CRP 的检测来诊断急性感染和判断治疗[16]。CRP 主要的生物学功能是识别病原并形成抗感染屏障，以及清除源于凋亡和坏死细胞的抗原[17]，保护机体免遭炎症和内毒素血症等的损伤[15]。CRP 非常重要的一个特征就是能够结合补体成分 C1q 而激活补体经典途径，杀灭病原微生物，间接保护机体免遭病原菌侵害[17]。已经确定 CRP 除了能调控急性炎症的基本作用外，同时还具有促炎和抗炎的功效，而 CRP 的抗炎或促炎功效受其表达水平的影响[17]。在转基因小鼠中，通过与 FcγR 的相互作用，高表达的 CRP 在一定程度上能保护机体免遭 LPS 的致死性攻击[18]。各组动物肝组织 *CRP* 基因相对表达量的检测结果表明，气分证模型动物具有显著但不强烈的急性期反应，但白虎汤和芩连液的治疗作用使机体发生更强烈的急性期反应，说明两种药物能通过提高 CRP 的表达来消除 LPS 致病攻击，对 CRP 介导的先天免疫保护具有显著增强作用，且两种药物的作用差异不显著。

4 结果

静脉注射 LPS 建立的家兔气分证模型中，作为先天免疫屏障的补体系统能通过经典途径活化。芩连液的治疗作用对 C3 表达的抑制作用直接导致补体活化受到抑制，但同时可显著提高 CRP 表达，从而增强 CRP 介导的先天免疫保护作用。阻断气分证动物的补体活化，而增强 CRP 的保护功效可能是芩连液的作用机制之一。白虎汤通过保证 C3 和 *CRP* 基因的高表达对介导补体活化和消除 LPS 的致病攻击具有显著的正向效应；而芩连液可抑制 C3 的表达与补体的活化，仅以提高 CRP 的表达量起到一定的保护作用。所以，白虎汤对增强机体先天免疫保护力具有更显著和更全面的功效。

参考文献

[1] 郭选贤，张华锴，闫俊峰. 温病气分证辨证规律初探［J］. 中国中医基础医学杂志，2010，16（9）：

742－743.
[2] 孙守芳．解毒白虎汤治疗温病气分证的实验研究［D］．福建：福建中医学院，2004.
[3] 张保国，程铁锋，刘庆芳．白虎汤药效及现代临床研究［J］．中成药，2009，31（8）：1 272－1 275.
[4] 刘建．白虎汤加减灌肠治疗温病气分热证的研究［D］．广州：广州中医药大学，2006.
[5] 李英，刘宏杰，邹万盛，等．褪黑素对内毒素血症山羊肝功能损伤的影响［J］．中国兽医学报，2009，29（1）：86－89.
[6] JANSSEN B J，GROS P. Structural insights into the central complement component C3 [J]. *Mol Immunol*，2007，44：3－10.
[7] CARROLL M C. The complement system in regulation of adaptive immunity [J]. *Nat Immunol*，2004，5：981－986.
[8] LAMPING N，SCHUMANN R R，BURGER R. Detection of two variants of complement component C3 in C3－deficient guinea pigs distinguished by the absence and presence of a thiolester [J]. *Mol Immunol*，2000，37：333－341.
[9] BEXBORN F，ANDERSSON P O，CHEN H，*et al*. The tick－over theory revisited：Formation and regulation of the soluble alternative complement C3convertase（C3（H_2O）Bb）[J]. *Mol Immunol*，2008，45：2 370－2 379.
[10] MASTELLOS D，PRECHL J，LÁSZLÓ G，*et al*. Novel monoclonal antibodies against mouse C3interfering with complement activation：description of fine specificity and applications to various immunoassays [J]. *Mol Immunol*，2004，40：1 213－1 221.
[11] JANSSEN B J，HUIZINGA E G，RAAIJMAKERS H C，*et al*. Structures of complement component C3 provide insights into the function and evolution of immunity [J]. *Nature*，2005，437：505－511.
[12] MOLLNES T E，JOKIRANTA T S，TRUEDSSON L，*et al*. Complement analysis in the 21st century [J]. *Mol Immunol*，2007，44：3 838－3 849.
[13] EKDAHL K N，NORBERG D，BENGTSSON A A，*et al*. Use of serum or buffer－changed EDTA－plasma in a rapid，inexpensive，and easy－to－perform hemolytic complement assay for differential diagnosis of systemic lupus erythematosus and monitoring of patients with the disease [J]. *Clin Vaccine Immunol*，2007，14，549－555.
[14] KOLKHOF P，GEERTS A，SCHAFER S，*et al*. Cardiac glycosides potently inhibit C－reactive protein synthesis in human hepatocytes [J]. *Biochem Biophys Res Commun*，2010，394：233－239.
[15] VOLANAKIS J E. Human C－reactive protein：expression，structure，and function [J]. *Mol Immunol*，2001，38：189－197.
[16] RIDKER P M. Clinical application of C－reactive protein for cardiovascular disease detection and prevention [J]. *Circulation*，2003，107：363－369.
[17] MARNELL L，MOLD C，DU CLOS T W. C－reactive protein：Ligands，receptors and role in inflammation [J]. *Clin Immunol*，2005，117：104－111.
[18] MOLD C，RODRIGUEZ W，RODIC－POLIC B，*et al*. C－reactive protein mediates protection from lipopolysaccharide through interactions with FcγR [J]. *J Immunol*，2002，169：7 019－7 025.

（发表于《畜牧兽医学报》）

射干地龙颗粒的安全药理学分析

王贵波[1]，罗永江[3]，罗超应[2]，李锦宇[2]，郑继方[1]，谢家声[1]，辛蕊华[1]

（1. 中国农业科学院兰州畜牧与兽药研究所，兰州 730050；2. 甘肃省中兽药工程技术研究中心，兰州 730050；3. 农业部兽用药物创制重点实验室，兰州 730050）

摘　要：为了进一步探讨射干地龙颗粒临床用药的安全性，首次通过无创方式对射干地龙颗粒开展了安全药理学研究。分别按体质量应用高剂量（2.0g/kg）、中剂量（1.0g/kg）、低剂量（0.5g/kg，相当于推荐剂量）的射干地龙颗粒和生理盐水对犬进行灌胃，并观察麻醉犬的平均动脉压、心率、体温、心电标准Ⅱ导联、呼吸频率、潮气量、血氧饱和度、呼吸曲线、尿液 11 项指标和尿液增质量等相关指标；同时对昆明系小鼠按照上述药物剂量进行灌胃，并观察小鼠的自主活动；通过以上指标的观测来说明药物对心血管系统、呼吸系统、泌尿系统和中枢神经系统的影响。结果表明，按照 2.0g/kg 将射干地龙颗粒进行灌胃，对麻醉犬的心血管系统、呼吸系统和泌尿系统无显著影响；小鼠自主活动结果表明，小鼠活动的总路程和时间及各边、角的路程和时间组间差异不显著（$P>0.05$），且小鼠的一般行为活动、姿势、步态均表现正常，无肌颤、不安、奔跑、嘶叫、蜷缩竖毛等异常，说明在试验剂量范围内药物对小鼠的中枢神经系统不表现显著影响。结果表明，至少在 0.5～2.0g/kg 范围内给药，射干地龙颗粒对动物心血管系统、呼吸系统、泌尿系统及中枢神经系统均无明显影响，表明其不良反应小，是一种适合于临床应用的、较为安全的药物。

关键词：射干地龙颗粒；安全药理；犬；小鼠

药物的安全药理学是涉及生理学、药理学和毒理学的综合性药理学科，在新药临床前安全性评价中的意义越来越受到相关管理部门的重视，在新药研发过程中具有不可替代性和重要意义。有报道表明药物不良反应是美国人口死亡的第 4 大原因，而安全药理学试验则可减少这种死亡的发生[1]。因此，有必要对临床前药物或新药进行安全药理学试验研究，以避免药物在较短时间内危及机体的重要系统。在早期的临床试验中，药物发生的不良反应主要发生于中枢神经系统、心血管系统、呼吸系统和泌尿系统[2]，在本试验中也主要对这四大系统的相关指标进行了观测。在国内，兽药的安全性评价主要侧重于急性毒性试验和亚慢性毒性试验，而对兽用药物的安全药理学研究鲜见报道。射干地龙颗粒是中国农业科学院兰州畜牧与兽药研究所研制的新型中兽药制剂，以射干、紫菀和地龙为主要成分，具有调节机体免疫力、解热镇痛、抑制病毒复制、抗菌消炎和高效低毒的综合功效，对防治鸡传染性呼吸道病具有良好的效果，避免了当前存在的疫苗免疫失败和病原耐药性问题，而且符合当前人们对畜产品安全性的需求。近年来，本课题组对该药的组方筛选、制剂工艺和药效学等进行了研究[3~4]。目前，对射干地龙颗粒的安全性评价体系研究，已经完成了对小鼠的急性毒性试验和靶动物的亚慢性毒性试验[5~7]，但尚缺乏相关的安全药理学研究。本试验旨在研究射干地龙颗粒对犬心血管系统、呼吸系统、泌尿系统和小鼠神经系统的影响，探讨射干地龙颗粒临床用药的安全性，以期为临床扩大试验及临床用药提供重要的参考。

1 材料与方法

1.1 主要仪器设备

BIOPAC MP150 多导生理记录仪，BIOPAC NIBP100 血压测量器，BIOPAC TSD137 呼吸换能器，ECG100C 心电信号放大器，均为美国 BIOPAC 公司产品；G3D 多参数监护仪，为深圳市杰纳瑞医疗仪器有限公司产品；KNF－100 型尿液分析仪，由扬州市凯达医疗设备有限公司生产；鼠博士八鼠集成旷场及鼠博士行为学分析系统，购于上海移数信息科技有限公司。

1.2 试验材料

射干地龙颗粒，由中国农业科学院兰州畜牧与兽药研究所提供，批号：100301；盐酸塞拉嗪注射液，由青岛汉河动植物药业有限公司生产，批号：2009022301。尿液分析试纸，由苏州弘益生物科技有限公司生产，批号：110808。

1.3 实验动物

健康成年杂种犬，由中国农业科学院兰州畜牧与兽药研究所标准化试验动物场饲养；昆明系小鼠，购于兰州大学医学实验中心。

1.4 试验方法

1.4.1 试验犬分组及给药

健康杂种犬 24 只，体质量 4.8～6.1kg，随机分为 4 组，每组 6 只：射干地龙颗粒高、中、低 3 个药物组和对照组，于试验开始前驱虫并适应性饲喂一周。试验时，按照体质量以 0.10ml/kg 的剂量注射盐酸赛拉嗪对犬进行麻醉。30min 后，射干地龙颗粒高、中、低剂量组犬分别按照体质量以 20、1.0 和 0.5g/kg 的剂量（按生药计）给药，其中，0.5g/kg 为推荐剂量；对照组犬以 2.0g/kg 的剂量给予生理盐水；各组均通过胃导管灌胃给药。

1.4.2 犬心血管系统、呼吸系统和泌尿系统指标的测定

用十六通道生理记录仪记录犬的前臂血压、呼吸节律曲线和心电标准Ⅱ导联的波动曲线，并计算呼吸频率和潮气量；同时用心电监护仪测定犬的肛门体温。分别记录灌胃前及试验 40、50、70、90 和 120min 等时间点（即给药后 10、20、40、60 和 90min 时）的血压、心率、呼吸频率、潮气量、血氧饱和度和体温等值，并记录呼吸节律曲线和心电标准Ⅱ导联曲线。其中血压数值以平均动脉压表示。

用尿液分析仪对灌胃前及试验结束时犬的尿液进行分析，同时收集试验过程中的尿液，观察尿量的变化。

1.4.3 小鼠神经系统指标的测定

昆明系小鼠 40 只，雌雄各半，体质量 18～22g，随机分为 4 组，每组 10 只：射干地龙颗粒高、中、低剂量组分别按照体质量以 2.0、1.0 和 0.5g/kg 的剂量（按生药计）灌胃；对照组小鼠则按 2.0g/kg 的剂量灌胃给予生理盐水。给药 55min 后，将小鼠置于集成旷场内适应 5min，后开启行为学分析系统观察并记录小鼠在 5min 内的自主活动。并在灌药后的 3h 内观察小鼠的一般行为活动、姿势、步态以及有无肌颤、不安等神经系统方面的异常。

1.5 统计分析

数据用 $\bar{x} \pm S_{\bar{x}}$ 表示，用 SPSS14.0 软件进行方差分析。平均值间比较 $P>0.05$ 时，组间差异不显著。

2 结果与分析

2.1 射干地龙颗粒对麻醉犬心血管系统的影响

试验期间，各组麻醉犬的平均动脉压组间差异不显著（$P>0.05$），心率和体温组间差异亦不显著（$P>0.05$），但是，各组麻醉犬的心率和体温在试验开始后，均出现了降低。通过对各组麻醉犬的心电图描记结果分析，各组麻醉犬的 P 波和 T 波为圆立波，QRS 波群、QT 间期、PR 间期和 ST 间期各组均基本正常。麻醉犬心律稳定，为窦性心律，心电图波形正常，未出现心电异常现象。提示按照 2.0g·kg^{-1}的剂量给药，射干地龙颗粒对犬的心血管系统（Ⅱ导联）无明显影响（表1，表2，表3 和图1，图2，图3，图4）。

表1 射干地龙颗粒对麻醉犬平均动脉压的影响（$n=6$） （mmHg）

组别	0min	40min	50min	70min	90min	120min
对照组	123.00±3.58	155.83±2.43	152.33±14.44	136.60±10.21	141.93±9.40	139.49±10.00
低剂量组	95.57±1.90	133.02±2.95	129.67±19.81	119.00±7.02	122.23±12.51	120.18±8.29
中剂量组	101.23±1.63	96.87±1.35	99.67±3.99	102.53±3.79	97.27±3.20	98.22±4.65
高剂量组	104.26±3.07	105.75±1.08	94.47±5.40	97.67±2.70	103.53±7.74	102.34±3.78

* 同列数据相比，差异均不显著（$P>0.05$）。表2~6同

表2 射干地龙颗粒对麻醉犬心率的影响（$n=6$） （Beats/min）

组别	0min	40min	50min	70min	90min	120min
对照组	68.32±2.65	61.00±1.44	51.40±3.46	45.97±1.92	54.95±10.10	49.32±7.98
低剂量组	57.92±1.99	52.53±2.06	52.50±6.18	50.47±6.30	44.82±3.57	50.09±3.97
中剂量组	66.96±1.04	65.49±1.26	60.68±6.38	54.32±3.19	50.38±3.84	56.87±4.66
高剂量组	59.47±0.77	48.29±1.65	52.04±6.18	42.78±3.91	41.16±3.32	47.52±4.39

表3 射干地龙颗粒对麻醉犬体温的影响（$n=6$） （℃）

组别	0min	40min	50min	70min	90min	120min
对照组	37.04±0.32	37.60±0.26	37.32±0.61	36.76±0.27	35.88±0.64	35.32±0.62
低剂量组	38.00±0.18	38.34±0.20	38.04±0.25	37.40±0.36	36.93±0.58	36.25±0.61
中剂量组	38.62±0.20	37.66±0.41	37.24±0.41	36.50±0.44	35.78±0.44	34.84±0.43
高剂量组	37.90±0.41	36.56±0.32	36.70±0.37	36.56±0.65	36.60±1.36	35.94±0.80

2.2 射干地龙颗粒对麻醉犬呼吸系统的影响

试验期间各观测点，各组间麻醉犬的呼吸频率、潮气量和血氧饱和度差异不显著（$P>0.05$）。呼吸节律曲线各组均基本稳定，除低剂量组 50min 和高剂量组 40min 时麻醉犬的血

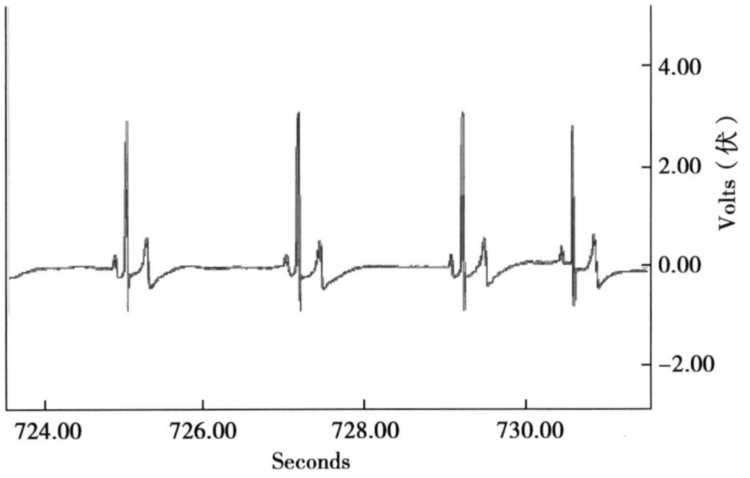

图 1　对照组犬心电图

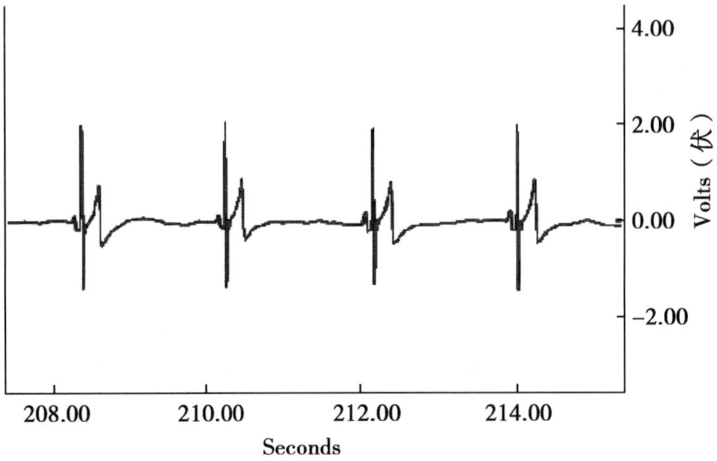

图 2　低剂量组犬心电图

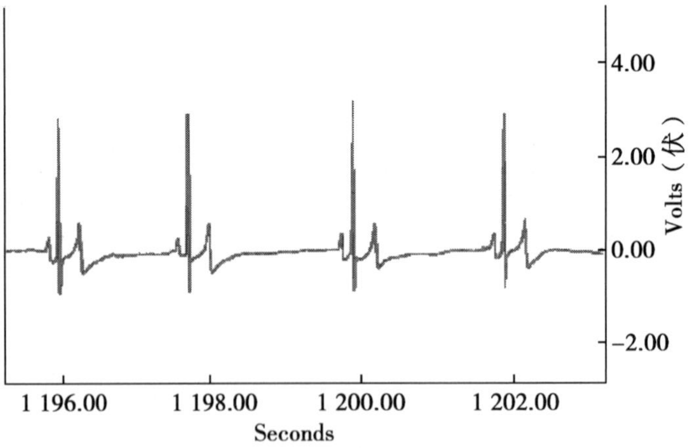

图 3　中剂量组犬心电图

氧饱和度在 90% 以下，其余各组各时间点均在 90% 以上，结合各组间差异不显著的统计结

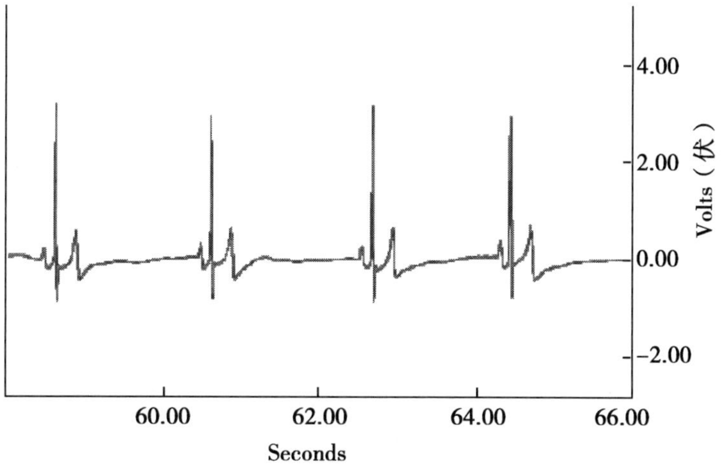

图4　高剂量组犬心电图

果及对照组在0min时血氧饱和度为90.20%的数据，并不能说明中剂量组和高剂量组的药物对麻醉犬的血氧饱和度造成了影响。结果表明，按照2.0g·kg^{-1}的剂量给药，射干地龙颗粒对麻醉犬的呼吸系统无明显影响（表4，表5，表6和图5，图6，图7，图8）。

表4　射干地龙颗粒对麻醉犬呼吸频率的影响（*n*=6）　（Times·min^{-1}）

组别	0min	40min	50min	70min	90min	120min
对照组	15.57±0.44	15.98±0.22	13.18±1.30	10.56±2.07	8.00±1.63	9.64±1.62
低剂量组	13.91±0.56	14.55±0.27	11.34±2.11	11.28±2.13	11.90±2.22	11.88±2.35
中剂量组	12.63±0.64	14.49±0.60	11.52±2.07	10.20±1.96	10.94±2.29	11.02±2.63
高剂量组	13.26±0.66	19.72±0.86	10.86±1.07	10.85±0.74	9.44±0.84	11.25±0.49

表5　射干地龙颗粒对麻醉犬潮气量的影响（*n*=6）　（ml）

组别	0min	40min	50min	70min	90min	120min
对照组	289.98±10.54	311.14±15.62	335.58±29.07	344.17±34.70	352.60±27.02	320.04±27.91
低剂量组	349.26±25.68	399.34±21.84	339.26±10.52	353.82±26.94	379.16±30.03	376.66±25.09
中剂量组	317.11±10.76	346.28±20.56	334.02±31.68	337.57±26.74	344.53±31.72	322.27±26.01
高剂量组	290.09±12.58	292.36±13.22	291.04±11.26	290.24±11.09	290.24±21.09	329.54±36.42

表6　射干地龙颗粒对麻醉犬血氧饱和度的影响　（%）

组别	0min	40min	50min	70min	90min	120min
对照组	90.20±3.64	92.40±3.36	96.40±1.29	96.80±0.86	96.40±0.75	94.60±1.25
低剂量组	91.80±1.98	91.40±2.84	93.60±2.66	96.80±1.11	95.20±1.39	96.40±1.03
中剂量组	92.60±4.92	93.40±3.04	89.40±4.48	94.20±2.11	96.20±0.92	96.80±0.58
高剂量组	92.40±2.68	89.00±3.08	92.20±2.97	93.80±1.20	96.75±0.85	95.80±1.11

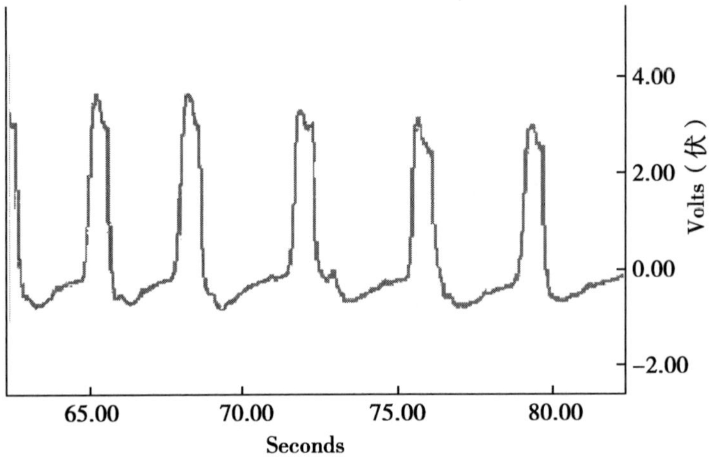

图 5　对照组犬呼吸节律图

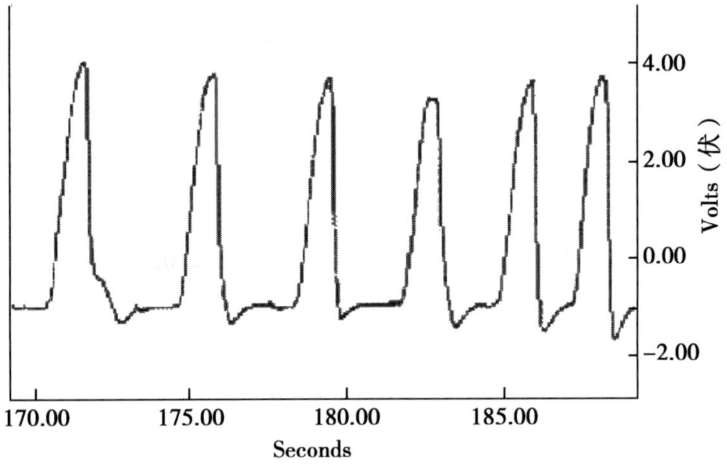

图 6　低剂量组犬呼吸节律图

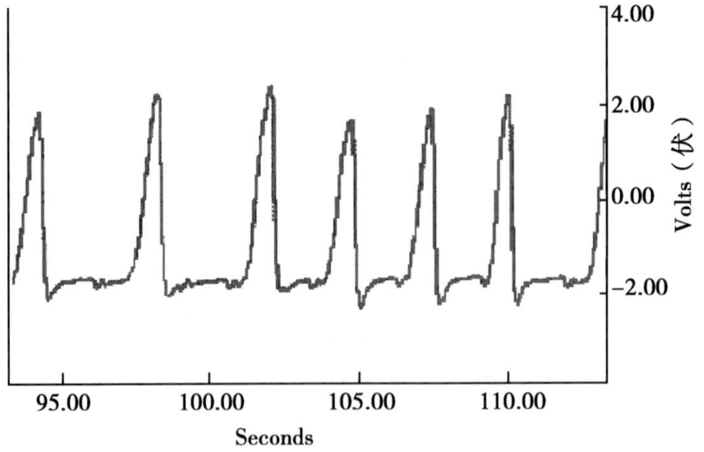

图 7　中剂量组犬呼吸节律图

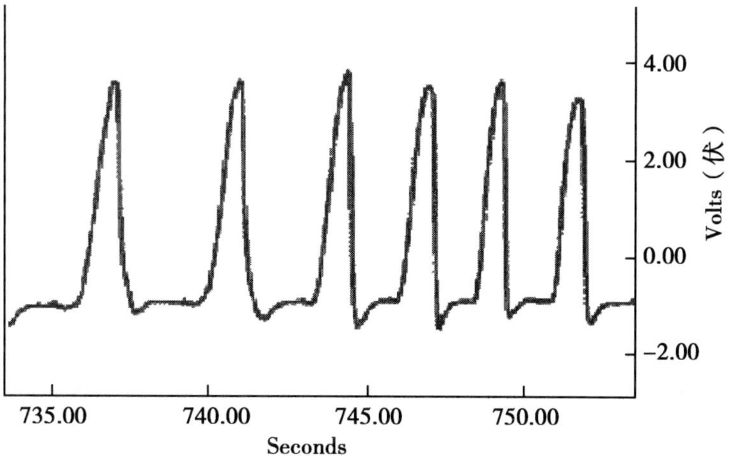

图 8　高剂量组犬呼吸节律图

2.3　射干地龙颗粒对麻醉犬泌尿系统的影响

试验前后各组麻醉犬的尿液白细胞、亚硝酸盐、蛋白质、潜血、酮体、胆红素和葡萄糖均呈现阴性；而尿胆原仅在低剂量组试验后出现可疑阳性，其余均为阴性，结合中剂量药物组和高剂量药物组试验后的阴性结果并不能说明药物对肝脏或肾脏造成了损伤。而各组尿液中的维生素 C 在试验前后均为阳性，组间并无差异。尿液增质量、pH 值和尿液相对密度在试验前后组间差异不显著（$P>0.05$）。综合以上结果，在 $2.0g \cdot kg^{-1}$ 给药剂量范围内，该药对麻醉犬泌尿系统无明显影响（表 7）。

表 7　射干地龙颗粒对麻醉犬泌尿系统参数的影响（$n=6$）

检测项目	对照组	低剂量组	中剂量组	高剂量组	
白细胞	试验前	-	-	-	-
	试验后	-	-	-	-
亚硝酸盐	试验前	-	-	-	-
	试验后	-	-	-	-
尿胆原	试验前	-	-	-	-
	试验后	-	+ -	-	-
蛋白质	试验前	-	-	-	-
	试验后	-	-	-	-
潜血	试验前	-	-	-	-
	试验后	-	-	-	-
维生素 C	试验前	+	+	+	+
	试验后	+	+	+	+
酮体	试验前	-	-	-	-
	试验后	-	-	-	-
胆红素	试验前	-	-	-	-
	试验后	-	-	-	-
葡萄糖	试验前	-	-	-	-
	试验后	-	-	-	-
尿液增质量（g）		2.02 ±0.73	3.90 ±1.57	2.40 ±3.96	3.75 ±1.55

（续表）

检测项目	对照组	低剂量组	中剂量组	高剂量组	
pH	试验前	6.80 ±0.20	7.60 ±0.48	7.70 ±0.54	7.50 ±0.58
	试验后	6.90 ±0.29	6.60 ±0.29	7.10 ±0.51	6.80 ±0.46
尿液相对密度	试验前	1.03 ±0.0012	1.01 ±0.0042	1.01 ±0.0063	1.01 ±0.0071
	试验后	1.03 ±0.0016	1.02 ±0.0045	1.02 ±0.0058	1.22 ±0.20

* 同行数据相比，尿液增质量、pH 和尿液相对密度在试验前后组间差异显著均不显著（$P>0.05$）；“－”表示测定结果为阴性，“＋”表示测定结果为阳性，“＋－”表示可疑阳性

2.4 射干地龙颗粒对小鼠神经系统的影响

结果表明，射干地龙颗粒高、中、低剂量组在给药后 1h 小鼠活动的总路程、平均速度，中央活动路程，中央活动时间，各角、边活动的路程与活动时间，组间比较并无显著差异（$P>0.05$）。同时，在给药后连续观察 3h，各组小鼠的一般行为活动、姿势、步态均表现正常，无肌颤、不安、奔跑、嘶叫、蜷缩竖毛等异常。提示，射干地龙颗粒高、中、低 3 个给药剂量对小鼠的自主活动没有抑制作用（表 8）。

表 8　射干地龙颗粒对小鼠自主活动的影响（$n=10$）

检测项目	对照组	低剂量组	中剂量组	高剂量组
总路程（m）	38.72 ±16.46	31.19 ±12.88	34.77 ±21.46	29.89 ±13.58
平均速度（$m \cdot s^{-1} \times 10^{-2}$）	12.85 ±5.47	10.36 ±4.28	11.55 ±7.13	9.94 ±4.51
中央活动路程（cm）	129.54 ±18.54	150.69 ±40.47	120.37 ±26.68	143.26 ±21.07
中央活动时间（s）	16.56 ±5.47	18.62 ±5.90	21.53 ±11.37	17.81 ±2.87
左上角活动路程（cm）	196.02 ±102.89	148.47 ±42.07	181.41 ±82.71	250.88 ±150.10
左上角活动时间（s）	40.98 ±10.13	55.43 ±9.21	62.73 ±17.95	45.99 ±10.10
右上角活动路程（cm）	99.54 ±23.19	80.54 ±11.12	146.37 ±80.17	101.85 ±25.14
右上角活动时间（s）	37.58 ±7.24	31.11 ±7.46	45.83 ±19.51	36.20 ±6.28
左下角活动路程（cm）	105.84 ±9.76	119.53 ±27.78	82.31 ±16.65	101.26 ±19.98
左下角活动时间（s）	39.56 ±3.36	39.12 ±7.58	25.71 ±5.10	39.28 ±6.75
右下角活动路程（cm）	180.15 ±31.12	118.32 ±13.09	82.96 ±19.11	101.26 ±19.98
右下角活动时间（s）	62.13 ±13.73	39.53 ±5.16	31.39 ±11.48	30.80 ±6.05
左边活动路程（cm）	120.22 ±17.64	119.64 ±14.82	124.08 ±25.72	186.13 ±53.98
左边活动时间（s）	22.46 ±4.44	20.36 ±2.13	40.28 ±14.23	29.97 ±6.96
上边活动路程（cm）	141.13 ±38.14	147.33 ±18.99	102.78 ±21.05	122.12 ±26.20
上边活动时间（s）	34.13 ±6.00	36.75 ±5.20	25.66 ±4.68	34.65 ±26.20
右边活动路程（cm）	121.95 ±13.82	131.49 ±18.80	74.24 ±20.29	102.30 ±16.85
右边活动时间（s）	23.04 ±4.01	28.81 ±5.61	16.16 ±4.45	26.76 ±6.60
下边活动路程（cm）	136.00 ±14.61	128.14 ±16.77	103.16 ±22.21	119.78 ±14.94
下边活动时间（s）	23.04 ±2.68	22.24 ±2.02	21.42 ±6.30	25.67 ±5.03

* 同行数据相比，组间差异均不显著（$P>0.05$）

3 讨论

3.1 射干地龙颗粒的药理作用机制

射干地龙颗粒是以张仲景创制的古方“射干麻黄汤”和“小青龙汤”为基础进行加减组合，并根据鸡传染性支气管炎的发病特点，而开发的新型复方中兽药制剂。该制剂治疗产蛋鸡呼吸型传染性支气管炎的效果显著；能够对抗组胺、乙酰胆碱所致的气管平滑肌收缩作用，从而起到松弛气管平滑肌和宣肺的功效；同时能明显减少咳嗽的次数，并能增强支气管的分泌作用，表现出镇咳、平喘、祛痰、抗过敏的作用。

射干地龙颗粒以射干、地龙和紫菀为主药。其中，射干对炎症早、晚期均有明显抑制作用，同时还具有抗菌和抗病毒的作用。异黄酮类化合物是射干中的主要抗炎成分，包括鸢尾黄素和鸢尾素。研究表明，鸢尾黄素和鸢尾素通过抑制炎性细胞中 COX-2 蛋白的诱导作用，来抑制前列腺素的产生，从而达到抗炎活性[8]；射干对伤寒、副伤寒杆菌、金黄色葡萄球菌、大肠杆菌、甲型、乙型链球菌、肺炎球菌、流感嗜血杆菌均有不同程度的抑制作用；其水煎剂中的野鸢尾苷元对流感病毒、疱疹病毒和腺病毒等有抑 制作用[9]。地龙具有解热镇痛、增强免疫和平喘的作用。地龙液能明显提高巨噬细胞活化率，因此，具有良好的抗炎能力[10]，同时地龙肽在体外均可明显增强小鼠巨噬细胞的吞噬活性[11]，使得机体的非特异性免疫增强；由地龙中分离出的酪氨酸衍生物可作用于体温调节中枢，使散热增加，起到很好的解热作用[12~13]；研究表明，地龙粉剂有明显的镇痛作用，与扑热息痛合用有协同作用[12]；地龙能扩张支气管，缓解咳喘时的支气管痉挛，具有很好的平喘效果[14]。紫菀中含有的紫菀酮、紫菀皂苷、表木栓醇及豆甾醇等具有抗菌、消炎、止咳、祛痰等功效[15~16]；同时紫菀可抑制由组织胺和乙酰胆碱引起的气管收缩[17]。

遍历所查资料表明，射干、地龙与紫菀的毒副作用很小，在大剂量服用时偶或出现。

3.2 射干地龙颗粒对心血管系统的影响

心血管系统属于维持生命的重要系统。按照兽用中药、天然药物安全药理学研究技术指导原则，必须对心血管系统开展临床前安全药理学试验研究，并完成一般观察。本试验中，各组犬的心率和体温在试验开始后均出现了降低。而研究也表明，QFM 麻醉合剂（盐酸塞拉嗪注射液）对犬心率和体温的确有降低作用[18]。结合各组犬在试验中，应用麻醉药的事实，推测各组心率和体温的普遍性下降主要是由麻醉药引起的。有研究表明射干提取物中的草夹竹桃苷[19]，对心脏可产生刺激性作用和毒性作用导致心脏传导阻滞；但在本试验中，心电图未见异常，推测主要是由于射干中草夹竹桃甙的含量较低，未达到可引起心脏传导阻滞的程度。试验过程中，各给药组犬的血压呈不规律变化，组间差异亦不显著，这与先前报道[20]中麻醉犬静脉注射地龙热浸液或乙醇浸出液后，血压降低的结果不一致；可能主要是由于射干地龙颗粒为复方且口服给药导致。

3.3 射干地龙颗粒对呼吸系统的影响

在新药的安全药理学评价中，呼吸系统仅依靠临床观察的方法不足以准确地反映其功能，因此，对呼吸功能指标的检测采用了定量观测的方法。本试验中通过检测麻醉犬的呼吸频率、潮气量、血氧饱和度和呼吸节律曲线来说明射干地龙颗粒对呼吸系统的影响。结果表明，各给药组麻醉犬的呼吸频率在试验 40min（给药 10min）时均出现了升高，推测主要是由于射干地龙颗粒扩张支气管作用所致；而在试验的 50、70、90 和 120min 呈现先降低后升高的变化趋势，与 QFM 麻醉合剂的研究结果基本一致[18]。而试验过程中检测呼吸功能的其他指标组间差异不显著（$P>0.05$），且并无明显的规律性，说明射干地龙颗粒对呼吸功能无明显影响。

3.4 射干地龙颗粒对泌尿系统的影响

药物的原形及代谢产物在排泄过程中会对泌尿系统造成影响。按照兽用中药、天然药物安全药理学研究技术指导原则，可以通过观察受试物引起肾参数的变化来说明药物对泌尿系统功能的影响。试验过程中，仅低剂量组的尿胆原试验后出现了可疑阳性，各组麻醉犬的其他相关指标在试验前后均呈现阴性；而各组犬尿液中的维生素 C 在试验前后均为阳性，组间并无差异，可能是与饲喂的犬粮有关，导致尿液中存在较多的还原剂致使出现了阳性结果。研究表明，地龙中蚯蚓素具有溶血作用，而且紫菀中含皂甙有强烈的溶血作用[21]，但本试验中各组犬的尿潜血在试验前后均为阴性，可能是射干地龙颗粒为复方制剂且应用剂量较小的原因。

3.5 射干地龙颗粒对神经系统的影响

在安全药理学研究中，心血管系统、呼吸系统和中枢神经系统短时间内可因药物作用而危及机体生命功能，作为其中的核心组合试验。对神经系统相关指标的检测主要包括运动功能、协调功能和行为改变等。试验过程中各组小鼠在活动、姿势、步态和自主活动方面差异不显著，说明射干地龙颗粒对神经系统无明显影响。

4 结果

本试验按照兽用中药、天然药物安全药理学研究技术指导原则，首次通过无创的方式测定了药物对麻醉犬心血管系统、呼吸系统和泌尿系统的影响，有效的减少了应激反应和手术对实验动物相关生理指标的干扰，保证了各项生理指标能够更准确的反映实验动物的真实状态；并通过测定小鼠的自主活动，确定了药物对神经系统的影响。试验结果表明，射干地龙颗粒在 $0.5\sim2.0g\cdot kg^{-1}$（根据体质量计算）的给药剂量范围内，对麻醉犬心血管系统、呼吸系统和泌尿系统均无明显影响，对小鼠的中枢神经系统也没有明显的抑制作用。说明此临床指导用量是安全可靠的，至于加大剂量是否影响临床治疗效果还应做进一步的探索。

参考文献

[1] REDFERN WS，WAKEFIELD ID，PRIOR H，*et al*. Safety pharmacology – aprogressive approach [J]. Fundam Clin Pharmacol，2002，16（3）：161–173.

[2] 王玉珠，王海学，王庆利. 新药临床试验前安全药理学研究的发展过程 [J]. 中国临床药理学杂志，2011，27（7）：557–560.

[3] 谢家声，王东升，李锦宇，等. 射干麻黄地龙散治疗鸡传染性支气管炎临床筛选试验 [J]. 中兽医学杂志，2009，增刊：134–137.

[4] 谢家声，罗超应，王贵波，等. 射干麻黄地龙颗粒治疗鸡呼吸型传染性支气管炎效果观察 [J]. 湖北农业科学，2011，50（5）：978–979.

[5] 罗永江，郑继方，辛蕊华，等. 射干麻黄地龙散对小白鼠的急性毒性试验 [J]. 黑龙江畜牧兽医，2011，（16）：106.

[6] 王东升，罗超应，王贵波，等. 射干麻黄地龙散对鸡脏器和血液生理指标的影响 [J]. 中国家禽，2010，32（11）：23–26.

[7] 王贵波，谢家声，郑继方，等．射干麻黄地龙散对鸡肝肾功能及脏器剖检的影响［J］．中国畜牧兽医，2010，37（9）：172－175．
[8] KIM YP，YAMADA M，LIM SS，et al. Inhibition by tectorigenin and tectoridin of prostaglandin E2production and cyclooxygenase－2induction in rat peritoneal macrophages［J］. Biochimica et Biophysica Acta，1999，1438（3）：399－407．
[9] 张明发，沈雅琴．射干药理研究进展［J］．中国执业药师，2010，7（1）：14－18．
[10] 刘英姿．地龙注射液免疫活性的体外研究［J］．安徽农业科学，2009，37（35）：17 514－17 517．
[11] 傅炜昕，董占双，李铁英，等．免疫活性地龙肽的制备及其对小鼠 NK 细胞活性的影响［J］．中国医科大学学报，2007，36（6）：650－652．
[12] 陈斌艳，张蕾，虞礼敏，等．地龙粉针对大鼠、小鼠与兔的解热镇痛作用［J］．上海医科大学学报，1996，13（3）：115－118．
[13] 朱道辰，张芩花，王运吉．地龙抗炎有效部位的分离筛选［J］．大连轻工业学院学报，2004，23（1）：35－37．
[14] 林建海，刘宝裕．中药对致敏性哮喘豚鼠气道的作用［J］．上海医学，1996，19（11）：638．
[15] LI YP，WANG YM. Evaluation of tussilagone：a cardiovascular－respiratory stimulant isolated from Chinese herbal medicine［J］. Gen Pharmacol，1988，19（2）：261－263．
[16] UKIYAM，AKIHISA T，TOKUDA H，*et al*. Constituents of Compositae plants Ⅲ． Anti－tumor promoting effects and cytotoxic activity against human cancer cell lines of triterpene diols and triols from edible Chrysanthemum flowers［J］. Cancer Lett，2002，177（1）：7－12．
[17] 李岩，王丽华．紫菀与甘草对豚鼠气管作用研究［J］．中医药信息，1999，16（4）：47．
[18] 刘焕奇，王洪斌，霍慧君．QFM 麻醉合剂对犬麻醉效果的观察［J］．中国兽医杂志，2005，41（12）：33－35．
[19] 刘延吉，吴波，张阳，等．中药射干毒性成分分析［J］．沈阳农业大学学报，2011，42（4）：491－493．
[20] 徐叔云，彭华民，邢文鑅．广地龙的降压作用和降压机制的探讨［J］．药学学报，1963，10（1）：15－21．
[21] 国家药典委员会．中国药典，Ⅰ部［S］．北京：化学工业出版社，2005：273．

（发表于《畜牧兽医学报》）

我国部分地区奶牛乳房炎源大肠杆菌生物学特性及耐药性分析

徐继英[1]，刘俊林[1]，李先波[2]，霍生东[1]，杨志强[3]

（1. 西北民族大学生命科学与工程学院，兰州 730030；2. 四川农业大学动物医学院，雅安 625000；3. 中国农业科学院兰州畜牧与兽药研究所，兰州 730050）

摘　要：调查奶牛乳房炎源大肠杆菌的某些生物学特性及其耐药状况，以提高药物疗效，减少牛乳中药物的残留。本研究从国内 7 个省、市、自治区部分地区患乳房炎奶牛的乳样中分离纯化与鉴定出 95 株大肠杆菌（*Escherichia coli*），并对大肠杆菌分离株进行 O 血清型鉴定、小鼠（*Mus musculus*）致病性试验以及抗菌药物敏感性分析。研究结果显示，95 株大肠杆菌共鉴定出 37 种血清型，覆盖了 54 株分离株，另有 2 株自凝，39 株未鉴定出型，较常见血清型为 093 和 09；大肠杆菌分离株接种小白鼠剖检可见明显病变；95 株大肠杆菌对 16 种抗菌药物中的 8 种药物耐药率超过 50%，青霉素的耐药率甚至达到 100%，同一菌株最多耐药 14 种，最少耐药 2 种，耐药 6 种以上菌株占到 51.58%。表明，奶牛乳房炎源大肠杆菌分离株血清型比较复杂，且对多种药物产生不同程度的耐药性，存在着严重的多重耐药情况。本研究为奶牛乳房炎疫苗的研制和乳房炎的临床治疗提供理论依据。

关键词：奶牛乳房炎；大肠杆菌；血清型；致病性；耐药性

奶牛乳房炎一直困扰着奶牛业的发展，是造成奶牛业经济损失最严重的常见、多发性疾病之一。乳房炎的严重程度是一个细菌与宿主免疫防御之间相互作用的结果（Burvenich *et al.*，2003）。由于其病原菌十分复杂，因而给防治工作带来很大的困难。迄今为止，人们从奶牛乳室中分离到的微生物以金黄色葡萄球菌（*Staphyloccocus aureus*）、链球菌（*Streptococcus*）和大肠杆菌（*Escherichia coli*）为主，由这三种细菌引起的乳房炎占病例数的 90% 以上，但各地奶牛乳房炎病原菌感染情况也有所差异（卜仕金等，1999. 兽药与饲料添加剂，4（3）：14～16）。

近年来由于国内外以菌苗预防为主的综合防治措施在生产中不断推广应用，当接触性乳房炎被成功控制的同时，大肠杆菌引起的环境性乳房炎在许多国家和地区有增多的趋势（Burvenich *et al.*，2003；Green *et al.*，2005；Rangel and Marin，2009）；在印度，大肠杆菌是引起奶牛乳房炎的第二大最常见的病原菌，仅次于金黄色葡萄球菌（Sumathi *et al.*，2008）；在中国哈尔滨市 3 个大型奶牛场和南京的规模化养殖场，大肠杆菌与金黄色葡萄球菌同为引起奶牛乳房炎最常见的病原菌（王娜和高学军，2011；金兰梅等，2010. 黑龙江畜牧兽医，42（21）：101～103）。大肠杆菌性奶牛乳房炎的高发生率可能源于糟糕的环境卫生或是乳房炎控制时抗菌药物的过度使用（Green *et al.*，2005）。

目前，抗菌药物是治疗奶牛乳房炎的常用药，但由于抗菌药物的大量使用和滥用，易导致病原菌产生抗药性，从而使得主要病原菌对抗菌药物的敏感性可能会发生变化。因此，定期对主要病原菌进行相应药物的敏感性试验，有针对性的选择治疗药物，可有效地提高药物的疗效，减少牛乳中抗菌药物的残留。为此，本实验对中国部分地区奶牛乳房炎乳样进行大肠杆菌的分离、血清型鉴定及药物敏感性分析，以期为疫苗研制、筛选敏感药物预防和临床治疗奶牛乳房炎提供依据。

1 结果与分析

1.1 分离菌生长培养与形态学特征

从530份乳样中分离出95株大肠杆菌，分离率为17.9%。分离菌在各种培养基上的生长培养特性、菌落特征、菌体形态及染色特性均符合大肠杆菌的特征。

1.2 生化鉴定

经三糖铁琼脂斜面生长、葡萄糖、乳糖、麦芽糖、甘露醇、蔗糖、侧金盏花醇、吲哚、M-R、V-P、枸橼酸盐利用等生化试验鉴定，95株分离菌的24项生化反应结果均符合大肠杆菌的生化特性（表1）。

表1　分离菌株生化鉴定结果

项目	阳性菌（占总菌数的%）	项目	阳性菌（占总菌数的%）
葡萄糖	95（100）	氧化酶	0（0）
乳糖	95（100）	接触酶	95（100）
麦芽糖	95（100）	产硫化氢	5（5.26）
甘露醇	95（100）	明胶液化	0（0）
蔗糖	33（34.73）	尿素分解	8（8.42）
侧金盏花醇	8（8.42）	V-P	0（0）
卫茅醇	50（52.63）	M-R	90（94.73）
肌醇	0（0）	水解七叶苷	0（0）
山梨醇	81（85.26）	硝酸盐还原	95（100）
阿拉伯糖	74（77.89）	吲哚试验	90（94.73）
棉子糖	74（77.89）	动力	33（34.73）
水杨苷	8（8.42）	枸橼酸盐利用	0（0）

1.3 血清型定型

O血清型鉴定结果显示，95个大肠杆菌分离株共计鉴定出37种血清型，覆盖了54株分离株，另有2株自凝，39株未鉴定出型，分别占送检菌株的56.85%、2.1%和41.05%，见表2。54个定型菌株中，较常见血清型为O93、O9、O146、O7和O74，各有8、6、5、3和3株，占已定型菌株的比率依次为14.81%、11.11%、9.26%、5.55%和5.55%。

本实验中，有些菌株能与多种单因子血清发生凝集反应，表现出多种血清型。云南有一株菌检出7种混合血清型，北京有一株菌检出4种混合血清型，内蒙古、北京各有2株菌，甘肃有3株菌都检出2种混合血清型，但在同一地区，不同分离株携带的多种血清型并没有完全或部分相同。

1.4 致病性试验

实验 95 个大肠杆菌分离株对昆明系小鼠（*Mus musculus*）均有致病性，高度致病者占 49.5%，中度致病者占 29.5%，低致病者占 21%（恽时峰等，1997）（表 2）。接种小鼠主要表现精神沉郁，饮食欲下降或废绝，高致病性分离株引起接种小鼠死亡高峰在 12～36h，从死亡小鼠或产生严重病变小鼠肝、脾、心血均能分离到接种菌，对照组小鼠全部健活。

表 2 95 株大肠杆菌血清型鉴定与致病性试验结果

地区	大肠杆菌株数	定型菌株数	未定型菌株数	自凝菌株数	致病性（高/中/低/无）
内蒙古	39	11	28	0	18/12/9/0
北京	19	15	4	0	9/6/4/0
甘肃	13	10	3	0	7/4/2/0
贵州	9	6	2	1	6/1/2/0
云南	6	6	0	0	4/1/1/0
重庆	5	4	0	1	2/2/1/0
四川	4	2	2	0	1/2/1/0
合计	95	54	39	2	47/28/20/0

* 致病性是指分离菌接种昆明系（KM）小鼠后 72h 内死亡情况。判定标准为 3 只全部死亡为高毒力，致死 2 只为中等毒力，致死 1 只为低毒力，不致死为无毒菌株（恽时峰等，1997）

1.5 药物敏感性试验

95 株分离大肠杆菌对 16 种抗菌药物的敏感性试验结果见表 3。16 种抗菌药物中，有 8 种药物的耐药率超过 50%，其中，克林霉素、青霉素、羧苄青霉素和红霉素的耐药率超过 80%，青霉素的耐药率甚至达到 100%；对诺氟沙星、氧氟沙星、头孢哌酮、氯霉素和头孢西丁等 5 种药物的高度敏感率均超过 70%，尤其以头孢西丁为最高，高度敏感率达到 94.74%。

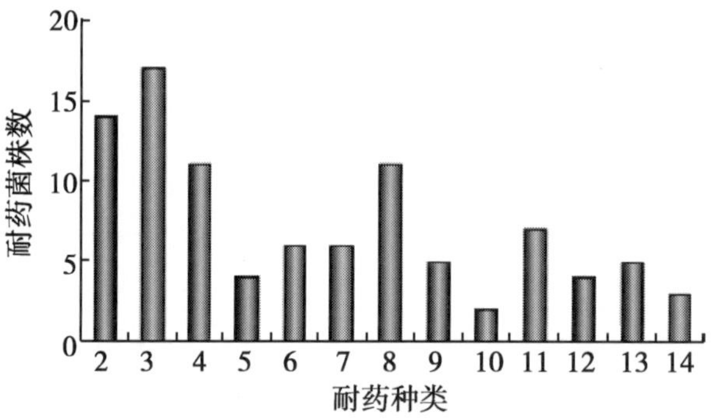

图 大肠杆菌分离株对 16 种常用抗菌药物的多重耐药性

此外，大肠杆菌分离株存在严重的多重耐药情况（图）。95 株大肠杆菌中，无一株菌对 16 种药物均有敏感性或仅对一种药物有耐药性，最多耐药 14 种，最少耐药 2 种，耐药 6 种以上菌株占到 51.58%。同时对氟喹诺酮类、青霉素类、头孢类、磺胺类和四环素类药物都产生耐药性的菌株也存在。

表 3　大肠杆菌分离株对 16 种常用抗菌药物的敏感性测定

抗菌药物	药敏性（株）			药率（%）	高敏率（%）
	耐药	中度敏感	高度敏感		
卡那霉素	11	37	47	11.58	49.47
诺氟沙星	21	0	74	22.11	77.89
氧氟沙星	21	0	74	22.11	77.89
克林霉素	84	11	0	88.42	0
阿莫西林	59	11	25	62.11	26.32
头孢唑啉	60	16	19	63.16	9.47
头孢哌酮	21	0	74	22.11	77.89
四环素	59	0	36	62.11	37.9
青霉素	95	0	0	100	0
羧苄青霉素	84	11	0	88.42	0
红霉素	90	5	0	94.74	0
复方新诺明	53	0	42	55.79	44.21
氯霉素	26	0	69	27.37	72.63
庆大霉素	37	4	54	38.95	56.84
呋喃妥因	26	32	37	27.37	38.95
头孢西丁	0	5	90	0	94.74

2 讨论

本实验取自国内 7 个省、市、自治区 530 份临床型和隐性奶牛乳房炎乳样中，分离出 95 株大肠杆菌，分离率为 17.9%。此结果虽然高于之前吉林地区大肠杆菌分离率 12.50%（金兰梅等，2011）、内蒙古地区大肠杆菌分离率 12.9%（邓海平等，2009. 安徽农业科学，37（2）：595～598）、无锡地区大肠杆菌分离率 15.79%（郭海军和周斌，2011）的结果，而低于哈尔滨地区 19.6% 的大肠杆菌分离率（王娜和高学军，2011），这种分离率的地域差异性可能与大肠杆菌为环境性致病菌有关，但也显示出奶牛乳房炎源大肠杆菌分离率近年来有着明显升高的趋势，因此，对于大肠杆菌引起的奶牛乳房炎应给予足够的重视。

实验结果显示，内蒙古地区奶牛乳房炎源大肠杆菌分离株最多，但未定型菌株所占比例也最高（约 71.8%），可能与奶样多来源于农户因而分散性较强有关。已定型 54 株大肠杆菌的较常见血清型为 093、09、0146、07 和 074，其中 093 型分布率最高，占 14.81%，这与王桂琴等（2006）报道内蒙古呼和浩特地区奶牛乳房炎源大肠杆菌 093 型分布率为 11.32% 的结果基本一致，但其余较常见血清型则不同，与禽源大肠杆菌分离株的优势血清型差异更大（高崧等，1999）。该结果除血清型 0146 与美国报道一致外，其余较常见血清型与美国（Wenz *et al.*，2006）、印度（Kausar *et al.*，2009）、巴西（Bueris *et al.*，2002）等国的报道也完全不同，可能与大肠杆菌作为环境性致病菌的地域差异性有很大关系。此外，同一株分离菌株能与多种单因子血清发生凝集反应而表现出多种血清型，可能与大肠杆菌表面存在多种 O 抗原有关，王桂琴等（2006）、恽时峰等（1997）也有相同报道，同一分离菌株表现出多种血清型对于临床上研制有效的奶牛乳房炎大肠杆菌疫苗具有重要的意义。

实验中所有大肠杆菌分离株对昆明系小白鼠均有致病性，但大肠杆菌分离株的致病性存在明显差异，所有致死小白鼠体内均能分离到大量大肠杆菌，而处死小白鼠体内只能分离到少量大肠杆菌。受试分离株来自国内7个省、市、自治区，覆盖了37个血清型，但不同地区大肠杆菌分离株的致病性并无明显差异，这表明分离株的致病性不受地区、血清型的影响，也显示出我国的奶牛源大肠杆菌病的复杂性，多种致病血清型的存在，给奶牛乳房炎的防制工作增加了难度。

目前，国内外针对奶牛乳房炎的治疗仍然以抗生素为主，抗菌药物的广泛应用，在消除敏感菌的同时，耐药菌株也得以生存繁殖下来，最终导致牛奶中抗生素残留超标及耐药菌株不断增加。这在危害人类健康的同时也给奶牛乳房炎的治疗带来了新的难题。本实验通过国内7个省、市、自治区部分地区奶牛乳房炎源大肠杆菌对16种常用抗菌药物的敏感性测定表明，目前临床上使用的多种抗生素对奶牛乳房炎源大肠杆菌均产生了不同程度的耐药性，尤其是克林霉素、羧苄青霉素、红霉素和青霉素4种药物的耐药率非常高，而对诺氟沙星、氧氟沙星、头孢哌酮、氯霉素和头孢西丁等5种药物仍保持着较高的敏感性。这与倪春霞等（2010）报道内蒙古、甘肃、四川地区乳样中分离大肠杆菌对氟哌酸、左氟沙星、氧氟沙星和氯霉素高度敏感，对青霉素、红霉素、复方新诺明等有广泛耐药性；王正兵等（2011. 贵州农业科学，39（4）：139~141）报道对奶牛乳房炎大肠杆菌产生耐药性的药物主要有青霉素类（青霉素G、氨苄青霉素、羧苄青霉素）、链霉素、四环素、红霉素、林可霉素和复方新诺明等；Sumathi 等（2008）报道大多数乳样中分离大肠杆菌对氨苄青霉素、多粘菌素E、新霉素、呋喃唑酮耐药性较高的实验结果基本一致。

此外，实验发现，许多大肠杆菌分离株存在严重的多重耐药情况，最多耐药14种，最少耐药2种，耐药6种以上菌株超过半数，同时对氟喹诺酮类、青霉素类、头孢类、磺胺类、四环素类药物都产生耐药性的菌株也存在，这种现象可能与奶牛乳房炎大肠杆菌对长期、大量、频繁使用的老抗生素的耐药性都较高（王桂琴等，2006），大肠杆菌多重耐药菌株的出现或许是对抗生素滥用所造成的选择性压力的一种应答表现（Bueris *et al.*，2002）有关。本实验乳样采自国内7个省、市、自治区部分地区，基本上能够反映该地区目前奶牛乳房炎源大肠杆菌对常用抗菌药物的耐药状况，而对多种药物不敏感以及如此高的多重耐药率应引起兽医工作者的高度重视。

据研究报道，许多编码抗生素耐药性的基因与某些毒力因子一样，也都位于细菌质粒上，而细菌耐药性可以通过质粒水平传播到另一株细菌，使得这些耐药菌株有可能在动物和人类之间互相传播（Martínez and Baquero，2002；Werckenthin *et al.*，2001）。应用于动物的抗菌药物其抗药性的发展可能会给人类带来危害，耐药菌能够引起人的疾病，并且能够经由食物进行传递（McKellar，1998），这将对人类的饮食健康和食品卫生安全带来很大的威胁。本研究可为中国奶牛乳房炎的预防和治疗提供一定的理论参考，对临床上研制有效的奶牛乳房炎大肠杆菌疫苗具有重要的指导意义，对保障人类食品安全和公共卫生安全也具有积极的意义。

3 材料与方法

3.1 材料

3.1.1 奶样来源

无菌采集自北京、内蒙古、甘肃、四川、重庆、云南、贵州等7个省、自治区、直辖市部分地区（2008—2009）临床型和隐性奶牛乳房炎奶样，共计530份。

3.1.2 培养基与试剂

麦康凯（Macconkcy）琼脂、伊红美蓝（Eosin Methylene Blue）琼脂、水解酪蛋白（Mueller - Hinton）培养基、普通肉汤、普通琼脂购自杭州天和微生物试剂有限公司；各种微量生化鉴定管、甲基红试剂，V - P试剂、吲哚试剂、硝酸盐还原试剂、氧化酶试剂均购自兰州麦迪生物试剂有限公司；大肠杆菌（Escherichia coli）O抗原定型多价血清、单因子血清由中国兽药监察所提供；药敏纸片购自杭州天和微生物试剂有限公司。

3.1.3 标准菌株

大肠杆菌O抗原阳性标准菌株由中国兽药监察所（CIVDC）提供；药敏试验标准质控大肠杆菌株ATCC25922购自中国食品药品检定研究院（NIFDC）。

3.1.4 小鼠

18 ~ 22g昆明系（KM）小鼠（Mus *musculus*）295只，雌雄不限，购自兰州大学实验动物中心。

3.2 方法

3.2.1 细菌分离培养

将奶样接种普通肉汤，37℃增菌培养24h后用铂耳圈挑取培养物划线接种于麦康凯琼脂平板，置于37℃温箱培养24h后，然后挑取麦康凯琼脂上玫瑰红色单个菌落，划线接种于伊红美蓝琼脂平板，37℃培养24h，再挑取伊红美蓝平板上黑紫色，对光观察具有金属光泽的单个菌落，接种普通肉汤及普通琼脂斜面，37℃培养18 ~ 20h，观察生长状况及菌落形态。同时挑取单个纯培养菌落涂片镜检，观察菌体形态及染色特性，普通琼脂斜面纯培养则以胶塞密封，4℃保存备用。

3.2.2 生化鉴定

分离的细菌按文献（甘肃农业大学主编，1980）的方法进行24项生化鉴定试验。

3.2.3 血清型鉴定

按Oie等（2002）和于学辉等（2008）的方法对分离菌中的95株大肠杆菌进行O抗原鉴定，定型标准为血清的试管凝集价不小于1：640。

3.2.4 致病性试验

经37℃ 24h培养的大肠杆菌分离株普通肉汤培养物，分别腹腔接种3只18 ~ 22g普通级昆明系（KM）小鼠，接种量0.2ml/只（约含109 CFU）（Iwahi *et al.*, 1982；恽时峰等，1997）；对照组10只小鼠，接种普通肉汤0.2ml/只。接种后观察72h，记录小鼠死亡情况，72h后全部扑杀，观察心、肝、脾、肺和肾脏病变，并从其心血、肝、脾分离接种菌。

3.2.5 药物敏感性测定

按照WHO推荐的纸片扩散法（K - B法），参考文献（李新圃等，2001）测定大肠杆菌分离株对16种常用抗菌药物的敏感性，根据卫生部抗菌药物耐药性检测中心颁布的《抗菌药物药敏试验判断标准》判定结果。16 种抗菌药物为青霉素（PG）、头孢哌酮（CFP）、氯霉素（CMP）、羧苄青霉素（CB）、头孢西丁（CFX）、丁胺卡那霉素（AMK）、诺氟沙星（NOR）、阿莫西林（AMO）、头孢唑啉（CEZ）、庆大霉素（GEN）、四环素（TET）、红霉素（ERY）、克林霉素（CL）、复方新诺明（SXT）、氧氟沙星（OFL）和呋喃妥因（FT）。

参考文献

[1] Bueris M V, Correa G P, Marin J M. Antimicrobial susceptibility of Escherichia coli isolated from mastitic bovine milk. Ars Veterinaria, Jaboticabal, SP, 2002, 18 (2): 125 ~ 129.

[2] Burvenich C, van Merris V, Mehrzad J, Diez – Fraile A and Duchateau L. Severity of E. coli mastitis is mainly determined by cow factors. Veterinaria Research, 2003, 34: 521 ~ 564.

[3] Gansu Agricultural University, ed. Veterinary Microbiology Experiment Guidance. Agriculture Press, Beijing, China, pp: 212 ~ 217（甘肃农业大学主编，1980. 兽医微生物学实验指导. 农业出版社，中国，北京，1980，pp: 212 ~ 217）

[4] Gao S, Liu X F, Zhang R K, Jiao X A, Wen Q Y, Wu C C, Tang Y M, Zhu X B, Li Z, Chen J, Cui L B and Cui H P. The isolation and identification of pathogeny Escherichia coli of chicken origion from some regions in China. Chinese Journal of Animal and Veterinary Sciences, 1999, 30 (2): 164 ~ 171（高崧，刘秀梵，张如宽，焦新安，文其乙，吴长新，唐一鸣，朱晓波，李琮，陈娟，崔力兵，崔洪平，1999. 我国部分地区禽病原性大肠杆菌的分离与鉴定. 畜牧兽医学报，30 (2): 164 ~ 171）

[5] Green M J, Green L E, Bradley A J, Schukken Y H and Medley G F. Prevalence and associations between bacterial isolates from dry mammary glands of dairy cows. Veterinaria Record, 2005, 156 (10): 71 ~ 77.

[6] Guo H J, Zhou B. Isolation and identification on pathogens from cow mastitis and drug resistance of common pathogenic bacteria. Shanghai Journal of Animal Husbandry and Veterinary Medicine, 2011, 3: 12 ~ 14（郭海军，周斌，2011. 奶牛乳房炎病原菌的分离鉴定与主要致病菌耐药性分析. 上海畜牧兽医通讯，3: 12 ~ 14）

[7] Iwahi T, Abe Y, Tsuchiy K. Virulence of Escherichia coli in ascending urinary – tract infection in mice. Journal Medicine Microbiology, 1982, 15: 303 ~ 316.

[8] Kausar Y, Chunchanur S K, Nadagir S D, Halesh L H, Chandrasekhar M R. Virulence factors, serotypes and antimicrobial suspectibility pattern of Escherichia coli in urinary tract infections. Al Ameen Journal Medicine Science, 2009, 2 (1): 47 ~ 51.

[9] Li X P, Yu J, Li H S, Luo J Y. Susceptibility test of major pathogenic bacterium from bovine mastitis in individual cow farm. Chinese Journal of Veterinary Science and Technology, 2001, 31 (11): 41 ~ 43（李新圃，郁杰，李宏胜，罗金印，2001. 个体奶牛场乳牛乳腺炎主要病原菌及药敏试验. 中国兽医科技，31 (11): 41 ~ 43）

[10] Martínez J, Baquero F. Interactions among strategies associated with bacterial infection: Pathogenicity, epidemicity, and antibiotic resistance. Clinic Microbiology Review, 2002, 15 (4): 647 ~ 679.

[11] McKellar Q A. Antimicrobial resistance: A veterinary perspective – antimicrobials are important for animal welfare but need to be used prudently. British Medical Journal, 1998, 317: 610 ~ 611.

[12] Ni CX, Pu WX, Hu YH, Deng HP, Wang L, Meng XQ. Isolation, identification and drug sensitive test of pathogenic bactcria causing dairy cattle mastitis. Acta Agriculturae Boreali – occidentalis Sinica, 2010, 19 (2): 20 ~ 24（倪春霞，蒲万霞，胡永浩，邓海平，王玲，孟晓琴. 奶牛乳房炎病原菌的分离鉴定及耐药性分析. 西北农业学报，2010，19 (2): 20 ~ 24）

[13] Oie S, Kawakami M, Kamiya A, Tomita M. In vitro susceptibility of four serotypes of enterohaemorrhagic Escherichia coli to antimicrobial agents. Biological and Pharmaceutical Bulletin, 2002, 25 (5): 671 ~ 673.

[14] Rangel P M, Marin J M. Antimicrobial resistance in Brazilian isolates of Shiga toxin – encoding Escherichia coli from cows with mastitis. Ars Veterinaria, Jaboticabal, SP, 2009, 25 (1): 18 ~ 23.

[15] Sumathi B R, Amitha R G, Krishnappa G. Antibiogram profile based dendrogram analysis of Escherichia coli serotypes isolated from bovine mastitis. VeterinaryWorld, 2008, 1 (2): 37 ~ 39.

[16] Wang G Q, Wu C M, Sun L H, Chen X, Shen J Z. Studies on serotype drug resistance of Escherichia coli from cow mastitis. Chinese Journal of Veterinary Medicine, 2006, 42 (12): 19 ~ 20 (王桂琴, 吴聪明, 宋丽华, 陈霞, 沈建忠. 奶牛乳房炎大肠杆菌的血清型及耐药性调查研究. 中国兽医杂志, 2006, 42 (12): 19 ~ 20)

[17] Wang N, Gao X J. Isolation and identification of pathogenic bacteria of bovine subclinical mastitis in Harbin. Journal of Northeast Agricultural University, 2011, 42 (2): 29 ~ 32 (王娜, 高学军. 哈尔滨地区奶牛隐性乳房炎病原菌的分离鉴定. 东北农业大学学报, 2011, 42 (2): 29 ~ 32)

[18] Wenz J R, Barrington G M, Garry F B, Ellis R P, Magnuson R J. Escherichia coli isolates' serotypes, genotypes, and virulence genes and clinical coliform mastitis severity. Journal Dairy Science, 2006, 89: 3 408 ~ 3 412.

[19] Werckenthin C, Cardoso M, Martel J, Schwarz S. Antimicrobial resistance in Staphylococci from animals with particular reference to bovine Staphylococcus aureus, porcine Staphylococcus hyicus, and canine Staphylococcus intermedius. Veterinaria Research, 2001, 32: 341 ~ 362.

[20] Yu X H, Cheng A C, Wang M S, Wang Y, Wang Y W, Tang C. Serotype identification and virulence – associated genes analysis of pathogenic E. coli isolated from ducklings. Acta Veterinaria et Zootechnica Sinica, 2008, 39 (1): 53 ~ 59 (于学辉, 程安春, 汪铭书, 王英, 王远微, 汤承. 鸭源致病性大肠杆菌的血清型鉴定及其相关毒力基因分析. 畜牧兽医学报, 2008, 39 (1): 53 ~ 59)

[21] Yun S F, Lan Z R, Wang WW, Zheng M Q, Cai B X. Characterization of avian Escherichia coli in Jiangsu. Acta Agriculture Shanghai, 1997, 13 (4): 7 ~ 10 (恽时峰, 兰邹然, 王伟武, 郑明球, 蔡宝祥. 江苏境内鸡源大肠埃希氏菌的某些生物学特性. 上海农业学报, 1997, 13 (4): 7 ~ 10)

(发表于《农业生物技术学报》)

Ⅰa型和Ⅱ型牛源无乳链球菌 *sip* 基因的遗传进化分析

王旭荣，张世栋，杨　峰，王国庆，杨志强，李建喜，李宏胜

（中国农业科学院兰州畜牧与兽药研究所，甘肃省中兽药工程技术研究中心，中国农业科学院临床兽医学研究中心，农业部兽用药物创制重点实验室，兰州 730050）

摘　要：为分析 8 株Ⅰa 型和 10 株Ⅱ型牛源无乳链球菌的 *sip* 基因特征和遗传进化关系，采用 PCR 方法扩增目的基因并克隆入 pGEM－T Easy 载体后测序，进行了同源性分析和构建了遗传进化树。结果显示，18 株菌株的 *sip* 基因全长均为 1 305bp，无基因缺失。8 株Ⅰa 型菌株之间的 *sip* 基因核苷酸同源性和推导的氨基酸同源性分别为 99．8%～100%和 99．5%～100%；而 10 株Ⅱ型菌株之间的 *sip* 基因核苷酸同源性和推导的氨基酸同源性均为 100%。18 株菌株与其他不同来源、不同血清型参考菌株相应序列的核苷酸同源性和氨基酸同源性分别为 97．8%～100%和 96．8%～99．8%。*sip* 基因的遗传进化树分析显示 18 株菌株处于 2 个不同的大分支上，其中Ⅱ型菌株处于同一分支上，与中国菌株 Mf32 和美国菌株 GB01068、GB01089 的亲缘关系最近，而Ⅰa 型菌株处于另一个分支上，与中国菌株 Ly2 和美国菌株 GB00549 的亲缘关系最近。说明 8 株Ⅰa 型菌株和 10 株Ⅱ型菌株分别来源于不同的菌株，但它们的 *sip* 基因同源性很高，而且与参考菌株的 *sip* 基因相比，目前仍属于保守基因。

关键词：牛；无乳链球菌；*sip* 基因；Ⅰa 型；Ⅱ型；遗传进化

（发表于《中国兽医科学》）

Ⅰa 型牛源无乳链球菌 M7 菌株 *sip* 基因的分子特征分析

王旭荣，张世栋，杨　峰，杨志强，李宏胜，李建喜

（中国农业科学院兰州畜牧与兽药研究所，甘肃省中兽药工程技术研究中心，农业部兽用药物创制重点实验室，中国农业科学院临床兽医学研究中心，兰州 730050）

摘　要：通过 PCR 方法扩增Ⅰa 型牛源无乳链球菌地方菌株 M7 的 *sip* 基因，将目的片段克隆入 pGEM－T Easy 载体并进行测序，采用多种生物软件对 *sip* 基因及其表达的蛋白质进行分子特征分析。试验结果表明，M7 菌株的 *sip* 基因为 1 305bp，未出现基因缺失；与 GenBank 中发表的不同血清型的无乳链球菌菌株的相应核苷酸序列同源性为 98.0%～100.0%，推导的氨基酸同源性为 97.2%～99.8%。M7 的 *sip* 基因与中国菌株 Ly2（FJ808732）和美国菌株 GB00549（FJ752159）的相应基因的核苷酸同源性和氨基酸同源性最高，核苷酸同源性均为 100.0%，氨基酸同源性均为 99.8%。该 *sip* 基因表达的蛋白质是一种稳定的分泌性外膜蛋白，疏水性强；其 N－末端的第 52－95 位氨基酸残基之间含有 1 个 LysM 超家族的保守结构域；存在 1～25 位氨基酸的信号肽，剪切位点在第 25～26 位氨基酸之间；存在多个 B 细胞和 T 细胞表位。说明 M7 菌株的 *sip* 基因是未缺失 LysM 超家族结构域的比较保守的免疫蛋白。

关键词：牛；无乳链球菌；*sip* 基因；Ⅰa 型

（发表于《中国畜牧兽医》）

二喹噁啉羟酸全抗原偶联比 HPLC 测定方法的建立

张景艳，杨志强，李建喜，张　凯，王　磊，王学智，孟嘉仁

（中国农业科学院兰州畜牧与兽药研究所，农业部新兽药工程重点实验室，
甘肃省中兽药工程技术研究中心，兰州 730050）

摘　要：为建立一种准确、可靠的测定人工抗原偶联比方法，本试验应用高效液相色谱法方法测定全抗原二喹噁啉羟酸－牛血清白蛋白（QCA－BSA）合成中游离二喹噁啉羟酸的含量。采用 Waters 2695 分离系统、2489 紫外检测系统；流动相：甲醇：水(60：40)；柱温：30℃；流速：1ml/min；检测波长 320nm。试验结果表明，游离二喹噁啉羟酸的线性范围为 0.2～128μg/ml，线性方程为 y＝17526x＋4648，R^2＝1，RSD 为 0.28%，全抗原 QCA－BSAⅠ、Ⅱ、Ⅲ的偶联比分别为 1.76：1.28.60：1、65.63：1。建立的方法准确、可靠，适用于喹噁啉类人工抗原偶联比的测定。

关键词：二喹噁啉羟酸；高效液相色谱法；偶联比

（发表于《中国畜牧兽医》）

防制胚泡着床障碍中药的筛选及药效研究

王东升，张世栋，荔　霞，董书伟，李世宏，严作廷

（中国农业科学院兰州畜牧与兽药研究所，农业部兽用药物创制重点实验室，
甘肃省中兽药工程技术研究中心，兰州 730050）

摘　要：为了筛选防制胚泡着床障碍中药，本试验利用皮下注射米非司酮所致的小鼠胚泡着床障碍模型，以妊娠率、着床胚泡数、雌激素和孕激素及其比值为指标，将丹参、黄芪、川芎、黄芩、白术、菟丝子、桑寄生、续断、党参、当归等中药组成复方进行筛选和药效评价。筛选试验结果表明，方 3 组妊娠率显著高于方 1 组、菟丝子组和续断组（$P<0.05$）也高于方 2 组和对照组，说明方 3 有利于小鼠胚泡的着床。方 3 的药效试验表明，保胎组妊娠率极显著高于对照组和高剂量组（$P<0.01$）显著高于低剂量组（$P<0.05$）。保胎组孕酮水平、孕酮和雌二醇比值极显著高于高、低剂量组、对照组和正常组（$P<0.01$）孕酮和雌二醇比值显著高于中剂量组（$P<0.05$），而保胎组、低剂量组和正常组雌二醇水平显著低于高剂量组和对照组（$P<0.05$），中剂量组孕酮水平显著高于对照组（$P<0.05$）。这表明方 3 能提高小鼠的孕酮水平及孕酮和雌二醇比值，有利于小鼠胚泡着床。

关键词：胚泡着床障碍；中药；雌激素；孕激素

（发表于《中国畜牧兽医》）

利用 iCODEHOP 设计简并引物克隆益生菌 FGM 通透酶基因片段

王　龙，张　凯，王旭荣，张景艳，孟嘉仁，郝桂娟，杨志强，李建喜

（中国农业科学院兰州畜牧与兽药研究所，甘肃省中兽药工程技术研究中心，兰州 730050）

摘　要：通过 iCODEHOP 在线设计细菌通透酶的简并引物，以发酵黄芪菌 FGM 基因组 DNA 为模板进行 Touchdown PCR 扩增，得到 740 bp PCR 产物，将产物经 pGEM – T Easy 载体连接，转化至 JM109 中，筛选阳性株并测序。序列通过 BLAST x 检索与 GenBank 进行同源性比对后，结果表明，此 DNA 产物序列与其他菌属来源的通透酶蛋白序列具有相似性，所克隆的序列即为 FGM 通透酶基因片段。用 iCODEHOP 在线设计的简并引物可信性强，阳性率高。FGM 通透酶基因的成功克隆为细菌发酵黄芪机理研究提供了依据。

关键词：益生菌；iCODEHOP；通透酶

（发表于《中国畜牧兽医》）

全抗原 MQCA – BSA 耦联比 HPLC 测定方法的建立

张景艳，杨志强，李建喜，王学智，张　凯，孟家仁，王　磊

（中国农业科学院兰州畜牧与兽药研究所，甘肃省新兽药工程重点实验室，甘肃省中兽药工程技术研究中心，兰州 730050）

摘　要：为了建立一种准确、可靠的测定人工抗原耦联比的方法，试验应用高效液相色谱（HPLC）法采用 Waters 2695 分离系统、2489 紫外检测系统，流动相为甲醇、水之比为 60：40，柱温为 30℃，流速为 1.00ml/min，检测波长为 320nm，测定全抗原合成中游离 3 – 甲基喹啉 – 2 – 羧酸（MQCA）的含量。结果表明：游离 MQCA 的线性范围为 0.2 ~ 128.0μg/ml，线性方程为 A = 17 100C + 16 667（R^2 = 0.9995）。说明本方法准确、可靠，适用于人工抗原偶联比的测定。

关键词：3 – 甲基喹啉 – 2 – 羧酸（MQCA）；高效液相色谱（HPLC）法；耦联比

（发表于《黑龙江畜牧兽医》）

体液防御在奶牛乳腺组织先天性免疫中的作用

王小辉，李建喜，王旭荣，李宏胜

（中国农业科学院兰州畜牧与兽药研究所，中国农业科学院临床兽医学研究中心，兰州 730050）

摘　要：深入研究乳腺组织的免疫防御对制定控制乳腺炎的措施非常重要。乳腺的先天性免疫是一个非常广泛的研究领域，尽管经过多年的研究，但目前对乳腺先天性防御的相关知识仍旧非常缺乏。本文综述了近年来关于奶牛乳腺组织的体液防御在其先天性免疫中的功能和作用机制的研究结果。

关键词：体液防御；乳房炎；先天性免疫

（发表于《中国奶牛》）

Effects of *Genhuasg* Dispersible Tablets Onpart of the Physiological and Biochemical Function in Broilers

XIN, Ruihua, LUO Yongjiang, ZHENG JIfang, LUO Chaoying, LI Jinyu, WANG Guibo, XIE Jiasheng

(1. Lanzhou Institute of Ifusbandry and Pharmaceutical Sciences of CAAS. Lanzhou, China; 2. Key laboralllry of Vctcrinan Pharmaccutlcal Devclopment. Ministry of Agriculture, Lanzhou, China)

Abstract: In order to make clear the safety of Genhuang Dispersible Tablets and provide date support for the clinical medication, 120 chickens were randomly divided into four groups including the low dose group (1g /kg. bw, according to the original medicinal materials), medium dose group (5g/kg. bw, according to the original medicinal materials), high dose group (10g/kg. bw, according to the original medicinal materials) and the control group. Every group had thirty broilers. The drug was continuously administrated for 30 days. On the 10Th, 20th and 30th days after administration, we randomly selected 10 broilers to examine the weight, organ index, liver and kidney function indicators. Observation and testing indicators are as the following. The first indicators are daily observation of chicken health, behavior, poisoning and death. Then the body weight gain of each dose group was also calculated. While 10 chickens were randomly selected from each group and the heart blood was collected for determination 0fbiochemical indices. And the heart, liver, spleen, lung and kidney and other organs were weighed and organ coefficient was calculated. Blood biochemical parameters. including aspartate aminotransferase (AST), alanine aminotransferase (ALT). albumin (ALB), urea (UREA) and creatJnine (CREA) were determined by Mindray BS-420 automatic biochemical analyzer. In this experiment, there Were no rules changes between the dose of the drug and the heart and lung coefficient twenty days after administration. So the drug affected organ index of broiler chickens was less.

The ALT and AST levels of every dose group compared with control group increased slightly in this experiment. 30 days at, er the administration in high-dose group, there was no significant dj fl0rence of ALB ($P > 0.05$), although some effects of tile drug ell the liver cell, but it did not affect liver function in broilers. 10 days after tile administration, the serum creatininc level of lhe high dose group was significantly higher ($P < 0.05$), but the difrerences between thc group of urea content was not significant ($P > 0.05$). So it was a Iow toxic and safe traditional Chinese medicine preparation which was suitable for clinical use.

Key words: Gcnhuang Dispersible Tablet; broiler; physiological and biochemical indicators

(Published the article in The Conference ASTVM)

Study on Safety Pharmacology of Shegan Dilong Particles

WANG, Guibo[1], LUO Yongjiang[2], LUO Chaoying[2], LI Jinyu[2], ZHENG Jifang[1], XIE Jiasheng[1], XIN Ruihua[1]

(1. Lanzhou Institute of Husbandry and Pharmaceutical Sciences of CAAS, Lanzhou, China; 2. Key Laboratory of Veterinary Pharmaceutical Development, Ministry of Agriculture, Lanzhou, China)

Abstract: Safety pharmacology of the drug is related to the physiology, pharmacology and toxicology of Pharmacology. The significance of the preclinical safety evaluation is attracted more and more attention of the relevant administrative departments. Its role in drug safety evaluation system is very important. In order to study clinical safety of Shegan Dilong particles, a safety pharmacology studies were carried out for the first time by non – invasive way.

Three doses of *shegan Dilong* particles including high dose (2g/ (kg · bw)), medium dose (1g/ (kg · bw)) and low dose (0. 5g/ (kg · bw), which is equivalent to the recommended dose) were given to the anestbetized dogs, there was also a group of anesthetized dogs were given saline. Then the anesthetized dogs were observed. BIOPAC MPI50 data acquisition and analysis system was used to collect the data of the mean arterial pressure, heart rate, body temperature, standard II lead ECG, respiratory rate, tidal volume, oxygen saturation and respiration curve of the dogs. The Emiction Analysis Instrument was used to determine dogs´urine of eleven indicators while the urine weight gain was also calculated. Meanwhile Kllnming strain mice were given the particles in accordance with the above different doses, and the spontaneous activity of mice was determined by behavioral analysis system of integrated open field. The above indicators observed were used to illustrate the impact on the cardiovascular system, respiratory system, urinary system and central nervous system ofthe particles.

The results showed that *Shegan Dilong* particles were given by oral according to the dose of 2. 0g/kg. bw had no significant effect on the cardiovascular system, respiratory system and urinary system of anesthetized dogs. The results of the mice's spontaneous activity in the testing process also sbowed that the total distance, the side of distance, the angle of distance and time among the groups had no significant differences ($P > 0.05$). And the general conduct of activities, posture and gait of the mice performed normal, while muscle trembling, anxiety, running, screamed, curled up on the vertical hair and other aboormalities in mice dido't exit. It indicated that in the range of the test dose the particles had no effect on the central nervous system in mice.

The results suggested that at least in the range of 2g/kg. bw the SheganDiiang particles had no significant effect on animal cardiovascular system, respiratory system, urinary system and central nervous system, and it also suggested that the adverse reactions was small, so it's a safe drug suitable for the clinical application.

(Published the article in The Conference ASTVM)

Wonderful Usage of Fu Zi (Radix Aconiti Lateralis Preparata) and Complexity Science Characteristic of TCM

LUO Chaoying, ZHENG Jifang, XIE Jiasheng, LUO Yongjiang, LI Jinyu, WANG Guibo, XIN Ruihua

(Engineering & Technology Research Center of Traditional Chinese Veterinary Medicine of Gansli Province. Lanzhou Institute of Husbandry & Pharmaceuties Science, Chinese Academy of Agricultural Sciences, Lanzhou, 730050)

Abstract: In view of the difficult understanding on the dosage variability of Fu Zi (Radix AconitiL; HcralisPreparata) in practical TCM clinic from the viewpoint of CM and one kind of the most common Traditollal Chinese Medicine (TCM) in all ages in China. the dosage variability and Nonlinear Characteristic of Zheng Differentiation and Treaonent of Traditional Chinese Medicine (TCM) were discussed from the viewpoint of complexity science. and it is thought that it is very difficult to understand the dosage variability of Fu Zi (Radix AconitiLaleralisPreparata) from lhe viewpoint of CM, but the wonderful usage of it and nonlinear action characteristic of Zheng Differentiation and Treatment of TCM could be seen clearly from the viewpoint of complexity science. It is very important not only to knowing and understanding of TCM, but also its curative effect and safe.

Key words: Fu Zi (Radix Aconiti Latcralis Preparata); Dosage: Complexity sCIence; Zheng DilTerentiation and Treatment/Condition analysis and treatment: TCM

(Published the article in The Conference ASTVM)

藏兽药蓝花侧金盏对兔螨的抑杀作用研究

尚小飞[1]，潘　虎[1]，苗小楼[1]，王东升[1]，唐木克[2]，
达　哇[2]，王　瑜[1]，杨耀光[1]
（1. 中国农业科学院兰州畜牧与兽药研究所，农业部兽用药物创制重点实验室，甘肃省
中兽药工程技术中心，兰州 730050；2. 四川省若尔盖县畜牧兽医局）

摘　要： 探讨藏兽药蓝花侧金盏对兔螨的抑杀作用。应用体外培养杀螨实验，比较了蓝花侧金盏水提取物、甲醇提取物、乙酸乙酯提取物及石油醚提取物的杀螨活性，并对活性较强的提取物的毒力和治疗兔螨病效果进行了评价。4 种提取物中，乙酸乙酯提取物具有较强的体外杀螨活性，高剂量（500mg/ml）在体外培养 6h 内全部杀死螨虫，4 种浓度（500、250、125 和 62.5mg/ml）的半数致死时间分别为 0.743、2.73、5.919 和 22.536h，并且其能有效治疗兔螨病，总有效率为 90%。蓝花侧金盏乙酸乙酯提取物具有良好的杀兔螨活性。

关键词： 蓝花侧金盏；兔螨；杀螨作用；乙酸乙酯提取物

（发表于《中兽医医药杂志》）

大黄末中蒽醌含量的测定

苗小楼，潘　虎，尚小飞，李宏胜
（中国农业科学院兰州畜牧与兽药研究所，农业部兽用药物创制重点实验室，甘肃省
中兽药工程技术研究中心，兰州 730020）

摘　要： 采用 PR－PHLC 法测定大黄末中五种成分的含量。wtares 2695 PHLC，PDA 检测器，检测波长 453mn，柱温 25℃，色谱柱：Hypersil ODS－2 C_{18} 柱（250mm×4.6mm，5μm），Kromasil ODs 保护柱（4.6m×10mm，5μm），流动相：甲醇－0.1%磷酸（85：15），流速 1.0ml/min。芦荟大黄素，大黄酸，大黄素，大黄酚，大黄素甲醚的回归方程分别为：$Y=1.37\times10^5X-3.52\times10^4$，r=0.9998；$Y=8.22\times10^4X-4.19\times10^4$，r=0.9999；$Y=1.11\times10^5X-3.24\times10^4$，r=0.9999；$Y=1.85\times10^5X-5.36\times10^4$，r=0.9999；$Y=7.17\times10^4X-2.41\times10^4$，r=0.9999，线性范围分别为：6.8～680，8.8～880，6.2～620，7.1～710，6.0～600μg。该法灵敏、分离度好、准确，可用于大黄末的质量控制。

关键词： 大黄末；含量测定；RP－HPLC

（发表于《第三届中国兽医临床大会》）

大黄末中芦荟大黄素等5种蒽醌类成分含量测定

苗小楼，潘　虎，尚小飞，李宏胜

（中国农业科学院兰州畜牧与兽药研究所，农业部兽用药物创制重点实验室，甘肃省中兽药工程技术研究中心，兰州 730050）

摘　要：采用RP－HPLC法测定大黄末中5种成分的含量。Waters 2695 HPLC，PDA检测器，检测波长435nm，柱温25℃，色谱柱：Hypersil ODS－2 C_{18}柱（250mm×4.6mm，5μm），Kromasil ODS保护柱（4.6mm×10mm，5μm），流动相：甲醇－0.1%磷酸（85∶15），流速1.0ml/min。结果：芦荟大黄素、大黄酸、大黄素、大黄酚、大黄素甲醚回归方程分别为：$Y=1.37\times10^5X-3.52\times10^4$，$r=0.9998$；$Y=8.22\times10^4X-4.19\times10^4$，$r=0.9999$；$Y=1.11\times10^5X-3.24\times10^4$，$r=0.9999$；$Y=1.85\times10^5X-5.36\times10^4$，$r=0.9999$；$Y=7.17\times10^4X-2.41\times10^4$，$r=0.9999$，线性范围分别为6.8－680、8.8－880、6.2－620、7.1－710、6.0－600μg。该法准确灵敏、分离度好，可用于大黄末的质量控制。

关键词：大黄末；含量测定；RP－HPLC

（发表于《中兽医医药杂志》）

蛋白质组学及其在奶牛蹄叶炎研究中的应用前景

董书伟[1,3]，李　巍[2]，严作廷[1]，王旭荣[1]，高昭辉[1,3]，荔　霞[1]

（1. 中国农业科学院兰州畜牧与兽药研究所，农业部兽用药物创制重点实验室，甘肃省中兽药工程技术中心，兰州 730050；2. 郑州牧业高等专科学校，郑州 450000；3. 甘肃省牦牛繁育重点实验室，兰州 730050）

摘　要：蛋白质组学技术已经成为筛选重大疾病的特异生物标志物和研究发病机制的有效手段，日益受到科研人员的重视。奶牛蹄叶炎是影响奶业健康快速发展的主要疾病之一，论文总结了奶牛蹄叶炎和蛋白质组学的主要研究概况，探讨了蛋白质组学在该病研究中的应用前景。

关键词：蹄叶炎；蛋白质组学；奶牛

（发表于《动物医学进展》）

蛋白质组学研究进展及其在中兽医学中的应用探讨

董书伟，荔　霞，刘永明，王胜义，王旭荣，刘世祥，齐志明

（中国农业科学院兰州畜牧与兽药研究所，农业部兽用药物创制重点实验室，
甘肃省中兽药工程技术中心，兰州 730050）

摘　要：随着生命科学研究进入后基因组时代，蛋白质组学作为重要的实验技术，已经成为筛选重大疾病的特异生物标志物和研究发病机理的新途径。文章总结了蛋白质组学研究的主要内容和方法，简述了蛋白质组学在生命科学的应用概况，并探讨了其在中兽医学研究中的应用前景。

关键词：蛋白质组；蛋白质组学；生物标志物；中兽医学

（发表于《中国畜牧兽医》）

发酵型党参提取物对肉鸡生产性能及生化指标的影响

张　凯，李建喜，杨志强，王学智，
孟嘉仁，张景艳，王　龙

（中国农业科学院兰州畜牧与兽药研究所，甘肃省新兽药工程重点实验室，
甘肃省中兽药工程技术研究中心，兰州 730050）

摘　要：为了研究发酵型党参提取物对肉鸡生产性能及生化指标的影响，试验对饲喂正常日粮（对照组）、含党参生药日粮（RC 组）和含发酵型党参提取物日粮（FRCE 组）黄羽肉鸡的全期饲料利用率、日增重等进行了测定；并测定血清乳酸脱氢酶、丙氨酸氨基转移酶、碱性磷酸酶、天冬氨酸氨基转移酶、γ－谷氨酸转肽酶活性及尿酸含量；捕杀后测定器官指数，并做组织切片镜检。结果表明：RC 组肉鸡的平均日增重和日采食量显著降低（$P<0.05$），FRCE 组饲料转化率显著降低（$P<0.05$）；与对照组相比，FRCE 组的丙氨酸氨基转移酶活性显著降低（$P<0.05$），RC、FRCE 组碱性磷酸酶活性显著下降（$P<0.05$），RC 组和 FRCE 组尿酸含量显著上升（$P<0.05$）；各组肉鸡内脏指数均差异不显著（$P>0.05$）。说明在试验期内发酵型党参提取物对肉鸡生产性能和健康状况有明显促进作用。

关键词：发酵型党参提取物；血清生化指标；肉鸡；生产性能

（发表于《黑龙江畜牧兽医》）

肝纤维化过程中抑制肝星状细胞活性的影响因素

秦　哲，杨志强，李建喜，张　凯，张锦艳，王　磊，
郝桂娟，邓慧媛，王国庆
（中国农业科学院兰州畜牧与兽药研究所，730050）

摘　要：叶纤维化是肝损物之厂的一种病理修复反应，表现为肝内胶原纤维异常增生，是慢性肝病发展至肝硬化的必经阶段。HSCs 的活化和增殖对一于肝纤维化的发生、发展起到关键作用。抑制 HSCs 的活性及促进其凋亡的发生是切断肝纤维化的重要途径。本文对抑制 HSCs 活化，促进其凋亡等因素的研究进展进行综述。

关键词：肝纤维化；月十早状细胞；凋亡

（发表于《中国毒理学会兽医毒理学与饲料毒理学学术讨论会暨兽医毒理专业委员会第四次全国代表大会会议论文录》）

根黄分散片的安全药理学研究

王贵波[1]，罗永江[3]，罗超应[2]，李锦宇[2]，郑继方[1]，谢家声[1]，辛蕊华[1]
（1. 中国农业科学院兰州畜牧与兽药研究研究所，兰州 730050；2. 甘肃省中兽药工程技术研究中心，兰州 730050；3. 农业部兽用药物创制重点实验室，兰州 730050）

药物的安全药理学是涉及生理学、药理学和毒理学的综合性药理学科，在新药临床前安全性评价中的意义越来越受到相关管理部门的重视，在药物安全性评价体系中的作用十分重要。为了进一步探讨根黄分散片临床用药的安全性，首次通过无创方式对根黄分散片开展了安全药理学研究。

（发表于《第三届中国兽医临床大会文集》）

根黄分散片的含量测定及制剂稳定性研究

辛蕊华[1]，罗永江[1]，郑继方[1]，李　维[2]，王贵波[1]，
罗超应[1]，李锦宇[1]，谢家声[1]
（1. 中国农业科学院兰州畜牧与兽药研究所，甘肃省新兽药工程重点实验室，甘肃省
中兽药工程技术研究中心，兰州 730050；2. 兰州市
疾病预防控制中心，兰州 730030）

摘　要：本试验旨在建立根黄分散片中黄芩苷的含量测定方法以及考察该制剂的稳定性，为制定该制剂质量标准中含量测定方法及保质期提供依据。采用反相高效液相色谱（RP－HPLC）法，色谱柱为 YWG－C18（150mm×4.6mm，10μm），流动相：甲醇：0.2%磷酸水溶液（47：53），流速：1ml/min，检测波长：280nm，柱温：30℃，进样量：10μl；根据药典，采用强光照射试验、加速试验及部分长期稳定性试验考察分散片中黄芩苷的稳定性。结果表明，黄芩苷的含量在 1.018～50.90μg/ml 范围内与峰面积呈良好的线性关系，$r=0.9999$（n＝6），平均加样回收率为 97.27%，RSD 为 0.75%（n＝9）；强光照射试验、加速试验及部分长期稳定性试验结果表明该产品基本稳定；此方法简便、灵敏、重现性好，可用于根黄分散片中黄芩苷的含量测定，且该制剂的稳定性良好。

关键词：反相高效液相色谱法；根黄分散片；黄芩苷；稳定性

（发表于《中国畜牧兽医》）

海藻糖生物特性及其在 ELISA 技术研发中的应用

王　磊，崔东安，张景艳，李建喜

（中国农业科学院兰州畜牧与兽药研究所，农业部新兽药工程重点实验室，甘肃省中兽药工程技术研究中心，兰州 730050）

摘　要：海藻糖属非还原性双糖，在生物体内作为一种贮藏性糖类，可提供能量来源，同时它也是一种重要的应激代谢产物，对应激状态具有高度抗性。外源性海藻糖对细胞、抗体等生物活性物质同样具有非特异性生物保护作用。文章综述了海藻糖的能量储备、抗应激、细胞稳定剂等生物学特性，以及对细胞、抗体和 ELISA 中包板抗原或抗体的保护作用，并对其在 ELISA 技术改善中的应用前景做了推测与展望。

关键词：海藻糖；生物学特性；ELISA

（发表于《中国畜牧兽医》）

寒痢宁口服液的薄层鉴别

王海军，王胜义，齐志明，刘世祥，荔　霞，刘永明

（中国农业科学院兰州畜牧与兽药研究所，农业部兽用药物创制重点实验室，甘肃省中兽药工程技术研究中心，兰州 730050）

摘　要：为了建立寒痢宁口服液的质量控制标准，采用薄层色谱法对寒痢宁口服液中的黄连、厚朴、陈皮、补骨脂进行定性鉴定。试验结果表明，黄连、厚朴、陈皮和补骨脂的特征成分在与对照品和对照药材色谱相应位置上显相同颜色的斑点，阴性对照无干扰。该方法专属性强，稳定性和重现性好，为寒痢宁口服液的质量控制建立了定性鉴定方法。

关键词：寒痢宁口服液；薄层色谱法；定性鉴定

（发表于《中国畜牧兽医》）

航天搭载对中草药品质的影响研究进展

王华东[1]，冯晓春[2]

（1. 中国农业科学院兰州畜牧与兽药研究所，兰州 730050；
2. 兰州理工大学技术工程学院）

伴随着“神州八号”宇宙飞船返回舱安全着陆及其搭载的植物种子、微生物、线虫等实验材料顺利交接，航天育种再次成为人们注目的焦点。本项目总体分析航天搭载对中草药品质的影响及其研究进展。航天育种也称为太空育种或空间技术育种，是指利用火箭、飞船等可返回式航天器或高空气球等将植物种子或胚胎送入太空环境，经宇宙射线、微重力、太空辐射、高真空、超低温、交变磁场等复杂空间环境因素诱变产生有益突变，再返回地面进行选育的新品种培育技术。目前，仅有美国、俄罗斯、中国 3 个掌握可返回式卫星技术的国家能够开展航天育种研究，我国已在农业航天诱变育种方面开展了广泛而系统的研究，用于航天育种的种质资源有农作物、蔬菜、花卉、药用植物等高附加值作物以及线虫、病毒、细菌、酵母等微生物试验用材[1]。与传统地面育种方法相比，航天育种具有育种范围广、育种程序简便、育种周期短、变异几率高、后期变异稳定时间快（可缩短 3 个世代左右）等优势。尤其是中草药航天育种研究为我国独创，研究成果与技术水平居世界领先地位，继 1999 年“神州一号”飞船首次成功搭载野生灵芝、红花、黄茂、柴胡、板蓝根等 30 多种中草药种子进入太空以来，先后经 10 次太空搭载，搭载自术、黄芬、菊花、青蒿、冬虫夏草、知母、远志、枸杞子、金钱草、砂仁、沙参、人参、五味子、射干、刺五加、杜仲、地黄、贝母、丹参、麻黄、桔梗、徐长卿、肉苁蓉、甘草、决明子等多种药用植物进行了航天诱变育种，搭载后效应研究涉及生物学性状、化学成分含量测定与分析、生理生化分析、药理活性、安全性、基因组与遗传性状等多个方面[1]，成为缓解我国优质中草药资源过度采挖、资源日益匮乏、种质退化、药材产地生态失衡、人畜争药现象加重等不利局面的有效手段之一，值得进一步进行研究分析。

（发表于《中兽医医药杂志》）

狐狸细小病毒病的诊治

董书伟[1,3]，严作廷[1,3]，刘姗姗[2]，高昭辉[1,3]，齐志明[1,3]，
刘世祥[1,3]，刘永明[1,3]，荔　霞[1,3]

（1. 中国农业科学院兰州畜牧与兽药研究所，农业部兽用药物创制重点实验室，
兰州 730050；2. 兰州工业研究院，兰州 730050；3. 甘肃省
中兽药工程技术中心，兰州 730050）

狐狸是特种经济毛皮动物，狐皮在国际市场上被公认为“软钻石”、“软黄金”，其市场价格也逐年上涨，狐狸养殖具有投资少、产出快、效益好的优势，是被人们看好的朝阳养殖业。

细小病毒病是犬科动物的一种高度接触性急性传染病，由犬细小病毒（CPV）感染引起，以呕吐、腹泻、便血为特征。本病多发于夏、秋季节，常发于 3 岁以下的动物，尤以 6 月龄内的犬科动物最为易感；管理不善、气温骤变、拥挤、环境卫生差等应激因素常诱发该病，并伴发或继发其他病原混合感染，死亡率极高。细小病毒病可使动物频繁呕吐，严重腹泻，迅速脱水；还可通过胎盘垂直传播至胎儿，引起孕兽产弱仔或流产、死胎、木乃伊等；携带病原的动物可通过口、鼻分泌物及粪便等将病毒排出体外，污染兽舍，成为主要传染源。该病常见于犬，而细小病毒感染狐狸的报道在国内并不多见，现将笔者近期诊治的狐狸细小病毒病例研究报道。

（发表于《畜牧与兽医》）

家禽肠道健康导向的功能性添加剂研究进展

崔东安[1,2]，王　磊[1,2]，程海鹏[2]

（1. 中国农业科学院兰州畜牧与兽药研究所，兰州 730050；
2. 北京康牧兽医药械中心药厂）

摘　要：本文从家禽肠道功能及微生态菌群对家禽健康的影响，探讨了基于肠道为靶标的功能性饲料添加剂的种类及其功效，并介绍现代复方制剂“菌物药”——肠泰的药效学原理。

关键词：肠道健康；功能性饲料添加剂；菌群平衡；菌物药

（发表于《家禽科学》）

金石翁芍散的亚慢性毒性试验

李锦宇，郑继方，罗超应，王东升，王贵波，汪晓斌
（中国农业科学院兰州畜牧与兽药研究所，农业部兽用药物创制重点实验室，
甘肃省中兽药工程技术研究中心，兰州 730050）

摘　要：为了评价连续使用中草药复方金石翁芍散而产生毒副反应和严重程度，以及停药后的发展和恢复情况为拟定临床安全用药剂量提供参考方法采用金石翁芍散高、中、低 3 个剂量组及空白对照组通过混饲给药方式在 90 只 18～22g 昆明系小鼠上进行亚慢性毒性试验。各组小鼠的血象及血清 GPT 均在正常范围。组织学检查，肝，肾，脾等均无中毒病变。饲喂 30d，也未见毒副作用。说明该制剂的临床用药剂量是非常安全的。

关键词：金石翁芍散；亚慢性毒性试验

（发表于《第三届中国兽医临床大会》）

喹乙醇单克隆抗体的制备

王　磊，李建喜，张景艳，详尔忽强，王学智，张　凯，孟嘉仁
（中国农业科学院兰州畜牧与兽药研究所，兰州 730050）

摘　要：喹乙醇（Olaquindox，OLA）常被用于饲料一添加剂，但过量或不规范使用有潜在的致畸和致突变性。欧盟、美国等禁止其用于饲料添加剂，我国规定只能添加于体重低于35kg的猪饲料中。本研究旨在制备一种效价高、特异性强的喹乙醇单克隆抗体，为饲料中喹乙醇快速检测方法的建立奠定基础。首先，采用活泼酯化法合成喹乙醇免疫抗原（OLA－NHS－BSA）、检测抗原（OLA－NHS－OVA）和鉴别抗原（OVA－NHS），利用BCA法和紫外分光光度法检测OLA－NHS－BSA和OLA－NHS－OVA的蛋白含量和偶联比。然后，采用皮下和腹腔注射的方法，分别用剂量为50μg/只、100μg/只和200μg/只的OLA－NHS－BSA免疫Balb/C小鼠，每隔2周免疫一次，免疫4次后，断尾采血，分离血清：分别使用OLA－NHS－OVA和OVA－NHS包被的检测板，通过间接ELISA法测定血清效价，确定免疫方案和融合备用小鼠：融合前4大，通过腹腔注射，冲刺免疫融合备用小鼠。融合前一大取健康小鼠的腹腔细胞，制备饲养细胞：融合当大取冲刺免疫的小鼠制备脾细胞，在PEG诱导下与SP2/0骨髓瘤细胞按8∶1的比例进行融合，37℃，5% CO_2 条件下，在含有饲养细胞的HAT选择培养基中进行培养。11天左右，再分别使用OLA－NHS－OVA和OVA－NHS包被的检测板，通过间接ELISA法筛选阳性杂交瘤细胞，并利用有限稀释法对阳性杂交瘤细胞进行克隆：最后采用体内诱生腹水法制备大量喹乙醇抗体，通过间接ELISA法测定抗体效价。全抗原OLA－N H S－BSA和OLA－NHS－OVA的蛋白含量分别为4.69mg/ml和8.85 mg/ml,偶联比分别为1∶4.9和1∶6.4，全抗原、半抗原和载体蛋白在200～500nm的扫描图谱显示平抗原和载体蛋白偶联成功。第4次免疫后，三个免疫剂量组的小鼠血清效价均达1∶16 000以上，且以100μg/只剂量组的免疫效果最好，为最佳免疫方案。细胞融合后共培养576孔细胞，11天后观察筛选阳性杂交瘤细胞，有杂交瘤细胞生长的共513孔，融合率为89.06%，阳性孔有84孔，阳性率为14.58%：通过反复筛选、克隆最终得到一株分泌高效价和特异性抗体的杂交瘤细胞株3B6，细胞培养上洁效价为 $1:5.12\times10^4$,制备的腹水效价为 $1:1.6\times10^7$，且均与OVA－NHS无交叉。【结论】制备的单克隆抗体效价高、特异性强，是针对喹乙醇的特异性抗体，可以用于建立喹乙醇的快速免疫学检测方法。

关键词：吟乙醇；单克隆抗体；制备

（发表于《中国毒理学会兽医毒理学与饲料毒理学学术讨论会暨兽医毒理专业委员会第四次全国代表大会会议论文录》）

麻杏石甘汤作用机制及其在兽医临床上的应用

刘晓磊，郑继方，罗永江，王贵波，罗超应，谢家声，李锦宇，辛蕊华

（中国农业科学院兰州畜牧与兽药研究所，甘肃省中兽药工程技术研究中心，
农业部兽用药物创制重点实验室，兰州 730050）

摘　要：麻杏石甘汤擅清宣肺热，镇咳平喘。鉴于该方剂在治疗畜禽呼吸道疾病上应用广泛，就麻杏石甘汤的作用机制研究进展及其在兽医临床上的应用方面进行了论述，并对其未来的发展进行了展望，以期其在动物疾病的防治上发挥更大的作用。

关键词：麻杏石甘汤；作用机制；研究进展；兽医临床；应用

（发表于《湖北农业科学》）

免疫失败诱发一起狐狸犬瘟热的诊治

董书伟[1,2]，严作廷[1,2]，荔　霞[1,2]，高昭辉[1,2]，刘永明[1,2]

（1. 中国农业科学院兰州畜牧与兽药研究所，农业部兽用药物创制重点实验室，
兰州 730050；2. 甘肃省中兽药工程技术中心，兰州 730050）

狐狸是特种经济毛皮动物，狐皮在国际市场上被公认为“软钻石”、“软黄金”，其市场价格也逐年上涨，狐狸养殖具有投资少、产出快、效益好的优势，是一种朝阳养殖业（聚焦三农，2010）。犬瘟热是由犬瘟热病毒引起的一种传染性极强的犬科动物传染病，主要临床表现为双相热型，消化道、呼吸道卡他性炎症，发病后期发生非化脓性脑炎，出现四肢甚至全身抽搐的神经症状（谭菊，2006；张立恒等，2010）；有些动物还有皮疹和硬足蹄的症状（秦佳芸，2006）。该病常见于犬，而犬瘟热感染狐狸的报道并不多见，现将笔者近期诊治的狐狸感染犬瘟热的病例报道。

该狐狸养殖场位于甘肃省定西地区，刚建好 3 年，前期一直处于自养扩繁阶段，现已有 635 只狐狸，有银狐和蓝狐两个品种，周围无其他犬科动物养殖场，也未发生过其他传染病。主诉：前几天狐场免疫了犬用五联苗（犬瘟热、细小病毒病、传染性肝炎、传染性支气管炎、犬窝咳），其中，犬瘟热为弱毒苗。免疫的当晚和次日，天气骤变寒冷，幼龄狐狸于第 5 天开始发病，随后每天都有新发病例。当地兽医采用抗生素治疗无效，疫情继续蔓延至种狐。笔者获悉到狐场时，已经有 200 多只狐狸发病，死亡 30 多只。为此，笔者进行及时诊治取得较好效果。

（发表于《中国畜牧兽医》）

纳米铜对大鼠肝脏毒性的蛋白质组 2 – DE 图谱分析

高昭辉[1,2,4]，董书伟[1,4]，薛慧文[2]，申小云[3]，荔　霞[1]

（1. 中国农业科学院兰州畜牧与兽药研究所，农业部兽用药物创制重点实验，
甘肃省中兽药工程技术中心，兰州 730050；2. 甘肃农业大学动物医学院，兰州 730070；
3. 贵州省毕节学院，贵州毕节 551700；4. 甘肃省牦牛繁育工程重点实验室，兰州 730050）

摘　要：为获取纳米铜对大鼠肝脏毒性的蛋白质组表达图谱，试验通过构建大鼠纳米铜中毒模型，采用双向凝胶电泳技术获取了大鼠纳米铜中毒组和对照组肝脏的蛋白质 2 – DE 图谱，结合 PDQuest 8. 0 软件分析发现 200mg/kg 纳米铜组大鼠肝脏 2 – DE 图谱中有 27 个蛋白质斑点表达上调，16 个蛋白质斑点表达下调，共有 43 个差异蛋白质点。试验利用蛋白质组学技术研究大鼠纳米铜肝脏蛋白质组学，为在蛋白质水平阐明纳米铜的毒性作用机制奠定基础。

关键词：蛋白质组学；2 – DE；纳米铜；大鼠

（发表于《中国畜牧兽医》）

奶牛临床型乳房炎的细菌分离鉴定与耐药性分析

王旭荣，李宏胜，李建喜，王小辉，孟嘉仁，杨　峰，杨志强
（中国农业科学院兰州畜牧与兽药研究所，甘肃省中兽药工程技术研究中心，
中国农业科学院临床兽医学研究中心，农业部兽用药物创制重点实验室，兰州 730050）

摘　要：2011 年山西省多个奶牛场发生了较严重的乳房炎，对 76 份采集的奶样进行细菌分离鉴定并采用药敏纸片法检测主要分离菌的抗生素耐药情况。所分离细菌以革兰氏阳性菌为主，革兰氏阳性球菌和其他革兰氏阳性菌分别占 60.67% 和 23.59%。分离出多种病原菌和机会致病菌，主要的病原菌有链球菌、金黄色葡萄球菌、大肠杆菌等，检出率为 2.24% ~11.24%，其中无乳链球菌的检出率最高；机会致病菌有粪链球菌、微球菌、克雷伯菌、凝固酶阴性葡萄球菌等，检出率为 1.12% ~11.24%，其中粪链球菌和微球菌的检出率较大，分别为 11.24% 和 6.74%。药敏试验检测结果显示，在所选的 15 种药物中，主要分离菌均对丁胺卡那霉素、氟哌酸和恩诺沙星 3 种药物敏感；大肠杆菌、克雷伯菌、凝固酶阴性葡萄球菌对青霉素类和 β－内酰胺/β－内酰胺酶抑制剂类药物产生了极强的耐药性，耐药率均为 100%；乳房链球菌对该类药物也产生不同程度的耐药，耐药率为 20% ~100%；对链霉素产生 100% 耐药的细菌有大肠杆菌、克雷伯菌、无乳链球菌、乳房链球菌、停乳链球菌等细菌；部分分离菌对卡那霉素、庆大霉素、链霉素、四环素、红霉素、先锋霉素Ⅴ、复方新诺明等药物产生不同程度耐药。被检奶牛场混合感染较为严重，应进一步加强环境卫生管理，临床治疗应合理有效用药。
关键词：临床型乳房炎；细菌分离；耐药性

（发表于《中国畜牧兽医》）

奶牛蹄叶炎与血液生理生化指标的相关性分析

董书伟，荔　霞，高昭辉，严作廷，王胜义，刘世祥，齐志明，刘永明

（中国农业科学院兰州畜牧与兽药研究所农业部兽用药物创制重点实验室，甘肃省中兽药工程技术中心，甘肃省牦牛繁育重点实验室，兰州 730050）

摘　要：为研究奶牛蹄叶炎与其血液生理生化指标的相关性，采集健康组奶牛 15 头和蹄叶炎患病组奶牛 36 头的血样，检测其生理生化指标。结果：患病组奶牛血浆中生化指标总胆固醇（TC）和高密度脂蛋白胆固醇（HDL－C）含量和健康组相比有显著差异（$P<0.05$）；而 2 组间奶牛血液生理指标和其他生化指标均无显著性差异（$P>0.05$）。表明奶牛发生蹄叶炎可能与脂质代谢紊乱有密切关系，TC 和 HDL－C 可作为蹄叶炎发病监测的候选指标，并建议通过调节奶牛的脂质代谢平衡来防治蹄叶炎。

关键词：蹄叶炎；血液生理生化指标；相关性；奶牛

（发表于《畜牧与兽医》）

奶牛微量元素营养舔砖对奶牛生产性能和健康的影响

王胜义，刘永明，齐志明，刘世祥，王　慧，王海军，荔　霞

（中国农业科学院兰州畜牧与兽药研究所，中国农业科学院临床兽医学研究中心，甘肃省中兽药工程技术研究中心，兰州 730050）

摘　要：选择 60 头不同产奶水平荷斯坦泌乳牛进行试验，在保持原有饲养管理和日粮水平不变的情况下，研究补饲奶牛微量元素营养舔砖对奶牛产奶性能、乳品质及牛体机能的影响。结果表明，补饲微量元素营养舔砖可以显著提高各水平奶牛产奶量（$P<0.05$）；奶牛异食癖消失，被毛状况得到明显改善，乳房炎、胎衣不下病例分别较对照组减少 17% 和 25.98%。

关键词：奶牛微量元素营养舔砖；产奶量；健康

（发表于《中兽医医药杂志》）

内毒素对奶牛子宫内膜细胞的毒性初探

张世栋，王旭荣，王东升

（中国农业科学院兰州畜牧与兽药研究所，农业部兽用药物创制重点实验室，中国农业科学院临床兽医学研究中心，甘肃省中兽药工程技术研究中心，兰州，730005）

摘　要：为探究奶牛子宫内膜炎发病过程中细菌内毒素的致病机理。运用细胞学技术分离纯化了原代中国荷斯坦奶牛子宫内膜上皮细胞（Endometrial Epihtelial Cell，EnEpC），并进行体外培养。经免疫组化鉴定培养的子宫内膜上皮细胞纯度较高、性质稳定。利用体外细胞毒理学的方法，以不同浓度的细菌内毒素（LPS）作用细胞，在一定作用时间后，以噻唑蓝（3－（4，5）dimethylthiahiazo（－z－y1）－3，5－diphentyetrazoliumromide，MTT）法和磺酰罗丹明B（Sulforhodamine B，SRB）法检测了LPS对细胞活力和细胞增殖的影响。MT试验结果显示，10ng/ml至6.52μg/ml的LPS对EnEpC的活力促进率具有剂量依赖效应，最高达到56.36%，6.25～25μg/ml的LPS对EnEpC的活力促进率在（54.94±1.30）%：50μg/ml的LPS开始对细胞活力产生显著的抑制作用，抑制率为5.12%，至250μg/ml时抑制率达到36.60%，根据线性回归方程计算出LPS对EnEpC的活力抑制率达到50%时的作用剂量为350μg/ml。SRB试验结果显示，0.781μg/ml、1.562μg/ml、3.125μg/ml的LPS对EnEpC的增值促进率依次增大为8.55%、10.88%和14.85%；6.25μg/ml、12.5μg/ml、25μg/ml的LPS对EnEpC的增值促进率依次降低为11.39%、9.87%和5.39%。50μg/ml的LPS开始对细胞增殖产生显著的抑制作用，抑制率为20.09%，至250μg/ml时增殖抑制率达到48.89%，根据线性回归方程计算出LPS对EnEpC的增殖抑制率达到50%时的作用剂量为250μg/ml。低剂量的内毒素（≤6.52μg/ml）对奶牛子宫内膜上皮细胞具有显著的促增殖作用，而高剂量的内毒素（≥50μg/ml）对奶牛子宫内膜上皮细胞具有显著的增殖抑制作用。

关键词：奶牛子宫内膜上皮细胞；内毒素；毒性测试

（发表于《中国毒理学会兽医毒理学与饲料毒理学学术讨论会暨兽医毒理专业委员会第四次全国代表大会会议论文录》）

牛源性无乳链球菌血清型分布及抗生素耐药性研究

李宏胜[1]，郁　杰[2]，罗金印[1]，李新圃[1]，徐继英[3]，王旭荣[1]，张礼华[1]

（1. 中国农业科学院兰州畜牧与兽药研究所，农业部兽用药物创制重点实验室，甘肃省中兽药工程技术中心，中国农业科学院临床兽医学研究中心，兰州 730050；
2. 江苏畜牧兽医职业技术学院，泰州 225300；3. 西北民族大学，兰州 730000）

摘　要：本研究旨在查明牛源性无乳链球菌血清型分布及对常见抗生素的耐药情况，指导临床合理用药。对从中国部分地区奶牛场采集的临床型乳房炎病牛乳中分离鉴定出78株无乳链球菌地方菌株，制备沉淀反应抗原及6株标准血清型无乳链球菌单因子血清抗体，采用环状沉淀试验，对78株无乳链球菌地方菌株进行了血清学分型鉴定；同时采用K－B纸片法测定了这些菌株对抗生素的耐药情况。结果表明，引起奶牛乳房炎的无乳链球菌血清型主要为X型（60.26%），其次为Ⅲ型（10.26%）、R型（7.69%）、Ⅱ型（7.69%）和Ⅰb型（5.13%），Ⅰa型尚未发现。无乳链球菌对目前临床上使用的大部分抗生素，如头孢唑啉、头孢噻肟、丁胺卡那霉素、卡那霉素、庆大霉素、四环素、强力霉素、氟苯尼考、多黏菌素B、环丙沙星、氟哌酸和头孢他啶/棒酸均较敏感；但对氨苄青霉素、链霉素、恩诺沙星、阿莫西林/棒酸和复方新诺明，有一定的耐药性，其耐药率达50%～100%。本研究对进一步研制有效的药物及疫苗，指导临床合理用药具有重要的意义。

关键词：奶牛乳房炎；无乳链球菌；血清型；抗生素；敏感性

（发表于《中国畜牧兽医》）

芩连液与白虎汤对气分证家兔 T 细胞亚群和 6 种细胞因子的影响

张世栋，王东升，李世宏，李锦宇，李宏胜，陈炅然，严作廷

（中国农业科学院兰州畜牧与兽药研究所，中国农业科学院临床兽医学研究中心，农业部兽用药物创制重点实验室，兰州 730050）

摘　要：为评价自拟方芩连液的疗效，通过静脉注射脂多糖（LPS）建立了气分证家兔模型，并以芩连液和白虎汤分别进行了治疗，比较了两种组方对气分证家兔 T 淋巴细胞亚群分布和血清细胞因子水平的影响。结果显示，静脉注射 15μg/kg 的 LPS 可成功建立家兔气分证模型。病理组家兔体温、IL－10 显著升高，而 $CD4^+/CD8^+$ 比值、TNF－α 和 IL－6 显著降低，但 IFN－γ、IL－2 和 IL－4 无显著变化。经白虎汤和芩连液治疗后，家兔体温都显著降低，IL－10 显著降低并恢复至正常水平，TNF－α、IL－2、IL－4 和 IFN－γ 不受两组药物的影响，其中白虎汤的治疗可使 $CD4^+/CD8^+$ 比值恢复、IL－6 显著升高，而芩连液的治疗对 $CD4^+/CD8^+$ 比值和 IL－6 无显著影响。结果表明，气分证患兔的细胞免疫功能受到显著抑制，白虎汤和芩连液的治疗都具有显著的解热作用，同时对细胞因子间的失衡都具有一定的调节作用，但白虎汤对机体细胞免疫功能的调节明显优于芩连液。

关键词：气分证；T 细胞亚群；细胞因子；LPS；中药

（发表于《畜牧与兽医》）

芩连液与白虎汤对气分证家兔免疫调节及抗氧化活性的影响比较

张世栋[1]，严作廷[1]，王东升[1]，李世宏[1]，荔　霞[1]，
李锦宇[1]，李宏胜[1]，陈灵然[1]，龚成珍[2]

（1. 中国农业科学院兰州畜牧与兽药研究所，中国农业科学院临床兽医学研究中心，
农业部兽用药物创制重点实验室，兰州 730050；2. 甘肃农业大学
动物医学院，兰州 730070）

摘　要： 为比较自拟方芩连液和白虎汤对气分证的疗效差别，以静脉注射脂多糖（LPS）制作气分证家兔模型，在两种药物治疗前后分别检测了血清免疫球蛋白 IgG、IgM 和 IgA 的含量变化，以及血清超氧化物歧化酶（SOD）、丙二醛（MDA）和总抗氧化能力（T－AOC）的变化。结果显示，白虎汤对气分证家兔升高了的血清免疫球蛋白和 T－AOC 的含量变化无显著影响，但能使血清 SOD 下降，MDA 上升；芩连液可使得 IgM 和 IgG 含量下降，SOD 上升，MDA 无显著变化，而 T－AOC 含量下降。结果表明，白虎汤对气分证家兔免疫调节和总抗氧化能力的保持优于芩连液，但对 SOD 和 MDA 调节不及芩连液。

关键词： 气分证；免疫调节；抗氧化活性；脂多糖；中药

（发表于《中国畜牧兽医》）

芩连液与白虎汤对气分证家兔肾功能损伤的疗效比较

张世栋，王东升，荔　霞，李世宏，李锦宇，李宏胜，陈灵然，严作廷

（中国农业科学院兰州畜牧与兽药研究所，农业部兽用药物创制重点实验室，
中国农业科学院临床兽医学研究中心，兰州 730050）

摘　要：为比较自拟方芩连液和白虎汤对气分证的疗效差别，以静脉注射内毒素（LPS）制作了气分证家兔模型，在两种药物治疗前后分别检测了血清免疫球蛋白 A（IgA）、补体 3（C3）、肌酐（Cr）、尿素（Urea）、尿酸（UA）和碱性磷酸酶（ALP）的含量变化，并观察了肾脏组织病理变化。结果显示，模型动物的肾脏产生了明显的病理损伤，血清 Cr、IgA 和 C3 含量都显著升高，ALP 含量降低，Urea 和 UA 无显著变化；与治疗前比较，白虎汤和芩连液的治疗使家兔肾脏组织病理损伤明显转轻，血清 Cr、UA、IgA 和 C3 含量都显著降低，Urea 含量无显著变化，且白虎汤治疗组 ALP 显著降低，而芩连液治疗组 C3 显著低于对照组与模型组。结果表明，静脉注射 LPS 制作的气分证模型家兔肾脏组织收到明显的病理损伤，表现出 IgA 肾病的特征，白虎汤与芩连液对气分证家兔肾功能损伤具有一定的治疗作用，两者比较白虎汤具有更显著的疗效。

关键词：气分证；LPS；肾损伤；血清生化指标；中药

（发表于《中国兽医杂志》）

芩连液与白虎汤对气分证家兔胃肠黏膜的病理影响比较

张世栋，王东升，李世宏，荔　霞，李锦宇，李宏胜，陈炅然，严作廷

（中国农业科学院兰州畜牧与兽药研究所，中国农业科学院临床兽医学研究中心，甘肃省中兽药工程技术研究中心，兰州 730050）

摘　要：为了比较自拟方芩连液和白虎汤对气分证动物模型的治疗效果，试验通过静脉注射内毒素（LPS）复制家兔气分证模型，并在2种药物治疗前后分别观察了家兔胃肠道黏膜组织的病理变化以及测定了血清免疫球蛋白（IgA）含量。结果表明：静脉注射LPS能致家兔胃肠道黏膜组织产生明显病理损伤，血清IgA含量上升；经白虎汤和芩连液治疗后胃肠道黏膜组织病理损伤显著减轻，但血清IgA含量仍然显著高于正常水平。说明白虎汤和芩连液对LPS引起的胃肠道黏膜组织损伤都具有显著的治愈效果，而对血清IgA的含量无显著影响。

关键词：内毒素（LPS）；气分证；黏膜；免疫球蛋白（IgA）

（发表于《黑龙江畜牧兽医》）

桑杏平喘颗粒对大鼠部分生化指标的影响

刘晓磊，辛蕊华，郑继方，王贵波，罗超应，谢家声，李锦宇，胡振英，罗永江

（中国农业科学院兰州畜牧与兽药研究所，甘肃省中兽药工程技术研究中心，农业部兽用药物创制重点实验室，兰州 730050）

摘　要：本试验旨在了解桑杏平喘颗粒的安全性，为临床用药提供数据支持。选用健康Wistar大白鼠80只，随机分成4组（1个对照组和3个试验组），每组20只，雌雄各半，对照组灌服生理盐水，试验组分别以桑杏平喘颗粒煎剂，相当于原药材3g/kg、15g/kg、30g/kg体重灌胃给药，连续给药30d，在给药10d、20d、30d采血，测定血清中部分生化功能指标。试验结果表明，桑杏平喘颗粒对大鼠肝脏、肾脏、心脏功能的影响较小，是一种适合临床应用的低毒、安全的中药制剂。

关键词：桑杏平喘颗粒；大鼠；生化指标；安全性评价；猪支原体肺炎

（发表于《中国畜牧兽医》）

桑杏平喘颗粒对大鼠肝脏、肾脏和心脏功能的影响研究

刘晓磊[1]，辛蕊华[1]，郑继方[2]，王贵波[1]，罗超应[1]，谢家声[1]，罗永江[3]

（1. 中国农业科学院兰州畜牧与兽药研究研究所，兰州 730050；2. 甘肃省中兽药工程技术研究中心，兰州 730050；3. 农业部兽用药物创制重点实验室，兰州 730050）

猪支原体肺炎（mycoplasmal pneumonia of swine，MPS），是猪的一种接触性、慢性呼吸道疾病，以咳嗽和喘气为特征性症状，解剖可见肺脏组织肉变或大理石样病变，发病率高，死亡率低；患猪生长、增重缓慢，饲料利用率降低，易继发其他疾病，导致相关产品的竞争力降低，严重制约了养猪业 的健康快速发展。因此，如何有效地防制该病的发生就成为当前兽医科研工作者的一项紧迫的工作任务。桑杏平喘颗粒是由中国农业科学院兰州畜牧与兽药研究所自主研发，采用了中医辨证施治原则和传统用药经验，结合现代中药药理研究及循证医学的思想组方，由桑白皮、杏仁、石膏、甘草等中药组成，具有止咳平喘、清热解毒、抗菌、抗病毒、抗应激、提高机体非特异性免疫力、调节机体新陈代谢的综合功效，能够祛除 MPS 引起的咳嗽和喘气症状，对 MPS 有良好的疗效，从而可以很好的保护猪群，促进养猪业的健康快速发展。本试验旨在对桑杏平喘颗粒的部分安全性做出客观评价，为临床用药提供数据支持。

（发表于《第三届中国兽医临床大会论文集》）

射干地龙颗粒对小白鼠止咳祛痰作用研究

罗永江，谢家声，辛蕊华，郑继方，罗超应，胡振英，邓素平

（中国农业科学院兰州畜牧与兽药研究所，农业部兽用药物创制重点实验室，甘肃省新兽药工程重点实验室，兰州 730050）

摘　要：考察射干地龙颗粒对动物发病模型的止咳祛痰效果。采用小白鼠浓氨水诱咳法和酚红祛痰法。生理盐水对照组和药物组引起半数小白鼠咳嗽所需的浓氨水喷雾时间（ET_{50}）分别为 19.2s 和 30.43s，R 值为 158.50%；以酚红标准溶液（X）对其吸收度（A 值）（Y）作标准曲线为：$Y=0.0747X-0.0036$，在 0.1－10.0μg/ml 浓度范围内呈线性关系，其相关系数为 0.9998，3 个药物剂量组小鼠气管酚红排泌量与生理盐水对照组比较均呈显著性差异；3 个药物剂量组之间无显著性差异。射干地龙颗粒对小白鼠具有明显的止咳作用；对小白鼠气管分泌功能有显著增强作用。

关键词：射干麻黄地龙散；小白鼠；止咳作用；祛痰作用

（发表于《中兽医医药杂志》）

射干地龙颗粒防治蛋鸡传染性支气管炎效果

谢家声，罗超应，王贵波，罗永江，辛蕊华，李锦宇，郑继方

（中国农业科学院兰州畜牧与兽药研究研究所甘肃省中兽药工程技术研究中心，农业部兽用药物创制重点实验室，兰州 730050）

摘　要：为进一步验证射干地龙颗粒防治鸡传染性支气管炎的临床疗效和安全性，试验选用临床自然发病鸡 1 080羽，分为试验组、药物对照组和空白对照组，试验期 7d，观察患病鸡死亡率、产蛋率以及破软蛋等。结果表明，试验组死亡率较药物对照组、空白对照组降低 0.8 个和 3.9 个百分点，经统计分析，试验组与空白对照组间差异显著（$P<0.05$）；试验组平均产蛋率较药物对照组和空白对照组分别提高 2.5 和 11.67 个百分点，经统计分析，试验组与空白对照组间差异显著（$P<0.05$）；试验组破壳蛋率分别较药物对照和空白对照下降 0.05 和 0.54 个百分点，经统计分析，试验组与空白对照组间差异不显著（$P>0.05$）。说明射干地龙颗粒可扩大应用于规模化养鸡场，防治鸡传染性支气管炎安全有效。

关键词：射干地龙颗粒；鸡传染性支气管炎；死亡率；产蛋率

（发表于《中兽医医药杂志》）

蹄叶炎奶牛血浆蛋白质组学 2 – DE 图谱的构建及分析

高昭辉[1,2]，荔　霞[1]，严作廷[1]，王旭荣[1]，阎　萍[1]，董书伟[1]

（1. 中国农业科学院兰州畜牧与兽药研究所，农业部兽用药物创制重点实验室，甘肃省中兽药工程技术中心，甘肃省耗牛繁育重点实验室，兰州 730050；2. 甘肃农业大学动物医学院）

摘　要：为研究蹄叶炎奶牛血浆蛋白质组表达变化，采集患病牛和健康牛血浆，用人白蛋白和 IgG 杭体去除其高丰度蛋白后，通过 2 – DE 技术和 PDQuest 分析软件比较两组间血浆蛋白组表达差异。结果显示，蹄叶炎奶牛血浆 2 – DE 图谱中有 39 个蛋白斑点表达上调，23 个蛋白斑点表达下调，共检测到 62 个差异蛋白点，表明奶牛发生蹄叶炎时，其血浆蛋白组表达模式发生了改变，这种改变可能与蹄叶炎的发病机制密切相关。

关键词：蛋白质组学；2 – DE；血浆；蹄叶炎；奶牛

（发表于《中兽医医药杂志》）

我国奶牛乳房炎无乳链球菌抗生素耐药性研究

李宏胜，罗金印，王旭荣，李新圃，王　玲，杨　峰，张世栋，苗小楼

（中国农业科学院兰州畜牧与兽药研究所，农业部兽用药物创制重点实验室，中国农业科学院临床兽医学研究中心，甘肃省中兽药工程技术研究中心，兰州 730050）

摘　要：为了查明我国奶牛乳房炎无乳链球菌对抗生素耐药情况，指导临床合理用药，从我国部分地区奶牛场采集的临床型奶牛乳房炎病乳中分离鉴定出无乳链球菌 115 株，采用 K－B 纸片法测定了这些菌株对抗生素的耐药情况。结果表明，无乳链球菌对目前临床上使用的大部分抗生素，如头孢唑啉、头孢噻肟、丁胺卡那霉素、卡那霉素、庆大霉素、四环素、强力霉素、麦迪霉素、林可霉素、氟苯尼考、多黏菌素 B、环丙沙星、氟哌酸、氨苄青霉素/舒巴坦、头孢噻肟/棒酸和头孢他啶/棒酸均比较敏感；但对青霉素 G、氨苄青霉素、链霉素、恩诺沙星、阿莫西林/棒酸和复方新诺明等有一定耐药性，其耐药率达 50%～100%。

关键词：奶牛乳房炎；无乳链球菌；抗生素；敏感性

（发表于《中兽医医药杂志》）

以复杂性科学观念指导奶牛疾病防治与中西兽医药学结合

罗超应，郑继方，谢家声，罗永江，李锦宇，辛蕊华，王贵波

（中国农业科学院兰州畜牧与兽药研究所，甘肃省中兽药工程技术研究中心，兰州 730050）

摘　要：面对当前奶牛疾病防治的日益困难，虽然不乏众多新进展与新成果，而其临床防治却依旧是步履维艰；在用复杂性科学理论对其进行分析的基础上认为，其主要原因是传统科学“单因素线性分析”的简单化认识观念与方法不能适应临床疾病的复杂多变性特点，而转变科学观念，以复杂性科学理论为指导，加强其临床疾病复杂多变性特点和规律的认识与把握，也许才是突破其发展瓶颈的关键所在。其次，要实现奶牛疾病防治的复杂性科学转变，离不了中兽医药学辨证施治（状态分析与处理）在认识方法上的支撑与帮助；而奶牛疾病防治，有可能成为中西兽医药学结合发展的最重要用武之地之一。

关键词：奶牛疾病；防治；中西兽医药学结合；科学观念转变；复杂性科学；传统科学

（发表于《中国兽医杂志》）

淫羊藿总黄酮提取方法的比较研究

王东升[1,2,3]，张世栋[1,2,3]，李世宏[1,2,3]，严作廷[1,2,3]

（1. 中国农业科学院兰州畜牧与兽药研究所，兰州 730050；2. 农业部兽用药物创制重点实验室，兰州 730050；3. 甘肃省中兽药工程技术研究中心，兰州 730050）

摘　要：为了比较不同提取方法对中药淫羊藿总黄酮含量的影响，试验采用水煎、超声波和回流法进行总黄酮的提取，采用紫外分光光度法测定其含量。结果表明：回流组和超声 3 组总黄酮含量极显著高于超声 2 组、超声 1 组和水煎组（$P<0.01$），超声 2 组总黄酮含量极显著高于超声 1 组和水煎组（$P<0.01$），超声 1 组总黄酮含量极显著高于水煎组（$P<0.01$）。说明采用回流法（用 80% 乙醇回流）提取淫羊藿总黄酮的提取率最高，水煎法提取率最低。

关键词：总黄酮；回流；超声波；紫外分光光度法

（发表于《黑龙江畜牧兽医》）

正交试验法优化催情助孕液制备工艺

王东升，张世栋，李世宏，苗小楼，尚小飞，严作廷

（中国农业科学院兰州畜牧与兽药研究所，农业部兽用药物创制重点实验室，甘肃省中兽药工程技术研究中心，兰州 730050）

摘　要：为了优化催情助孕液的水提醇沉工艺，采用正交试验法设计，用高效液相色谱法测定制剂中淫羊藿苷含量并以其为评价指标，考察了加水量、煎煮时间、煎煮次数、醇沉浓度、醇沉次数对制剂工艺的影响。结果表明：最佳制剂工艺为加 10 倍量水，煎煮 3 次，每次 1h，乙醇醇沉浓度为 75%，醇沉 1 次。用该工艺制备的催情助孕液淫羊藿苷含量为 0.98892mg/ml，说明该优选工艺稳定可行。

关键词：淫羊藿苷；正交设计；HPLC；含量

（发表于《中国农学通报》）

中兽药穴位注射疗法治疗猪温热病研究概况

李新圃，李剑勇，杨亚军，罗金印

（中国农业科学院兰州畜牧与兽药研究所，农业部兽用药物创制重点实验室，甘肃省新兽药工程重点实验室，甘肃省中兽药工程中心，兰州 730050）

猪温热是以发热为主的急性外感热病，具有流行性。其病因十分复杂，许多与急性传染病有密切关系，发病率高达到50% ~100%，死亡率达30% ~90%。患猪主要表现为体温升高至41℃以上，高热稽留，大多数病猪呼吸困难，部分流涕、打喷嚏、咳嗽；部分患猪粪便秘结，或出现腹泻；病程稍长的病猪消瘦、贫血，部分耳朵发绀呈紫红色，皮肤有紫斑、甚至出血。对于猪温热病的治疗，临床常采用降温、抗菌、抗病毒、抗炎等方法，主要是在饲料、饮水中添加此类中、西药物，或进行肌肉注射，但治疗效果都不理想[1,2]，临床迫切需要寻找新的治疗方法。本研究对中兽药穴位注射法治疗猪温热病进行全面分析，并提出一些有效的成功经验。

（发表于《中兽医医药杂志》）

中药子宫灌注剂治疗奶牛不孕症研究进展

王东升，严作廷，张世栋，谢家声，李世宏

（中国农业科学院兰州畜牧与兽药研究所，中国农业科学院临床兽医学研究中心，甘肃省中兽药工程技术研究中心，兰州 730050）

摘　要：中药子宫灌注剂在奶牛子宫内膜炎和卵巢疾病等繁殖疾病的防治中发挥着巨大的作用，临床用于防治奶牛主要繁殖障碍疾病的子宫灌注剂主要有溶液型、混悬型和乳浊型3种。本文对这3种剂型的组方、疗效及作用机理等的研究概况进行综述，以期为新型中药子宫灌注剂的研制及其在临床上的应用提供参考。

关键词：奶牛；中药；子宫灌注剂；子宫内膜炎；卵巢疾病

（发表于《中国奶牛》）

中医药抗感染研究的困惑与复杂性科学分析

罗超应[1]，罗盘真[2]，郑继方[1]，谢家声[1]，罗永江[1]，李锦宇[1]，辛蕊华[1]，王贵波

（1. 中国农业科学院兰州畜牧与兽药研究所，甘肃省中兽药工程技术研究中心，
甘肃兰州 730500；2. 户县中医医院，陕西户县 10300）

摘　要：在对中医药抗病原体最低有效浓度普遍太高，而其对机体免疫功能调节及其诱导干扰素生成的结果又与临床脱节等困惑进行综述分析的基础上，采用复杂性科学“非线性理论”对病原微生物感染多因素综合作用、辨证施治与复杂性科学“初始条件”等进行了探讨，认为中医药乃至整个生物医药抗感染研究都要转变科学观念，以复杂性科学为指导，重视各种因素的综合作用及其各自作用的初始条件；中医药辨证施治（状态分析与处理）实质上就是一种多因素综合分析法，应该强化对它的认识与临床研究。

关键词：中医药；抗感染；复杂性科学；初始条件；辨证施治；状态分析与处理

（发表于《第三届中国兽医临床大会》）

奶牛繁殖障碍的综合防制技术

严作廷，王东升

（中国农业科学院兰州畜牧与兽药研究所，兰州 730050）

奶牛繁殖障碍是指生殖机能紊乱和生殖器官畸形以及由此引起的生殖活动异常的现象。在奶牛养殖业中，繁殖性能是奶牛产后泌乳和畜群扩大的关键，繁殖性能的高低直接影响到产业的经营效益，因此，提高奶牛繁殖性能是奶牛业生产中一项重要工作。目前，在国内认为，牛超过始配年龄的或产后经过三个发情周期（65d 以上）仍不发情，或者繁殖适龄母牛经过三个发情周期的配种仍不受孕或不予配种的（管理利用性不育），即为繁殖障碍。而在国外，则以产犊间隔 365d 计算，产后超过 85d 未孕者定为繁殖障碍。我国奶牛不孕症的发病率在 20% ~30%，对奶牛业的健康发展有较大的影响。本文综合论述当前奶牛繁殖障碍及有效的综合防制技术。

（发表于《兽医导刊》）

奶牛子宫内膜炎的预防和治疗

严作廷，王东升，李世宏，张世栋
（中国农业科学院兰州畜牧与兽药研究所，兰州 730050）

奶牛子宫内膜炎是子宫黏膜的黏液性或化脓性炎症。是奶牛最常见的产科病之一，是引起奶牛不孕的主要原因。据报道，每年美国因奶牛不孕症造成的经济损失近 2.5 亿美元；英国报道不孕牛中约 95% 是由于子宫内膜炎引起。瞿自明等 1983—1985 年对北京、上海、南宁、兰州等 16 个城市 41 个奶牛场调查了 9 754 头适龄奶牛，有 1 684 头发生子宫内膜炎（17.26%），占不孕牛的 68.34%。患子宫内膜炎的奶牛不能正常发情、配种、受孕、产犊，从而延长了胎间距，据统计，胎间距每延长一天，导致少产鲜奶和犊牛损失约合 10 元人民币（尚未计入饲养管理、饲草料、人工、医药等费用）。因此在奶业生产中，必须重视奶牛子宫内膜炎的预防和有效的治疗工作，否则将会给奶业生产带来重大经济损失。本文概述奶牛子宫内膜炎的预防和治疗情况，并提出有益建议。

（发表于《兽医导刊》）

清宫助孕液治疗奶牛子宫内膜炎临床试验

严作廷[1]，王东升[1]，李世宏[1]，张世栋[1]，谢家升[1]，王雪郦[2]，
杨明成[2]，朱新荣[3]，陈道顺[3]
（1. 中国农业科学院兰州畜牧与兽药研究所，甘肃省中兽药工程技术研究中心，兰州 730084；
2. 兰州市城关奶牛场，兰州 730020，3. 甘肃荷斯坦奶牛繁育示范中心，兰州 730086）

摘　要：为研究“清宫助孕液”治疗奶牛子宫内膜炎的临床疗效，本试验将临床确诊为子宫内膜炎的奶牛 80 头，随机分为“清宫助孕液”治疗组和药物对照组，每组分别为 60 头和 20 头，治疗组采用直肠把握法将清宫助孕液子宫灌注，每次 100ml，隔日一次，4 次为一疗程；对照药物将青霉素 100 万单位、链霉素 100 万单位，用生理盐水 50ml 稀释后，子宫灌注，隔日一次，4 次为一疗程。结果表明，试验组治疗 60 头，治愈 52 头，治愈率 86.67%，显效 3 头，显效率 5.0%；有效 2 头，有效率 3.33%；无效 3 头，无效率 5.0%，总有效率 95.0%。三个情期受胎率 86.67%。对照组治疗 20 头，治愈 15 头，治愈率 75.0%，显效 1 头，显效率 5.0%；有效 2 头，有效率 10%；无效 2 头，无效率 10%，总有效率 90.0%。三个情期受胎率 75.0%。可以得出，“清宫助孕液”对奶牛子宫内膜炎具有较好的治疗效果，可以提高情期受胎率。

关键词：奶牛；清宫助孕液；子宫内膜炎

（发表于《中国奶牛》）

草业学科

A Comparative Study of Different Methods on Quality Assessment of Soil Environment Polluted by Zinc in Agricultural Production Areas

——A Case Study in Shulan City of Jilin Province

LI Runlin[1,2,3,4], YAO Yanmin[3,4], YU Shikai[3,4]

(1. The Lanzhou Scientific Observation and Experiment Field Station, Ministry of Agriculture for Ecological System in Loess Plateau Areas, Lanzhou 730050, China; 2. Lanzhou Institute of Husbandry and Pharmaceutical, Chinese Academy of Agricultural Sciences, Lanzhou 730050, China; 3. Key Laboratory of Agri – informatics, Ministry of Agriculture, Beijing 100081, China; 4. Institute of Agricultural Resources and Regional Planning, Chinese Academy of Agricultural Sciences, Beijing 100081, China)

Abstract: The aim was to explore evaluated precision on quality of soilenvironment polluted with zinc in agricultural production areas and to provide references for verification of production area. In Shulan City in Jilin Province, soils were sampled and analyzed in a laboratory using single – factor pollution index and GIS based spatial interpolation. The quality of environment polluted with zincwas assessed and related methods were compared according to Environment Quality Standard of Green Food Production Area. Spatial interpolation of zinc in soils based on GIS proved more precise than traditional methods; cokriging method with co – factors was higher in precision than common cokriging; cokriging method with zinc and organic matter was higher in precision than cokriging with zinc alone. Quality assessment on environment polluted with zinc based on GIS interpolation is more scientific and reasonable than traditional methods.

Key words: Agricultural production area; Soil Environment Quality Assessment; Cokriging; Ordinary kriging

(Published the article in Agricultural Science and Technology)

Study on the Biome Classification of Helophytes at Maqu Wetland

ZHANG Huaishan[1,2], ZHAO Guiqin[1], ZHANG Jiyu[3]

(1. College of Prataculture, Gansu Agricultural University, Lanzhou 730070, China; 2. Lanzhou Institute of Animal&Veterinary Pharmaceutics Science, Chinese Academy of Agriculture Science, Lanzhou 730050, China; 3. College of Pastoral Agriculture Science and Technology, Lanzhou University, Lanzhou 730000, China)

Abstract: This study aimed to accomplish a biome classification of helophytes at Maqu, the first bend of the Yellow River. Helophgtes in the Maqu wetland were investigated using quadrat sampling method with references to plant specimens. The helophyte communities at Maqu wetland could be divided into two categories: sedge marshes and non – sedge marshes, which can be further subdivided into 4 biomes. The constructive species mainly included *Blysmus sinocompressus*, *Blysmocarex nudicarpa*, *Eleocharis valleculosa* and *Polygonum amphibian*. The sub – constructive species consisted mainly of *Carex brunnescens*, *Catabrosa aquatica*, *Kobresia kansuensis*, *Polygonum amphibium* and *Leontopodium alpinum*. The total coverage of communities ranged from 5% to 90%, which were commonly found in areas permanently ponded with water, such as watercourse depressions, floodplains, valley depressions, terrace scarp depressions and riverhead depressions, with the underground water depth of 20 ~ 30cm. The biome classification of helophyte communities provided scientific basis for the ecological restoration and control of Maqu wetland prairie.

Key words: Wetland; Helophytes; Biome classification; Maqu County

(Published the article in Agricultural Science and Technology)

封育对玛曲高寒沙化草地生态位特征的影响

陈子萱[1]，周玉雷[2]，田福平[3]，胡　宇[3]，白　璐[4]，时永杰[3]

（1. 甘肃省农业科学院生物技术研究所，兰州 730070；2. 仲恺农业工程学院园艺园林学院，广州 510225；3. 中国农业科学院兰州畜牧与兽药研究所/农业部兰州黄土高原生态环境重点野外科学观测试验站，兰州 730050；4. 内蒙古鄂尔多斯市草地牧业综合开发办，鄂尔多斯 017000）

摘　要：封育是玛曲高寒沙化草地重要的管理方式。为研究封育对玛曲高寒沙化草地主要植物种生态位的影响，对玛曲高寒沙化草地主要植物种进行生态位特征调查，计算主要植物的重要值、生态位宽度和生态位重叠值。结果表明：对玛曲高寒沙化草地封育 2 年后，草地优势种为禾本科牧草异针茅，其重要值平均为 0.29。乳浆大戟作为高寒沙化草地的主要杂类草，生态位宽度为 0.973。紫羊茅、优势种异针茅与杂类草乳浆大戟生态位重叠较大。说明高寒沙化草地经过 2 年的围栏封育，禾本科牧草有了较大的竞争力，草地植物群落发生良性演替，可食牧草的比例上升，草地经济价值有所提高。

关键词：高寒沙化草地；封育；重要值；生态位宽度；生态位重叠值

（发表于《江苏农业科学》）

施肥对玛曲高寒沙化草地地上生物量的影响

田福平[1]，陈子萱[2]，石　磊[3]

（1. 中国农业科学院兰州畜牧与兽药研究所，农业部兰州黄土高原生态环境重点野外科学观测试验站，兰州 730050；2. 甘肃省农业科学院生物技术研究所，兰州 730070；3. 中国科学院遗传与发育生物学研究所农业资源研究中心，石家庄 050021）

摘　要：为研究施肥对玛曲高寒沙化草地地上生物量的影响，设置不施肥（CK）、施有机肥（O）、施氮肥（N）、施磷肥（P）、施氮磷肥（NP）、施有机肥和氮肥（ON）、施有机肥和磷肥（OP）及施有机肥和氮磷肥（ONP）等 8 个处理，采用样方法对玛曲高寒沙化草地植被盖度和地上生物量进行研究。结果表明：在 2006 年和 2007 年，3 种肥料混施处理（ONP）比对照盖度增加了 7%，地上生物量均为最高。2007 年施肥处理的地上生物量在不同生长季均比 2006 年显著增加。氮、磷肥和有机肥组合的处理（ONP）对增加高寒沙化草地的地上生物量效果最显著。

关键词：施肥；高寒沙化草地；盖度；地上生物量

（发表于《中国农学通报》）

我国草田轮作的研究历史及现状

田福平[1,2,3]，师尚礼[2]，洪绂曾[2]，时永杰[1]，余成群[3,4]，张小甫[1]，胡　宇[1]

（1. 中国农业科学院兰州畜牧与兽药研究所，农业部兰州黄土高原生态环境重点野外科学观测试验站，兰州 730050；2. 甘肃农业大学草业学院，兰州 730070；3. 西藏高原草业工程技术研究中心，拉萨 850001；4. 中国科学院地理科学与资源研究所，北京 100101）

摘　要：草田轮作是提高作物产量和改良土壤性状、实现土地可持续生产的重要措施。我国的草田轮作从魏晋南北朝开始就有记载。近年来的研究主要集中在草田轮作潜力及重要性、轮作效益、轮作模式及对土壤影响等方面。草田轮作既提高了粮食产量，又为家畜提供了充足的饲料，促进了养分循环，是实现农牧民增收及我国生态、经济和社会可持续发展的有效途径。

关键词：草田轮作；历史；现状；重要性

（发表于《草业科学》）

我国人工草地碳储量研究进展

田福平[1]，时永杰[1]，胡　宇[1]，陈子萱[2]，路　远[1]，张小甫[1]，李润林[1]

（1. 中国农业科学院兰州畜牧与兽药研究所，农业部兰州黄土高原生态环境重点野外科学观测试验站，兰州 730050；2. 甘肃省农业科学院生物技术研究所，兰州 730070）

摘　要：作为草地生态系统重要的组成部分，人工草地碳储量的研究对草地生态系统碳减排增汇措施的实施具有重要意义。综述了我国人工草地在生物量碳储量、土壤碳储量（包括土壤微生物量）及土壤呼吸等，为揭示人工草地固碳机制及准确评估人工草地碳源/汇提供科学依据。

关键词：人工草地；碳储量；土壤；有机碳

（发表于《江苏农业科学》）

燕麦与箭筈豌豆不同混播比例对生物量的影响研究

田福平[1]，时永杰[1]，周玉雷[2]，张小甫[1]，陈子萱[3]，胡　宇[1]，白　璐[4]

（1. 中国农业科学院兰州畜牧与兽药研究所/农业部兰州黄土高原生态环境重点野外科学观测试验站，兰州 730050；2. 仲恺农业工程学院园艺园林学院，广州 510225；3. 甘肃省农业科学院生物技术研究所，兰州 730070；4. 鄂尔多斯市草地牧业综合开发办，鄂尔多斯 017000）

摘　要：为了研究燕麦与箭筈豌豆不同混播模式的增产效应。在农业部兰州黄土高原生态环境重点野外科学观测试验站对一年生牧草燕麦和箭筈豌豆进行了混播试验。结果表明：丹麦444燕麦50%和333/A春箭筈豌豆50%的混播处理，地上生物量、种子产量及0～50cm的地上生物量均最高，分别为22 193.33、4 776.67、11 792.90kg/hm^2，且与其他混播处理的差异显著（$P<0.05$）。该混播模式具有较好的协同效应，可在黄土高原及类似地区推广应用。

关键词：燕麦；箭筈豌豆；混播；生物量

（发表于《中国农学通报》）

种子引发机理研究进展及牧草种子引发研究展望

赵　玥[1]，辛　霞[2]，王宗礼[1]，卢新雄[2]

（1. 中国农业科学院兰州畜牧与兽药研究所，兰州 730050；2. 中国农业科学院作物科学研究所，北京 100081）

摘　要：对30多年来种子引发处理在细胞膜修复、贮藏物质动员和能量转化、转录和翻译与诱导合成保护物质、细胞分裂的促进作用等方面的研究进展进行了综述，简述了近年来我国初步开展牧草种子引发研究的进展，提出了今后种子引发研究的重点命题及在牧草种子引发研究、技术应用与临亡牧草种质抢救保存中的展望。

关键词：种子引发；引发机理；研究进展；牧草种质；展望

（发表于《中国草地学报》）

紫花苜蓿航天诱变田间形态学变异研究

杨红善[1]，常根柱[1]，包文生[2]，柴小琴[3]，周学辉[1]

（1. 中国农业科学院兰州畜牧与兽药研究所，兰州 730050；2. 甘肃省航天育种工程技术中心，天水 741030；3. 天水市农业科学研究所，天水 741001）

摘　要：航天诱变育种是以高科技返回式卫星为背景的新型育种方法，以搭载于我国"神舟三号"飞船的4个紫花苜蓿材料：德宝、德福、阿尔冈金、三得利，种植而成的当代（SP_1）试验材料为研究对象。连续观测记载，初步确定了突变类型和单株，并进行了集团分类，大致分为以下7种类型：多叶单株（6株）、大叶单株（11株）、速生单株（14株）、白花突变单株（1株）、抗病单株（7株）、早熟单株（7株）和矮生、分蘖性强单株（2株），所选单株已分别挂牌标记。不同品种经航天诱变表现出不同的突变类型，德宝、德福以生长速度较为显著、三得利以多叶变异较为显著、德福以叶面积增大较为显著，可见航天诱变并非盲目、无规律变异，而与品种自身特性相关联，同时变异材料的遗传稳定性有必要下一步在分子标记和后续世代田间继续观测中进行确认。

关键词：紫花苜蓿；航天诱变；田间观察

（发表于《草业学报》）

The Effect of Soybean Meal Fermented by *Aspergillus usami* on Phosphor Metabolism in Growing Pigs

WANG Xiaoli[1], WANG Chunmei[1], ZHANG Huaishan[1], QIAO Guohua[1], ZHANG Qian[1], LU Yuan[1], WANG Xiaobin[1], SUN Qizhong[2]

(1. Lanzhou Institute of Husbandry and Pharmaceutical Sciences of CAAS, Lanzhou 730050, China; 2. Institute of Grassland Research of CAAS, Hohhot 010010, China)

Abstract: The experiment aimed to check the pigs' digestive utilization of phosphor and other nutrient components in fermented soybean meal. 15 8 week old two-way cross growing pigs (average weight was 23.6kg) were selected and divided randomly into 3 groups, 5 pigs per group. The method of total feces collection was adopted to compare the digestibility of phosphor and other nutrient components in growing pigs fed respectively by fermented soybean meal by *Aspergill ususami* and normal soybean meal. The experimental design were normal soybean group (NS group, P: 0.087%), fermented soybean meal group (FSgroup, P: 0.089%), normal soybean + fermented soybean meal mixing group (MS group, P: 0.089%). The results indicated that the digestibility of growing pigs in FS group was significantly higher than that of NS group. Meanwhile, the digestibility of crude protein was significantly improved. [Conclusion] The above results indicated that Feeding with soybean meal fermented by *Aspergillus usami* not only could increase the digestibility of phosphor and protein, but also decrease the excretion of phosphor and nitrogen.

Key words: Fermented soybean meal; Growing pigs; Phosphor; Digestibility

(Published the article in Animal Husbandry and Feed Science)

对我国苜蓿产业化及基地建设的分析与思考

常根柱，周学辉，杨红善

（中国农业科学院兰州畜牧与兽药研究所，兰州 730050）

摘　要：文章在对国内外苜蓿产业发展水平予以一定分析的基础上，论证提出了我国苜蓿产业尚处于“初级阶段”水平，走出“草粮争地”的误区，将苜蓿纳入农耕范畴和以企业为核心，带动我国苜蓿产业发展的观点；提出了在我国“三北地区”（西北、华北和东北）建立5大苜蓿基地（河西走廊荒漠绿洲区节水型苜蓿基地、黄土高原干旱、半干旱区苜蓿基地、内蒙古鄂尔多斯高原草地农业苜蓿基地、南疆温润盆地低产田提升改造苜蓿基地、东北低湿盐碱区土地改造苜蓿基地），种植33.33万 hm^2 苜蓿，与“三北防护林工程”共同构建我国北方荒漠化林草生态防治体系的建议。

关键词：苜蓿；产业化；基地建设

（发表于《中国草食动物科学》）

苜蓿种子生产及其研究进展

杨　晓[1]，李锦华[1]，余成群[2]，乔国华[1]，朱新强[1]

（1. 中国农业科学院兰州畜牧与兽药研究所，兰州 730050；
2. 中国科学院地理科学与自然资源研究所）

摘　要：从苜蓿种子前期处理、影响苜蓿种子产量的因素（气候条件、密度、灌溉、施肥、授粉、生长调节剂、杂草防治）以及种子后期收获处理等3方面论述了我国苜蓿种子生产的研究现状，以期为今后苜蓿种子生产提供参考。

关键词：苜蓿；种子生产；种子处理

（发表于《中国草食动物科学》）

其　他

科技论文中数据资料结论正误判断方法的原理和应用

——“差比系数判断法”简介

魏云霞，程胜利，肖玉萍，杨保平，李东海

（中国农业科学院兰州畜牧与兽药研究所，兰州 730050）

摘　要：“差比系数判断法”是判断科技论文中数据资料结论正误的一种方法。凡是在论文中采用随机分组 t 检验或单因素方差分析多重检验的数据分析结果，都可以用该方法快速准确地判断其正误。文中对该判断方法的统计学原理作了简要说明，着重介绍了判断的具体方法和注意事项，具有很强的操作性。

关键词：差比系数判断法；t 检验；单因素方差分析

（发表于《中国草食动物科学》）

修购专项实施成效分析

杨志强，袁志俊，肖　堃，邓海平

（中国农业科学院兰州畜牧与兽药研究所，兰州 730050）

摘　要：2006 年财政部正式立项设立中央级科学事业单位修缮购置专项资金项目以来，到 2011 年年底兰州畜牧与兽药研究所共获得修购专项支持项目 17 项，总经费 5 286万元。修购专项的实施使该所的基础设施和科研条件得到了极大的改善，科研能力得到显著提升。依托修购项目支持获得国家和省、部级科技成果 9 项，获得重大科研项目立项 14 项，形成研发机构 8 个。修购专项的实施，为该所的快速发展奠定了坚实的基础。

关键词：修购专项；实施；成效

（发表于《农业科研经济管理》）

作者简介

杨志强（1957—），中共党员，学士，二级研究员，博士生导师。甘肃省优秀专家，甘肃省第一层次领军人才，中国农业科学院跨世纪学科带头人，甘肃省“555”创新人才，《中兽医医药杂志》主编。兼任中国毒理学兽医毒理学分会会长，中国畜牧兽医学会常务理事，中国兽医协会常务理事，中国畜牧兽医学会动物药品学分会副会长，中国畜牧兽医学会毒物学分会副会长，中国畜牧兽医学会中兽医学分会副会长，中国畜牧兽医学会西北地区中兽医学会理事长，农业部兽药评审委员会委员。长期从事中兽医药学、兽医药理毒理、动物营养代谢与中毒病等研究工作，是该领域内的知名专家，先后主持和参加国家、省、部级科研课题33项，其中，主持20项，获奖9项，自主和参与研发新产品8个，获授权专利3项。先后培养硕士研究生20名，培养博士研究生10名。在国内和国际学术刊物上共发表学术论文100余篇，其中主笔发表论文80篇。主编和参与编写《微量元素与动物疾病》等学术专著13部。

刘永明（1957—），中共党员，大学文化程度，三级研究员，硕士研究生导师。先后担任中国农业科学院中兽医研究所党委办公室副主任、主任，中国农业科学院兰州畜牧与兽药研究所人事处处长、副所长、党委副书记和纪委书记等职务，2001年7月至今任中国农业科学院兰州畜牧与兽药研究所党委书记、副所长、工会主席，兼任《中兽医医药杂志》和《中国草食动物科学》杂志编委会主任、中国农业科学院思想政治工作研究会理事、中国兽医协会会员和兰州市科学技术奖励委员会委员等职务。主要从事动物营养与代谢病研究工作。先后主持国家科技支撑计划、公益性行业专项、科技成果转化基金项目、948项目以及省级科研课题或子专题12项，主持基本建设项目4项，获授权专利8项，取得新兽药证书1个、添加剂预混料生产文号5个；主编（主审）、副主编著作6部，参与编写著作6部，

张继瑜（1967—），中共党员，博士研究生，三级研究员，博（硕）士生导师，中国农业科学院三级岗位杰出人才，中国农业科学院兽用药物研究创新团队首席专家，国家现代农业产业技术体系岗位科学家。兼任中国兽医协会中兽医分会副会长，中国畜牧兽医学会兽医药理毒理学分会副秘书长，农业部兽药评审委员会委员，农业部兽用药物创制重点实验室常务副主任，甘肃省新兽药工程重点实验室常务副主任，中国农业科学院学术委员会委员，黑龙江八一农垦大学和甘肃农业大学兼职博导。主要从事

兽用药物及相关基础研究工作，重点方向包括兽用化学药物的研制、药物作用机理与新药设计、细菌耐药性研究。带领的研究团队在动物寄生虫病、动物呼吸道综合症防治药物研究上取得了显著进展。在肠杆菌耐药机理、血液原虫药物作用靶标筛选的研究处于领先地位。先后主持完成国家、省部重点科研项目20多项，获得科技奖励6项，研制成功4个兽药新产品，其中，国家一类新药一个，以第一完成人申报10项国家发明专利，取得专利授权5项。培养研究生21名，发表论文170余篇，主编出版《动物专用新化学药物》和《畜牧业科研优先序》等著作2部。

阎萍（1963—）中共党员，博士，三级研究员，博士生导师。2012年享受国务院特殊津贴，是中国农业科学院三级岗位杰出人才，甘肃省优秀专家，甘肃省“555”创新人才，甘肃省领军人才。曾任畜牧研究室副主任、主任等职务，2013年3月任研究所副所长职务。兼任国家畜禽资源管理委员会牛马驼品种审定委员会委员，中国畜牧兽医学会牛业分会副理事长，全国牦牛育种协作组常务副理事长兼秘书长，中国畜牧兽医学会动物繁殖学分会常务理事和养牛学分会常务理事等。阎萍研究员主要从事动物遗传育种与繁殖研究，特别是在牦牛领域的研究成绩卓越，先后主持和参加完成了科技部支撑计划、科技部基础性研究项目、科技部“863”计划、“948”计划、农业部行业科技项目、国家肉牛产业技术体系岗位专家、人事部回国留学基金项目、科技部成果转化项目、甘肃省科技重大专项计划、甘肃省农业生物技术项目等20余项课题。现为国家肉牛牦牛产业技术体系牦牛选育岗位专家，甘肃省牦牛繁育工程重点实验室主任。作为高级访问学者多次到国外科研机构进行学术交流。培育国家牦牛新品种1个，填补了世界上牦牛没有培育品种的空白。获国家科技进步奖1项，省部级科技进步奖5项及其他科技奖励3项。培养研究生15名，发表论文180余篇，出版《反刍动物营养与饲料利用》《现代动物繁殖技术》《牦牛养殖实用技术问答》《Recend Advances in Yak Reproduction》《中国畜禽遗传资源志—牛志》等著作。

包鹏甲（1980—），男，甘肃武威人，助理研究员，动物遗传育种与繁殖专业硕士。2007年毕业于甘肃农业大学，主要从事生物技术与草食动物遗传繁育研究工作，主持完成中央级公益性科研院所基本科研业务费2项，参加完成国家、省部级及其他科研课题多项，参编著作3部，第一作者发表国内外文章10余篇，获专利十余项。

常根柱（1956—），甘肃甘谷县人。研究员，中共党员，硕士研究生导师，中国草学会理事，甘肃省一层次领军人才，甘肃省草品种审定委员会委员。先后参加、主持完成国家、省部级课题12项；获省、部级奖3项，地、厅级奖4项；主编、副主编学术专著4部，发表论文68篇；获国家授权发明专利1项（第二）。选育成功牧草新品种4个（2个第一，2个第二），通过甘肃省草品种审定委员审定登记，其中，2个通过国家草品种审定委员会评审，参加全国区域试验。独立培养硕士研究生5名，合作培养博士研究生1

名。研究方向：牧草育种。现承担项目：甘肃省科技支撑计划－牧草航天诱变品种选育。

程富胜（1971—）博士，副研究员，硕士生导师。主要从事天然药物活性成分免疫药理学研究及其制剂开发研制。先后主持和参加国家、省、部级科研课题 20 项。参加完成的国家支撑项目“中草药饲料添加剂‘敌球灵’的研制”获农业部科技进步二等奖；国家自然基金项目“免疫增强剂－8301 多糖的研究与应用”成果获中国农业科学院科技进步二等奖；“蕨麻多糖免疫调节作用及其机理研究与临床应用”成果已分别获中国农科院、兰州市可进步二等奖。主持完成的“富含活性态微量元素酵母制剂的研究”获兰州市科技进步二等奖。在国内和国际学术刊物上共发表学术论文 50 余篇，其中，主笔发表论文 30 余篇。

褚敏（1982—），中共党员，在读博士，助理研究员。主要从事牦牛分子遗传与育种研究，先后参加国家科技支撑计划、农业部行业科技、甘肃省重大科研项目、“863”“948”项目、甘肃省生物技术和甘肃省科技支撑等重大课题十余项，现主持中央级公益性科研院所基本科研业务费专项资金项目 1 项。主笔发表 SCI 学术论文 2 篇，中文 10 多篇，参与编写著作《牦牛养殖实用技术问答》和《适度规模肉牛场养殖技术示范》2 部，翻译并编辑第五届国际牦牛会议英文学术论文集 1 部，发明实用新型专利 6 项。

崔东安（1981—），博士，主要从事奶牛胎衣不下方证代谢组学研究（奶牛胎衣不下血瘀证的代谢组学研究 No. 1610322015006），发表文章 4 篇，其中，3 篇 SCI 文章。

丁学智（1979—），博士、副研究员，甘肃省杰青年金获得者、2012 年度中国农业科学院“青年文明号”。主要从牦牛高寒低氧适应方面的研究工作。目前，主持国家自然基金青年科学基金、甘肃省杰出青年科学基金、国家自然科学基金国际（地区）合作重大项目等，发表 SCI 收录论文 10 余篇、国家发明专利了 2 项、实用新型专利 1 项。

董书伟（1980—），男，汉族，在读博士，助理研究员，毕业于西北农林科技大学。主要从事奶牛营养代谢病与中毒病的蛋白质组学研究。先后主持参加国家科技支撑计划、中央级公益性科研院所基本科研业务费专项资金项目、中国科学院西部之光项目、国家自然科学青年基金，发表学术论文 20 余篇，申请专利 18 项。

郭健（1964 —），硕导、学士、副研究员。中国畜牧兽医学会养羊学分会常务理事，甘肃省“三区”科技人才，国家科技型创新基金管理中心项目评审专家，甘肃省财政厅项目评审专家，甘肃省科技厅项目评审验收专家。现主持“十二五”国际科技支撑计划项目子课题“优质细毛羊新品种（系）繁育及相关技术研究”、“十二五”甘肃省重大专项项目“甘肃超细毛羊新品种培育及产业化研究示范”。主编出版著作2部，副主编出版3部，发表文章20余篇。以第一完成人取得星星实用专利4项。累计获得甘肃省科技进步二等奖3项，中华科技奖三等奖1项，中国农科院科技成果一等奖2项，农牧渔业丰收奖1项，兰州市科技创新成果一等奖1项。

郭天芬（1974—），学士、副研究员。主要从事动物营养、毛皮质量检测及标准研究制定工作。期间参加十余项国家级省部级科研项目及撰写制定十余项标准制定项目，主持完成一项国家标准制定项目和二项中央公益类基本科研项目。以第一完成人获发明专利1项、实用新型专利6项，获甘肃省科学技术进步奖1项（第3完成人），酒泉市科技进步奖1项（第4完成人）。正式发表学术期刊论文100余篇，其中，第一作者的30余篇。参与撰写专著5部，其中，副主编的2部。

郭宪（1978—），中国农业科学院硕士生导师，博士，副研究员。中国畜牧兽医学会养牛学分会理事，中国畜牧业协会牛业分会理事，全国牦牛育种协作组理事。主要从事牛羊繁育研究工作。先后主持国家自然科学基金、国家支撑计划子课题、中央级公益性科研院所基本科研业务费专项资金项目等5项。科研成果获奖3项，参与制定农业行业标准3项，授权专利3项。主编著作2部，副主编著作5部，参编著作2部。主笔发表论文30余篇，其中，SCI收录5篇。

郭志廷（1979—），男，内蒙古人，助理研究员，执业兽医师，九三学社兰州市青年委员会委员，中国畜牧兽医学会中兽医分会理事。2007年毕业于吉林大学，获中兽医硕士学位。近年主要从事中药抗球虫、免疫学和药理学研究。先后主持或参加国家、省部级科研项目5项，包括中央级公益性科研院所专项基金。作为参加人获得兰州市科技进步一等奖1项，兰州市技术发明一等奖1项，完成甘肃省科技成果鉴定4项，授权国家发明专利5项（1项为第一完成人），参编国家级著作2部。在国内核心期刊上发表学术论文80余篇（第一作者30篇）。

韩吉龙（1987—），中国农业科学院在读博士研究生，动物遗传育种与繁殖专业，主要开展绵羊重要经济性状分子机理研究，第一作者发表论文3篇，其中，SCI文章1篇。

郝宝成（1983—），甘肃古浪人，硕士研究生，助理研究员，研究方向为新型天然兽用药物研究与创制。先后主持和参与了中央级公益性科研院所基本科研业务费专项资金、国家支撑计划、863项目子课题等项目6项，以第一作者发表论文23篇（其中，SCI论文2篇，一级学报2篇），参与编写著作3部《中兽药学》、《兽医中药学及试验技术》、《天然药用植物有效成分提取分离与纯化技术》，以第一发明人获得国家发明专利3项，荣获2012年度兰州市九三学社参政议政先进工作者。

孔晓军（1982—），研究实习员，主要从事中兽药、藏兽药的药理与毒理学研究，发表代表性SCI文章2篇。申请专利：一种用于组织切片或涂片烘干装置，201420204678．8，获得中国农业科学院二等奖一项：重金属镉/铅与喹乙醇抗原合成、单克隆抗体制备及ELISA检测技术研究。

郎侠（1976—），博士，现为中国畜牧兽医学会养羊学分会理事，青藏高原研究会会员。主要从事绵、山羊育种，动物遗传资源保护利用和现代生态畜牧业方面的研究以及现代畜牧生产咨询工作。主持及参与国家、省部级科研项目10余项。主编出版著作6部，副主编著作4部，参编著作3部，在国内外学术期刊发表学术论文100余篇，获得专利4项。主持项目获中国农业科学院科技进步二等奖1项，参与项目获中国农业科学院科技进步一等奖1项、甘肃省科技进步二等奖1项、中华神农科技奖1项。

李宏胜（1964—），九三学社社员，博士，研究员，硕士生导师，甘肃省“555”创新人才。中国畜牧兽医学会家畜内科学分会常务理事。多年来主要从事兽医微生物及免疫学工作，尤其在奶牛乳房炎免疫及预防方面有比较深入的研究。先后主持和参加完成了国家自然基金、国家科技支撑计划、国际合作、甘肃省、兰州市及企业横向合作等20多个项目。先后获得农业部科技进步三等奖1项；甘肃省科技进步二等奖2项、三等奖2项；中国农业科学院技术成果二等奖3项；兰州市科技进步一等奖2项、二等奖2项。获得发明专利4项，实用新型专利14项，培养硕士研究生5名，在国内外核心期刊上发表论文160余篇，其中，主笔论文60余篇。

李剑勇（1971—），研究员，博士学位，硕士和博士研究生导师，国家百千万人才工程国家级人选，国家有突出贡献中青年专家。现任中国农业科学院科技创新工程兽用化学药物创新团队首席专家，农业部兽用药物创制重点实验室副主任，甘肃省新兽药工程重点实验室副主任，甘肃省新兽药工程研究中心副主任，农业部兽药评审专家，甘肃省化学会色谱专业委员会副主任委员，中国畜牧兽医学会兽医药理毒理学分会理事，国家自然基金项目同行评议专家，《黑龙江畜牧兽医杂志》常务编委，《PLOS ONE》、《Medicinal Chemistry Research》等SCI杂志审稿专家。多年来一直从事兽用药物创制及与之相关的基础和应用基础研究工作。曾先后完成药物研究项目40多项，主持16项。获省部级以上奖励10项，2011年度获第十二届中国青年科技奖；2011年度获第八届甘肃青年科技奖；获2009年度兰州市职工技术创新带头人称号。获国家一类新兽药证书，均为第2完成人。申请国家发明专利22项，获授权9项。发表科技论文200余篇，其中SCI收录22篇，第一作者和通讯作者15篇。出版著作4部，培养研究生15名。

李锦宇（1973—），学士学位，副研究员。主要从事中兽医针灸研究、中兽药新药研发工作。现主持科技部成果转化项目“抗禽感染疾病中兽药复方新药‘金石翁芍散’的推广应用”和甘肃省科技支撑课题，参加其它国家科研课题15项。获甘肃省科技进步奖二等奖一项（第二完成人），农科院科技进步二等奖一项。申报获取国家三类新药－金石翁芍散（第一完成人）；获取专利16项（第一完成人3项），参与编写著作10部（副主编2部）；在国内外正式刊物发表科技论文公开发表各种学术论文50篇。

李润林（1982—），硕士，研究实习员。主要从事农业资源遥感监测研究，主持项目2项目，参加项目5项，发表科技论文13篇，其中第一作者发表论文5篇。

李维红（1978—），博士，副研究员。主要从事畜产品质量评价技术体系、畜产品检测新方法及其产品开发利用研究等。先后主持和参加了中央级公益性科研院所基本科研业务费专项资金项目、甘肃省自然基金项目等。主笔发表论文20余篇。主编《动物纤维超微结构图谱》，副主编《绒山羊》，参加编写《动物纤维组织学彩色谱》和《甘肃高山细毛羊的育成和发展》著作2部。获得专利4项，作为参加人获甘肃省科技进步一等奖1项、三等奖1项。

李新圃（1962—），博士，副研究员。主要从事兽医药理学研究工作。已主持完成省、部、市级科研项目7项。参加完成国际合作、省、部、市级科研项目三十余项。参加项目“奶牛重大疾病防控新技术的研究与应用”获2010年甘肃省科技进步二等奖；“奶牛乳房炎主要病原菌免疫生物学特性的研究”在2008年获兰州市科技进步一等奖、中国农科院科学技术成果二等奖和甘肃科技进步三等奖；“绿色高效饲料添加剂多糖和寡糖的应用研究”获2005年兰州市科技进步一等奖；“奶牛乳房炎综合防治配套技术的研究及应用”获2004年甘肃省科技进步二等奖。已发表研究论文40余篇，其中，5篇被SCI收录。获实用新型专利授权3个。

梁春年（1973—），硕士生导师，博士，副研究员，副主任，兼任全国牦牛育种协作组常务理事，副秘书长，中国畜牧兽医学会牛学会理事，中国畜牧业协会牛业分会理事，中国畜牧业协会养羊学分会理事等职。主要从事动物遗传育种与繁殖方面的研究工作。现主持国家科技支撑计划子课题“甘肃甘南草原牧区牦牛选育改良及健康养殖集成与示范”和国家星火计划项目子课题“牦牛高效育肥技术集成示范”等课题4项；参加完成国家及省部级科研项目30余项，获得省部级科技奖励7项。参与制定农业行业标准5项。参加国内各类学术会议30余次，国际学术会议4次。主编著作2本，副主编著作5本，参编著作6本，发表论文90余篇，其中，SCI文章5篇。

刘希望（1986—），硕士，助理研究员。2007年毕业西北农林科技大学环境科学专业，2010年获西北农林科技大学化学生物学专业硕士学位，同年参加工作至今。曾主持农业部兽用药物创制重点实验室开放基金项目1项，现主持中央公益性科研院所基本科研业务费1项。以第一作者发表SCI论文4篇，主要从事新兽药研发，药物合成方面的研究工作。

刘宇（1981—），硕士，助理研究员。2007年7月来中国农业科学院兰州畜牧与兽药研究所工作至今，主要从事有机及天然药物化学研究。先后主持及参与国家级省部级药物研究项目10余项，包括“十二五”农村领域国家科技计划项目、国家支撑计划项目、甘肃省科技支撑计划项目、甘肃省自然科学基金项目及中央级公益性科研院所基本科研业务费专项资金项目等。获省部级科技进步奖1项，院厅级奖2项，国家发明专利4个。在国内外学术刊物上发表论文20余篇，出版著作2部。

罗超应（1960 —），学士，研究员，中西兽医药学结合研究（主持科技部科研院所开发研究专项“新型中兽药射干地龙颗粒的研究与开发”、科技部基础工作专项“华东区传统中兽医学资源抢救与整理”与国家“十一五”科技支撑子项目“中兽药中试及其生产工艺研究”等）。主编、副主编与参编出版《牛病中西医结合治疗》等著作 16 部，共计 679 余万字；主笔发表“Variability of the Dosage, Effects and Toxicity of Fu Zi (Aconite) From a Complexity Science Perspective”、“以复杂科学理念指导中西医药学结合”、“奶牛乳房炎的复杂性及其对传统科学观念的挑战”等学术论文近 100 篇，其中英文期刊文章 5 篇；专利 12 项，成果奖励 5 项。

罗永江（1966—），硕士生导师，学士，副研究员。主要从事中兽药的研制与开发工作。主持或参加了“十一五”国家科技支撑项目子课题、甘肃省自然科学基金、兰州市农业科技攻关课题、中央级公益性科研院所基本科研业务费专项以及横向课题等。发表论文 40 余篇，参加编写专著 8 部，获得发明专利 11 项，实用新型专利 5 项。获得过农业部科技进步三等奖 1 次，中国农科院科技进步一等奖、二等奖各 1 次，甘肃省科技进步二等奖 1 次、三等奖 2 次，兰州市技术发明一等奖 1 次，北京市科学技术奖三等奖 1 次。

苗小楼（1972—），学士，副研究员。主要从事兽药研发、传统兽医药物研究工作，主持参加多个省部课题，发表论文 20 余篇，获得授权专利 3 项，主持研制的中兽药“益蒲灌注液”获得国家三类新兽药证书，参与研制的一类兽药“喹烯酮”曾先后获国家科技进步二等奖和甘肃省科技进步一等奖。

秦哲（1983 —），博士，助理研究员。从事益生菌发酵中药，中药药理等研究工作。先后参加各类研究课题 10 余项，2011 年参加的“新型中兽药饲料添加剂‘参芪散’的研制与应用”获中国农业科学院科学技术成果二等奖；2013 年参加的“非解乳糖链球菌发酵黄芪转化多糖的研究与应用”获得甘肃省科技进步三等奖；2014 年参加的“重金属镉/铅与喹乙醇抗原合成、单克隆抗体制备及 ELISA 检测技术研究”获中国农业科学院科学技术成果二等奖；2014 年参加的“奶牛乳房炎联合诊断和防控新技术研究及示范”获甘肃省农牧渔丰收奖一等奖；发表科技论文 10 余篇，SCI 论文 1 篇，授权实用新型 2 项。

尚若锋（1974—），博士，副研究员。主要从事兽用药物研发工作。主持或参与国家支撑计划、“863”计划以及其他国家级和省部级的科研项目20余项。获得省部级科研奖励4项，国家发明专利12项，以第一作者或通讯作者发表文章30余篇，其中，SCI文章12篇。

尚小飞（1986—），硕士，助理研究员。主要从事藏兽医药的现代化研究。目前，主持中央级公益性院所基本科研业务费项目一项，参与多项国家及省部级课题。在*Journal of Ethnopharmacology*，*Veterinary Parasitology*等SCI杂志发表文章8篇，参与编写著作一部。

时永杰（1961—），硕士生导师，学士，三级研究员，中国农业科学院三级岗位杰出人才。从事牧草育种栽培、草地培育改良和草地生态环境治理等方面研究工作30余年，先后主持和参加各类研究课题30余项；获省、部级科技成果奖12项，院、厅级奖5项。作为主要撰稿人编写并出版了《甘肃省种草区划》、《我国西部荒漠化生态环境及其治理》、《优质牧草高产栽培及加工利用技术》等学术著作9部，发表学术论文120余篇，培养硕士研究生6名。现主持国家重点基础研究发展计划（973）项目、国家公益性行业科研专项项目和国家科技支撑计划项目等研究课题。兼任中国草学会理事，中国畜牧业协会草业分会理事，甘肃省草原学会理事，美国北美苜蓿协会（NAAIC）会员等。

孙晓萍（1962—），硕士生导师，学士，副研究员。主要从事绵羊遗传育种工作，先后主持的项目有：甘肃省自然科学基金：绵羊毛生长机理研究；甘肃省农委推广项目：绵羊双高素推广应用研究；甘肃省支撑计划项目：奶牛产奶量的季节性变化规律研究；甘肃省星火项目：肉羊高效繁殖技术研究；甘肃省支撑计划项目：肉用绵羊高效饲养技术研究。先后发表学术论文40余篇，主编参编著作11部，实用新型专利3个。获甘肃省畜牧厅科技进步二等奖1项，甘肃省科技进步二等奖1项，中国农业科学院科技进步一等奖1项，中华农业科技二等奖1项。

田福平（1976—），甘肃武山人，副研究员。中国草学会会员，2003年获草业科学王栋奖学金，中国草学会青年工作委员会理事，一直从事草地生态、牧草种质资源与育种、牧草栽培等方面的研究。先后参加国家、部、省级相关研究项目40项，主持项目6项。获甘肃省科技进步奖二等奖2项，中国农业科学院科学技术成果奖二等奖1项。发表学术论文60余篇，其中，SCI论文4篇。参编著作7部，其中主编著作4部。

王东升（1979—），农学硕士，助理研究员。主要从事奶牛繁殖疾病的研究。主持“奶牛子宫内膜中天然抗菌肽的分离、鉴定及其生物学活性研究”和“狗经穴靶标通道及其生物效应的研究”2个课题，参加“十二五”国家科技支撑计划项目“奶牛健康养殖重要疾病防控关键技术研究”和“十一五”国家科技支撑计划项目“奶牛主要繁殖障碍疾病防治药物研制”等10多个项目。参与申请并获得发明专利4项，实用性新专利5项，取得三类新兽药证书1个，参加的成果获甘肃省科技进步二等奖2项和兰州市科技进步二等奖2项。参编著作《兽医中药配伍技巧》《兽医中药学》《奶牛围产期饲养与管理》和《奶牛常见病综合防治技术》等5部，主笔发表论文20余篇。

王贵波（1982—），临床兽医学硕士，助理研究员，中国生理学会会员，主要从事兽医针灸及中兽医药学研究。主持完成了甘肃省青年基金项目“针刺镇痛对犬脑内Jun蛋白表达的影响研究”，中央级公益性事业单位专项资金项目“针刺镇痛对中枢Fos蛋白表达的影响研究”、“电针对犬痛阈及中枢强啡肽基因表达水平的研究”及横向委托项目“银翘双解颗粒/饮与灵丹草饮临床疗效验证委托试验”；还参与完成了国家自然科学基金、财政部中央级公益性科研院所基本科研业务费项目、“十一五”国家科技支撑项目和农业科技成果转化资金项目等。第一作者发表论文20余篇，副主编著作一部，参编著作一部，第一发明人获得的专利7项。

王宏博（1977—），博士，助理研究员，博士。主要从事动物营养与饲料科学。先后主持甘肃省科技厅项目2项，中央级公益性科研院所科研业务费专项资金项目3项，先后参加农业部公益性行业（农业）专项3项，“948”项目1项，获得国家发明专利2项，实用新型专利1项，参与完成国家发明专利和实用新型专利总计10余项。参与制定国家和农业部标准10余项。主编1部，副主编著作1部，参编著作3部。发表学术论文30余篇。

王华东（1979—），硕士，助理研究员。主要从事科技期刊编辑工作，先后出版著作3部（副主编1部，参编2部），主笔发表学术论文5篇。编辑、出版的《中兽医医药杂志》发行范围涵盖中国、北美、澳洲、西欧、韩国、日本、新加坡等海内外15个国家和地区。随着中兽医药在畜禽疫病防治及公共卫生安全等方面的独特作用的日益凸显，作为中兽医领域唯一的国家级学术刊物，《中兽医医药杂志》全面、详实地报道了近年来该领域的最新研究成果，并力促其推广应用与转化，为继承和传播祖国兽医学诊疗技术、促进中西兽医结合、繁荣学术、服务“三农”等做出了重大贡献，取得了显著的经济效益、社会效益和生态效益。

王慧（1985—），硕士，研究实习员。主要从事动物营养代谢病研究。现主持“中央级科研院所基本科研业务费专项资金项目（NO. 1610322013003）。2013年获农业部全国农牧渔业丰收二等奖（第7完成人）；兰州市科学技术进步二等奖（第9完成人）。目前第一作者发表论文16篇，其中SCI收录8篇。申请专利2项。

王磊（1985—），硕士，研究实习员。主要从事中兽医药理学、奶牛疾病防治和药物残留研究。先后参加各类研究课题8项，主持课题1项。2013年参加的“非解乳糖链球菌发酵黄芪转化多糖的研究与应用”获甘肃省科技进步三等奖、2014年参加的“重金属镉/铅与喹乙醇抗原合成、单克隆抗体制备及ELISA检测技术研究”获中国农业科学院科学技术成果二等奖；主笔发表科技论文4篇、其中SCI收录1篇，获得授权专利2项。

王玲（1969—），预防兽医学硕士，副研究员，研究方向为兽医微生物与免疫学，主要从事创新兽药的基础研究与应用。工作期间先后参加国家重点科技支撑项目、国家自然科学基金、农业部“跨越计划”、农业部和科技部专项资金项目、中央级公益性科研院所专项资金项目、甘肃省及兰州市科技攻关项目等课题40余项。主笔发表论文40余篇，其中核心期刊20余篇，主笔发表SCI论文1篇，署名发表SCI 2篇。工作以来作为主要完成人获得研究成果5项，发明专利4项，实用新型专利3项，甘肃省四类新兽药证书3项，转让成果1项。

王胜义（1981—），硕士学位，助理研究员。主要从事中兽药新药研发和微量元素代谢病研究，先后主持和参与了国家科技支撑计划项目、科技部农业成果转化项目、公益性行业（农业）科研专项、农业部“948”项目等。发表中文核心21篇，参与著作1部，以第一完成人或参与人获得专利13个，以第二完成人获得全国农牧渔业丰收奖二等奖1项。

王晓力（1965—），副研究员。主要从事牧草资源开发利用方面的研究工作。近年来参加省部级课题多项，主持农业行业科研专项子课题2项，获全国农牧渔业丰收二等奖1项、内蒙古自治区农牧业丰收一等奖1项、2011年贵州省科学技术成果转化二等奖1项。发表论文30余篇，出版著作4部，参编4部，授权专利5项。

王旭荣（1980—），博士，助理研究员。主要从事分子病毒学和分子细菌学方面的研究。主持完成“犬瘟热病毒”的基本科研业务费项目1项；现主持“奶牛乳房炎病原菌“方面的农业行业标准项目1项和甘肃省农业生物技术项目1项；参与国家奶牛产业技术体系项目、948项目、国家科技支撑项目、国家自然基金、科技基础性工作专项等10余项。参与的研究项目在2011—2014年期间获得奖项3个，其中，甘肃省科技进步三等奖1项，甘肃省农牧渔丰收奖一等奖1项、中国农业科学院科学技术成果二等奖1项。发表科技论文40余篇，主笔15篇；参编著作1部；实用新型授权20余项，第一发明人授权5项；申请发明专利（第一发明人）3项。

王学红（1975—），硕士，高级兽医师。主要从事天然药物提取研究。先后主持及参加国家级省部级药物研究项目20余项，包括“十二五”农村领域国家科技计划项目、国家支撑计划项目、甘肃省科技支撑计划项目及中央级公益性科研院所基本科研业务费专项资金项目等。获省部级科技进步奖4项，院厅级奖2项，国家发明专利10余个。在国内外学术刊物上发表论文20余篇，出版著作4部。

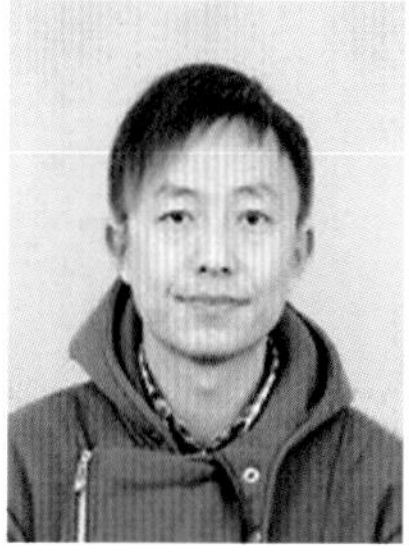

魏云霞（1965—），博士，副研究员。主要从事学术杂志的编辑出版工作。《中国草食动物科学》副主编。其主编的杂志先后获得中国农学会和中国期刊协会“第四届全国优秀农业学术期刊二等奖”、中国畜牧兽医学会期刊编辑学分会“第四届全国畜牧兽医优秀期刊二等奖”等奖项。作为副主编编辑出版的《中国草食动物》杂志从2011年起被世界著名检索机构美国《化学文摘》收录。《中国草食动物》杂志在2004年、2006年、2008年、2010年连续4次被评为“中国科技核心期刊”。作为第一作者和通讯作者共发表论文13篇，其中，在科技核心期刊上发表8篇，在中文核心期刊上发表2篇。作为主编出版专著1部，作为副主编出版专著2部，参编4部；作为副主编出版8部，参编6部。

吴晓云（1986—），在读博士，主要从事牦牛低氧适应机制和牦牛肉品质形成遗传机制的研究。目前，以第一作者发表论文9篇，其中，5篇被SCI收录。

席斌（1981—），硕士，助理研究员。主要从事畜产品质量及安全监测工作。主持项目2项，甘肃省星火计划及中央级公益性科研院所基本科研业务费专项资金项目。参加完成科研项目4项，参加国家标准6项，参与发明专利2项。以第一作者发表论文30余篇，其中核心期刊20余篇。参编出版专著1部（《动物纤维组织学彩色图谱》2007），并获第十六届2008年度中国西部地区优秀科技图书二等奖。

谢家声（1956—），大学本科学历，高级实验师。主要从事猪、禽、奶牛等动物疾病的中兽医临床防治研究，主持完成了甘肃省自然基金—“复方中草药防治畜禽病毒性传染病新制剂的研究”、甘肃省科技攻关项目—“治疗奶牛胎衣不下天然药物的研制”等项目。完成国家“七五”、“八五”攻关项目、国家“十五”及“十一五”奶业专项、国家“十一五”、“十二五”支撑计划以及省市各类研究项目30余项。发表论文60余篇，合编专著8部，获得国家发明专利3项。1990年以来先后获甘肃省科技进步三等奖1项；甘肃省科技进步二等奖3项；中国农业科学院科技进步二等奖4项；兰州市科技进步二等奖3项；甘肃省农牧厅二等奖1项；甘肃省天水市科技进步二等奖2项。

辛蕊华（1981—），硕士，助理研究员。主要从事中兽医药物学工作，先后主持和参与过中央级公益性科研院所基本科研业务费专项资金项目、公益性行业（农业）科研专项、国家科技支撑项目等；发表论文 12 篇，参与著作两部。参加的“富含活性态微量元素酵母制剂的研究”科研成果获得兰州市科学技术二等奖。

肖玉萍（1979—），动物遗传育种与繁殖专业硕士，助理研究员，主要从事《中国草食动物科学》的编辑出版工作。发表论文 15 余篇，副主编出版专著 2 部。

熊琳（1984—），硕士，助理研究。主要从事农产品质量安全的研究。发表相关科研论文 4 篇，其中，SCI 收录 1 篇，授权专利 8 项，参与制定国家标准 1 项。

严作廷（1962—），硕导，博士，研究员。现为第五届农业部兽药评审专家、中国畜牧兽医学会家畜内科学分会常务理事、中国畜牧兽医学会中兽医学分会理事。主要从事工作奶牛疾病和中兽医学研究（先后主持国家自然科学基金、国家科技支撑计划等项目 10 多项，现主持“十二五”国家科技支撑计划课题“奶牛健康养殖重要疾病防控关键技术研究”课题）。主要成就（主编或副主编《家畜脉诊》《奶牛围产期饲养与管理》等著作 3 部，参编《中兽医学》《奶牛高效养殖技术及疾病防治》和《犬猫病诊疗技术及典型医案》等著作 6 部。发表论文 70 余篇，获省部级科技进步奖 6 项，院厅级奖 7 项。获得国家新兽药证书 2 个，国家发明专利 10 个，实用新型专利 5 个）。

杨红善（1981—），硕士，助理研究员。主要从事牧草种质资源搜集与新品种选育研究工作，主持在研项目 3 项，其中，甘肃省青年基金项目 1 项、甘肃省农业生物技术研究与应用开发项目 1 项、中央级公益性科研院所基本科研业务费专项资金项目 1 项，参加国家支撑计划子课题等各类项目共计 3 项。工作期间以第一完成人或参加人审定登记甘肃省牧草新品种 4 个。获中国农业科学院科技进步二等奖 1 项（第三完成人）。参加编写《高速公路绿化》著作 1 本（副主编）。在各类期刊发表论文 20 篇，其中，主笔 13 篇。

杨晓（1985—），硕士，助理研究员。主要从事牧草栽培和草原生态研究，先后参与国家科技支撑计划项目、公益行业专项项目等。发表代表性文章5篇，其中SCI文章1篇。

岳耀敬（1980—），在读博士。主要从事绵羊繁殖、羊毛和高原适应性等重要经济、抗逆性状的分子调控机制研究。国家绒毛用羊产业技术体系分子育种岗位团队成员，兼任中国畜牧兽医学会养羊学分会副秘书长、世界美利奴育种者联盟成员，曾先后到法国、澳大利亚、新西兰等国学习考察细毛羊育种工作。现主持国家自然青年基金、甘肃省青年基金等项目3项；申报发明专利5项，授权发明专利2项、实用新型2项；参编著作5部，发表学术论文38篇，其中，SCI5篇。

张怀山（1969—），博士，助理研究员。长期从事草类植物种质资源及育种研究。参与完成国家、省部级、市级科研项目16项，主持完成省部级、市级、院所级科研项目4项，发表论文49篇，出版专著2部，获得国家发明专利5项，获得省部级、市级科技奖3项。

张景艳（1980—），硕士，助理研究员。从事兽药残留快速检测技术和细胞药理学等研究工作。先后参加各类研究课题10余项，主持课题1项。2011年参加的“新型中兽药饲料添加剂‘参芪散’的研制与应用”获中国农业科学院科学技术成果二等奖；2013年参加的“非解乳糖链球菌发酵黄芪转化多糖的研究与应用”获得甘肃省科技进步三等奖；2014年参加的“重金属镉/铅与喹乙醇抗原合成、单克隆抗体制备及ELISA检测技术研究”获中国农业科学院科学技术成果二等奖；2014年参加的“奶牛乳房炎联合诊断和防控新技术研究及示范”获甘肃省农牧渔丰收奖一等奖；发表科技论文10余篇，授权发明专利1项、实用新型2项。

张凯（1983—）江西新余人，2008年毕业于西北民族大学，获得临床兽医学硕士学位，现就读于意大利墨西拿大学临床兽医学博士学位。同时现为兰州畜牧与兽医研究所助理研究员，研究方向为中兽药生物转化与畜禽传染病防控。发表论文20余篇，其中，主笔7篇。曾获得中国农业科学院发明成果“三等奖”。

张茜（1980—），博士，助理研究员。主要从事西北干旱荒漠地区牧草科研和生产、资源保护和利用，以及植物分了生物和遗传学方面及抗逆功能基因方面的研究。先后主持国家自然基金、甘肃省自然基金、中央级公益性科研院所基本科研业务费专项资金项目等项目。发表代表性文章5篇，其中，SCI文章3篇。

张世栋（1983—），硕士，助理研究员。主要从事奶牛疾病与中兽医药研究工作。近年来，主持中央级基本科研业务费及其增量项目共3项，国家自然基金1项；参与国家“十一五”、“十二五”课题3项，甘肃省、兰州市科技项目多项。参与获得兰州市科技进步二等奖1项。发表论文10篇，获得专利6项，参与申请专利10多项。

周绪正（1971—），副研究员，硕士生导师。先后主持和参加国家、省、部级科研课题近40项，其中主持3项；参与项目获得国家科技进步二等奖1项，甘肃省科技进步一等奖2项，兰州市技术发明一等奖1项、临夏回族自治州科技进步一等奖1项、兰州市科技进步二等奖1项、中国农业科学院科学技术成果二等奖2项；参与研制成功国家一类新药（喹烯酮原料及预混剂）2个，三类兽药新制剂4个（多拉菌素、替米考星、黄霉素），取得8项国家发明专利；在国内和国际学术刊物上共发表学术论文100余篇，其中，主笔发表论文40余篇；主编、参编著作10部。

周学辉（1964—），助理研究员。主要从事中兽医药、草食动物营养与饲料、草业科学研究工作。主持及参加的各类科研项目42项，其中，主持3项（甘肃省自然基金项目、世行全球环境基金项目、中国农科院科技经费项目各1项）；主笔发表论文50余篇，参加50余篇；主编著作2部、论文集1册，副主编3部，参编5部；获国家发明专利1项（第1）；审定登记甘肃省草新品种4个；获农业部科技进步三等奖1项，甘肃省科技进步二等奖2项，甘肃省科技进步三等奖1项，全国农牧渔业丰收奖二等奖1项，甘肃省畜牧厅科技进步一等奖1项，中国农业科学院科技成果二等奖2项，中国农业科学院科技进步一等奖1项，兰州市科技进步二等奖1项。

朱新书（1957—），学士，副研究员，主要从事牦牛藏羊品种资源开发和利用工作，主要主持和参加的项目有：公益性行业（农业）科研专项《放牧牛羊营养均衡需要研究与示范》；农业部重点攻关项目《大通牦牛选育与杂交利用》；甘肃省重大科技项目《甘南牦牛改良与选育技术研究示范》；国家肉牛牦牛产业技术体系《牦牛繁育技术研究与示范岗位科学家团队》等。发表学术论文20余篇，主编参编著作3部，荣获农业部科技进步三等奖1项。